Advances in Intelligent Systems and Computing

Volume 939

The series "Advances in Intelligent Systems and Computing" contains publications on theory, applications, and design methods of Intelligent Systems and Intelligent Computing. Virtually all disciplines such as engineering, natural sciences, computer and information science, ICT, economics, business, e-commerce, environment, healthcare, life science are covered. The list of topics spans all the areas of modern intelligent systems and computing such as: computational intelligence, soft computing including neural networks, fuzzy systems, evolutionary computing and the fusion of these paradigms, social intelligence, ambient intelligence, computational neuroscience, artificial life, virtual worlds and society, cognitive science and systems, Perception and Vision, DNA and immune based systems, self-organizing and adaptive systems, e-Learning and teaching, human-centered and human-centric computing, recommender systems, intelligent control, robotics and mechatronics including human-machine teaming, knowledge-based paradigms, learning paradigms, machine ethics, intelligent data analysis, knowledge management, intelligent agents, intelligent decision making and support, intelligent network security, trust management, interactive entertainment, Web intelligence and multimedia.

The publications within "Advances in Intelligent Systems and Computing" are primarily proceedings of important conferences, symposia and congresses. They cover significant recent developments in the field, both of a foundational and applicable character. An important characteristic feature of the series is the short publication time and world-wide distribution. This permits a rapid and broad dissemination of research results.

** Indexing: The books of this series are submitted to ISI Proceedings, EI-Compendex, DBLP, SCOPUS, Google Scholar and Springerlink **

More information about this series at http://www.springer.com/series/11156

Ajith Abraham · Niketa Gandhi ·
Millie Pant
Editors

Innovations in Bio-Inspired Computing and Applications

Proceedings of the 9th International
Conference on Innovations in Bio-Inspired
Computing and Applications (IBICA 2018)
held in Kochi, India during
December 17–19, 2018

 Springer

Editors
Ajith Abraham
Machine Intelligence Research Labs
Auburn, WA, USA

Niketa Gandhi
Machine Intelligence Research Labs
Auburn, WA, USA

Millie Pant
Department of Applied Science
and Engineering
Indian Institute of Technology
Roorkee, India

ISSN 2194-5357   ISSN 2194-5365   (electronic)
Advances in Intelligent Systems and Computing
ISBN 978-3-030-16680-9   ISBN 978-3-030-16681-6   (eBook)
https://doi.org/10.1007/978-3-030-16681-6

Library of Congress Control Number: 2019936144

This Springer imprint is published by the registered company Springer Nature Switzerland AG
The registered company address is: Gewerbestrasse 11, 6330 Cham, Switzerland

Preface

Welcome to the Proceedings of the ninth International Conference on Innovations in Bio-Inspired Computing and Applications (IBICA 2018) and seventh World Congress on Information and Communication Technologies (WICT 2018). Conferences are held at Toc H Institute of Science and Technology (TIST) during December 17–19, 2018. Last year, IBICA 2017 was held in Marrakech, Morocco, and WICT 2017 was held in New Delhi, India.

The aim of IBICA is to provide a platform for world research leaders and practitioners, to discuss the full spectrum of current theoretical developments, emerging technologies, and innovative applications of bio-inspired computing. Bio-inspired computing is currently one of the most exciting research areas, and it is continuously demonstrating exceptional strength in solving complex real-life problems. WICT 2018 provides an opportunity for the researchers from academia and industry to meet and discuss the latest solutions, scientific results, and methods in the usage and applications of ICT in the real world. Innovations in ICT allow us to transmit information quickly and widely, propelling the growth of new urban communities, linking distant places and diverse areas of endeavor in productive new ways, which a decade ago was unimaginable. Thus, the theme of this World Congress is "Innovating ICT For Social Revolutions."

IBICA–WICT 2018 brings together researchers, engineers, developers, and practitioners from academia and industry working in all interdisciplinary areas of intelligent systems, nature-inspired computing, big data analytics, real-world applications and to exchange and cross-fertilize their ideas. The themes of the contributions and scientific sessions range from theories to applications, reflecting a wide spectrum of the coverage of intelligent systems and computational intelligence areas. IBICA 2018 received submissions from 10 countries, and each paper was reviewed by at least five reviewers in a standard peer-review process. Based on the recommendation by five independent referees, finally 25 papers were accepted for the conference (acceptance rate of 37%). WICT 2018 received submissions from 12 countries, and each paper was reviewed by at least five reviewers in a standard peer-review process. Based on the recommendation by five independent referees, finally 25 papers were accepted for the conference (acceptance rate of 35%).

Many people have collaborated and worked hard to produce the successful IBICA–WICT 2018 conference. First, we would like to thank all the authors for submitting their papers to the conference, for their presentations and discussions during the conference. Our thanks go to program committee members and reviewers, who carried out the most difficult work by carefully evaluating the submitted papers. Our special thanks to Oscar Castillo, Tijuana Institute of Technology, Tijuana, Mexico; Florin Popentiu Vladicescu, University Politehnica of Bucharest, Romania; and Sheng-Lung Peng, National Dong Hwa University, Taiwan, for the exciting plenary talks. We express our sincere thanks to the session chairs and organizing committee chairs for helping us to formulate a rich technical program.

Ajith Abraham
Preethi Thekkath
General Chairs - (IBICA-WICT 2018)

Millie Pant
Simone Ludwig
Antonio J. Tallón-Ballesteros
N. Vishwanath
Rasmi P. S.
Program Co-chairs

Organization

Patrons

K. Varghese   Toc H Public School Society, India
Alex Mathew   Toc H Public School Society, India

General Chairs

Ajith Abraham   Machine Intelligence Research Labs (MIR Labs), USA
Preethi Thekkath   Toc H Institute of Science and Technology, India

Program Co-chairs

Simone Ludwig   North Dakota State University, USA
Millie Pant   Indian Institute of Technology, Roorkee, India
Antonio J. Tallón-Ballesteros   University of Seville, Spain
N. Vishwanath   Toc H Institute of Science and Technology, India
Rasmi P. S.   Toc H Institute of Science and Technology, India

International Advisory Board

Achuthsankar S. Nair   University of Kerala, Thiruvananthapuram, India
Albert Zomaya   University of Sydney, Australia
Andre Ponce de Leon F. de Carvalho   University of Sao Paulo at Sao Carlos, Brazil
Bruno Apolloni   University of Milan, Italy

Francisco Herrera   University of Granada, Spain
Imre J. Rudas   Óbuda University, Hungary
Janusz Kacprzyk   Polish Academy of Sciences, Poland
Marina Gavrilova   University of Calgary, Canada
N. Krishnan   Manonmaniam Sundaranar University, Tamil
                                        Nadu, India
Patrick Siarry   Université Paris-Est Créteil, France
Ronald Yager   Iona College, USA
Salah Al-Sharhan   Gulf University of Science and Technology,
                                        Kuwait
Sankar Kumar Pal   ISI, Kolkata, India
Sebastian Ventura   University of Cordoba, Spain
Vincenzo Piuri   Universita' degli Studi di Milano, Italy
Vrinda V. Nair   APJ Abdul Kalam Technological University,
                                        India

Publication Chairs

Azah Kamilah Muda   UTeM, Malaysia
Niketa Gandhi   Machine Intelligence Research Labs (MIR Labs),
                                        USA

Web Service

Kun Ma   Jinan University, China

Publicity Committee

Atta Rahman   University of Dammam, Dammam, Saudi Arabia
Chinmay Chakraborty   Birla Institute of Technology, Jharkhand, India
G. Sudha Sadasivam   PSG College of Technology, Coimbatore, India
Meera Ramadas   University College of Bahrain,
                                        Kingdom of Bahrain
Marjana Prifti Sk'nduli   University of New York, Tirana
Mayur Rahul   C.S.J.M. University, Kanpur, India
Neeraj Rathore   Jaypee University of Engineering
                                        and Technology, Guna, MP, India
Nesrine Baklouti   University of Sfax, Tunisia
Sanju Tiwari   National Institute of Technology, Kurukshetra,
                                        Haryana, India

| Shikha Mehta | Jaypee Institute of Information Technology, Noida, India |
| Sourav Banerjee | Kalyani Government Engineering College, West Bengal, India |

Organizing Chairs

Babu John	Toc H Institute of Science and Technology, India
Sreela Sreedhar	Toc H Institute of Science and Technology, India
Rasmi P. S.	Toc H Institute of Science and Technology, India
N. Vishwanath	Toc H Institute of Science and Technology, India

Local Organizing Committee

Jesna Anver	Toc H Institute of Science and Technology, India
Saira Varghese	Toc H Institute of Science and Technology, India
Leda Kamal	Toc H Institute of Science and Technology, India
Elsaba Jacob	Toc H Institute of Science and Technology, India
Abin Oommen Philip	Toc H Institute of Science and Technology, India
Ceira Sara Cherian	Toc H Institute of Science and Technology, India
Mima Manual	Toc H Institute of Science and Technology, India
Ashly Joseph	Toc H Institute of Science and Technology, India
Rinu Rose George	Toc H Institute of Science and Technology, India
Mithu Mary George	Toc H Institute of Science and Technology, India
Anju Kuriakose	Toc H Institute of Science and Technology, India
Anuraj C. K.	Toc H Institute of Science and Technology, India

International Program committee

Abid Hussain Wani	University of Kashmir, India
Alberto Cano	University of Córdoba, Spain
Andries Engelbrecht	University of Pretoria, South Africa
Arun Kumar Sangaiah	Vellore Institute of Technology, India
Aswani Kumar Cherukuri	Vellore Institute of Technology, India
Azah Muda	UTeM, Malaysia
Cesar Hervas	University of Córdoba, Spain
Christian Veenhuis	HELLA Aglaia Mobile Vision GmbH, Germany
Daniel Valcarce	University of A Coruña, Spain
Daniela Zaharie	West University of Timisoara, Romania
Denis Felipe	Federal University of Rio Grande do Norte, Brazil

Sylvain Piechowiak Université de Valenciennes et du
  Hainaut-Cambrésis, France
Terry Trowbridge York University, Canada
Thanasis Daradoumis Open University of Catalonia, Greece
Thatiana C. N. Souza Federal University Rural Semi-Arid, Brazil
Thomas Hanne University of Applied Sciences Northwestern
  Switzerland, Switzerland
Varun Ojha ETH Zurich, Switzerland

Contents

Dynamic Parameter Adaptation Based on Using Interval Type-2 Fuzzy Logic in Bio-inspired Optimization Methods

Oscar Castillo[✉]

Tijuana Institute of Technology, Calzada Tecnologico s/n,
Tomas Aquino, 22379 Tijuana, Mexico
ocastillo@tectijuana.mx

Abstract. In this paper we perform a comparison of the use of type-2 fuzzy logic in two bio-inspired methods: Ant Colony Optimization (ACO) and Gravitational Search Algorithm (GSA). Each of these methods is enhanced with a methodology for parameter adaptation using interval type-2 fuzzy logic, where based on some metrics about the algorithm, like the percentage of iterations elapsed or the diversity of the population, we aim at controlling their behavior and therefore control their abilities to perform a global or a local search. To test these methods two benchmark control problems were used in which a fuzzy controller is optimized to minimize the error in the simulation with nonlinear complex plants.

Keywords: Interval type-2 fuzzy logic · Ant Colony Optimization · Gravitational Search Algorithm · Dynamic parameter adaptation

1 Introduction

Bio-inspired optimization algorithms can be applied to most combinatorial and continuous optimization problems, but for different problems need different parameter values, in order to obtain better results. There are in the literature, several methods aim at modeling better the behavior of these algorithms by adapting some of their parameters [18, 19], introducing different parameters in the equations of the algorithms [4], performing a hybridization with other algorithm [17], and using fuzzy logic [5–9, 14, 16].

In this paper a methodology for parameter adaptation using an interval type-2 fuzzy system is presented, where on each method a better model of the behavior is used in order to obtain better quality results.

The proposed methodology has been previously successfully applied to different bio-inspired optimization methods like BCO (Bee Colony Optimization) in [1], CSA (Cuckoo Search Algorithm) in [3], PSO (Particle Swarm optimization) in [5, 7], ACO (Ant Colony Optimization) in [6, 8], GSA (Gravitational Search Algorithm) in [9, 16], DE (Differential Evolution) in [10], HSA (Harmony Search Algorithm) in [11], BA (bat Algorithm) in [12] and in FA (Firefly Algorithm) in [15].

The algorithms used in this research are ACO (Ant Colony Optimization) from [8] and GSA (Gravitational Search Algorithm) from [9], each one with dynamic parameter

© Springer Nature Switzerland AG 2019
A. Abraham et al. (Eds.): IBICA 2018, AISC 939, pp. 1–12, 2019.
https://doi.org/10.1007/978-3-030-16681-6_1

adaptation using an interval type-2 fuzzy system. Fuzzy logic proposed by Zadeh in [20–22] help us to model a complex problem, with the use of membership functions and fuzzy rules, with the knowledge of a problem from an expert, fuzzy logic can bring tools to create a model and attack a complex problem.

The contribution of this paper is the comparison between the bio-inspired methods which use an interval type-2 fuzzy system for dynamic parameter adaptation, in the optimization of fuzzy controllers for nonlinear complex plants. The adaptation of parameters with fuzzy logic helps to perform a better design of the fuzzy controllers, based on the results which are better than the original algorithms.

2 Bio-inspired Optimization Methods

ACO is a bio-inspired algorithm based on swarm intelligence of the ants, proposed by Dorigo in [2], where each individual helps each other to find the best route from their nest to a food source. Artificial ants represent the solutions to a particular problem, where each ant is a tour and each node is a dimension or a component of the problem. Biological ants use pheromone trails to communicate to other ants which path is the best and the artificial ant tries to mimic that behavior in the algorithm.

Artificial ants use probability to select the next node using Eq. 1, where with this equation calculate the probability of an ant k to select the node j from node i.

$$P_{ij}^k = \frac{[\tau_{ij}]^\alpha [\eta_{ij}]^\beta}{\sum_{l \in N_i^k} [\tau_{ij}]^\alpha [\eta_{ij}]^\beta}, \qquad if\ j \in N_i^k \tag{1}$$

The components of Eq. 1 are: P^k is the probability of an ant k to select the node j from node i, τ_{ij} represents the pheromone in the arc that joins the nodes i and j and η_{ij} represents the visibility from node i to node j, with the condition that node j must be in the neighborhood of node i. Also like in nature the pheromone trail evaporates over time, and the ACO algorithm uses Eq. 2 to simulate the evaporation of pheromone in the trails.

$$\tau_{ij} \leftarrow (1 - \rho)\tau_{ij}, \qquad \forall (i,j) \in L \tag{2}$$

The components of Eq. 2 are: τ_{ij} representing the pheromone trail in the arc that joins the nodes i and j, ρ represents the percentage of evaporation of pheromone, and this equation is applied to all arcs in the graph L.

There are more equations for ACO, but these two equations are the most important in the dynamics of the algorithm, also these equations contain the parameters used to model a better behavior of the algorithm using an interval type-2 fuzzy system.

GSA proposed by Rashedi in [13], is a population based algorithm that uses laws of physics to update its individuals, more particularly uses the Newtonian law of gravity and the second motion law. In this algorithm each individual is considered as an agent, where each one represent a solution to a problem and each agent has its own mass and can move to another agent. The mass of an agent is given by the fitness function, agents with bigger mass are better. Each agent applies some gravitational force to all other agents, and is calculated using Eq. 3.

$$F_{ij}^d(t) = G(t) \frac{M_{pi}(t) \times M_{aj}(t)}{R_{ij}(t) + \varepsilon} \left(x_j^d(t) - x_i^d(t) \right) \tag{3}$$

The components of Eq. 3 are: F_{ij}^d is the gravity force between agents i and j, G is the gravitational constant, M_{pi} is the mass of agent i or passive mass, and M_{aj} is the mass of agent j or active mass, R_{ij} is the distance between agents i and j, ε is an small number used to avoid division by zero, x_j^d is the position of agent j and x_i^d is the position of agent j.

The gravitational force is used to calculate the acceleration of the agent using Eq. 4.

$$a_i^d(t) = \frac{F_i^d(t)}{M_{ii}(t)} \tag{4}$$

The components of Eq. 4 are: a_i^d is the acceleration force of agent i, F_i^d is the gravitational force of agent i, and M_{ii} is the inertial mass of agent i.

In GSA the gravitational constant G from Eq. 3, unlike in real life here it can be variable and is given by Eq. 5.

$$G(t) = G_0^{-\alpha t/T} \tag{5}$$

The components of Eq. 5 are: G is the gravitational constant, G_0 is the initial gravitational constant, α is a parameter defined by the user of GSA and is used to control the change in the gravitational constant, t is the actual iteration and T is the total number of iterations. To control the elitism GSA uses Eq. 6 to allow only the best agents to apply their force to other agents, and in initial iterations all the agents apply their force but Kbest will decrease over time until only a few agents are allowed to apply their force.

$$F_i^d(t) = \sum_{j \in Kbest, j \neq 1} rand_i F_{ij}^d(t) \tag{6}$$

The components of Eq. 6 are: F_i^d is the new gravity force of agent i, $Kbest$ is the number of agents allowed to apply their force, sorted by their fitness the best $Kbest$ agent can apply their force to all other agents, in this equation j is the number of dimension of agent i.

3   Methodology for Parameter Adaptation

The optimization methods involved in this comparison have dynamic parameter adaptation using interval type-2 fuzzy systems, and each of these adaptations are described in details for ACO in [8] and for GSA in [9]. The way in which this adaptation of parameters was performed is as follows: first a metric about the perfor-mance of the algorithms needs to be created, in this case the metrics are a percentage of iteration elapsed described by Eq. 7 and the diversity of individuals described by Eq. 8, then after the metrics are defined we need to select the best parameters to be dynamically adjusted, and this was done based on experimentation with different levels of all the parameters of each optimization method.

$$Iteration = \frac{Current\ Iteration}{Maximum\ of\ Iterations} \tag{7}$$

The components of Eq. 7 are: *Iteration* is a percentage of the elapsed iterations, *current iteration* is the number of elapsed iterations, and *maximum of iterations* is the total number iterations set for the optimization algorithm to find the best possible solution.

$$Diversity(S(t)) = \frac{1}{n_s} \sum_{i=1}^{n_s} \sqrt{\sum_{j=1}^{n_x} \left(x_{ij}(t) - \bar{x}_j(t)\right)^2} \tag{8}$$

The components of Eq. 8 are: *Diversity(S)* is a degree of dispersion of the population S, n_s is the number of individuals in the population S, n_x is the number of dimensions in each individual from the population, x_{ij} is the j dimension of the individual i, *tested* x_j is the j dimension of the best individual in the population. After the metrics are defined and the parameters selected, a fuzzy system is created to adjust just one parameter, and with this obtain a fuzzy rule set to control this parameter, and for all the parameters we need to do the same, and at the end only one fuzzy system will be created to control all the parameters at the same time combining all the created fuzzy systems. The proposed methodology for parameter adaptation is illustrated in Fig. 1, where it has the optimization method, which has an interval type-2 fuzzy system for parameter adaptation.

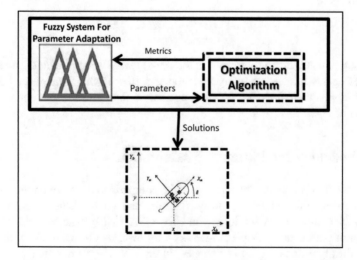

Fig. 1. General scheme of the proposal for parameter adaptation

Figure 1 illustrates the general scheme for parameter adaptation, in which the bio-inspired optimization algorithm is evaluated by the metrics and these are used as inputs for the interval type-2 fuzzy system, which will adapt some parameters of the optimization algorithm based on the metrics and the fuzzy rules. Then this method with parameter adaptation will provide the parameters or solutions for a problem, in this case the parameters for the fuzzy system used for control. The final interval type-2 fuzzy systems for each optimization method are illustrated in Figs. 2 and 3 respectively, for ACO and GSA correspondingly. Each of these fuzzy systems has iteration and diversity as inputs, with a range from 0 to 1 using the Eq. 7 and Eq. 8 correspondingly to each input, and two outputs but these differs from each optimization method because each one has its own parameters to be dynamically adjusted.

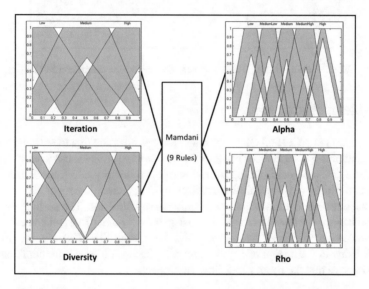

Fig. 2. Interval type-2 fuzzy system for parameter adaptation in ACO

The interval type-2 fuzzy system from Fig. 2 has two inputs and two outputs, the inputs are granulated into three type-2 triangular membership functions and the outputs into five type-2 triangular membership functions, and nine rules, in this case the parameters to be dynamically adjusted over the iterations are α *(alpha)* and ρ *(rho)* from Eq. 1 and Eq. 2 respectively, both with a range from 0 to 1.

The interval type-2 fuzzy system from Fig. 3 has *iteration* and *diversity* as inputs with three type-2 triangular membership functions and two outputs, which are the parameters to be adjusted in this case, α *(alpha)* with a range from 0 to 100 and *Kbest* from 0 to 1, each output is granulated into five type-2 triangular membership functions with a fuzzy rule set of nine rules. The parameters α *(alpha)* and *Kbest* are from Eq. 5 and Eq. 6 respectively.

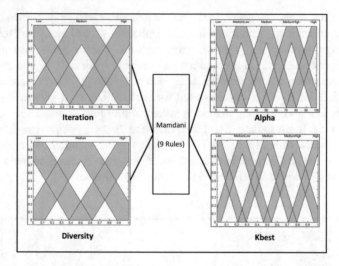

Fig. 3. Interval type-2 fuzzy system for parameter adaptation in GSA

4 Problems Statement

The comparison of ACO and GSA is through the optimization of a fuzzy controller from two different non-linear complex plants, where these two problems use a fuzzy system for control. The first problem is the optimization of the trajectory of an autonomous mobile robot and the objective is to minimize the error in the trajectory, the robot has two wheeled motors and one stabilization wheel, it can move in any direction. The desired trajectory is illustrated in Fig. 4, where first the robot must start from point (0, 0) and it needs to follow the reference using the fuzzy system from Fig. 5 as a controller. The reference illustrated in Fig. 4 helps in the design of a good controller because it uses only nonlinear trajectories, to assure that the robot can follow any trajectory.

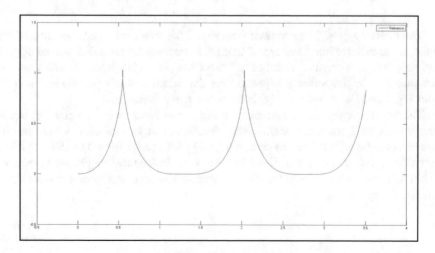

Fig. 4. Trajectory for the autonomous mobile robot

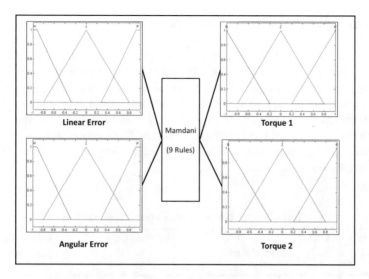

Fig. 5. Fuzzy controller for the autonomous mobile robot

The fuzzy system used for control illustrated in Fig. 5, and uses the linear and angular errors to control the motorized wheels of the robot. In this problem the optimization methods will aim at finding better parameters for the membership functions, using the same fuzzy rule set. The second problem is the automatic temperature control in a shower, and the optimization method will optimize the fuzzy controller illustrated in Fig. 6, which will try to follow the flow and temperature references. The fuzzy system used as control is illustrated in Fig. 6 and has two input variables, temperature and flow; a fuzzy rule set of nine rules and two outputs cold and hot is presented. The fuzzy system uses the inputs and with the fuzzy rules to control the open-close mechanism of the cold and hot water.

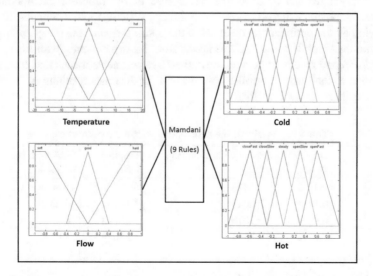

Fig. 6. Fuzzy controller for the automatic temperature control in a shower

5 Simulations, Experiments and Results

The optimization methods were applied to the optimization of the membership functions of the fuzzy system used as controllers for the two problems described in Sect. 4. Using the parameters from Table 1, each method was applied to both problems. In the case of the problem of the trajectory of an autonomous mobile robot there are 40 points to be search for all the membership functions, and in the problem of the automatic temperature control in a shower there are 52 points. The methods to be compared are: the original ACO method, ACO with parameter adaptation, original GSA method and GSA with parameter adaptation.

Table 1. Parameters for each optimization method

Parameter	Original ACO	ACO with parameter adaptation	Original GSA	GSA with parameter adaptation
Population	30	30	30	30
Iterations	50	50	50	50
α (Alpha)	1	Dynamic	40	Dynamic
β (Beta)	2	2		
ρ (Rho)	0.1	Dynamic		
Kbest			Linear decreasing from 100% to 2%	Dynamic
G_0			100	100

The parameters from Table 1 are a challenge for the optimization methods, because there are only 50 iterations to found the best possible fuzzy controller for each problem. This is a good manner to show the advantages of the proposed methodology for parameter adaptation using an interval type-2 fuzzy system. Table 2 contains the results of applying all the optimization methods to the optimization of the fuzzy controller for an autonomous mobile robot, the average is from 30 experiments (with Mean Square Error (MSE)) and results in bold are best, the 30 experiments means that each method was applied to the fuzzy controller optimization for 30 times resulting in 30 different fuzzy controllers for each method.

Table 2. Results of the simulations with the robot problem

MSE	Original ACO	ACO with parameter adaptation	Original GSA	GSA with parameter adaptation
Average	0.4641	**0.0418**	36.4831	15.4646
Best	0.1285	**0.0048**	10.4751	3.2375
Worst	0.9128	**0.1276**	76.0243	30.8511
Standard deviation	0.2110	**0.0314**	15.8073	8.6371

From Table 2 the optimization method that obtains better results is ACO with parameter adaptation using the proposed methodology with an interval type-2 fuzzy system, also it can be seen that the results of GSA with parameter adaptation are better that the original GSA, but ACO is better. The results in Table 3 are from applying all the methods to optimize the fuzzy controller for the automatic temperature control in a shower, the average is from 30 experiments (with the Mean Square Error (MSE)) and also the results in bold are best, same as the first problem the 30 experiments means that each method was applied to the optimization of the fuzzy controller and obtaining 30 different fuzzy controller for each method.

Table 3. Results of the simulations with the shower problem

MSE	Original ACO	ACO with parameter adaptation	Original GSA	GSA with parameter adaptation
Average	0.6005	0.4894	3.8611	**0.1151**
Best	0.5407	0.3980	1.9227	**0.0106**
Worst	0.9036	0.5437	6.5659	**0.3960**
Standard deviation	0.0696	0.0378	1.0860	**0.0913**

From the results in Table 3 in this case GSA with parameter adaptation using the proposed methodology using an interval type-2 fuzzy system can obtains better results than the other methods. Also it can be seen that ACO with parameter adaptation can obtain better results than the original ACO method and the original GSA method.

6   Statistical Comparison

The Z-test is a tool to prove that the methods with parameter adaptation can obtain on average better results than its counterparts the original methods, also to know what method is better on certain problem by comparing its results with all of the other methods. The comparison between the methods is using the statistical test Z-test, using the parameters from Table 4 and the results of the comparisons are in Tables 5 and 6 for the robot and shower problems, respectively.

Table 4. Parameters for the statistical Z-test

Parameter	Value
Level of significance	95%
Alpha (α)	5%
Alternative hypothesis (H_a)	$\mu_1 < \mu_2$ (claim)
Null hypothesis (H_0)	$\mu_1 \geq \mu_2$
Critical value	-1.645

The results in Table 5 are using the parameters in Table 4 for the Z-test, where it claims that a method (μ_1) has on average better results (we are comparing errors, so minimum is better) than the other method (μ_2), in Tables 5 and 6 the first column correspond to the methods as μ_1 and the first row correspond to the methods as μ_2, also we are not comparing the same method with itself, results in bold means that there are enough evidence to reject the null hypothesis.

Table 5. Results of the Z-test for comparison in the robot problem

μ_2 μ_1	Original ACO	ACO with parameter adaptation	Original GSA	GSA with parameter adaptation
Original ACO		10.8415	**−12.4795**	**−9.5098**
ACO with parameter adaptation	**−10.8415**		**−12.6269**	**−9.7803**
Original GSA	12.4795	12.6269		6.3911
GSA with parameter adaptation	9.5098	9.7803	**−6.3911**	

From the results in Table 5, which correspond to the optimization of a fuzzy controller for the trajectory of an autonomous mobile robot, there is enough evidence that ACO method with parameter adaptation can obtain on average better results than all of the other methods. There is enough evidence that the original ACO method can obtain on average better results than the original GSA and GSA with parameter adaptation. There is also enough evidence that GSA with parameter adaptation can obtain on average better results than the original GSA method.

Table 6. Results of the Z-test for comparison in the shower problem

μ_2 μ_1	Original ACO	ACO with parameter adaptation	Original GSA	GSA with parameter adaptation
Original ACO		7.6813	**−16.4115**	23.1516
ACO with parameter adaptation	**−7.6813**		**−16.9950**	20.7332
Original GSA	16.4115	16.9950		18.8264
GSA with parameter adaptation	**−23.1516**	**−20.7332**	**−18.8264**	

From the results in Table 6, which correspond to the optimization of a fuzzy controller for the automatic temperature control in a shower, there is enough evidence that GSA with parameter adaptation can obtain on average better results than all of the other methods. There is enough evidence that ACO with parameter adaptation can obtain on average better results than the original ACO method and the original GSA method. There is also enough evidence that the original ACO method can obtain on average better results than the original GSA method.

7 Conclusions

The optimization of fuzzy controllers is a complex task, because it requires the search of values for several parameters in a set of infinite possibilities in the range of each input or output variables. The bio-inspired optimization methods help in the search because they are guided by some kind of intelligence, from swarm intelligence or from laws of physics and can make a better search of parameters. With the inclusion of a fuzzy system in this case an interval type-2, the bio-inspired methods can search even in a better way, because is guided by the knowledge of an expert system that model a proper behavior in determined states of the search, in the beginning improves the global search or exploration of the search space and in final improves the local search or the exploitation of the best area found so far of the entire search space. From the results with the MSE there is clearly that ACO with parameter adaptation has the best results in the robot problem, and GSA with parameter adaptation has the best results in the shower problem, but with the statistical test it confirm these affirmations. The statistical comparison shows that the methods with parameter adaptation are better than their counterparts the original methods. Also ACO is a better method with the robot problem, but GSA is better in the shower problem.

References

1. Amador-Angulo, L., Castillo, O.: Statistical analysis of type-1 and interval type-2 fuzzy logic in dynamic parameter adaptation of the BCO. In: 2015 Conference of the International Fuzzy Systems Association and the European Society for Fuzzy Logic and Technology (IFSA-EUSFLAT-15). Atlantis Press, June 2015
2. Dorigo, M.: Optimization, learning and natural algorithms. Ph.D. thesis, Dipartimento di Elettronica, Politechico di Milano, Italy (1992)
3. Guerrero, M., Castillo, O., Garcia, M.: Fuzzy dynamic parameters adaptation in the cuckoo search algorithm using fuzzy logic. In: 2015 IEEE Congress on Evolutionary Computation (CEC), pp. 441–448. IEEE, May 2015
4. Hongbo, L., Ajith, A.: A fuzzy adaptive turbulent particle swarm optimization. Int. J. Innov. Comput. Appl. 1(1), 39–47 (2007)
5. Melin, P., Olivas, F., Castillo, O., Valdez, F., Soria, J., Garcia, J.: Optimal design of fuzzy classification systems using PSO with dynamic parameter adaptation through fuzzy logic. Elsevier Exp. Syst. Appl. 40(8), 3196–3206 (2013)
6. Neyoy, H., Castillo, O., Soria, J.: Dynamic fuzzy logic parameter tuning for ACO and its application in TSP Problems. In: Studies in Computational Intelligence, vol. 451, Springer, pp. 259–271 (2012)
7. Olivas, F., Valdez, F., Castillo, O., Melin, P.: Dynamic parameter adaptation in particle swarm optimization using interval type-2 fuzzy logic. Soft. Comput. 20(3), 1057–1070 (2016)
8. Olivas, F., Valdez, F., Castillo, O., Gonzalez, C., Martinez, G., Melin, P.: Ant colony optimization with dynamic parameter adaptation based on interval type-2 fuzzy logic systems. Appl. Soft Comput. 53, 74–87 (2016)
9. Olivas, F., Valdez, F., Melin, P., Sombra, A., Castillo, O.: Interval type-2 fuzzy logic for dynamic parameter adaptation in a modified gravitational search algorithm. Inf. Sci. 476, 159–175 (2019)

10. Ochoa, P., Castillo, O., Soria, J.: Differential evolution with dynamic adaptation of parameters for the optimization of fuzzy controllers. In: Recent Advances on Hybrid Approaches for designing intelligent systems, pp. 275–288. Springer International Publishing (2014)
11. Peraza, C., Valdez, F., Castillo, O.: An improved harmony search algorithm using fuzzy logic for the optimization of mathematical functions. In: Design of Intelligent Systems Based on Fuzzy Logic, Neural Networks and Nature-Inspired Optimization, pp. 605–615. Springer International Publishing (2015)
12. Perez, J., Valdez, F., Castillo, O., Melin, P., Gonzalez, C., Martinez, G.: Interval type-2 fuzzy logic for dynamic parameter adaptation in the bat algorithm. Soft Comput. 1–19 (2016)
13. Rashedi, E., Nezamabadi-pour, H., Saryazdi, S.: GSA: a gravitational search algorithm. Inf. Sci. **179**(13), 2232–2248 (2009)
14. Shi, Y., Eberhart, R.: Fuzzy adaptive particle swarm optimization. In: Proceeding of IEEE International conference on evolutionary computation, Piscataway, NJ: IEEE Service Center, Seoul, Korea, pp. 101–106 (2001)
15. Solano-Aragon, C., Castillo, O.: Optimization of benchmark mathematical functions using the firefly algorithm with dynamic parameters. In: Fuzzy Logic Augmentation of Nature-Inspired Optimization Metaheuristics, pp. 81–89. Springer International Publishing (2015)
16. Sombra, A., Valdez, F., Melin, P., Castillo, O.: A new gravitational search algorithm using fuzzy logic to parameter adaptation. In: 2013 IEEE Congress on Evolutionary Computation (CEC), pp. 1068–1074. IEEE, June 2013
17. Taher, N., Ehsan, A., Masoud, J.: A new hybrid evolutionary algorithm based on new fuzzy adaptive PSO and NM algorithms for distribution feeder reconfiguration. Energy Convers. Manag. **54**, 7–16 (2012)
18. Valdez, F., Melin, P., Castillo, O.: Evolutionary method combining particle swarm optimization and genetic algorithms using fuzzy logic for decision making. In: IEEE International Conference on Fuzzy Systems, pp. 2114–2119 (2009)
19. Wang, B., Liang, G., Chan Lin, W., Yunlong, D.: A new kind of fuzzy particle swarm optimization fuzzy_PSO algorithm. In: 1st International Symposium on Systems and Control in Aerospace and Astronautics, ISSCAA 2006, pp. 309–311 (2006)
20. Zadeh, L.: Fuzzy sets. Inf. Control **8**, 338–353 (1965)
21. Zadeh, L.: Fuzzy logic. IEEE Computer, pp. 83–92 (1965)
22. Zadeh, L.: The concept of a linguistic variable and its application to approximate reasoning —I. Inform. Sci. **8**, 199–249 (1975)

Effect of Derivative Action
on Back-Propagation Algorithms

Ahmet Gürhanlı[1]([envelope]), Taner Çevik[2], and Nazife Çevik[3]

[1] Department of Computer Engineering, İstanbul Aydın University,
İnönü Cad. 38, Sefaköy Küçükçekmece, 34295 Istanbul, Turkey
`ahmetgurhanli@aydin.edu.tr`
[2] Department of Software Engineering, İstanbul Aydın University,
İnönü Cad. 38, Sefaköy Küçükçekmece, 34295 Istanbul, Turkey
`tanercevik@aydin.edu.tr`
[3] Department of Computer Engineering, İstanbul Arel University,
Türkoba Mahallesi, Erguvan Sokak No: 26/K, Tepekent – Büyükçekmece,
34537 Istanbul, Turkey
`nazifecevik@arel.edu.tr`

Abstract. Multilayer neural networks using supervised training try to minimize the error between a given correct answer and the ones produced by the network. The weights in the neural network are adjusted at each iteration and after adequate epochs, adjusted weights give results close to correct answers. Besides the current error, accumulated errors from past iterations are also used for updating weights. This resembles the integral action in control theory, but the method took the name momentums in machine learning. Control theory uses one more technique for achieving faster tracking: the derivative action. In this research, we added the missing derivative action to the training algorithm and obtained promising results. The training algorithm with derivative action achieved 3.8 times speedup comparing to the momentum method.

Keywords: Machine learning · Neural networks · Deep learning ·
Artificial intelligence · PID controllers

1 Introduction

Training of multilayer neural networks is done by adjusting the weights between the nodes. It is often a time-consuming process and needs high performance computing. So, developing faster training algorithms is a crucial point in machine learning applications. Back-propagation algorithm [2–4] is the basis for training a network using supervised learning. Momentum [5–7] approach is widely used to minimize the error between learned outputs and correct ones. Ruder gives a good summary of gradient descent optimization algorithms that are trying to reach a minimum error level [8]. Numerous unsupervised training methods and reinforcement learning methods are also developed. Schmidhuber provides a good summery of these methods in his work [9]. The Momentum approach is very similar to the PI control (proportional and integral) which is very widely used to track a feedback signal in industry. However, in PID

© Springer Nature Switzerland AG 2019
A. Abraham et al. (Eds.): IBICA 2018, AISC 939, pp. 13–19, 2019.
https://doi.org/10.1007/978-3-030-16681-6_2

control, the derivative action is the one that accelerates the error tracking. It is usually omitted in real time control, because it may lead to instability issues. But, in machine learning applications the focus is on shortening the training time. Stability can be ensured by proper parameter tuning. So, the derivative action should not be omitted. Frequently, it might take hours or days to train an artificial neural network. Derivative action allows using higher proportional gains by avoiding the overshoots, and higher proportional gains lead to faster learning. The aim of this research is to evaluate the possible contributions of derivative action in various machine learning applications. In this paper, the initial research outcomes are reported. The background information will be given in Sect. 2 as introduced in Kim's deep learning book [1]. Proposed algorithm will be explained in Sect. 3. Simulation results will be discussed in Sect. 4. Conclusion remarks will be given in Sect. 5.

2   Background

2.1   Back-Propagation Algorithm

For sake of simplicity, let us consider a neural network having 2 layers as depicted in Fig. 1. This network has 1 input layer, 1 output layer and 1 hidden layer. The input data propagates through the nodes from left to right, being scaled by the weights of each connection. Each node has a non-linear activation function that makes multilayers contribute to the learning process. In this research, the sigmoid function is employed as the non-linear component at the nodes. Let us call the inputs to the network as x_1 and x_2. The inputs to the hidden nodes are $v_1^{(1)}$ and $v_2^{(1)}$. In our notation the superscripts in parenthesis denote the layer number for scalar numbers. The layer number is given as a subscript for matrices and vectors which are written in bold font. Input layer is not counted, because there is no operation at this layer. The outputs of the hidden layer or layer 1 are $y_1^{(1)}$ and $y_2^{(1)}$. A similar notation is used for the output layer.

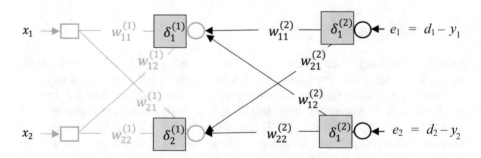

Fig. 1. Back-propagation algorithm

Now let's derive the outputs from the inputs. First the inputs to the hidden layer will be as in Eq. 1.

$$\begin{bmatrix} v_1^{(1)} \\ v_2^{(1)} \end{bmatrix} = \begin{bmatrix} w_{11}^{(1)} & w_{12}^{(1)} \\ w_{21}^{(1)} & w_{22}^{(1)} \end{bmatrix} \begin{bmatrix} x_1 \\ x_2 \end{bmatrix} \triangleq \mathbf{W}_1 \mathbf{x} \tag{1}$$

The output of the first hidden layer is given by the sigmoid function. Some other non-liner functions can also be employed, but in this paper, we will keep using the sigmoid function.

$$\begin{bmatrix} y_1^{(1)} \\ y_2^{(1)} \end{bmatrix} = \begin{bmatrix} \varphi(v_1^{(1)}) \\ \varphi(v_2^{(1)}) \end{bmatrix} \tag{2}$$

The same equations operate at the output layer, as well.

$$\begin{bmatrix} v_1^{(2)} \\ v_2^{(2)} \end{bmatrix} = \begin{bmatrix} w_{11}^{(2)} & w_{12}^{(2)} \\ w_{21}^{(2)} & w_{22}^{(2)} \end{bmatrix} \begin{bmatrix} y_1^{(1)} \\ y_2^{(1)} \end{bmatrix} \triangleq \mathbf{W}_2 \mathbf{y}_1 \tag{3}$$

$$\begin{bmatrix} y_1^{(2)} \\ y_2^{(2)} \end{bmatrix} = \begin{bmatrix} \varphi(v_1^{(2)}) \\ \varphi(v_2^{(2)}) \end{bmatrix} \tag{4}$$

When training the neural network, the error between the correct output and the calculated one is propagated back through the network. Here, a variable called delta is used to move back through a node. Delta is defined as the product of the error at the output of the node by the derivative of the activation function.

$$e_1^{(2)} = d_1 - y_1^{(2)}$$
$$\delta_1^{(2)} = \varphi'\left(v_1^{(2)}\right)e_1^{(2)}$$

$$e_2^{(2)} = d_2 - y_2^{(2)}$$
$$\delta_2^{(2)} = \varphi'\left(v_2^{(2)}\right)e_2^{(2)} \tag{5}$$

The errors of the hidden nodes are defined as the weighted sum of back-propagated deltas from the layer on the right-side. Once the error is calculated, the delta of the inner node is obtained the same way as in Eq. 5.

$$e_1^{(1)} = w_{11}^{(2)}\delta_1^{(2)} - w_{21}^{(2)}\delta_2^{(2)}$$
$$e_2^{(1)} = w_{12}^{(2)}\delta_1^{(2)} - w_{22}^{(2)}\delta_2^{(2)}$$

$$\begin{bmatrix} e_1^{(1)} \\ e_2^{(1)} \end{bmatrix} = \begin{bmatrix} w_{11}^{(2)} & w_{21}^{(2)} \\ w_{12}^{(2)} & w_{22}^{(2)} \end{bmatrix} \begin{bmatrix} \delta_1^{(2)} \\ \delta_2^{(2)} \end{bmatrix}$$

$$\mathbf{e}_1 = \mathbf{W}_2^{\mathbf{T}}\boldsymbol{\delta}_2 \tag{6}$$

Equation 6 tells that we can obtain the error vector of the hidden node as the product of the transposed weight matrix and the delta vector. After that we can calculate weight update matrices as previously defined and adjust the weights according to the following learning rule:

$$\Delta \mathbf{W_k} = \alpha \delta_k \mathbf{x_k^T}$$
$$\mathbf{W_k} \leftarrow \mathbf{W_k} + \Delta \mathbf{W_k} \tag{7}$$

In Eq. 7, $\mathbf{W_k}$ is the weight matrix of Layer k. α is the scaling parameter for weight adjustment at each iteration. Usually it is taken as 0.9, but it might need to be adjusted for different applications. $\mathbf{x_k^T}$ is the transpose of the input vector of the Layer k. This method is a form of stochastic gradient descent (SGD).

2.2   Momentums

To improve the speed of learning, we can add the scaled sum of previous weight update matrices to the current weight update matrix [1]:

$$\Delta \mathbf{W_k} = \alpha \delta_k \mathbf{x_k^T}$$
$$\mathbf{M} = \Delta \mathbf{W_k} + \beta \mathbf{M^-}$$
$$\mathbf{W_k} = \mathbf{W_k} + \mathbf{M} \tag{8}$$
$$\mathbf{M^-} = \mathbf{M}$$

$\mathbf{M^-}$ denotes the momentum in the previous iteration, it is added to the weight adjustment matrix after scaling with β and its values are updated using the last adjustment matrix after each iteration. This method uses the error history, so that a faster and more stable learning process is obtained.

3   Back-Propagation with Derivative Action

The momentum concept in previous section is very similar to the integral action in PID (proportional, integral, derivative) control. Integral action sums up the error history and leads to a zero error after the settling time of the system. So, we can say back-propagation plus momentum acts like a PI controller.

However derivative action is missing in this technique. Derivative action takes the derivative of the error signal and adds it to the weight adjustment signal after scaling with a control parameter. Derivative action should accelerate the learning time considerably. When the error is away from zero the differential signal has higher values and as error approaches to zero it will have values close to zero. Deep learning calculations take very long time in most applications and accelerating the process is very crucial for building a good algorithm.

We propose the following learner:

1. Initialize the weights using estimated values
2. Get the input from the training data.
3. Obtain neural network's output.
4. Calculate the error between obtained output and correct output of the training data.

$$\mathbf{e} = \mathbf{d} - \mathbf{y}$$

5. Calculate the delta of the output nodes.

$$\boldsymbol{\delta} = \varphi'(\mathbf{v}) \odot \mathbf{e}$$

Note: When using the sigmoid function as the non-linear element its derivative can be placed in this equation as below.

$$\boldsymbol{\delta} = \mathbf{y} \odot (1 - \mathbf{y}) \odot \mathbf{e}$$

6. Propagate the output node delta backwards and calculate the deltas of the left nodes one by one.

$$\mathbf{e_k} = \mathbf{W}_{k+1}^{T} \boldsymbol{\delta}_{k+1}$$

$$\boldsymbol{\delta}_k = \varphi'(\mathbf{v_k}) \odot \mathbf{e}_k$$

7. Repeat 6 until reaching the layer on the immediate right of the input layer.
8. Adjust the weights according to the PID learning rule.
   a. Calculate the proportional term.

$$\mathbf{P_k} = \boldsymbol{\delta}_k \mathbf{x}_k^T$$

   b. Calculate the integral term.

$$\mathbf{I_k} \leftarrow \mathbf{I_k} + \mathbf{P_k}$$

   c. Calculate the derivative term.

$$\mathbf{D_k} \leftarrow \mathbf{P_k} - \mathbf{P_k^-}$$
$$\mathbf{P_k^-} \leftarrow \mathbf{P_k}$$

   d. Update weights using parameters to control the PID activity.

$$\Delta \mathbf{W_k} = \alpha_k \mathbf{P_k} + \beta_k \mathbf{I_k} + \gamma_k \mathbf{D_k}$$
$$\mathbf{W_k} \leftarrow \mathbf{W_k} + \Delta \mathbf{W_k}$$

9. Repeat steps 2–8 for each training data point.
10. Repeat steps 2–9 until the network is properly trained.

4 Simulation Results

The SGD, Momentum and proposed learning algorithms are implemented in MATLAB. The machine is trying to learn a supervisor function as shown in the simulation setup in Fig. 2. The settling time is defined as the time accomplishing an error level less than 0.005. The performance of SGD, Momentum, and proposed algorithms are plotted on Fig. 3. Settling time of each method is marked with a star sign on their plots. Proportional method reaches the desired level after 141 iterations. Momentum approach performs much better with a settling iteration of 80 times. Proposed method's learning is much faster and arrives to the settling level after 21 iterations. After adding derivative action, the new algorithm gives 6.71 times speedup with respect to the SGD method and 3.8 times speedup when compared to the Momentum approach.

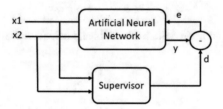

Fig. 2. Simulation setup

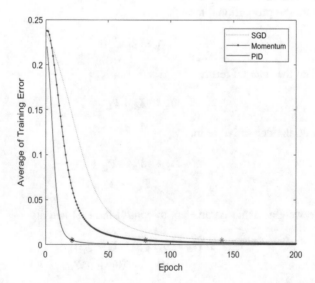

Fig. 3. Performance comparison of SGD, Momentum and proposed learning algorithms

5   Conclusion

Derivative action improved the learning speed considerably. When the difference between the correct output and calculated one is big, the derivate term adds more correction to the network weights. As the output error becomes smaller, the correction signal gets smaller, as well. Obtained speedup over momentum method is 3.8 times. The tuning of learning parameters is subject of our next research. The method can be used in most of the machine learning applications to reduce the training time. As our future work, we will evaluate the algorithm using several datasets and formal comparison results will be reported. The algorithm will also be tested in a real-world application.

References

1. Kim, P.: MATLAB Deep Learning with Machine Learning, Neural Networks and Artificial Intelligence, 1st edn. Apress, Soul (2017)
2. Nitta, T.: A back-propagation algorithm for complex numbered neural networks. In: Proceedings of 1993 International Conference on Neural Networks, Nagoya, pp. 1649–1652 (1993)
3. Kim, M.S.: Modification of backpropagation networks for complex-valued signal processing in frequency domain. In: Proceedings of IJCNN vol. 3, pp. 27–31 (1990)
4. Nitta, T., Fruya, T.: A complex back-propagation learning. Trans. Inf. Process. Soc. Japan **32** (10), 1319–1329 (1991)
5. Wiegerinck, W., Komoda, A., Heskes, T.: Stochastic dynamics of learning with momentum in neural networks. J. Phys. A Math. Gen. **27**, 4425–4437 (1994)
6. Swanston, D.J., Bishop, J.M., Mitchell, R.J.: Simple adaptive momentum: new algorithm for training multiplayer perceptrons. Electron. Lett. **30**, 1498–1500 (1994)
7. Scarpetta, S., Rattray, M., Saad, D.: Natural gradient matrix momentum. In: Proceedings of the Ninth International Conference on Neural Networks, pp. 43–48. The Institution of Electrical Engineers, London (1999)
8. Ruder, S.: An overview of gradient descent optimization algorithms. arXiv preprint arXiv: 1609.04747 (2016)
9. Schmidhuber, J.: Deep learning in neural networks: An overview. Neural Networks **61**, 85–117 (2015)

Quantum Cryptography: A Survey

Lav Upadhyay$^{(\boxtimes)}$

Madhav Institute of Technology and Science, Gwalior, India
lavupadhyay@gmail.com

Abstract. This paper represents the overview of Quantum Cryptography. Cryptography is the art of secrecy and it is the use of quantum mechanical properties to perform cryptographic tasks. It is a way of securing the channel using quantum mechanics properties. There are so many examples of quantum cryptography but the most important example is Quantum Key Distribution, which provides a solution to the breaking of various popular public key encryption and signature schemes (e.g. RSA and ElGamal). This helps to solve the security problems and also makes the communication channel is more secure. There are so many advantages of quantum cryptography, one thing is that the quantum computer gives the quadratic speed up on the general problems and second thing is that the quantum cryptography lies in the fact it allows the completion of various cryptographic tasks. That is proven to be impossible using classical communication.

Keywords: Quantum cryptography · Quantum cryptography protocols ·
Post Quantum Cryptography

1 Introduction

When quantum cryptography came in early 1970s, so many concept or techniques came like conjugate coding, bit commitment etc. The quantum key cryptography follows the principle of Heisenberg uncertainty and the concept of photon polarization. In this section, firstly describe why we need of this type of cryptography as being classical computing cryptography is there, in simple way quantum cryptography is more secure than classical cryptography. There is no possibility of copying the data encoded in quantum cryptography. This can be helped as detecting the eavesdropping in cryptography. When Conjugate coding concept came, then after some times, it was rejected by information theory society. After this SIGACT news published then a method is proposed for secure communication which is based on Conjugate Observables, which is now called as BB84. After this, published some approaches like Entanglement which is known as Quantum Entanglement. There have some quantum protocols like BB84 protocol, E91 protocol, BB92 protocol, SARG04 protocol, S09 protocol, S13 protocol, COW protocol, DPS protocol, KMB09 protocol. These protocols are used in QKD. The most important example of Quantum Cryptography is Quantum Key Distribution. In this paper, discussed some phase of quantum computing. As after quantum computing, move to quantum cryptography. There are some fundamental concepts of quantum cryptography.

© Springer Nature Switzerland AG 2019
A. Abraham et al. (Eds.): IBICA 2018, AISC 939, pp. 20–35, 2019.
https://doi.org/10.1007/978-3-030-16681-6_3

Quantum Computers
It is that one whose operation is governed by the laws of quantum mechanics. It can be differentiated from the classical computer i:e, 0 -> Bit, 1 -> Bit but in quantum computer $|0>$ -> state, $|1>$ -> are showing the states.

Qubit
Qubit is the fundamental unit of quantum computing. Sometimes it is called quantum bits or quantum numbers. Qubit has two states like horizontal and vertical polarization. Quantum mechanics allow the both states of qubit in superposition at the same time. For describing the quantum states Di-rac or Bra-ket notation required. In this notation angel brackets ("<and>") and vertical bars ("|"). In such terms another representation is very important in quantum mechanics that is: $<\phi \mid \psi>$, consisting left part: $<\phi|$ called Bra and a right part $|\psi>$ called Ket and overall notation is called Bra ket notation.

Quantum State
The linear polarization photon is quantum state. Generally the value of the variable cannot be determined certainty. If focus on quantum mechanics then the value of the polarization angle can be in range of $(0°, 180°)$, although only Boolean predicates (States) can be measured about this variable [1].

Superposition
Quantum Superposition is the very important feature of quantum mechanics, it tells about the quantum states. It states that any two quantum states can be added together or superposed and finally the result will be another valid quantum state and if we say opposite from that any quantum state can be shown as sum of two or more other distinct states. Quantum mechanics follows the Schrödinger equation from this it can be solved easily.

Heisenberg Uncertainty Principle
In this principle there is only one property to be known among conjugate pair of particle with certainty. Like momentum of a particle and the position of the particle, if position is known then momentum has to be shown uncertain. For example in quantum cryptography if we take two basis like horizontal and vertical and once the information have been made in diagonal direction then all the information in horizontal direction will be disappeared so at the same time it is impossible to find the horizontal and diagonal direction of particle [2], this shows that the Heisenberg Principle follow.

No Cloning Theorem
In this theorem, a single quantum cannot be cloned [4], there has no copy of quantum state. It is impossible to create multiple copies of quantum states. It follows the Heisenberg Uncertainty principle.

Quantum Entanglement
This is the quantum phenomenon [5], the pair of states cannot do measure anything independently. The first prior condition is both states should be twisted or entangled. The measurement of physical properties like polarization, spin etc. they cannot measure separately. The measurement is done simultaneously at both states. Like an example if we measure the spin of the pair of particles and the sum of the spin of particles is 0 and

one particle found in clockwise direction then another particle found in anticlockwise direction simultaneously as shown in Fig. 1.

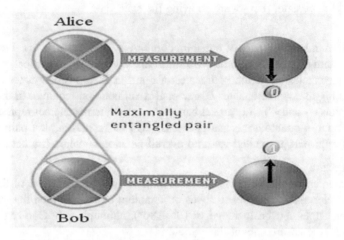

Fig. 1. Entangled states

2 Quantum Key Distribution

This scheme allows the one to distribute the sequence of random bits whose randomness and privacy are guaranteed by the laws and concept of quantum mechanics. These sequences can be used as secret keys and also surety about the confidentiality data transfer. There have some practical limitations of quantum key distribution; QKD requires an optical environment for transmission like optical fiber. The range of QKD is limited about 60 miles or 100 kms and research is going up to 250 kms [21].

One important thing is that QKD is to used to identify the third party between users during processing. Quantum key is only used to produce and distribute key only not for message and any type of data.

There are some symbols which are used in quantum cryptography for polarization of photon:

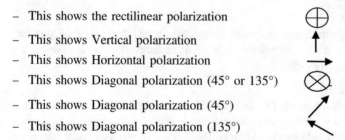

- This shows the rectilinear polarization
- This shows Vertical polarization
- This shows Horizontal polarization
- This shows Diagonal polarization (45° or 135°)
- This shows Diagonal polarization (45°)
- This shows Diagonal polarization (135°)

Quantum Key Distribution is secure [21, 22], but if multi photon particles generated on channel then could be possibility of Eve. A PNS attack which is known as Photon number Splitting attack more possibility to occur on the channel.

In QKD distance has major important role, experimentally 100 km are done but researchers trust that they are able to send till 250 km [21, 22].

In quantum mechanics can be observed with the help of operators, like commutative operator where A.B = B.A, does not followed by quantum mechanics. It states that A is happened before B or B is happened before A.

Quantum Key Distribution follows non commuting operator property means that A. B ≠ B.A, this operator shows the property of Heisenberg phenomenon. When measuring the polarization angle between basis 1 and basis 2, if Eve finds the angle in basis 1 before Bob must be different if Bob finds the angle in basis 1 before Eve in basis 2.

3   Quantum Distribution Key Parameters

There are some parameters to analysis the quantum computing i:e distance, error bit rate, data rate etc.

There are many theoretical and practical experiment was done on the long distance transmission over an optical fiber cable [20], but distance is a big challenge in quantum computing. Efficient entanglement distribution can be done over 200 km. Quantum Error bit rate can be analyzed with distance as, QBER $\propto$ 1/distance (Table 1).

Table 1.   Quantum computing parameters

Fiber distance	20 km	100 km
Secure key rate	1.02 Mbit/s	10.1 Kbit/s
Quantum bit error rate	Less	More
Security	More	Experimentally less

4   Quantum Distribution Protocols

BB84, proposed in 1984 by Bennett and Brasssard, In this protocol the pair of states are used and these pair of states are orthogonal to each other. In it Alice chooses a random bit like 0 or 1 then set with the basis, basis are two types one is rectilinear and diagonal basis. These are denoted by some specific symbols. The Rectilinear and Diagonal basis are shown in Fig. 2 respectively. Alice sends a single photon in that state to Bob till all the bits would not be send. Bob is also randomly go with basis. If bit have same basis correspondingly at both ends then successfully shared key. There is chance to present evesdropping, if we take an example such as Alice generate random bits 0 or 1 and chooses a random bit then send to Bob with basis. Bob chooses also basis randomly, if Alice sends a bit with basis as horizontally but Bob don't choose same so conflicting will be occur and the probability of matching the state of both is 1 (horizontally), but the probability of D/A is (45°) and −0.5 for D/A (for 135°) but if Eves is not present between then the probability is 0.75 for accept the bits. BB84 Protocol has some basic phenomenon as given below.

(1) Alice will choose (4 * n + $) number of bits, will encode randomly and send to Bob.
(2) Bob also measures random bases.
(3) In the likelihood probability at least 2 * n cases based on basis would be agree.
(4) Bob discards all those bits where the basis would not be matched.
(5) In case of error is small then they will use the remaining n bits as a one time pad for secure communication (Fig. 3).

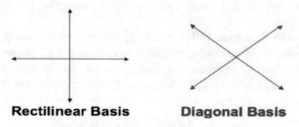

Rectilinear Basis   **Diagonal Basis**

Fig. 2. Bit encoding

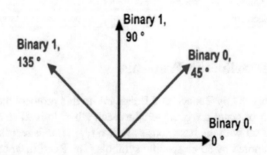

Fig. 3. Photon polarizations

This can be understood by the given tables below, there are two cases like:

(1) Table 2 represents that only Alice and Bob are there, No Eve is there between them. Alice wants to send message to Bob, First Alice chooses random basis and bits and send to Bob, Bob also chooses random basis if basis are matched then Bob receives the same bits as Alice has chosen.

Table 2. Bit sending without error

Alice	Bit	1	0	0	1	1	0	0	0
	Bases	$\oplus$	$\otimes$	$\otimes$	$\oplus$	$\otimes$	$\oplus$	$\oplus$	$\otimes$
Bob	Bit	1	0	0	0	1	0	0	0
	Bases	$\oplus$	$\oplus$	$\otimes$	$\otimes$	$\oplus$	$\otimes$	$\oplus$	$\otimes$
	Secured Key	1	-	0	-	-	-	0	0

(2) Table 3 represents that there are three Alice, Bob and Eve which is a eaves-dropper. Alice wants to send message to Bob. When Alice chooses random bits and basis and sends on the channel then between Alice and Bob, Eve is there; she also chooses the random basis and steals the bits and basis. Now the final secured key is based on all three basis matched.

Table 3. Bit sending with error

Alice	Bit	1	0	0	1	1	0	0	0
	Bases	$\oplus$	$\otimes$	$\otimes$	$\oplus$	$\otimes$	$\oplus$	$\oplus$	$\otimes$
Eve	Bit	1	1	0	0	0	1	0	1
	Bases	$\oplus$	$\oplus$	$\otimes$	$\otimes$	$\otimes$	$\oplus$	$\otimes$	$\oplus$
Bob	Bit	1	0	0	0	1	0	0	0
	Bases	$\oplus$	$\oplus$	$\otimes$	$\otimes$	$\oplus$	$\otimes$	$\oplus$	$\otimes$
	Secured Key	1	-	0	-	-	-	-	-

E91, This is also known as Eckert protocol, this scheme is based on entangled pairs of photons. The EPR pair will be used to detect the presence of Eve in the system. In this, anybody can create these photons like Alice, Bob even evesdropper Eve. There are two situations here, one is both Alice and Bob measure same result with 100% probability either polarization is horizontal or vertical and second is if any attacked by evesdropper occur then both can detect it.

B92, the scheme is similar to BB84 protocol but difference is, here only two states out of four [4]. In this protocol Alice encodes her basis in a predetermined way – encoding 0 in computational basis, but bit 1 in any basis not orthogonal to this. Considering the case of diagonal basis.

$$|\psi> \; = \begin{cases} |0> & \textbf{If bit a} = \textbf{0} \\ (|0> + |1>)/\sqrt{2} & \textbf{If bit a} = \textbf{1} \end{cases}$$

Suppose Alice wants to encode bits like that then –
a = 0: computational basis |0>
a = 1: Diagonal basis (|0> + |1>)/$\sqrt{2}$

It converts the classical bits in two non orthogonal states. There is no measurement can separate out two non orthogonal states so this have to create the issue that is there is no possibility of identifying the bits with certainty. This coding scheme allows to the one party who is receiver to learn whenever the bits get send without any discussion with Alice. Bob uses a coin toss to decide the basis for measurement. Assume Bob chooses bit 1 then Alice cannot choose the bit and if Bob chooses bit 1' then Alice cannot choose bit 1 as. There is no measurement when Bob chooses bit 0.

4.1   Different QKD Protocols

Table 4 describes the QKD protocols, it also differentiates the protocols with each other. The first protocol BB84 has four states and follow the Heisenberg Uncertainty principles but later we see that E91 protocol follows the Quantum Entanglement as mentioned in table. B92 protocol has two non orthogonal states and more secure than BB84 and E91. As mentioned in above table all the protocols are classified according to their features, application etc.

Table 4. QKD protocols

QKD protocol	Phases/features	Principles	Advantage
BB4 protocol (1984)	It is the first QKD protocol, use 4 quantum states, classical post processing phase is not there [39]	Heisenberg Uncertainty Principles	Uses photon polarization, 4 states with difference of 45° (0°–135°)
E91 protocol (1991)	In this protocol both users can detect the attacks is present [39]	Quantum Entanglement	Uses ERP
B92 protocol (1992)	B92 uses 2 non orthogonal quantum states, more secure than BB84 and SARG04, Strong phase – reference coherent light factor present [39]	Heisenberg Uncertainty Principles	Only 2 states are required (0°, 45°)
SSP protocol (1999)	Six State Protocol is like BB84 protocol with an additional basis [32, 39]	Heisenberg Uncertainty Principles	Uses 6 states in all 3 directions ± (x, y, z)

(continued)

Table 4. (*continued*)

QKD protocol	Phases/features	Principles	Advantage
DPS protocol (2003)	Experimentally set up is easy and number of PNS attacks are less and due to this high efficiency in terms of secret bits per qubit [33, 39]	Quantum Entanglement	Robustness against PNS attacks
SARG04 protocol (2004)	SARG04 protocol has 4 states, can achieve higher secret key rate and greater source distance than BB84 [39]	Heisenberg Uncertainty Principles	More secure than BB84, QBER = 2 * BB84 (QBER)
COW protocol (2004)	Coherent one way protocol is the new one protocol	Quantum Entanglement	Reduced PNS attacks, efficient rate
KMB09 protocol (2009)	KMB09 is the another alternative protocol in QKD, this scheme is more secure due to minimum index transmission error rate (ITER) and QBER, which created by Eve [36, 39]	Heisenberg Uncertainty Principles	Minimum ITER and QBER
S09 protocol (2012)	This protocol is based on public and private key cryptography [37, 39]	Public private key Cryptography	More robust but implementation is harder due to exchanging qubits multiple times
S13 protocol (2013)	It is also a new protocol, it is similar to BB84 protocol and different from if by use of Private Reconciliation from a random see and Asymmetric cryptography [38, 39]	Heisenberg Uncertainty Principles	S13 protocol can be created in existing device without any modification

5 Quantum Key Distribution Networks

Quantum networks are responsible for transmission of message in physical channel. These networks are used to provide the transportation of quantum information from one end to other end which is separated by distance. While in distributed computing the transmission of quantum information can be done through quantum gates in networks. Secure communication can be done through quantum networks in quantum key distribution. There are some specific quantum networks as:

Darpa [27], Secoqc [28, 29], Swiss Quantum [30], Tokyo Qauntum Network [31].

6   Security of Quantum Cryptography

The security of quantum cryptography depends on the quantum mechanics [2, 21]. This behavior depends on the photon polarization, which have rectilinear states and diagonal states, when message is send on the quantum channel then no 100% change to identify the direction of photons during transmission [23]. And clone is not possible in quantum mechanics; if anyone try to this getting same polarization then its hope is useless. It is not sure that is quantum cryptography is perfectly secured, still chances to attacks on the quantum channels. For example noise can be the cause of insecurity. There is unconditionally secure way of Quantum cryptography. If noise is not there on the channel then it is unconditionally secured [24].

The Security issues are there over long distance in quantum cryptography [24]. On the channel data is confidential over long distance is has important role in quantum cryptography [26]. The security can be seen in these ways [24, 25]:

(1)   Quantum Key Distribution with Noisy Channels (Privacy Amplification)
(2)   Quantum Key Distribution with practical Equipment (PNS Attack)

There are some other attacks on QKD and quantum cryptography.

- Intercept and Resend Attack
- Man in the Middle Attack
- Photon Number Splitting Attack
- Denial of services
- Trojan- horse Attack
- Time Channel Attacks
- Quantum Hacking is also possible

7   Applications

There are many applications and methods used earlier, these are following.

7.1   Quantum Money

It was first application of quantum mechanics to cryptography. It is proposed design of bank notes, no one can to found and this is used to further creation the Oblivious Transfer. In the oblivious transfer sender sends the message to receiver with probability 0.5 but sender is not aware for that message is received or not by receiver so called it is oblivious. Rabin's Oblivious Transfer is based on RSA system and more useful transfer is 1–2 oblivious transfer. Further shown that is a very important critical problem in cryptography. In general 1 out of n oblivious transfer the user get only one database information, no scared about the others users what they have got or not.

7.2   Randomness Generation

It tells about how to transfer the data through random number, the one important thing is that when a random number is generated then its way is random but not at time random so some deterministic algorithms are used for find the random number. So a new term like pseudo random generator is used in classical world. In quantum world random number generator creates many random numbers sequences but can't predict due to this more security and low performance also exists, the performance can be enhanced on the hardware related tools.

8   Quantum Programming Language and Tools

As earlier cryptography [6, 7] known as classical cryptography exists, there have no idea about data structures and operators to define the quantum properties for quantum cryptography, so the Quantum Computer programming language is needed, it is the collection of programming language which provide the phrase of quantum algorithms using high level constructs. The goal of quantum programming language is to design to identify and support the useful high level concepts.

Linear logic inventor Girard [7, 8], did the important and significant work in formation of quantum programming language, basically these based on calculus (Lamda calculus). The first practical proposal for quantum programming was Knill's step in 1996 [8]. In his proposed pseudo code defined the description of the quantum random access machine (QRAM) model. This model is based on assumption which would be probable accurate. It defines the number of operations which have to performed system hardware to form the quantum states etc.

There are so many different types [7] of tools and circuits are created with the help of basic programming languages i:e C, C++, Java, Mathematica and Matlab. At last it solves the so many problems in fastest exponentiation time [6]. It has to be divided in two phases which are (1) Imperative programming language and (2) Functional quantum programming language [9].

1. Imperative Programming Language
This language is also known as procedural language, used to change the global environment of the program and system variables. The more appropriate complete imperative language defined by Omer, Sanders and Zuliani [11] and Bettelli [12]. There are so many common classical imperative languages include C, C++, FORTAN and JAVA.

With argued the first quantum programming language was Quantum Computational Language (QCL) [6]. The most important feature of imperative language is to create and to optimize the quantum operators. There have some quantum operations i:e Quantum Fourier, Quantum Not, Quantum Swap, Quantum Hadamard etc.

Quantum Guarded Command Language (qGCL) [6, 11] is another imperative quantum programming language.

In October 2007 Mlnarik [13] described another quantum programming language i:e LanQ. It has C like syntax and it has a feature that it supports the classical and quantum both operations.

2. Functional Quantum Programming Language

It is also known as declarative language [7]. As compare to imperative language, functional programming language do not rely on global system state, depends on to perform the mathematical operations transformed by the mapping from input to output. The basic thing focused in this language is Lamda Calculus.

3. Other Quantum Programming Language

There are some other approaches, defined by Freedman, Kitaev, and Wong [14], refer to Topological quantum field theories (TQFT) by quantum computers. It includes the topological properties such as particles follow and paths.

Mauerer's [15] defined another specific of cQPL language based on Selinger's QPL. Some additional extension are added with it to support distribution system communication. Udrescu [16] defined the hardware language to create the quantum circuits.

4. Challenges in Quantum Programming Language

There are some pending challenges [6, 18] for designing and analysis the quantum programming language. Semantically and Syntax analysis is also a important issue for designing purpose. There are some other challenges such that quantum concurrency, Continuous classical Output, Infinite Data type and Higher Order Data type etc.

8.1   Tools and Simulator

There are many hardware tools for programming language to create the quantum circuits i:e JQuantum simulator, Quirk simulator, Davy Wybiral quantum circuit simulator, Quasi simulator etc. There have some example of quantum circuits which have to created by Jquantum simulator and Matlab Simulator. To make the quantum circuits need of the quantum gates. They can be differentiated from the classical gates as (Table 5):

Table 5. Difference between classical and quantum gate

Features	Classical gate	Quantum gate
Feedback loop	Yes	NIL
Fain-in	Yes (allow wires to be joined together)	NIL
Fan-out	Yes (several copies of bits are produced)	NIL (No cloning theorem)
Representation	Bits	Matrices

8.2   Quantum Gate and Circuits with JQuantum Simulator

Hadamard Gate (Fig. 4).

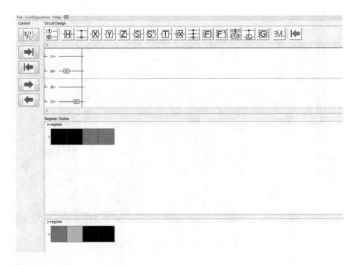

Fig. 4. Hadamard gate

Toffoli Gate (Fig. 5).

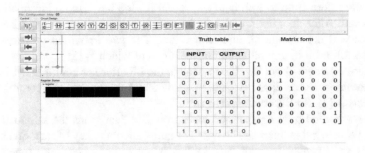

Fig. 5. Toffoli gate

9   Quantum Cryptography vs Post Quantum Cryptography

Quantum cryptography is [19] secure way to transmit the message from one party to another party. The question comes that still quantum cryptography is there then what is need of Post Quantum Cryptography. Answer is that Post Quantum Cryptography determines the some algorithms which are used to prevent the attacks through by the quantum computers. It is known to all that asymmetric key cryptography (public key cryptography) can be destroyed by the use of quantum computers. Also in quantum computing there have to some challenges for analysis and designing, another important approach is there called Post Quantum Cryptography. This cryptography is classified further (Fig. 6).

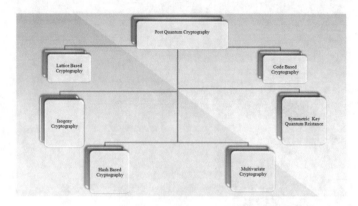

Fig. 6. Post Quantum Cryptography

10   Challenges

Bit Commitment
In this scheme Alice has a bit b but she wishes to commit to Bob but she does not want to send any information to Bob until she reveal it. Bob has also no information about the bit b until Alice does not reveal it. But a very important scenario is that Bob knows that Alice will not change bit after committed. The two important properties which should be followed in any bit commitment protocol which is [3]

Binding, Alice will not be able to change the bit after her commitment [3].

Concealing, Bob will not be able to know about the bit that she committed until she reveals it [3]. So Bit commitment as follows:

Alice writes bit b on a paper and locks in a safe then sends the safe to Bob. Bob receives the safe but at this stage he does not know about the key of safe. So he can not open this so b cannot be revealed. So this is showing Concealing property of bit commitment protocol and the second thing is that because off being safe at Bob's side, Alice cannot change the b value, so this is showing Binding property of bit commitment protocol. When she wants to reveal it she sends key to Bob.

There are two possibilities, one is secure commitment and second is unconditionally secure. If both the above schemes have done at same time then it is secure and it is unconditionally secure against the cheating of either Alice or Bob. There is a problem that how to secure unconditionally bit commitment scheme in classical world. But in quantum world there is hope of unconditionally secure bit commitment.

Quantum Rewinding
Zero knowledge proof that is in this system one end party can proof the other end party that this statement is true without transferring any information about this. So in quantum world. There has some problem because there is no copy of the script in order to return to it later on. The problem is how to store the quantum information and used as a secondary copy. After survey [10], there are some current challenges of quantum cryptography i:e Impossibility of quantum bit commitment, impossibility of secure two party computation using quantum communication, quantum rewinding, superposition access to oracles: quantum security notions, position based quantum cryptography. There are many other challenges in quantum cryptography.

   Current Challenges

- Long Distance Quantum Key Distribution
- Impossibility of Bit commitment
- Authentication Problem
- Confidentiality Problem over Long Distance
- Secure Quantum Key Distribution
- Zero Knowledge against Quantum Confliction

11   Conclusion

Quantum cryptography is too important and most advanced technology in area of quantum details. When it is used in quantum world then system become more secure and also have some stability. In quantum world when apply cryptography then it becomes more secure way to transmit the information. But this has faced to so many problems for convey the messages, so on this so many techniques are under research like security in internet of things and in networks generations like 5 g generations. At instant quantum cryptography is yet bounded distance so that's why performance issues are too much. But from last decade there are so many experiments are being to prove the high performance regarding this.

References

1. Brassard, G., Crepeau, C.: Quantum bit commitment and coin tossing protocols. In: CRYPTO 1990. LNCS, vol. 537, pp. 49–61 (1991)
2. Chhabra, N.: Secret key generation and eavesdropping detection using quantum cryptography 3(2), 3348–3354 (2012)
3. Watrous, J.: Impossibility of quantum bit commitment in university of calgary. In: CPSC 519/619: Quantum Computation
4. Wootters, W.K., Zurek, W.H.: A single quantum cannot be cloned. Nature **299**, 802–803 (1982). https://doi.org/10.1038/299802a0
5. Bennett, C.H., Brassard, G., Mermin, N.D.: Quantum cryptography without Bell's theorem. Phys. Rev. Lett. In **68**(5), 557–559 (1992). https://doi.org/10.1103/physrevlett.68.557

6. Selinger, P.: A brief survey of quantum programming languages. In: International Symposium on Functional and Logic Programming, pp. 1–6. Springer, Heidelberg, April 2004
7. Sofge, D.A.: A survey of quantum programming languages: history, methods, and tools. In: 2008 Second International Conference on Quantum, Nano and Micro Technologies, pp. 66–71. IEEE, February 2008
8. Girard, J.: Linear logic. Theor. Comput. Sci. **50**, 1–102 (1987)
9. Ambainis, A., Rosmanis, A., Unruh, D.: Quantum attacks on classical proof systems: the hardness of quantum rewinding. In: FOCS 2014, pp. 474–483 (2014). https://doi.org/10.1109/focs.2014.57
10. Broadbent, A., Schaffner, C.: Quantum cryptography beyond quantum key distribution. Des. Codes Cryptogr. **78**(1), 351–382 (2016)
11. Sanders, J.W., Zuliani, P.: Quantum programming. In: International Conference on Mathematics of Program Construction, pp. 80–99. Springer, Heidelberg (2000)
12. Bettelli, S., Calarco, T., Serafini, L.: Toward an architecture for quantum programming. Eur. Phys. J. D-Atomic Mol. Opt. Plasma Phys. **25**(2), 181–200 (2003)
13. Mlnarik, H.: Operational semantics and type soundness of quantum programming language LanQ. arXiv:quant-ph/0708.0890v1 (2007)
14. Freedman, M., Kitaev, A., Wong, Z.: Simulation of topological field theories by quantum computers. arXiv:quant-ph/0001071/v3 (2000)
15. Mauerer, W.: Semantics and simulation of communication in quantum computing. Master's thesis, University Erlangen-Nuremberg (2005)
16. Udrescu, M., Prodan, L., Vlâdutiu, M.: Using HDLs for describing quantum circuits: a framework for efficient quantum algorithm simulation. In: Proceedings of the 1st ACM Conference on Computing Frontiers. ACM Press (2004)
17. Gay, S.J.: Quantum programming languages: survey and bibliography. Math. Struct. Comput. Sci. **16**(04), 581–600 (2006)
18. Unruh, D.: Quantum programming languages. Informatik-Forschung und Entwicklung **21** (1–2), 55–63 (2006)
19. Shrestha, S.R., Kim, Y.S.: New McEliece cryptosystem based on polar codes as a candidate for post- quantum cryptography. In: 2014 14th International Symposium on Communications and Information Technologies (ISCIT), pp. 368–372. IEEE, September 2014
20. Dynes, J.F., Takesue, H., Yuan, Z.L., Sharpe, A.W., Harada, K., Honjo, T., Kamada, H., Tadanaga, O., Nishida, Y., Asobe, M., Shields, A.J.: Efficient entanglement distribution over 200 kilometers. Opt. Express **17**(14), 11440–11449 (2009)
21. Los Alamos National Laboratory. http://www.physorg.com/news86020679.html
22. Vignesh, R.S., Sudharssun, S., Kumar, K.J.: Limitations of quantum & the versatility of classical cryptography: a comparative study. In: Second International Conference on Environmental and Computer Science, ICECS 2009, pp. 333–337. IEEE, December 2009
23. Bennett, C.H.: Quantum cryptography: uncertainty in the service of privacy. Science **257**(7), 752–753 (1992)
24. Lo, H.K., Chau, H.F.: Unconditional security of quantum key distribution over arbitrarily long distances. Science **283**(5410), 2050–2056 (1999)
25. Gottesman, D., Lo, H.K., Lutkenhaus, N., Preskill, J.: Security of quantum key distribution with imperfect devices in Information Theory. In: Proceedings of the International Symposium on ISIT, p. 136. IEEE (2004)
26. Braun, J., Buchmann, J., Mullan, C., Wiesmaier, A.: Long term confidentiality: a survey. Des. Codes Cryptogr. **71**(3), 459–478 (2014)

27. Elliott, C., Colvin, A., Pearson, D., Pikalo, O., Schlafer, J., Yeh, H.: Current status of the DARPA quantum network. In: Defense and Security, pp. 138–149. International Society for Optics and Photonics, May 2005
28. Peev, M., Pacher, C., Alléaume, R., Barreiro, C., Bouda, J., Boxleitner, W., Debuisschert, T., Diamanti, E., Dianati, M., Dynes, J.F., Fasel, S.: The SECOQC quantum key distribution network in Vienna. New J. Phys. 11(7), 075001 (2009)
29. Poppe, A., Peev, M., Maurhart, O.: Outline of the SECOQC quantum-key-distribution network in Vienna. Int. J. Quant. Inf. 6(02), 209–218 (2008)
30. Stucki, D., Legre, M., Buntschu, F., Clausen, B., Felber, N., Gisin, N., Henzen, L., Junod, P., Litzistorf, G., Monbaron, P., Monat, L.: Long-term performance of the SwissQuantum quantum key distribution network in a field environment. New J. Phys. 13(12), 123001 (2011)
31. Sasaki, M., Fujiwara, M., Ishizuka, H., Klaus, W., Wakui, K., Takeoka, M., Miki, S., Yamashita, T., Wang, Z., Tanaka, A., Yoshino, K.: Field test of quantum key distribution in the Tokyo QKD Network. Opt. Express 19(11), 10387–10409 (2011)
32. Bechmann-Pasquinucci, H., Gisin, N.: Incoherent and coherent eavesdropping in the six-state protocol of quantum cryptography. Phys. Rev. Lett. A 59, 4238–4248 (1999)
33. Gisin, N., Ribordy, G., Zbinden, H., Stucki, D., Brunner, N., Scarani, V.: Towards practical and fast quantum cryptography. arXiv preprint arXiv:quant-ph/0411022 (2004)
34. Inoue, K., Waks, E., Yamanoto, Y.: Differential-phase- shift quantum key distribution using coherent light. Phys. Rev. A 68, 022317 (2003)
35. Waks, E., Takesue, H., Yamamoto, Y.: Security of differential-phase-shift quantum key distribution against individual attacks. Phys. Rev. A 73, 012344 (2006)
36. Khan, M.M., et al.: High error-rate quantum key distribution for long distance communication. New J. Phys. 11, 063043 (2009)
37. Esteban, E., Serna, H.: Quantum key distribution protocol with private-public key. arXiv: 0908.2146v4quant-ph, 12 May 2012
38. Serna, E.H.: Quantum Key Distribution from a random seed. arXiv:1311.1582v2quant-ph, 12 November 2013
39. Singh, H., Gupta, D.L., Singh, A.K.: Quantum Key Distribution protocols: a review. J. Comput. Inf. Syst. 8, 2839–2849 (2012)

Transfusion of Extended Vigenere Table and ASCII Conversion for Encryption Contrivance

Sakshi[1], Prateek Thakral[2]([⊠]), Karan Goyal[1], Tarun Kumar[1], and Deepak Garg[1]

[1] National Institute of Technology, Kurukshetra, Haryana, India
[2] Jaypee University of Information Technology, Waknaghat, Solan, India
18.prateek@gmail.com

Abstract. In the field of cryptography, to make cryptosystem more secure an evaluation of modulus operations on integral values followed by ASCII value generation on plain text characters have been deeply explored in this research paper. On the basis of this an extended algorithm has been proposed. In this, the Encryption technique consists of an extended combination of Vigenere and Caesar cipher which is the main key feature of this algorithm and then decryption of text along with ASCII algorithm and substitution methodology has been done. The Algorithm is initiated on the basis of inspection of various research papers, furthermore, reviews have been made for proving this system more reliable. In the proposed algorithm modified Vigenere table and ASCII values are taken into consideration for decreasing the steps to reduce complexity and making a more secure way of cryptography.

Keywords: Computer security · ASCII value · Cryptography · Cryptology · Multiplicative cipher · Random key · Symmetric key

1 Introduction

Network security comprises various strategy and application that are taken to avoid and check unintended access, alteration, misuse, or refusal of a network and other network resources that are accessible [1]. With each passing year, the security dangers confronting computer systems have turned out to be all the more, in fact, modern, better sorted out and harder to recognize. A great part of the data imparted every day must be kept private. Data, for example, monetary reports, worker information and therapeutic records should be conveyed in a way that guarantees secrecy and honesty [2]. The issue of unsecured correspondence is exacerbated by the way that quite a bit of this data is sent over people in general Internet and might be handled by outsiders, as in email or texting Presently, cryptography is recognized as a part of computer science as well as of science and mathematics, and is connected closely with, computer security, information concept, and engineering. Encryption simply means to change the message, also known as plaintext, to make it unreadable to any user who is not an authorized user [2]. In other words, we can say the whole procedure of encryption is to discover a protected

© Springer Nature Switzerland AG 2019
A. Abraham et al. (Eds.): IBICA 2018, AISC 939, pp. 36–45, 2019.
https://doi.org/10.1007/978-3-030-16681-6_4

mechanism by means of which just the valid or the authorized user has the access to the message sent by the sender. Cryptography is termed as "Hidden Secrets" and is mainly deals with encrypting the data [3]. It is useful for inspecting those conventions, which are related to various perspectives in information security, for instance, verifying data, ordering of data, non-dissent and data uprightness. The authorization secures and sustains the interest of both the parties i.e. the person who is sending message as well the authorized receiver [4].

There are various security aspects for a cryptographic technique including Authentication, Confidentiality, Integrity, and Non-Repudiation [5]. Authentication is only a procedure of demonstrating one's personality. Furthermore, if privacy or secrecy is taken into the picture it is tied in with confirming that no one else can read the information except for the intended user. To confirm that the received message has not been altered at all from its original content, is keeping up the Integrity. Non-Repudiation describes a system that shows that message or information received is from a valid user [6].

2 Related Work

The current work done in this field can be analyzed in this section briefly by experiencing the proposed model and various algorithm proposed by the different authors.

Mathur [7] proposed an algorithm that performs modulus operation on the plain text message and the secret key by using ASCII based conversion mechanism. The minimum value of both is stored in separate arrays. A binary conversion of the minimum key value is performed and right shifting is done. The shifted encrypted key is added to the minimum value of plain text characters and the final cipher text is produced. Saraswat et al. [8] proposed an algorithm that brings the new version for the vigenere table having the 26 alphabets (A–Z) as well as the 10 numeric digits (0–9). The alphanumeric cross-section of the rows and columns of the table provides the intersection text on which advance version of Caesar cipher is performed to get the final cipher text. Krishna [9] proposed a new algorithm that is totally different from symmetric, asymmetric or hashing function. The algorithm changes the input plain text into small sized packets of constant length and then stored in a binary matrix. The binary matrix has the 8 bit equivalent of every plain text characters' ASCII value. It uses rotational mathematical method and conversion of radix on the text repeatedly. The encrypted text is rotated by the cyclic function. The final cipher text is in the form of unprintable characters below ASCII value 32.

Bhargava et al. [10] proposed a new algorithm that incorporates the features of transposition and substitution cipher mechanism. It uses Multiplicative cipher techniques and Rail fence cipher techniques. In multiplicative cipher technique a secret key is selected such that taking its product with any other character would give another changed character for every plain text character. The final cipher text is a sequence of special characters that is formed by performing the algorithm on plain text. Gupta et al. [11] proposed a symmetric key encryption algorithm that takes plain text and two different keys. A modified Vigenere table is used that consists characters from ASCII table and values ranging from 33–126. An intersection text is produced by plain text and key 1 from the table. The Intersection text is then added with second key using modified Caesar cipher and modulo 26 arithmetic. It gives the final cipher text.

3 Proposed Algorithm

The algorithm that is being proposed here comes with the fusion of ASCII conversion and mathematical modification concept. Since the ASCII values are valid values for the numbers, characters, and other special symbols. The proposed algorithm will be coming up with the fulfillment of the shortcomings of the existing work and it will efficiently apply the ASCII values to produce the cipher text [12].

If we examine the ASCII table, we can draw conclusion that ASCII values from 33–126 depict the numeric values, alphabetical characters and special characters that are mostly used in day to day conversation while typing. These 94 characters are the printable characters that are available on every keyboard. In this algorithm the sender begins with the plain text message and a randomly generated key of length equivalent to the plain text.

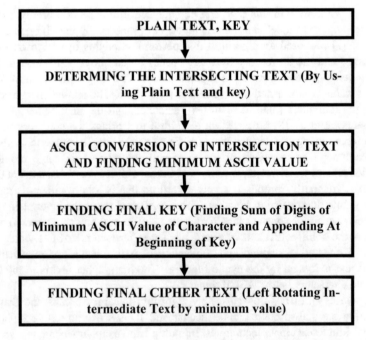

Fig. 1. Flowchart for the processing of the proposed algorithm

The entire operation can be understood by having gone through the proposed algorithm sequentially. The proposed algorithm can be viewed by the Fig. 1.

3.1 Step 1: Determining the Intersecting Text

This part is mainly affected by using the Vigenere table and is further extended to include the numeric values and the special characters. The below Fig. 1 presents the vigenere cipher table. The Vigenere cipher table contains only alphabets both in horizontal and vertical axis. The horizontal axis represents the plain text and the vertical axis represents the key. The intersection of these axis values i.e. the plain text character value and key character value gives the cipher text. The form of ordering used in the vigenere table will be used for ordering the characters for the table in the proposed algorithm (Fig. 2).

	A	B	C	D	E	F	G	H	I	J	K	L	M	N	O	P	Q	R	S	T	U	V	W	X	Y	Z
A	A	B	C	D	E	F	G	H	I	J	K	L	M	N	O	P	Q	R	S	T	U	V	W	X	Y	Z
B	B	C	D	E	F	G	H	I	J	K	L	M	N	O	P	Q	R	S	T	U	V	W	X	Y	Z	A
C	C	D	E	F	G	H	I	J	K	L	M	N	O	P	Q	R	S	T	U	V	W	X	Y	Z	A	B
D	D	E	F	G	H	I	J	K	L	M	N	O	P	Q	R	S	T	U	V	W	X	Y	Z	A	B	C
E	E	F	G	H	I	J	K	L	M	N	O	P	Q	R	S	T	U	V	W	X	Y	Z	A	B	C	D
F	F	G	H	I	J	K	L	M	N	O	P	Q	R	S	T	U	V	W	X	Y	Z	A	B	C	D	E
G	G	H	I	J	K	L	M	N	O	P	Q	R	S	T	U	V	W	X	Y	Z	A	B	C	D	E	F
H	H	I	J	K	L	M	N	O	P	Q	R	S	T	U	V	W	X	Y	Z	A	B	C	D	E	F	G
I	I	J	K	L	M	N	O	P	Q	R	S	T	U	V	W	X	Y	Z	A	B	C	D	E	F	G	H
J	J	K	L	M	N	O	P	Q	R	S	T	U	V	W	X	Y	Z	A	B	C	D	E	F	G	H	I
K	K	L	M	N	O	P	Q	R	S	T	U	V	W	X	Y	Z	A	B	C	D	E	F	G	H	I	J
L	L	M	N	O	P	Q	R	S	T	U	V	W	X	Y	Z	A	B	C	D	E	F	G	H	I	J	K
M	M	N	O	P	Q	R	S	T	U	V	W	X	Y	Z	A	B	C	D	E	F	G	H	I	J	K	L
N	N	O	P	Q	R	S	T	U	V	W	X	Y	Z	A	B	C	D	E	F	G	H	I	J	K	L	M
O	O	P	Q	R	S	T	U	V	W	X	Y	Z	A	B	C	D	E	F	G	H	I	J	K	L	M	N
P	P	Q	R	S	T	U	V	W	X	Y	Z	A	B	C	D	E	F	G	H	I	J	K	L	M	N	O
Q	Q	R	S	T	U	V	W	X	Y	Z	A	B	C	D	E	F	G	H	I	J	K	L	M	N	O	P
R	R	S	T	U	V	W	X	Y	Z	A	B	C	D	E	F	G	H	I	J	K	L	M	N	O	P	Q
S	S	T	U	V	W	X	Y	Z	A	B	C	D	E	F	G	H	I	J	K	L	M	N	O	P	Q	R
T	T	U	V	W	X	Y	Z	A	B	C	D	E	F	G	H	I	J	K	L	M	N	O	P	Q	R	S
U	U	V	W	X	Y	Z	A	B	C	D	E	F	G	H	I	J	K	L	M	N	O	P	Q	R	S	T
V	V	W	X	Y	Z	A	B	C	D	E	F	G	H	I	J	K	L	M	N	O	P	Q	R	S	T	U
W	W	X	Y	Z	A	B	C	D	E	F	G	H	I	J	K	L	M	N	O	P	Q	R	S	T	U	V
X	X	Y	Z	A	B	C	D	E	F	G	H	I	J	K	L	M	N	O	P	Q	R	S	T	U	V	W
Y	Y	Z	A	B	C	D	E	F	G	H	I	J	K	L	M	N	O	P	Q	R	S	T	U	V	W	X
Z	Z	A	B	C	D	E	F	G	H	I	J	K	L	M	N	O	P	Q	R	S	T	U	V	W	X	Y

Fig. 2. Vigenere table

Similar to this vigenere table, a new and extended table i.e. the extended version of previous Table is designed that not only contains the alphabetical characters but also the numeric and special characters. For Table 1 the horizontal axis depicts the plain text and the vertical axis depicts the key. The Table 1 is a matrix of 94 rows and columns for which the records start with the character whose value start with the ASCII value 33 with character "!" and it moves on till character whose ASCII value is 126 with character "~" in both the horizontal and vertical axis. This table consist of all 94 printable characters that contains all the capital letters (A–Z), small letters (a–z), numeric values (0–9) and remaining are the special printable characters. The following Table 1 i.e. the extended vigenere table will be used for the proposed algorithm.

The determination of the intersection text can be understood by the next example. Let us take the plain text that needs to be sent be "ABC12#" and the key be "SA4$#*". Now we will see how Intersecting Text is calculated.

Table 1. The extended vigenere table

	!	"	#	$	%	&	'	(	)	*	And up to ~
!	!	"	#	$	%	&	'	(	)	*	
"	"	#	$	%	&	'	(	)	*	+	
#	#	$	%	&	'	(	)	*	+	,	
$	$	%	&	'	(	)	*	+	,	-	
%	%	&	'	(	)	*	+	,	-	.	
&	&	'	(	)	*	+	,	-	.	/	
'	'	(	)	*	+	,	-	.	/	0	
(	(	)	*	+	,	-	.	/	0	1	
)	)	*	+	,	-	.	/	0	1	2	
*	*	+	,	-	.	/	0	1	2	3	
+	+	,	-	.	/	0	1	2	3	4	
,	,	-	.	/	0	1	2	3	4	5	
-	-	.	/	0	1	2	3	4	5	6	
.	.	/	0	1	2	3	4	5	6	7	
/	/	0	1	2	3	4	5	6	7	8	
0	0	1	2	3	4	5	6	7	8	9	
1	1	2	3	4	5	6	7	8	9	:	
2	2	3	4	5	6	7	8	9	:	;	
Up to till ~											

We can calculate Intersecting Text from table by finding the entry which is cross section point of plaint text character and key characters. Below gives the table showing the calculation of intersection text (Table 2).

Table 2. Working of proposed algorithm

Plain text	A	B	C	1	$	#
Key	S	A	4	$	#	*
Intersection text	S	B	V	4	&	,

Intersection text can also be calculated by using the mathematical formula. The formula for calculation intersection text is:

$$I.T = (PT - 33) + K \tag{1}$$

Where I.T = ASCII value of Intersection Text, PT = ASCII value of Plain Text, K = ASCII value of Key.

$$\text{IF (I.T > 126) THEN}$$
$$\{$$
$$\text{I.T} = \text{I.T} - 94; \qquad (2)$$
$$\}$$

This technique is applied so that the ASCII value of intersecting text is between the ranges of 33–126.

3.2 Step 2: Finding Minimum Value from the Intersection Text

In this step the ASCII values of the characters of the Intersection text are stored in an array and then sorted in ascending order. Now the smallest value among them is taken into consideration and stored in a separate variable (Table 3).

Table 3. Finding the minimum value from intersection text and sorting

Intersection text	S	B	V	4	&	,
ASCII value	115	98	86	52	38	44
Sorted ASCII values	38	44	52	86	98	115

Minimum (ASCII_VALUES) = 38, Store it in MIN. MIN = 38

Adding the digits of MIN recursively up to a single digit and store it in variable SUM.

SUM = 2.

3.3 Step 3: Final Cipher Text Determination

This step goes with determination of final cipher text. The ASCII values of the Intersection text are circularly sifted by the value of SUM (Table 4).

Table 4. Circular shifting

Intersection text	S	B	V	4	&	,
ASCII value	115	98	86	52	38	44
Circularly shifted ASCII values	86	52	38	44	115	98
Intermediate text	V	4	&	,	S	B

CIPHER TEXT: V4&,sb

The intermediate text obtained by the circular shift is the ultimate cipher text. Hence,

Plain Text	ABC1$#
Key	SA4$#*
Cipher Text	V4&,sb

4 Encryption and Decryption Algorithm

Encryption:
Input: Plain Text, Secret key. **Output**: Cipher Text
1. Begin: Read the Plain Text.
2. Generate a random Key equal to the length of Plain Text.
3. Determine the Intersecting text(I.T) – each character
4. Check the modified extended Vigenere table or
5. Calculate using formula
    a. I.T=Plain Input Text - 33+key
    b. If (I.T > 126)
    c. Then I.T = I.T - 94;
6. Take rotate= Smallest ASCII value from all characters of the Intersection Text.
    a. If (Rotate >9)
    b. Then Rotate=sum of all digits of Rotate.
7. Left Circular Shift the characters of Intersection text by the value of Rotate.
8. Append the value of rotate at the beginning of Intersection Text = Intermediate Text.
9. Final CIPHER TEXT = Intermediate Text.
10. End

Decryption:
1. Begin: Take the cipher Text.
2. Extract the rotate value from beginning of cipher text.
3. Right Circular Shift the characters of the Cipher Text by the value of rotate.
4. Find the Final Intersection Text (FIT)
    a) FIT = Cipher Text +33 – key
    b) If (FIT > = 126)
    c) Then FIT = FIT -94;
    d) If (FIT < = 33)
    e) Then FIT = FIT+94;
5. Convert the Final Intersection Text To Plain Text.
6. End

The implementation of the above algorithm can be in any language of our choice. A sample output screen of the encryption and decryption is given below (Fig. 3).

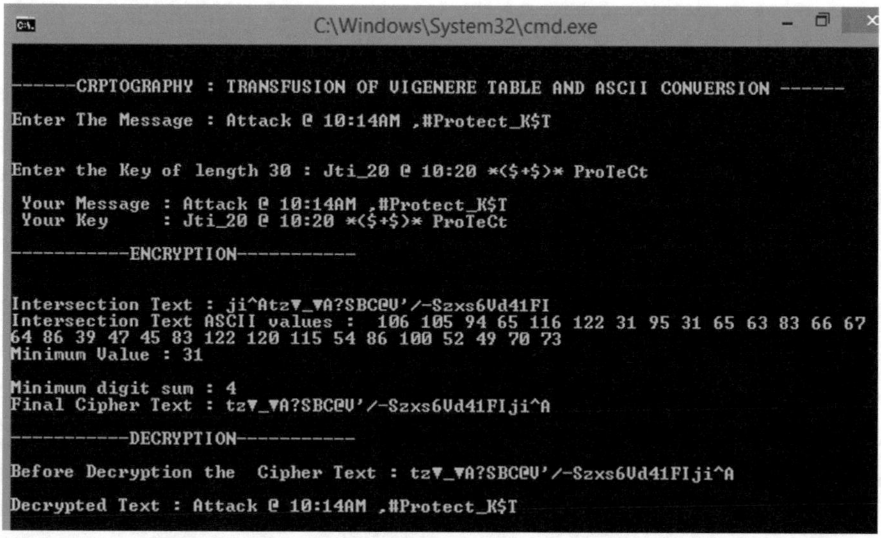

Fig. 3. Complete execution of the proposed algorithm

5 Comparative Analysis

While comparing the various existing and proposed algorithm till now, it has been found that the proposed algorithm is greatly secured in context of, uniqueness, security and it is more efficient for every printable characters of the plain text message. The encrypted text [13] produced by this algorithm takes very less execution time which is very high in the other existing papers.

The algorithm of Saraswat takes into consideration of only alphabetical and numeric characters whereas the proposed model works for every type of characters alphanumeric and special characters. This provides a range of security for the data. The algorithm execution time of Mathur increases with the length of character exponentially, whereas in the presented algorithm the execution time never exceeds the size of the smallest ASCII value in the intersection text. The work of Gupta works efficiently for all types of characters but the number of steps taken in the encryption is high and complex, the presented algorithm makes the encryption simpler and secure. Here is a comparison between the existing algorithm and proposed algorithm based on their execution time (in milliseconds) and size of the plain text and it can be referred from the graph that the proposed algorithm works far better than the existing algorithm (Fig. 4).

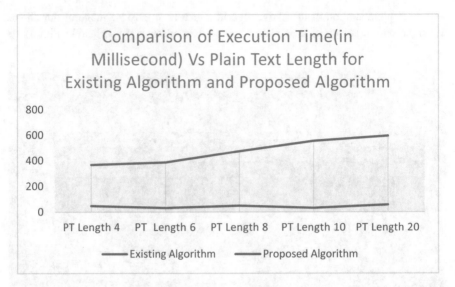

Fig. 4. Comparison graph of execution time (in Millisecond) vs plain text length for existing algorithm and proposed algorithm

6 Conclusion

In this research paper, we have introduced a more efficient and secure mechanism for the encryption of the plaint text. The algorithm addresses the major issue of security and privacy at the transmission phase of the data. The plain text throughout the process remains encrypted until the intended user provides the valid encryption key. From the previous section of the algorithm, we conclude that this new technique is robust and provides high level of security. The algorithm uses 4 steps to perform the encryption making it more secure. The range of characters that have been provided for encryption makes it tougher for cryptanalysis. Even the brute force attack will also take a large amount of time for decryption. The range of possibilities i.e. the 94 characters makes it tough to correctly reach to the plain text.

Even there are some limitations with the current system as the length of key and plain text must be same. The key security is also an issue. It's a belief that, this technique is a notable initiative towards the securing the data over network transmission. For future work our target would be to take a whole text file and then decrypt it with less time. Also to come up with more participation for data security of the cloud server such that we can secure the data storage over there.

References

1. Stallings, W.: Cryptography and Network Security Principles and Practice. Pearson Education Inc., Prentice Hall (2011)
2. Rivest, R.L.: Cryptography. In: Van Leeuwen, J. (ed.) Handbook of Theoretical Computer Science, vol. 1. Elsevier, Amsterdam (1990)
3. Justin, M.J., Manimurugan, S.: A survey on various encryption techniques. Int. J. Soft Comput. Eng. 2(1), 429–432 (2012)
4. Shinghe, S.R., Patil, R.: An encryption algorithm based on ASCII value. Int. J. Comput. Sci. Inf. Technol. 5(6), 7232–7234 (2014)
5. Sultana, R., Kamari, T.: An ASCII value based optimized text data encryption system. Int. J. Adv. Res. Electr. Electron. Instrum. Eng. 5(8) (2016)
6. Sukhraliya, V., Chaudhary, S., Solanki, S.: Encryption and decryption algorithm using ASCII values with substitution array approach. Int. J. Adv. Res. Comput. Commun. Eng. 2(8), 3094–3097 (2013)
7. Mathur, A.: An ASCII value based data encryption algorithm and its comparison with other symmetric data encryption algorithms. Int. J. Comput. Sci. Eng. 4(9), 1650–1657 (2012)
8. Saraswat, A., Khatri, C., Sudhakar, Thakral, P., Biswas, P.: An Extended Hybridization of Vigenere and Caeser cipher techniques for secure communication. Procedia Comput. Sci. 92, 355–360 (2016)
9. Krishna, Y.S.R.: Cryptographic algorithm based on ASCII conversions and a Radix function. Int. J. Sci. Eng. Res. 6(11), 1191–1194 (2015)
10. Bhargava, U., Sharma, A., Chawla, R., Thakral, P.: A new algorithm combining substitution & transposition cipher techniques for secure communication. In: ICOEI Tamil Nadu, pp. 619–624 (2017)
11. Gupta, C., Thakral, P.: ASCII conversion based two keys V4S scheme for encryption and decryption-a four step approach. In: IEEE International Conference on Computing, Communication & Networking ICCCNT 2017, IIT Delhi, pp. 1–6 (2017)
12. Liwandouw, V., Wowor, A.: The existence of cryptography: a study on instant messaging. Procedia Comput. Sci. 124, 721–727 (2017)
13. Joshi, A., Wazid, M., Goudar, R.H.: An efficient cryptographic scheme for text message protection against brute force & cryptanalytic attacks. Procedia Comput. Sci. 48, 360–366 (2015)

A Review on Human Action Recognition and Machine Learning Techniques for Suicide Detection System

V. Rahul Chiranjeevi[✉] and D. Elangovan[✉]

Computer Science and Engineering, Panimalar Engineering College,
Chennai 600123, India
rahulchiranjeevi777@gmail.com,
elangovan.durai@yahoo.com

Abstract. In current world about 800,000 people commit suicide every year. Mortality rate is increasing due to stress and depression. There are various types of suicide out of which hanging is the most common way of death. Though various systems are available for detecting hanging attempts their limitations results in inefficiency of the system. Numerous technologies are evolving everyday out of which an advanced system to detect hanging attempt can be established. This paper provides a comprehensive survey of human action recognition, machine learning techniques and various suicides prevention methods through which hanging attempts can be detected. Finally, an accuracy of various machine learning and human action recognition approaches is described.

Keywords: Hanging · Suicide · Machine learning · Convolution · Prison · SVM classifier

1 Introduction

The main source of mortality is Suicide by hanging. It is a demonstration of killing oneself by hanging from a stay point. It is most typically utilized suicide technique which has high passing rate. As shown by World Health Organization in perspective of the review of 56 country's mortality information found that hanging is the most widely recognized technique in numerous nations, out of which 59% are male suicides and 39% are female suicides. Suicide rates are rapidly increasing in young adults. Suicides occur mostly in Tech industries, colleges and prisons. There are many influential factors for suicide such as loneliness, depression, long non-ailing diseases, family problems, work pressure, financial problems, and psychiatric disorders. A few things can turn out badly while completing a hanging, prompting damage instead of death, and bringing about additional agony for the subject. The suspension mechanism may not be able to bear the full force needed to break the neck, resulting in injury rather than death due to a weak rope that was chosen, wrong knot or improperly tied rope, weak overhead beam and it breaks. To create various prevention endeavors, it is important to depict the qualities and personal conduct standards of the person. Different strategies

© Springer Nature Switzerland AG 2019
A. Abraham et al. (Eds.): IBICA 2018, AISC 939, pp. 46–55, 2019.
https://doi.org/10.1007/978-3-030-16681-6_5

have been proposed to reduce suicide endeavors. One of the techniques is the modernized discourse and movement examination. Facial feelings and speech give numerous courses in distinguishing the self-destructive reasoning. It is through mechanized investigation different analysts can discover the distinction in the self-destructive casualties. The recurrence of discourse in the self-destructive individual adjusts from low recurrence to high recurrence.

In 2005 [1] another suicide location gadget was outlined in which the individual would wear an ear piece to screen the beat and oxygen level of the individual. If the chance that the individual's vital signs are demonstrated as out of range an alert would go off and a crisis message will be sent. In another example, unique defensive dress like coverall has been planned which is comprised of high nylon texture that is hard to tear [2]. Although World Health association planned 'wearable bracelet' to record the mental parameters which distinguishes the individual's fundamental signs and, in the event, that they are out of range an alert is activated. A door alarm alert is proposed and it is activated when the entryway is utilized as a stay point for suicide. Despite the fact that different frameworks are intended to screen and counteract submitting the suicide, these frameworks are difficult to execute due to their high limitations. Truly these frameworks have different downsides as the framework needs wearing of different types of gear and they are even the hardware is expelled.

2 Human Action Recognition

A smart surveillance system [3] is developed for pedestrians to track and detect utilizing the action recognition. A track blob region in the image of pedestrian is captured and collected images are arranged in a sequence. Kalman filter technique is utilized for calculating the velocity and shape of the pedestrian. A signal of warning is issued when the person is entered into the restricted area or the speed of the pedestrian is exceeded. There are various motions of pedestrians like running, walking, falling etc. The drawback is that when illumination changes performance of the system is affected and the difference between the pedestrian with same speed cannot be identified for example a bicyclist and a runner.

A method is established for tracking [4] the person and recognizing the action of the person. A template captured previously is used to fund the region of interest and calculate the current action of the person. The Principle component analysis is utilized to calculate the activity recognition. The values from current frames and previous frames are compared and maximum likelihood is estimated. Various examinations are performed on sports like soccer and hockey out of which six actions are collected. In football various run like left run, right run, walk left walk right are estimated. For the examinations previously collected, ten templates are used for comparison.

Feifei et al. worked on human [5] pose recognition by utilizing marker less motion capture. A technique called background subtraction is utilized to find the interest of object moving in the videos. A new background model is established which works as a light invariant. Human tracking is performed by utilizing the particle filtering technique. A human body's 2D model is created for torso detection. The skin and non-skin color can be segmented by detecting the 2d model using various cameras in which one camera is used in front and one camera is used in corners of the room. Torso is obtained by combining the videos obtained from the cameras. Based on the positions obtained from cameras 20 features are relatively estimated and the nearest classifier is utilized for classification. As a future work, more number of features can be utilized thereby increasing the performance of the system.

In [6] action recognition in the videos is recognized by utilizing the spatio-temporal model and human object interaction. The Human object relationship is described by evaluating the spatial and temporal model based on the streams, obtained objects and humans are detected as well as the position of human and objects in the streams are relatively encoded and described. Using the detection window, features of humans and objects are extracted. The entire sequence interval between the videos is estimated to describe the entire system and Classification is performed by SVM classifier. Various examinations are evaluated on various datasets and accuracy of 96.3% and 98% is obtained on utilized dataset. The limitation is that the location poses and objects surrounding are not taken into consideration.

Table 1. Different types of technologies developed by various researchers on human action recognition

Author	Methodology	Dataset	Result	Limitations
Nguyen et al.	Principle component analysis and Independent subspace [7] analysis using three-layer convolution is utilized for performing the human recognition	UT-interaction	Based on the examinations an accuracy of 93% and 90% obtained is on set1 and set2	Spatial and temporal localization of activities are not possible using proposed approach as well as localization of spatial and temporal model cannot be performed
Zhang et al.	In this framework [8, 9] SVM is utilized for detecting the images, optical flow field And Fisher encoding technique is used	Human interaction	Experiments are performed on various datasets and an accuracy of 50% is obtained on public dataset	Accuracy does not meet the state of art results
Wang et al.	Dense trajectories technique is utilized to recognize the human action Optical flow technique is used. Bags of descriptors are utilized for classification. [10]	Hollywood-2	Examinations are performed and an accuracy of 50% is obtained	Accuracy does not meet the state of art results

(*continued*)

Table 1. (*continued*)

Author	Methodology	Dataset	Result	Limitations
Jiménez et al.	Various images from the video are obtained [11] Optical flow is computed for consecutive frames. Support Vector Machine is used for classification	Human interaction	Based on the performance of various datasets obtained results show an accuracy of 0.463%	High experimentation is needed to complete this approach
Kong et al.	This frame work utilized interaction model and Data driven [12] phase is used on training dataset	Collective Activity dataset and UT interaction dataset is utilized	Based on the examinations performed, an accuracy of 90.63% and 82.54% is obtained	Interactions may be confused with attributes and phrases
Yun et al.	Body pose feature [13] estimation is utilized for recognizing the human activity by obtaining the skeleton of the person	Microsoft and Kinect are utilized	Experiments are performed on various datasets and an accuracy of 87.3% is obtained on public dataset	Only based on a particular view videos are obtained

The Table 1 describes about different methodologies, datasets, limitations and results from various researchers based on the human action recognition.

Santhiya et al. [14] created a system in detecting the abnormal crowd tracking and motion analysis. It is difficult to detect the humans in videos obtained from crowded environment. An adaptive background model is utilized to perform on the obtained videos. A threshold value is set to the video to separate the background from foreground, Crowd is detected by utilizing the blob analysis. The density of the crowd is estimated by utilizing the crowd detection. Direct and indirect methods of crowd detection are performed. A new abnormal model is created based on the crowd activities. Experiments are performed on various public datasets.

Out of various view independent action recognition, a single view independent approach is adopted to perform the action recognition. Using this approach, the Videos are obtained from various cameras and actions are recognized. The proposed method outperforms the various views invariant representations [15–19] by utilizing a universal classifier which has training data corresponding to all the views and angles obtained through which various motions of the action can be recognized and templates for the actions are created and stored in the database [20–24].

3 Machine Learning

Weiss et al. proposed a [25] machine learning approach called transfer learning, where training and testing of data can be obtained from the field of conventional machine learning. The features obtained and distributed data can be identical, in many cases the process of obtaining the training and testing data is a tedious task. To overcome this problem, Transfer learning approach is utilized. Learning is performed using a high-performance learner from target domain, many applications and systems are based on this approach.

Deng et al. proposed a [26] machine learning approach called deep learning. Various ML approaches have structured learning which has a single layer of trans-formations for instance HMMs, GMMs etc. Deep leaning techniques utilize both supervised and unsupervised techniques in the deep nodes which contains various stages of nonlinear processing where the output is fed as input to an intermediate layer. High performance of recognition can be achieved through various neural network approaches.

An active learning approach is [27] proposed. Whenever, the videos are not labeled it become tedious task to label all videos in the database. Through active learning, the labeling can be performed through subset approach. Learning based on kernels is one of the approaches widely used to find the efficiency of the Support Vector Machines, Principle Component Analysis, etc.

Table 2. Different technologies developed by various researchers on machine learning

Author	Methodology	Datasets	Results	Limitations
Eisinger et al.	Multilayer Perceptron [27] and Bayesian Inference, is used for decision engine	Tom's Hardware website	Based on the examinations an accuracy of 83% is obtained	When the number of variables are increased, it is difficult to model the problem
Anishchenko	In this framework CNN [28] AlexNet, is used to detect the fall	Laboratory of Electronics and Imaging of the National Centre for Scientific Research	Experiments are performed on various datasets and an accuracy of 99.23% is obtained on public dataset	Classifier performance is not suitable for state of art methods
Ren et al.	Video summarization approach [29] based on machine learning which is used for explanation of videos	MINERVA	Examinations are performed and an accuracy of 87.1% is obtained	Temporal continuity of the video is affected if the frame rate is changed

(continued)

Table 2. (*continued*)

Author	Methodology	Datasets	Results	Limitations
Wei et al.	Various images from the [30] video are obtained. Semi supervised learning and support vector machine is used	TRECVID' 06	Based on the performance of various datasets obtained results show an accuracy of 76%	Labeling large set of data cannot be performed
Nawaratne et al.	This frame work utilized [31] incremental knowledge acquisition and self-learning	Action Recognition Dataset	Based on the examinations performed, an accuracy of 94.63% is obtained	Story line generation for autonomous video streams is not possible

Table 2 explains about methodologies, limitations, and the results obtained using the particular datasets with the accuracy by various researchers.

4 Suicide Detection

Lee proposed an image analysis [32] technique using 3-D modeling to detect the suicide by hanging in prisons. The classification is performed by utilizing a random forest and ensemble learning method based on tree classification algorithm. The joint positions in the human body are captured for instance elbows, hands knees, and feet. The data is normalized thereby generating the motion history of obtained data. A 3-D data is extruded from 2-D joints in the human action. Various examinations are performed on 75 positive and 75 negative data to achieve higher accuracy in predicting hanging attempts. However the system detects partial attempts and only particular type of hanging can be detected.

In [33] a suicide detection system is proposed to find suicidal words in twitter. PHP and My SQL are utilized to build a twitter API. The Proposed method searches keywords having suicide related tweets. Whenever a word is detected, an alert is sent to psychologist who uses the system to keep track of the person with suicidal tendency. The proposed system also detects the location where the tweet has been tweeted.

A mobile application for suicide [34] prevention is proposed. It uses patient's health status as a tool to detect the suicidal tendency. Based on the question, choice analysis is performed. The application has security requirements which have high user quality and friendliness. An EMA technique is used to monitor and collect huge data based on the frequency. The Patient's details from databases are extracted using data mining techniques, grouped based on similarity of characteristics between them and Supervised techniques are used to predict the patient behaviors. These tools help in building an effective health application.

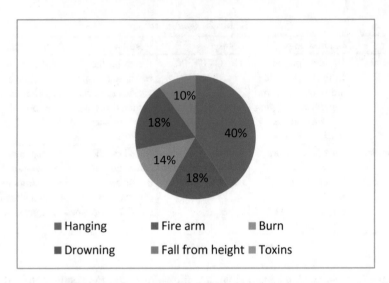

Fig. 1. Various types of suicides with occurrence rate.

Figure 1 show that the occurrence rate of different types of suicides which describes that hanging plays a major role.

Huang [35] proposed a novel dataset of various social media accounts which has high suicide rate. The location of the users is considered and text analysis is performed based on the longitudinal positions which have significant change in their death. Weibo is used to collect the data of the users who has died of confirmed suicide and are grouped based on the gender. The average age of death is below 30 in the proposed dataset. Location analysis is performed on the dataset. By examining the social media patterns, number of suicides can be reduced.

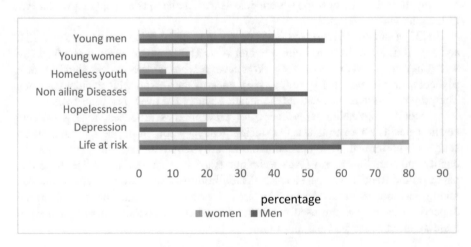

Fig. 2. Classification of people committing suicide

Figure 2 describes about the classification on various factors for suicide attempt by both men and women.

5 Conclusion

This paper discusses in detail about various human action recognition and machine learning techniques. The Various techniques have been portrayed in this paper for implementing a suicide detection system by recognizing the human actions. An analysis is performed using various techniques in the above-mentioned tables. From the analysis performed so far, a detection system of high accuracy can be established using the current action recognition techniques. The various studies suggest that current technologies and further improvements can be used to detect the suicide attempts more efficiently.

References

1. Cook, F.E.: Door suicide alarm. US Patent RE42,991 (2011)
2. Hayes, L.M.: Suicide prevention in correctional facilities: reflections and next steps. Int. J. Law Psychiatry **36**, 188–194 (2013)
3. Robert, B., Jackson, B., Papanikolopoulos, N.: Vision-based human tracking and activity recognition. In: Proceedings of the 11th Mediterranean Conference on Control and Automation, vol. 1 (2003)
4. Lu, W., Little, J.: Simultaneous tracking and action recognition using the PCA-HOG descriptor. In: IEEE 3rd Canadian Conference on Computer and Robot Vision, p. 6. IEEE (2006)
5. Feifei, H., Hendriks, E., Paclik, P., Oomes, A.: Markerless human motion capture and pose recognition. In: 10th Workshop on Image Analysis for Multimedia Interactive Services, pp. 13–16. IEEE (2009)
6. Victor, E., Niebles, J.: Spatio-temporal human-object interactions for action recognition in videos. In: Computer Vision Workshops (ICCVW), pp. 508–514. IEEE (2013)
7. Nguyen, N., Yoshitaka, A.: Human interaction recognition using independent subspace analysis algorithm. In: International Symposium on Multimedia (ISM), pp. 40–46. IEEE (2014)
8. Zhang, B., Yan, Y., Conci, N., Sebe, N.: You talkin' to me?: recognizing complex human interactions in unconstrained videos. In: International Conference on Multimedia, pp. 821–824. ACM (2014)
9. Iwata, S., Ohyama, W., Wakabayashi, T., Kimura, F.: Recognition and transition frame detection of Arabic news captions for video retrieval. In: 2016 23rd International Conference on Pattern Recognition (ICPR), Cancun, pp. 4005–4010 (2016)
10. Wang, H., Kläser, A., Schmid, C., Liu, C.: Action recognition by dense trajectories. In: Computer Vision and Pattern Recognition, pp. 3169–3176. IEEE (2011)
11. Jiménez, M., Blanca, N.: Human interaction recognition by motion decoupling. In: Pattern Recognition and Image Analysis, pp. 374–381. Springer, Heidelberg (2013)
12. Kong, Y., Jia, Y., Fu, Y.: Interactive phrases: semantic descriptions for human interaction recognition. Pattern Anal. Mach. Intell. IEEE **36**, 1775–1788 (2014)

13. Yun, K., Honorio, J., Chattopadhyay, D., Berg, T., Samaras, D.: Two-person interaction detection using body-pose features and multiple instance learning. In: Computer Society Conference on Computer Vision and Pattern Recognition Workshops, pp. 28–35. IEEE (2012)
14. Santhiya, G., Sankaragomathi, K., Selvarani, S.: Abnormal crowd tracking and motion analysis. In: International Conference Advanced Communication Control and Computing Technologies, pp. 1300–1304. IEEE (2014)
15. Yilmaz, A., Shah, M.: Recognizing human actions in videos acquired by uncalibrated cameras. In: ICCV (2005)
16. Shen, Y., Foroosh, H.: View-invariant action recognition using fundamental ratios. In: CVPR (2009)
17. Junejo, I.N., Dexter, E., Laptev, I., Perez, P.: View-independent action recognition from temporal self-similarities. IEEE TPAMI 33, 172–185 (2011)
18. Lewandowski, M., Makris, D., Nebel, J.C.: View and style-independent action manifolds for human activity recognition. In: ECCV (2010)
19. Wu, X., Jia, Y.: View-invariant action recognition using latent kernelized structural SVM. In: ECCV (2012)
20. Song, Y., Morency, L.P., Davis, R.: Multi-view latent variable discriminative models for action recognition. In: CVPR (2012)
21. Weinland, D., Ozuysal, M., Fua, P.: Making action recognition robust to occlusions and viewpoint changes. In: ECCV (2010)
22. Lv, F., Nevatia, R.: Single view human action recognition using key pose matching and viterbi path searching. In: CVPR (2007)
23. Zhu, F., Shao, L., Lin, M.: Multi-view action recognition using local similarity random forests and sensor fusion. Pattern Recogn. Lett. 24, 20–24 (2013)
24. Iosifidis, A., Tefas, A., Pitas, I.: View-invariant action recognition based on artificial neural networks. IEEE TNNLS 23(3), 412–424 (2012)
25. Weiss, K., Khoshgoftaar, T.M., Wang, D.: A survey of transfer learning. J. Big Data. 3 (2016). https://doi.org/10.1186/s40537-016-0043-6
26. Deng, L.: A tutorial survey of architectures, algorithms and applications for deep learning. APSIPA Trans. Sig. Inf. Process. 3, 1–29 (2014). https://doi.org/10.1017/atsip.2013.9
27. Eisinger, R., Romero, R.A.F., Goularte, R.: Machine learning techniques applied to dynamic video adapting. In: 2008 Seventh International Conference on Machine Learning and Applications (2008). https://doi.org/10.1109/icmla.2008.42
28. Anishchenko, L.: Machine learning in video surveillance for fall detection. In: 2018 Ural Symposium on Biomedical Engineering, Radioelectronics and Information Technology (USBEREIT) (2018)
29. Ren, W., Zhu, Y.: A video summarization approach based on machine learning. In: 2008 International Conference on Intelligent Information Hiding and Multimedia Signal Processing (2008)
30. Wei, S., Zhu, Z., Zhao, Y., Liu, N.: A cooperative learning strategy for interactive video search. In: 2007 6th International Conference on Information, Communications & Signal Processing (2007)
31. Nawaratne, R., Bandaragoda, T., Adikari, A., Alahakoon, D., De Silva, D., Yu, X.: Incremental knowledge acquisition and self-learning for autonomous video surveillance. In: 43rd Annual Conference of the IEEE Industrial Electronics Society, IECON 2017 (2017)
32. Lee, S., Kim, H., Lee, S., Kim, Y., Lee, D., Ju, J., Myung, H.: Detection of a suicide by hanging based on a 3-D image analysis. IEEE Sens. J. 14(9), 2934–2935 (2014). https://doi.org/10.1109/jsen.2014.2332070

33. Varathan, K.D., Talib, N.: Suicide detection system based on Twitter. In: 2014 Science and Information Conference (2014)
34. Berrouiguet, S., Billot, R., Lenca, P., Tanguy, P., Baca-Garcia, E., Simonnet, M., Gourvennec, B.: Toward e-health applications for suicide prevention. In: Connected Health: Applications, Systems and Engineering Technologies (CHASE). https://doi.org/10.1109/chase.2016.37
35. Huang, X., Xing, L., Brubaker, J.R., Paul, M.J.: Exploring timelines of confirmed suicide incidents through social media. In: 2017 IEEE International Conference on Healthcare Informatics (ICHI) (2017). https://doi.org/10.1109/ichi.2017.47

EEG Signal Analysis Using Wavelet Transform for Driver Status Detection

P. C. Nissimagoudar$^{(\boxtimes)}$, Anilkumar V. Nandi, and H. M. Gireesha

B.V.B. College of Engineering and Technology, Hubballi, Karnataka, India
pcngoudar@gmail.com

Abstract. The proposed work aims at providing an optimized method for determining driver status by analyzing Electroencephalogram (EEG) signals by time and frequency domain analysis using wavelet transforms. Human brain alertness level can be detected by direct measurement of electrical activity inside the brain using EEG Signals. These signals are acquired from electrodes placed on scalp. Such signal would be usually contaminated with various artifacts like muscle movements. Therefore the noise from the raw signals acquired from electrodes is removed using suitable band pass filter. Filtered signals are further subjected to discrete wavelet transform for isolating EEG rhythms. EEG rhythms include five frequency bands namely (delta (0.5–4 Hz), theta (4–8 Hz), alpha (8–12 Hz), beta (12–30 Hz) and gamma (>30 Hz). Debauches DB8 wavelet transform is used for decomposing the signal in to eight levels and lower five frequency bands are considered for the analysis. For the proposed work sleep data sets from Physionet are used for analysis.

Keywords: Wavelet transform · EEG · Driver status · Rhythm isolation

1 Introduction

The techniques to analyse biomedical signals such as electrocardiogram, electroencephalogram are significantly contributing in health care systems. Earlier methods of diagnosing human status and diseases by observations and by visual or hearing heuristic approaches have become obsolete. Various biological signals measured from human body represent the disorders or condition of human and further can be used for treatment [1, 2]. The analysis of such physiological signals involve various stages with complex mathematical computations such as,

- Acquisition of signals using sensors/electrodes
- Modelling of biomedical systems
- Analysis of non- stationary signals
- Filtering for removal of artifacts
- Event detection and characterization
- Frequency-domain characterization
- Pattern classification and diagnostic decision

In this work it is mainly focussed on analysis of biomedical signals applied to driver's vital information monitoring and analysis. Advanced Driver Assistance

© Springer Nature Switzerland AG 2019
A. Abraham et al. (Eds.): IBICA 2018, AISC 939, pp. 56–65, 2019.
https://doi.org/10.1007/978-3-030-16681-6_6

Systems (ADAS) have become integrated part in all the recent vehicles. These systems which will assist the driver to take better decisions and improve drivability are built with a smart integration of mechanical, electronic and software based systems. Driver alertness indication is one such system which belongs to advanced driver assistance systems. This is commonly known as driver's drowsiness detection. According to the survey there are around 30–40% accidents happen because of driver's drowsy status. Recently various methods have been proposed to distinguish between driver's alert status and drowsiness using physiological signals. The physiological signals like Electrooculogram (EOG) which measures basically the eye movement of driver, heart rate variability (HRV) were used as means of detecting driver's alertness level. These methods were considered as indirect methods for indicating driver's alertness. The EEG based analysis is considered as a quantitative and accurate approach which directly indicates the driver's consciousness level.

Electro encephalogram (EEG) the brain waves: The human brain consists of millions of brain cells called as neurons. These neurons communicate with each other using electrical signals. When millions of neurons communicate with each other, there would be huge amount of electrical activity takes place inside the brain. These signals are measured using electro encephalogram or EEG [3].

The brain state can also be measured using other techniques like Computed Tomography (CT), Magnetic resonance imaging (MRI), Positron Emission Tomography (PET) and functional Magnetic Resonance Imaging (fMRI). All these techniques have their own merits and demerits. EEG frequency analysis being a direct measure of the brain activity is found to be more effective [4].

EEG signals are classified based on the frequency ranges mainly as Alpha (8 Hz to less than 13 Hz), Beta (more than 13 Hz), Theta (4 Hz to Less than 8 Hz), and Delta (less than 4 Hz). The gamma waves with the frequency ranging from 40 Hz to 70 Hz represent the complete alert state of the human body. There is a direct correlation between high amounts of brain activity to the increased brain functioning ability. The beta waves with frequency ranging from 13 Hz to 30 Hz represent increased ability to focus on external reality. The alpha waves represent the relaxed status of brain with the frequency ranging from 8 Hz to 13 Hz. The alpha activity in the brain increases with eye closure. These waves are the best indicators of drowsy condition. Theta waves with the frequency 4 Hz to 10 Hz represent the early stage of sleep and delta waves with the frequency 1 Hz to 4 Hz represent deep sleep status. Different ranges representing different brain state and analysis of these signals help medical personnel in understanding functional and behavioral characteristics of human brain. Different brain disorders like autism, Alzheimer, epilepsy, memory loss can be diagnosed by analyzing EEG signals and also applications like drowsiness detection for automotives can be developed [5, 6].

For any EEG signal analysis application, signal acquisition and extraction of information is very important. The signal acquisition is done using either clinical wet electrode for healthcare related applications and by using dry head band type of electrodes for other applications like determining driver's status or for determining meditation levels. For both type of electrodes the acquired information will be contaminated with other noise signals which have to be removed to extract EEG information. Interpretation of EEG signals for different types of brain disorders or brain states is done using classification techniques like neuro-fuzzy or machine learning techniques. The study shows that the measurement of alpha and theta waves gives the direct indication of drowsy status. The EEG spectrum is a combination of different types of brain waves and extracting alpha and theta waves using the appropriate signal processing methods gives the status of the driver [7].

EEG signals are analyzed using four different types of techniques, time domain analysis, frequency domain analysis, time-frequency domain analysis and nonlinear analysis [11].

Time domain analysis method: For early EEG analysis time domain analysis technique is usually applied. The analysis of features like variance, histogram, peak detection, pass zero point detection are some of the major features which are done using time domain analysis technique [12].

Frequency domain analysis method: This is the most commonly used technique for EEG analysis. Here the power spectrum is analyzed to obtain different power distribution related features.

Time-Frequency domain analysis: Wavelet transforms is the most commonly used technique uses to perform time-frequency domain analysis. It mainly contributes in analyzing non-stationary signals.

Nonlinear analysis method: EEG represents highly nonlinear behavior. For analysis of such signals and to obtain non-linear features, techniques like complicity analysis, Lorenz scattered picture, kolmogorov entropy and Lypunov exponential are commonly used [13, 14].

2 Methodology

The methodology to analyze and interpret EEG information for determining the driver status is proposed using the following steps,

- EEG Data Collection
  - EEG Data from Physionet
  - EEG Sub set data creation.
- Preprocessing of EEG signals
  - Filtering and artifact removal
  - Epoch extraction (Splitting the data)
- Rhythm isolation
  - Extraction of EEG frequency bands.

2.1 EEG Data Collection

EEG data collection is done using Physionet website where stored data is available from different sources for research. Around 20 h each of sleep data is available for 61 different subjects representing different sleep stages. As per the requirement the sleep data which contains various sleep stages from Wakeup (W) stage, initial sleep or drowsy state S1 to Subsequent sleep stages named as S2, S3, S4 and Rapid Eye Moment (REM) which is in European Data Format (EDF) can be used for processing. To distinguish drowsy and wake up stages, the data from the regions W and S1 can be used.

The data consists of 61 polysomnograms (PSGs) in the form of *PSG.edf files and its related hypnograms (annotations of sleep stages by experts) in the form of *Hypnogram.edf files. *PSG.edf files are whole-night recordings of polysomnographic sleep recordings consisting of EEG from Fpz-Cz and Pz-Oz locations of electrodes. The sleep EEG data is obtained from Physionet database and each file consists of data with all sleep stages from W, S1-S4 and REM. One such data of *PSG. edf containing EEG, EOG, EMG and other data and *Hypnogram.edf with annotations is shown in Fig. 1. The files are viewed through the software Polyman.

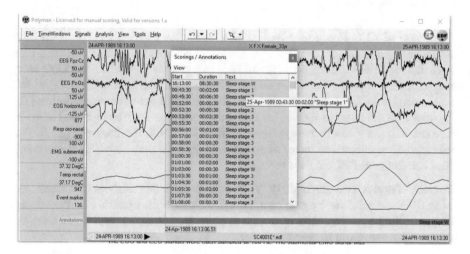

Fig. 1. *PSG.edf and *Hypnogram.edf file viewed through the software Polyman

Electrode Selection

According to neurophysiological literature the drowsiness is the transition from awake state to sleep state with the characteristics of decreased attention and slow movements. The sleep stage starts with activation and inhibition of neurons in different parts of brain. From EEG signals it is identified as,

(i) Decrease in beta rhythm activity (13–30 Hz) (ii) Increase and subsequent decrease in alpha rhythms (8–13 Hz) (iii) Increase in activity of theta rhythm (4–8 Hz) [7, 8].

The PSG database from Physionet contains the EEG recordings from Fpz-Cz, and Pz-Oz electrodes, only Fpz-cz electrode information is used in this experimentation. This is because EEG electrode positioned at prefrontal region gives ease placement of electrode for driver [9].

2.2   Pre-processing of EEG Signals

Filtering and Artifact Removal: The raw EEG data is expected to be noisy and non-stationary, so the data is filtered using band pass filter with the frequency band between 0.5 Hz to 90 Hz to remove DC shifts and other artifacts. This process also helps in minimizing the epoch limits. A 50 Hz notch filter is applied to remove 50 Hz line noise [12].

Epoch Extraction (Splitting the Data): Each wakeup and sleep data is separated in to different time regions as epochs for processing. Epochs considered can be continuous, separated or overlapping. Signals from each wake up and sleep data are epoched into desired number of samples [14].

2.3   Rhythm Isolation

Extraction of EEG Frequency Bands: EEG data after filtering and epoch selection has to be separated in to different frequency regions namely delta (0.5–4 Hz), theta (4–8 Hz), alpha (8–12 Hz), beta (12–30 Hz), and gamma (1–40 Hz). As EEG being non stationary signal, FFT is not suited. So, wavelet transforms (Daubechies db8) which translates the information into both frequency and time domain is chosen [13].

After removal of noise signals, the EEG information needs to be separated in to five different frequency bands, namely δ, θ, α, β and γ. From the literature [2, 7], [34] wavelets are an alternative and efficient methods used for analyzing EEG signals apart from conventional FFT analysis methods.

Discrete Wavelet Transform (DWT): The multi resolution analysis of a signal x(n) using DWT is carried out by passing it through a series of filter banks. First the samples are passed through a low pass filter with impulse response g(n) resulting in a convolution of the two given in below equation.

$$y(n) = x(n) * g(n) = \sum_{k=-\infty}^{k=+\infty} x(k)g(n-k)$$

The output of filter gives the approximate coefficients. The signal is also decomposed simultaneously using a high pass filter h(n) to get detail coefficients.

$$y(n) = x(n) * h(n) = \sum_{k=-\infty}^{k=+\infty} x(k)h(n-k)$$

Approximation coefficients give the characteristics of lower frequencies in the signal, whereas details give information about higher frequency characteristics. The Approximation coefficients at each level can be used for another level decomposition and this can be extended to multiple levels in order to get more frequency resolution. This is shown in Fig. 2. Hence both approximate and detail coefficients are considered. The signal is decomposed into L number of levels and approximate coefficients of level L and detail coefficients of 1 to L levels, which represents different frequency EEG bands.

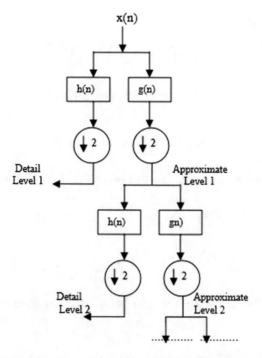

Fig. 2. Wavelet decomposition for EEG decomposition

Wavelets provide time-frequency domain analysis and are more suited for non-stationary signals like EEG. In the current work the discrete wavelet transform is used to recognize different brain rhythms.

Daubechies Wavelets

Daubechies has defined a type of wavelet for a given vanishing moment p to find the minimum size discrete filter. The vanishing moment defines how a function decays towards infinity. For example, the function sin t/t2 decays at a rate of 1/t2 as t approaches to infinity. We can estimate the rate of decay by the following integration,

$$\int_{-\infty}^{\infty} t^k f(t) dt$$

The parameter k indicates the rate of decay.

If we want the wavelet function with p vanishing moments, the minimum filter size is 2p. We can write the equation as,

$$H_0\left(e^{j\omega}\right) = \sqrt{2} \left(\frac{1 + e^{-j\omega}}{2}\right)^P R e^{j\omega}$$

The absolute-square of this function is,

$$\left|H_0\left(e^{j\omega}\right)\right|^2 = 2\left(\frac{\cos\omega}{2}\right)^{2p} P(\sin^2\omega/2)$$

A theorem in algebra, called Bezout theorem, can solve this equation. The unique solution is,

$$P(y) = \sum_{k=0}^{p=1} p - 1 + k/k)y^k$$

Taking p = 2 for an example. The obtained polynomial P(y) is,

$$P(y) = \sum_{k=0}^{1}\left(\frac{1+k}{k}\right)y^k = 1 + 2y \qquad (5.9)$$

$$\frac{P(2 - z - z^{-1})}{4} = 2 - \frac{1}{2z} - \frac{1}{2z} - 1$$

The roots are 2 + p3 and 2 − p3. After factorization, we have the low pass filter to be,

$$H_0\left(e^{j\omega}\right) = \sqrt{2} + \frac{\sqrt{6}}{8} + 3\sqrt{2} + \frac{\sqrt{6}}{8}e^{-j\omega} + 3\sqrt{2} - \frac{\sqrt{6}}{8}e^{-2jw} + \sqrt{2} - \frac{\sqrt{6}}{8}e^{-j\omega}$$

The discrete-time domain representation is,

$$h_\phi[n] = \sqrt{2} + \frac{\sqrt{6}}{8}\delta[n] + 3\sqrt{2} + \frac{\sqrt{6}}{8}\delta[n-2] + \sqrt{2} - \frac{\sqrt{6}}{8}\delta[n-3]$$

The result is the minimum size filter with 2 vanishing moments and the corresponding filter size is 4. Higher order Daubechies wavelets are derived at similar way [11].

dbN wavelets: The dbN wavelets are Daubechies wavelet's with N vanishing moments and 2N filters. For db8 wavelet it is with 8 vanishing moments and 16 filter levels [15].

3   Results and Discussion

To separate EEG signals in to five frequency bands, db8 Daubechies wavelet is used. The procedure is as follows,

- Input is wakeup and sleep S1 *PSG recordings in the edf format.
- The edf data is converted to .mat format.
- From each wakeup and sleep S1data, around 2000 samples are considered, shown in Fig. 2.

- Data is sampled at 500 Hz for next level of analysis.
- Wavelet function of type Daubechies db8 with 8 level decomposition is applied on each wakeup and sleep data.
- Level N (1–8) coefficients are extracted and the corresponding 1-D signals are reconstructed.
- Level 4 to 8 decomposition represents gamma, beta, alpha, and theta and delta frequency bands and time domain; shown in Fig. 3.
- FFT is applied to represent EEG rhythms in frequency domain; shown in Fig. 4.
- For each of the EEG rhythm maximum frequency point is detected; shown in Table 1.

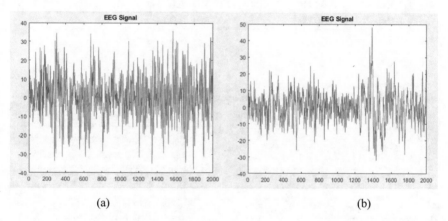

(a)   (b)

Fig. 3. (a) Raw EEG for wakeup signal; (b) Raw EEG for sleep S1 stage signal

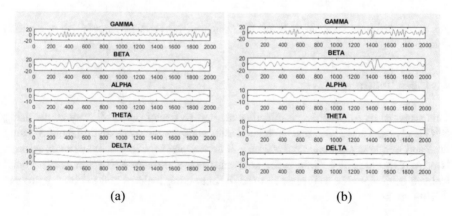

(a)   (b)

Fig. 4. (a) Time domain wake up EEG rhythms; (b) Time domain sleep S1 EEG rhythms

EEG rhythm extraction is an important phase in EEG signal analysis (Fig. 5). All the EEG frequency bands (gamma to delta) are extracted using db8 wavelet function.

Wavelet gives the signal representation in both frequency and time domain. These signals can further be used for feature extraction and classification.

Table 1. Maximum frequency for each EEG rhythm for wakeup signal and for sleep (S1) signal

EEG bands	Wakeup signal	Sleep signal
	Maximum frequency occurs at	Maximum frequency occurs at
Gamma	39.00 Hz	40.00 Hz
Beta	27.00 Hz	25.00 Hz
Alpha	10.00 Hz	9.00 Hz
Theta	5.00 Hz	6.00 Hz
Delta	3.00 Hz	3.00 Hz

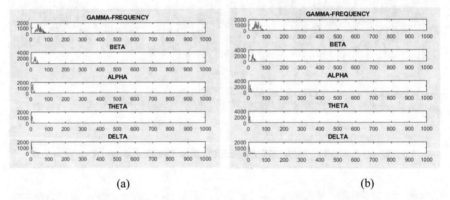

(a)   (b)

Fig. 5. (a) Frequency domain wake up EEG rhythms; (b) Frequency domain sleep S1 EEG rhythms

4   Conclusion

This paper aims to propose an effective method to detect driver drowsy status using EEG analysis. The EEG signals extracted after preprocessing are rhythm isolated using time and frequency domain analysis technique wavelet transforms. Wavelet transforms is proved to efficient technique for analyzing non-stationary EEG signals. The band isolated EEG signals can further be used for extracting features and to classify drowsy driving status.

References

1. Rangayyan, R.M.: Biomedical Signal Analysis. Wiley, Hoboken (2002)
2. Tompkins, W.J.: Biomedical Digital Signal Processing. Prentice-Hall, Upper Saddle River (1995)
3. Sun, Y., Ye, N., Wang, X., Xu, X.: The research of EEG analysis methods based on sounds of different frequency. In: IEEE/ICME International Conference on Complex Medical Engineering, pp. 1746–1751 (2007)
4. Kumar, J.S., Bhuvaneshwari, P.: Analysis of Electroencephalography (EEG) signals and its categorization-a study. In: International Conference on Modeling, Optimization and Computing (ICMOC 2012). Elsevier Publications (2012)
5. Rechtschaffen, A., Kales, A.E.: A Manual of Standardized Terminology, Techniques and Scoring Systems for Sleep Stages of Human Subjects, p. 10. UCLA Brain Information Service. Brain Research Institute, Los Angeles, (1968)
6. https://physionet.org/physiobank/database/sleep-edfx/
7. da Silveira, T., de Jesus Kozakevicius, A., Rodrigues, C.R.: Drowsiness detection for single channel EEG by DWT best m-term approximation. Res. Biomed. Eng. **31**(2), 107–115 (2015)
8. Blinowska, K., Durka, P.: Electroencephalography (EEG). Wiley, New York (2006)
9. Aboalayon, K.A.I., Faezipour, M., Almuhammadi, W.S., Moslehpour, S.: Sleep stage classification using EEG signal analysis: a comprehensive survey and new investigation. Entropy **18**, 272 (2016). https://doi.org/10.3390/e18090272
10. Ilyas, M.Z., Saad, P., Ahmad, M.I.: A survey of analysis and classification of EEG signals for brain-computer interfaces. In: 2nd International Conference on Biomedical Engineering (ICoBE), 30–31 March 2015, Penang (2015)
11. Chun-Lin, L.: A tutorial of the wavelet transforms, February 2010
12. Awais, M., Badruddin, N., Drieberg, M.: A hybrid approach to detect driver drowsiness utilizing physiological signals to improve system performance and wearability. Sensors **2017**, 17 (1991). https://doi.org/10.3390/s17091991
13. Sun, Y., Ye, N., Wang, X., Xu, X.: The research of EEG analysis methods based on sounds of different frequency. In: IEEE/ICME International Conference on Complex Medical Imaging. Information Science & Engineering College, Northeastern University, Shenyang (2007)
14. Blaiech, H., Neji, M., Wali, A., Alimi, A.M.: Emotion recognition by analysis of EEG signals. In: 2013 13th International Conference on Hybrid Intelligent Systems. Research Groups on Intelligent Machines University of Sfax, National Engineering School of Sfax (ENIS), Sfax (HIS) (2013)
15. Mantri, S., Agrawal, P., Patil, D., Wadhai, V.: Non invasive EEG signal processing framework for real time depression analysis. In: SAI Intelligent Systems Conference, 10–11 November 2015, London, UK (2015)

BMI Application: Accident Reduction Using Drowsiness Detection

Chirag Ghube$^{(\boxtimes)}$, Anuja Kulkarni, Chinmayi Bankar, and Mangesh Bedekar

Department of Computer Engineering, Maharashtra Institute of Technology,
Pune, India
chirag.ghube@gmail.com

Abstract. Among the numerous factors that are responsible for increasing road accidents, the second most common cause is drowsiness. In an attempt to reduce the rate of accidents, we propose a system which would efficiently handle the timely detection of drowsiness and would accordingly curb the speed of the vehicle being driven. As a proof of concept of the proposed method, we have trained the SVM classifier on the EEG (electroencephalogram) waves derived from "Analysis of a sleep-dependent neuronal feedback loop: the slow-wave micro continuity of the EEG" by Kemp et al. [1,2]. The data is obtained from a wireless EEG headset. The classification results will determine whether the EEG data corresponds to drowsiness or alertness. This level of drowsiness is then used to determine the maximum speed limit. As the work in [3] has stated, there is a strong correlation between the number of accidents and the speed limit. Hence altogether, the proposed system integrates EEG waves for sleep level detection, and speed lock as a preventive measure to reduce the number of plausible accidents.

Keywords: EEG · Brain machine interface · BCI ·
Drowsiness detection · SVM · Speed lock · Sleep level ·
Accident prevention

1 Introduction

Every 1 out of 25 adult drivers, as reported by The National Highway Traffic Safety Administration, report to have fallen asleep while driving [6,7]. They estimated that drowsy driving was responsible for 72,000 crashes, 44,000 injuries, and 800 deaths in 2013 [8]. But the actual number of accidents are estimated to reach up to 6,000 fatal crashes a year.

The Department of Mobility and Transport of the European Commission has provided a detailed report on road safety, which indicates a correlation between the driving speed and the associated risk of accident occurrence. Specifically, the report focuses on sudden increases in the normal speed and its corresponding probability of an accident. These two factors are observed to have an exponential

A. Abraham et al. (Eds.): IBICA 2018, AISC 939, pp. 66–72, 2019.
https://doi.org/10.1007/978-3-030-16681-6_7

relation with one another i.e. the higher the speed, the steeper is the increase in accident risk [3].

$$A_2 = A_1 \left(\frac{v_2}{v_1} \right)^2$$

The above formula developed by Nilsson [4] shows that the relationship between the number of injury accidents before the change in speed (A1) and after the change in speed (A2) depends on the new average speed (v1) and the old average speed (v2). This indicates that the speed changes and accident risks are closely related by a power function (Table 1).

Table 1. Relation of speed reduction and accident risk [3]

Hypothetical situation	Percent reduction in:		Average Delta V	Average crash energy
	No. of crashes	No. of persons injured		
10 km/h speed reduction	41.5	34.6	25.5	38.7
5 km/h speed reduction	15.0	13.1	16.1	23.6

More information about the above table and further explanation is given by Kloeden et al. [3]. From this table and by the formula developed by Nilsson [4], we can conclude that by reduction of speed i.e. by applying a speed-lock we can reduce the risk percentage associated with the crashes.

Now, the speed shows considerable variations when the driver undergoes a transition from wake state towards the subconscious/drowsy state. These sudden increases in speed which occur unintentionally, have to be prevented in order to ensure the safety of the drowsy driver. Hence we propose to limit the speed in proportion to the degree of drowsiness as a preventive measure.

2   Proposed Method

In this paper, we put forth a unique system which would be able to detect whether the driver is sleepy or not. The classifier used, will initially be trained on a previously labeled dataset. The EEG waves generated by the driver will be classified using an already trained classifier. The EEG waves will be captured and transmitted by a wireless EEG headset. Depending on the outcome of the classification, if the driver is found to be drowsy, a speed lock will be applied which will prohibit him from increasing the speed unknowingly. Also the extent

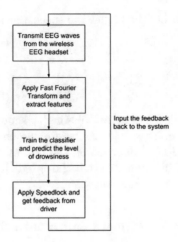

Fig. 1. Proposed method

of his/her drowsiness will be a contributing factor to the value of the speed lock applied. After applying the speed lock, feedback will be taken from the driver regarding the correctness of the output. Hence over time, the classifier will learn better and the misclassification rate will decrease. Thus, this would enhance safety. The system is explained in Fig. 1.

2.1 Dataset Used

The dataset used to train our classifier was "Analysis of a sleep-dependent neuronal feedback loop: the slow-wave micro continuity of the EEG" by Kemp et al. [2]. The sleep-edf database contains 197 whole-night PolySomnoGraphic (PSG) sleep recordings, containing EEG, EOG (electrooculography), chin EMG (electromyography), and event markers. Some records also contain respiration and body temperature. Corresponding hypnograms (sleep patterns) were manually scored by well-trained technicians according to the Rechtschaffen and Kales manual, and are also available. The EOG and EEG signals were each sampled at 100 Hz. The submental-EMG signal was electronically high-pass filtered, rectified and low-pass filtered after which the resulting EMG envelope expressed in uV rms (root-mean-square) was sampled at 1 Hz. For the purposes of this system we will be considering the EEG (from Fpz-Cz and Pz-Oz electrode locations) data from the SC*PSG.edf files. As this system is a prototype built with the purpose of serving as the proof of concept for the proposed system, we will be analyzing the sleep data of only 1 subject for 2 nights containing 7,000,000 entries.

2.2 Data Preprocessing

For the purposes of building a prototype which serves as a proof of concept for the system proposed, we will be using the EEG signals from channels Fpz-Cz and

Pz-Oz. The remaining data like the EOG, EMG, etc. will be pruned. The 20 h
sleep data for every night is grouped in 10-s epochs. A Fast Fourier Transform
(FFT) was performed on each channel of a 10-s EEG data epoch. The PSD
(Power Spectral Density) was divided into 4 segments: Delta (0–4 Hz), Theta
(4–8 Hz), Alpha (8–13 Hz), Beta (13–20 Hz).

After dividing into 4 power bands, the maximum PSD was used for each
band. Then the average of the PSDs was calculated. CGF (Center of Gravity
Frequency) was calculated with the formula:

$$CGF = \frac{[\sum_i P\left(f_i\right) \times f_i]}{[\sum_i P\left(f_i\right)]}$$

and Frequency Variability (FV) is calculated from the formula:

$$\frac{\sum_i P\left(f_i\right) \times f_i^2 - [\sum_i P\left(f_i\right) \times f_i]^2 \div \sum_i P\left(f_i\right)}{\sum_i P\left(f_i\right)}$$

where f is frequency and $P(f_i)$ is the estimated power spectral density.

As a result, the columns in the input matrix are converted from 2 to 32 (2
channels X 4 power bands X 4 features per power band). It was observed that
the data was highly skewed (class -1:class 1 = 125:16315). So, under-sampling
was used to adjust class distribution.

2.3   Classifiers Used

SVM: A Support Vector Machine (SVM) was used to classify the above data.
According to [5] it was found that SVM was an excellent classifier for EEG data
with the classification accuracy reaching upto 99.3%. We used the Radial Basis
Function (RBF) kernel. The probability estimate was calculated and the data
was classified accordingly. We have used Stratified 5-fold cross validation for
evaluating the performance of the classifier (Table 2).

Table 2. Cross validation results

Performance parameter	Accuracy
0.1	0.984
0.5	0.984
0.8	0.984
1	0.993
1.5	0.993

AdaBoost: AdaBoost was used to classify the EEG data. The accuracy measures according to the variations in learning rate and the number of estimators
were recorded (Table 3).

Table 3. Accuracy measures

Learning rate	Estimators	Accuracy
0.1	3	0.976
0.1	5	0.977
0.1	10	0.979
1	3	0.977
1	5	0.977
1	10	0.976
10	3	0.977
10	5	0.976
10	10	0.976

3   Results

After fitting the data to the trained models of SVM and AdaBoost, accuracy measures were achieved. The classification of data was done into two classes: alert and drowsy.

Following are the results of the confusion matrix and the required scores for SVM (Tables 4 and 5):

Table 4. Confusion matrix for SVM

Actual\predicted	Predicted alert	Predicted drowsy
Actual alert	65	28
Actual drowsy	0	6533

Table 5. Performance scores for SVM

Scores	Values
Accuracy	0.9958
Misclassification rate	0.0042
Precision	0.9957
Specificity	0.6989
F - Score	0.9978

Following are the results of the confusion matrix and the required scores for AdaBoost (Tables 6 and 7):

Table 6. Confusion Matrix for AdaBoost

Actual\predicted	Predicted alert	Predicted drowsy
Actual alert	65	28
Actual drowsy	91	6442

Table 7. Performance scores for AdaBoost

Scores	Values
Accuracy	0.9820
Misclassification rate	0.0179
Precision	0.9900
Specificity	0.9860
F - Score	0.9800

After the calculation of the probability of drowsiness, depending on user requirement speed locks can be applied. Thus we can summarize the approach in 4 stages:

1. Train the Classifier.
2. Retrieve EEG waves of the driver from the wireless EEG headset.
3. Classify the input at real time.
4. Map the classified output to the appropriate speed-lock value.

4   Conclusion

The main aim of the system is to reduce the accidents caused due drowsiness. The existing systems require heavy processing, take a lot of time to produce output and have many restrictions. This system has the potential to provide real-time results and to respond to drowsiness quickly in turn keeping the driver and the others on the road safe. The accuracy of this system can be improved by acquiring a better dataset with a balanced class distribution. This system has the potential of reducing the probability of accidents and thus satisfies our main goal.

5   Future Scope

The implementation done in this paper is just a proof of concept. The final system can include data from the car such as acceleration, breaking pattern and speed and emotions of the driver. The driver profiling based on acceleration and breaking patterns is explained in [9]. The emotions can be detected from data acquired by the wireless EEG headset. As the emotions change, our driving pattern changes with it. All these systems working together will give better outputs and as the driver uses these systems more the systems will increase their accuracy. The ultimate goal of the complete system is to allow the driver to give his or her undivided attention to the road, thus, keeping him or her safe.

References

1. Kemp, B., Zwinderman, A.H., Tuk, B., Kamphuisen, H.A.C., Oberyé, J.J.L.: Analysis of a sleep-dependent neuronal feedback loop: the slow-wave microcontinuity of the EEG. IEEE-BME **47**(9), 1185–1194 (2000)
2. Goldberger, A.L., Amaral, L.A.N., Glass, L., Hausdorff, J.M., Ivanov, P.C., Mark, R.G., Mietus, J.E., Moody, G.B., Peng, C.-K., Stanley, H.E.: PhysioBank, PhysioToolkit, and PhysioNet: components of a new research resource for complex physiologic signals. Circulation **101**(23), e215–e220 (2000). [Circulation Electronic Pages; http://circ.ahajournals.org/cgi/content/full/101/23/e215]
3. https://ec.europa.eu/transport/road_safety/specialist/knowledge/speed/speed_is_a_central_issue_in_road_safety/speed_and_accident_risk_en. Accessed 31 Aug 2018
4. Nilsson, G.: Traffic safety dimensions and the power model to describe the effect of speed on safety. Bulletin 221, Lund Institute of Technology, Lund (2004)
5. Yeo, M.V.M., Li, X., Shen, K., Wilder-Smith, E.P.V.: Can SVM be used for automatic EEG detection of drowsiness during car driving? Saf. Sci. **47**(1), 115–124 (2009). ScholarBank@NUS Repository
6. Wheaton, A.G., Chapman, D.P., Presley-Cantrell, L.R., Croft, J.B., Roehler, D.R.: Drowsy driving: 19 states and the District of Columbia, 2009–2010.[630 KB] MMWR Morb Mortal Wkly Rep. 2013; 61:1033
7. Wheaton, A.G., Shults, R.A., Chapman, D.P., Ford, E.S., Croft, J.B.: Drowsy driving and risk behaviors: 10 states and Puerto Rico, 2011–2012.[817 KB] MMWR Morb Mortal Wkly Rep. 2014; 63:557–562
8. https://www.cdc.gov/features/dsdrowsydriving/index.html. Accessed 31 Aug 2018
9. Dangra, B.S., Bedekar, M., Panicker, S.S.: User profiling of automobile driver and outlier detection, **3**(12) (2014). ISSN 2278 - 0211, (Special Issue)

Genre Based Classification of Hindi Music

Deepti Chaudhary[1,2(✉)], Niraj Pratap Singh[1], and Sachin Singh[3]

[1] Department of Electronics and Communication Engineering,
National Institute of Technology Kurukshetra, Kurukshetra 136119,
Haryana, India
deepti.nitk@gmail.com, nirajatnitkkr@gmail.com
[2] Department of Electronics and Communication Engineering, UIET,
Kurukshetra University, Kurukshetra 136119, Haryana, India
[3] Department of Electrical and Electronics Engineering,
National Institute of Technology, Delhi, Delhi 110040, India
sachinsingh.iitr@gmail.com

Abstract. The emotional content perceived from music has great impact on human beings. Research related to music is attaining more and more recognition not only in the field of musicology and psychology but also getting attention of engineers and doctors. The categorization of music can be carried out by considering various attributes such as genres, emotional content, mood, instrumental etc. In this work Hindi music signals belonging to four genres - Classical, Folk, Ghazal and Sufi are considered. Music signals belonging to these genres are divided into positive arousal, negative arousal, positive valence and negative valence by considering arousal and valence as parameters. Spectral features are calculated for the music clips using MIR toolbox. The classification is done by using K-nearest neighbor (K-NN), Naive Bayes (NB) and Support vector machine (SVM). The classification process is conducted for all the four genres and also for arousal and valence classes. The accuracy, precision and recall are considered as evaluation parameter in this work. The evaluation parameters of all the genres and classification results of all the classifiers used are compared in the proposed work. Results reveal that SVM classifier outperforms other two classifiers in terms of the parameters considered.

Keywords: Music emotion recognition · Human computer interaction ·
MIR toolbox · Arousal · Valence

1 Introduction

Music is the traditional method used for communication. Emotion can be induced as well as perceived from songs. All living organisms respond to music. Thus music is not only the method to communicate and heal human beings but it is also used by other living organisms to communicate. Music acts as fertilizer for plants [1]. Music therapy using Vedic chants is used in ayurveda since years to provide healing. Indian classical music is used for meditation which is used as naturopathy treatment for various diseases [2]. Music therapy can be used to treat physical as well as psychological problems of human beings [3, 4]. Thus to provide proper cure and healing to human beings

© Springer Nature Switzerland AG 2019
A. Abraham et al. (Eds.): IBICA 2018, AISC 939, pp. 73–82, 2019.
https://doi.org/10.1007/978-3-030-16681-6_8

music is to be classified in various categories. Manual classification of the music signals is very difficult with the large increase in digitally available music database. Thus automatic music classifier is required for classification of songs in different categories. The automatic music classifier system (MCS) based on human computer interaction is used to automatically detect the emotion of musical clips [5]. The MCS follows the five main procedures dataset collection, preprocessing, annotation, feature extraction and classification [6]. The categorization of songs can be done by using two methods categorical and dimensional approach. The categorical approach is proposed by Hevner in 1936 [7]. The dimensional approach is proposed by Russel [8] and Thayer [9]. In 2010 emotion classification of south Indian classical music is done in 2010 [10]. The categories of data are prepared in 2012 by Velankar and Sahasrabuddhe for mood classification of Hindi classical music [11]. In 2015, Ahsan et al. makes use of Indian classical music set and categorical approach to classify the songs. Researchers also consider 2-d approach to classify Hindi music on the basis of emotion. In 2012, Hampiholi, classified the bollywood music based on mood by using Thayer's model for categorizing the songs [12]. In 2012, Ujlambkar and Attar classified the Indian music in five categories of mood by using Thayer's model [13]. In 2014, Bhat et al. considered 100 songs from bollywood music and classified them in eight different moods [14]. In paper [15] Soleymani described the advantages of dimensional approach over categorical approach and adopted categorical approach to classify 1000 songs. Shakya et al. proposed a model to classify the songs according to genre in 2-d Arousal-Valence plane and proves that Mel-Frequency cepstrum coefficient gives best result among all features [16]. Patra et al. proposed a mood taxonomy for by considering western and Hindi songs [17].

In the proposed work Hindi songs related to four genres classical, ghazal, sufi and folk are considered to design the MCS. A MCS is proposed to categorize the songs on the basis of arousal and valence part of songs. Arousal is related to energy of the music and valence is related to type of emotion in the song. The songs related to all genres are further divided in positive-negative arousal and positive-negative valence as described in Thayer's model of music emotion perception. As proved in [16], MFCC from spectral features gives the best results among all features. Thus in this work spectral features are calculated using MIR toolbox. The classification process is done by using SVM, NB and K-NN.

2   Methodology

MCS has five basic steps as shown in Fig. 1. These steps include Data collection, Preprocessing, Annotation, Feature extraction and classification.

In the proposed work the Hindi music songs are four genres (Classical, Ghazal, Sufi and Folk) are collected from largely available database online. In the proposed work 400 songs belonging to all the four genres are considered out of which 280 songs are used for training and 120 songs are used for testing. In preprocessing 30-s song clips are extracted from complete songs by considering most representative part of song in standard format of 44100 Hz sample rate and 16 bits precision. Further windowing and framing techniques are applied to 30-s clips. Windowing is directly in co-operated with

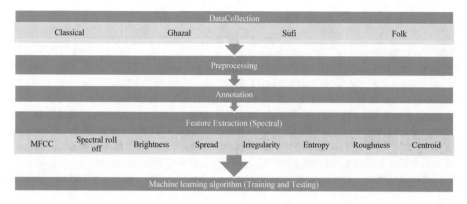

Fig. 1. Basic block diagram of MCS [6]

the Fourier transform function. Hamming window is used to preprocess the signal. The sound signals are non-stationary, thus the analysis of sound signals is carried out by considering short time signals. The process of transforming the sound signal in short time signals is framing. Frame length of 25 ms is considered in this work. Arousal-Valence emotion plane is used to categorize the songs during annotation process. Songs are annotated by a group of eight people into two categories of arousal and two categories of valence i.e. Positive – Negative arousal and Positive-Negative Valence by considering Thayer's model of emotion plane as shown in Fig. 2. After annotation the spectral features of songs are extracted by using MIR toolbox. After feature extraction of music clips, training and testing process is conducted for classifiers based on the feature values of various classes. In the proposed work, SVM, NB and K-NN are used as classifiers. All the songs belonging four genres are classified in two categories of arousal i.e. Positive arousal and Negative arousal. Songs are also classified on the basis of Valence i.e. Positive Valence and Negative Valence as represented in Fig. 2. Accuracy, precision and recall are considered as evaluation parameter in this work. The results for all the four genres and their classes for arousal and valence are compared.

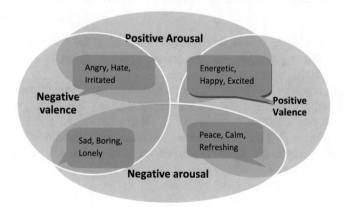

Fig. 2. A-V emotion model

3 Performance Evaluation

The implementation of the proposed work is carried out by using MATLAB. The spectral features are extracted by using MIRtoolbox 1.6.1 in MATLAB. The songs of four genres Classical, Ghazal, Sufi and Folk are considered. The database collected and sampled at a rate of 44100 Hz and 16 bit precision. 400 songs consisting of 100 songs of each genre are considered in this research work. The music clips considered are downloaded from freely available online songs. These songs are clipped to 30 s music clips by considering the most representative part of song. The training dataset consists of 280 songs and test data contains 120 songs. Spectral feature is used to differentiate the defined classes. The example of songs considered in this work is given in Table 1.

Table 1. Description of songs

Songs genre		Example
Ghazal	Positive arousal	Bahut pahle se un kadmo ki aahat
	Negative arousal	Pahle jo dard tha wahi pyara hai in dino
	Positive valence	Apni marji se hawao ka safar
	Negative valence	Ae khuda reeet k sehaaroon ko samandar kar de
Folk	Positive arousal	Aai aai basanti mela
	Negative arousal	Dharti kahe pukar ke
	Positive valence	Gaon Taraane mann ka
	Negative valence	Phir tumhari yaad
Sufi	Positive arousal	Allah Hoo Allah Hoo
	Negative arousal	Mere Sahiba
	Positive valence	Tu mane ya na mane
	Negative valence	Tumhain Dillagi Bhool
Classical	Positive arousal	Ajori badhavara
	Negative arousal	Gunaun gaun tumharo
	Positive valence	Bajan laagi bansri kaanhaki
	Negative valence	Kate na Biraha ki raat

Table 1 provides the lyrical description of songs of all the four genres considered. The positive and negative arousal and valence part of songs is considered while choosing the dataset.

Accuracy, precision and recall are considered as performance evaluation parameter for genre based classification. Table 2 shows the accuracy for all the genres individually. Accuracy is measured by using SVM, Naive Bayes and K-NN classification model. As proved in [18] polynomial kernel performs better then Radial basis kernel in SVM. Therefore polynomial kernel is used for SVM and value of parameter k = 5 are chosen for experiment.

Table 2. Accuracy comparison of SVM, NB and K-NN

Genres	Accuracy		
	SVM	NB	K-NN
Ghazal	0.875	0.825	0.825
Folk	0.85	0.8166667	0.783333
Sufi	0.883333	0.8333333	0.816667
Classical	0.908333	0.8583333	0.858333
Overall	0.7583	0.667	0.6417

The accuracy results represented in Table 2 are summarized by following points.

(i)   Accuracy achieved for ghazal, folk, sufi and classical songs is 87.5%, 85%, 88.3% and 90.8% correspondingly by using SVM.
(ii)   Accuracy achieved for ghazal, folk, sufi and classical songs is 82.5%, 81.67%, 83.33% and 85.83% correspondingly by using NB.
(iii) Accuracy achieved for ghazal, folk, sufi and classical songs is 82.5%, 78.33%, 81.67% and 85.83% correspondingly by using K-NN.
(iv)   Overall accuracy achieved for all the combined songs by using SVM, NB and K-NN is 75.83%, 66.7% and 64.17%.

Table 3. Precision comparison of SVM, NB and K-NN

Genres	Precision		
	SVM	NB	K-NN
Ghazal	0.758621	0.6666667	0.666667
Folk	0.7	0.6333333	0.566667
Sufi	0.766667	0.65625	0.625
Classical	0.806452	0.7096774	0.709677
Overall	0.5116	0.3913	0.3673

Table 3 shows the precision of all the genres using above mentioned classifiers. The precision results represented in Table 3 are summarized by following points.

(i)   Precision achieved for ghazal, folk, sufi and classical songs is 75.86%, 70%, 76.67% and 80.64% correspondingly by using SVM.
(ii)   Precision achieved for ghazal, folk, sufi and classical songs is 66.67%, 63.33%, 65.62% and 70.96% correspondingly by using NB.
(iii) Precision achieved for ghazal, folk, sufi and classical songs is 66.67%, 56.67%, 62.5% and 70.97% correspondingly by using K-NN.
(iv)   Overall precision achieved for all the combined songs by using SVM, NB and K-NN is 51.16%, 39.13% and 36.73%.

Recall is also considered as one of the evaluation parameter. Table 4 shows the recall of all the genres using above mentioned classifiers.

Table 4. Recall comparison of SVM, NB and K-NN

Genres	Recall		
	SVM	NB	K-NN
Ghazal	0.733333	0.6	0.6
Folk	0.7	0.6333333	0.566667
Sufi	0.766667	0.7	0.666667
Classical	0.833333	0.7333333	0.733333
Overall	0.7333	0.6	0.6

The recall results represented in table above are summarized by following points.

(i)   Recall achieved for ghazal, folk, sufi and classical songs is 73.33%, 70%, 76.67% and 83.33% correspondingly by using SVM.
(ii) Recall achieved for ghazal, folk, sufi and classical songs is 60%, 63.33%, 70% and 73.33% correspondingly by using NB.
(iii) Recall achieved for ghazal, folk, sufi and classical songs is 60%, 56.67%, 66.67% and 73.33% correspondingly by using K-NN.
(iv) Overall recall achieved for all the combined songs by using SVM, NB and K-NN is 73.33%, 60% and 60%.

Figures 3, 4 and 5 represents the graphical representation of the comparison among accuracy, precision and recall of the classifiers by considering different genres.

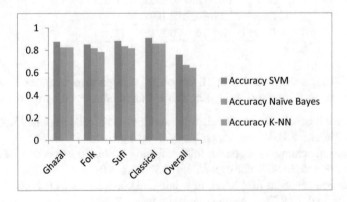

Fig. 3. Graphical representation of accuracy comparison

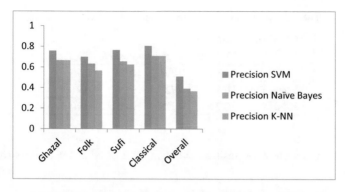

Fig. 4. Graphical representation of precision comparison

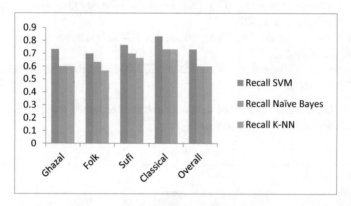

Fig. 5. Graphical representation of recall comparison

The results described above reveal that the accuracy achieved for classical songs is highest for all the three classifiers i.e. SVM, NB and K-NN (90.8%, 85.8% and 85.8%). The accuracy for SVM classifier is best among all the three classifiers and for NB and K-NN is same for classical songs.

It also reveals that the precision values attained are also best for classical songs using SVM, NB and K-NN i.e. 80.6%, 70.6% and 70.6%. The precision value attained for SVM is best among all the classifiers considered and precision value for NB and K-NN are same for classical songs.

It has also been predicted that the recall values achieved are best for classical songs using SVM, NB and K-NN i.e. 70.3%, 60% and 60%. The recall is also better for SVM than other two classifiers and similar for NB and K-NN for classical songs.

Music clips belonging to all the genres are further divided into two classes by considering arousal part and two classes by considering valence part as represented in Fig. 2. The SVM, Naive Bayes and K-NN are used to train the classes and then testing is performed by considering the spectral features of music clips. Accuracy for arousal and valence related classes by all the classifiers is shown in Table 5.

Table 5. Accuracy comparison of arousal and valence classes

Classifier	Arousal		Valence	
	Positive	Negative	Positive	Negative
SVM	0.86	0.79	0.83	0.76
Naive Bayes	0.78	0.72	0.76	0.71
K-NN	0.77	0.718	0.758	0.70

The comparative study of positive and negative arousal and valence is summarized as follows.

(i) The accuracy attained by SVM, NB and K-NN for positive arousal genre songs is 86%, 78% and 77%.

(ii) The accuracy attained by SVM, NB and K-NN for negative arousal genre songs is 79%, 72% and 71.8%.

(iii) The accuracy attained by SVM, NB and K-NN for positive valence genre songs is 83%, 76% and 75.8%.

(iv) The accuracy attained by SVM, NB and K-NN for negative valence genre songs is 76%, 71% and 70%.

The graphical representation of the comparison of accuracy represented in Table 5 for SVM, NB and K-NN is shown in Fig. 6.

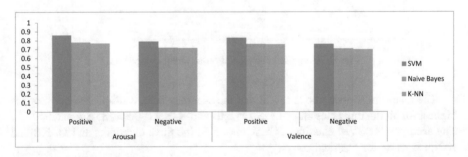

Fig. 6. Graphical representation of accuracy comparison of arousal and valence classes

The above results show that the accuracy, precision and recall (0.91, 0.81 and 0.83) for classical songs are better among all genres. It has also been revealed from Table 5 and Fig. 6 that the accuracy of positive classes of arousal and valence is 0.86 and 0.83 which is better than negative arousal and valence which is 0.79 and 0.76. It has also been experimentally proved from results that SVM outperforms Naive Bayes and K-NN in terms of the evaluation parameters considered in this approach.

4 Conclusion

In this work the classification of songs is done with respect to four genres. Based on the multi-genre dataset MCS is proposed by using SVM, NB and K-NN. The results of all the classifiers are compared in terms of accuracy, precision and recall. Results show that accuracy, precision and recall is attained maximum for classical songs among all genres and SVM classifier outperforms Naives Bayes and K-NN. The accuracy, precision and recall achieved by classical songs by using SVM are 0.91, 0.81 and 0.83. The songs are also categorized on the basis of arousal and valence. The accuracy achieved by positive arousal is more than negative arousal and accuracy achieved by positive valence is more than negative valence. The accuracy achieved by positive arousal is 0.86 and by positive valence is 0.83 by using SVM. Thus it has been experimentally shown that SVM outperforms Naives Bayes and K-NN for both the cases. In this paper authors make the comparison of genres of Hindi songs. The same techniques can be implemented on songs of different languages depending on user's choice.

References

1. Ramekar, U.V., Gurjar, A.A.: Empirical study for effect of music on plant growth. In: 10th International Conference on Intelligent Systems and Control, 7–8 January 2016, Coimbatore, India (2016)
2. Gurjar, A.A., Bardekar, A.A.: Study of Indian classical ragas structure and its influence on human body for music therapy. In; 2nd International Conference on Applied and Theoretical Computing and Communication Technology, iCATccT, 21–23 July 2016, Bangalore, India (2016)
3. Phasukkit, P., Mahrozeh, N., Kumngern, M.: A simple laboratory test of music therapy. In: Biomedical Engineering International Conference, 25–27 November 2015, Pattaya, Thailand (2015)
4. Raglio, A., et al.: Effects of music and music therapy on mood in neurological patients. World J. Psychiatry 5(1), 68–78 (2015)
5. Yang, Y.-H., Lin, Y.-C., Su, Y.-F., Chen, H.H.: A regression approach to music emotion recognition. IEEE Trans. Audio Speech Lang. Process. 16(2), 448–457 (2008)
6. Yang, Y.-H., Su, Y.-F., Lin, Y.-C., Chen, H.H.: Music Emotion Recognition. CRC Press, Boca Raton (2011)
7. Hevner, K.: Experimental studies of the elements of expression in music. Am. J. Psychol. 48(2), 246–268 (1936)
8. Russell, J.A.: A circumplex model of affect. J. Pers. Soc. Psychol. 39(6), 1161–1178 (1980)
9. Thayer, R.E.: The Biopsychology of Mood and Arousal. Oxford University Press, Oxford (1990)
10. Koduri, G.K., Indurkhya, B.: A behavioral study of emotions in south Indian classical music and its implications in music recommendation systems. In: Proceedings of the 2010 ACM Workshop on Social, Adaptive and Personalized Multimedia Interaction and Access, SAPMIA 2010, 29 October 2010, Firenze, Italy (2010)
11. Velankar, M.R., Sahasrabuddhe, H.V.: A pilot study of hindustani music sentiments. In: Proceedings of the 2nd Workshop on Sentiment Analysis where AI Meets Psychology, December 2012, Mumbai, pp. 91–98 (2012)

12. Hampiholi, V.: A method for music classification based on perceived mood detection for Indian bollywood music. Int. J. Comput. Electr. Autom. Control Inf. Eng. **6**(12), 507–5014 (2012)
13. Ujlambkar, A.M.: Mood classification of Indian popular music categories and subject descriptors. In: Sixth Asia Modelling Symposium, 29–31 May 2012, Bali, Indonesia, pp. 278–283 (2012)
14. Bhat, A.S., Amith, V.S., Prasad, N.S., Mohan, D.M.: An efficient classification algorithm for music mood detection in Western and Hindi music using audio feature extraction. In: Proceedings 5th International Conference on Signal and Image Processing, ICSIP, 8–10 January 2014, Bangalore, India, South Korea, pp. 359–364 (2014)
15. Soleymani, M., Caro, M.N., Schmidt, E.M., Sha, C.-Y., Yang, Y.-H.: 1000 songs for emotional analysis of music. In: Proceedings of the 2nd ACM International Workshop on Crowdsourcing for Multimedia, CrowdMM 2013, 22 October 2013, Barcelona, Spain, pp. 1–6 (2013)
16. Shakya, A., Gurung, B., Thapa, M.S., Rai, M.: Music classification based on genre and mood. In: International Conference on Computational Intelligence, Communications and Business Analytics, CICBA 2017. Communications in Computer and Information Science, vol. 776, pp. 168–183. Springer, Singapore (2017)
17. Patra, B.G., Das, D., Bandyopadhyay, S.: Multimodal mood classification of Hindi and Western songs. J. Intell. Inf. Syst. 1–18 (2018)
18. Song, Y.: The role of emotion and context in musical preference. Queen Mary, University of London (2016)

A Novel Meta-heuristic Differential Evolution Algorithm for Optimal Target Coverage in Wireless Sensor Networks

Chandra Naik$^{(\boxtimes)}$ and D. Pushparaj Shetty

Department of Mathematical and Computational Sciences,
National Institute of Technology Karnataka, Surathkal 575025, India
chandra.nitk2017@gmail.com, prajshetty@nitk.edu.in

Abstract. A wireless sensor network (WSN) faces various issues one of which includes coverage of the given set of targets under limited energy. There is a need to monitor different targets in the sensor field for effective information transmission to the base station from each sensor node which covers the target. The problem of maximizing the network lifetime while satisfying the coverage and energy parameters or connectivity constraints is known as the Target Coverage Problem in WSN. As the sensor nodes are battery driven and have limited energy, the primary challenge is to maximize the coverage in order to prolong network lifetime. The problem of assigning a subset of sensors, such that all targets are monitored is proved to be NP-complete. The Objective of this paper is to assign an optimal number of sensors to targets to extend the lifetime of the network. In the last few decades, many meta-heuristic algorithms have been proposed to solve clustering problems in WSN. In this paper, we have introduced a novel meta-heuristic based differential evolution algorithm to solve target coverage in WSN. The simulation result shows that the proposed meta-heuristic method outperforms the random assignment technique.

Keywords: Target coverage · Differential evolution ·
Wireless sensor networks

1 Introduction

Recent development in communication technology has enabled the development of low cost, low power, tiny devices, which communicate through short distances. These small devices consist of sensing, processing, and communicating components. The collaborative setting of these sensor nodes forms the WSN. It has seen a wide variety of applications like military application, fire detection in a forest or home appliances, flooding, earthquake, tsunami detection, health applications, and in surveillance applications. Variety of use in urban areas which include traffic monitoring, air pollution monitoring, water pipeline monitoring, different gas

© Springer Nature Switzerland AG 2019
A. Abraham et al. (Eds.): IBICA 2018, AISC 939, pp. 83–92, 2019.
https://doi.org/10.1007/978-3-030-16681-6_9

monitoring, precipitation monitoring in sewage, asset monitoring, temperature monitoring, power system monitoring, geo sensing applications, etc. [1].

Wireless sensor networks are facing challenges due to limited communication range, limited battery source and prone to failure by external events. The coverage problem plays a vital role in extending the lifetime wireless sensor networks. The coverage problem is centered around, how well sensors cover physical space in a deployed area [2].

2   Existing Work

To solve the target coverage problem, authors in [9] discussed a heuristic which produces a disjoint sensor cover. The authors in [2] proposed an approximation algorithm, where a sensor can take part in more than one sensor cover, and proved that the target coverage problem belongs to NP-Complete class. There have been several works on target coverage problem using many meta-heuristic algorithms. In [6] authors used artificial bee colony algorithm to solve target coverage problem. In [7] a 2-connected target cover solution was proposed. In [8] a GA-based algorithm is proposed to solve the target coverage problem, where authors try to prolong lifetime of the network by selecting sensors with highest residual energy. Further, in [4] authors solved both coverage and connectivity problem using GA approach, where for a given a set of points, it finds the minimum number of potential positions to place sensor nodes to achieve k-coverage of targets and m-connectivity with other sensors. Authors in [3] proposed a BBO-based scheme for solving target coverage problem, where optimal sensors positions are computed for achieving k-coverage and m-connectivity of the given wireless sensor network. In this paper, a novel differential evolution algorithm is proposed to solve the target coverage problem to prolong global network lifetime of the given WSN.

3   Classical Differential Evolution

Differential evolution (DE) is a widely used evolutionary algorithm in many of the applications. It has been used in many streams of engineering to solve optimization problems. It is divided into four stages which consists of initialization of population vector or chromosome, mutation, crossover, and selection. The algorithm starts with randomly generated population vectors of specified size. Each individual vector is a solution for the optimization problem. The fitness of each individual vector is calculated to know the quality of the solution. Once the population vectors are generated, the DE passes through, mutation, crossover, and selection process to optimize the solution in the population vector. Finally, depending on the fitness value best vector is selected as the best solution [10]. The various stages of classical DE are shown in Fig. 1a.

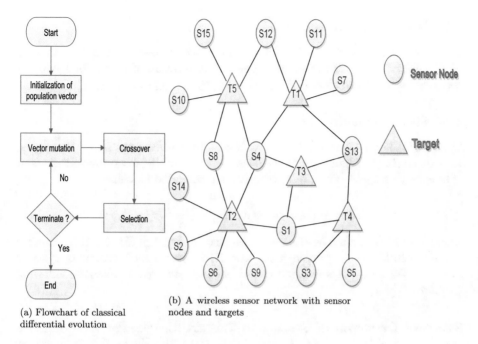

(a) Flowchart of classical differential evolution

(b) A wireless sensor network with sensor nodes and targets

Fig. 1. Classical DE method and a sample WSN

4   Proposed Differential Based Algorithm

In this paper, we discuss target coverage problem.

Definition 1 *Target Coverage Problem. Let S denote the set of n sensors, $S = \{s_1, s_2 \ldots, s_n\}$ deployed to cover m targets $T = \{t_1, t_2, \ldots, t_m\}$. The coverage requirement is that at any given moment, target t_k is covered by at least one sensor node and the total energy of the network is minimized.*

The objective of this study is to obtain a disjoint subset of sensors that covers each target. Later, these sensor may be scheduled to extend the global network lifetime of WSN.

4.1   Vector Encoding

In the proposed technique, each vector consist of two rows, and length equals the number of sensors. Sensors and targets are represented in the first row and the second row respectively. The ith component in vector represents a sensor assigned to a target as shown in Fig. 3. To illustrate, we take a sensor network which consists of 5 targets $T = \{t_1, t_2, \ldots, t_5\}$ and 15 sensors $S = \{s_1, s_2, \ldots, s_{15}\}$.

4.2   Initialization of the Population Vector

The scheme represents vector as follows. Each vector has the complete mapping of all sensors to one of the targets. The Gth generation of ith vector having N components is represented as $X_{i,G} = [x_{1,i,G}, x_{2,i,G}, x_{3,i,G}, \ldots, x_{N,i,G}]$.

4.3   Fitness Function

Our goal is to extend the lifetime of WSN such that at least one sensor is assigned to each target for a longer time. This can be achieved by an optimal assignment of sensors to targets. Thus, our objective is formulated as below

$$Maximize\ R_j = Maximize\ \frac{T_j(S)}{S}, \forall j \in m \qquad (1)$$

where $T_j(S)$ is the number of sensors assigned to target j, S is the total number of sensors in the field of deployment, and R_j is the ratio of sensors assigned to target T_j. We derive fitness function and evaluate each vector using the following method.

Standard Deviation of Sensors to Targets Assignment. Our objective is to maximize the lifetime of networks, in order to achieve this, each target should be assigned to an optimal number of sensors, which are in the communication range. Therefore, we compute the standard deviation of R_j. This can be formulated as below,

$$\sigma_T = \sqrt{\frac{1}{m} \sum_{j=1}^{m} (\mu_T - R_j)^2} \qquad (2)$$

where $\mu_T = \frac{1}{m} \sum_1^m R_j$. Since fitness $\propto \frac{1}{\sigma_T}$ in our formulation, we obtain following formula to compute fitness.

$$fitness = \frac{1}{\sigma_T} = \frac{1}{\sqrt{\frac{1}{m} \sum_{j=1}^{m} (\mu_T - R_j)^2}} \qquad (3)$$

4.4   Mutation

We adopted DE/best/1/bin scheme [10] for mutation and crossover operation. For each candidate vector of the population (called target vector), a mutation vector is created through DE mutation process. In this scheme, out of three vectors, a best vector and two random distinct vectors are selected. Let $X_{i,G}$, $X_{best,G}$ and $V_{i,G}$ be the target, best and donor vectors respectively. Then the mutation vector is obtained by,

$$V_{i,G} = X_{best,G} + F \times D_{i,G} \qquad (4)$$

where F is the scaling vector which lies in the interval [0.4, 1]. We set F as 1.0 and $D_{i,G} = X_{a,G} - X_{b,G}$ with $a, b \in [1, P]$, such that $a \neq b \neq best$. This classical mutation operation does not work for our scenario. This is because the subtraction of two components of the vectors gives a difference vector with negative values. Due to the fact that our vectors consists of 0 and 1, we adopted the scheme proposed in [5].

$$D_{j,i,G+1} = \begin{cases} 1 + (X_{j,a,G} - X_{j,b,G}) & if(X_{j,a,G} - X_{j,b,G} \leq 0) \\ X_{j,a,G} - X_{j,b,G} & otherwise \end{cases} \tag{5}$$

Again, the same problem may occur at the time of addition operation. Therefore, donor vectors are generated as mentioned in [5]

$$V_{j,i,G} = \begin{cases} X_{j,best,G} + F \times D_{j,a,G} - 1 & (if\ X_{j,best,G} + X_{j,a,G} > 1) \\ X_{j,best,G} + F \times D_{j,b,G} & otherwise \end{cases} \tag{6}$$

4.5 Crossover

A trial vector $U_{i,G+1}$ is derived from the target vector $X_{i,G}$ and the donor vector $V_{i,G}$, as shown below:

$$U_{j,i,G} = \begin{cases} V_{j,i,G} & if\ rand() \leq C_r \\ X_{j,i,G} & otherwise \end{cases} \tag{7}$$

where C_r is the crossover probability set to 0.5. To generate a jth component of a trial vector, a random number is obtained between 0 and 1. If the random number is less than or equal to C_r, then we select j^{th} component of donor vector as j^{th} component of the trial vector; otherwise it is selected from the target vector. The entire process of crossover is depicted in Fig. 2.

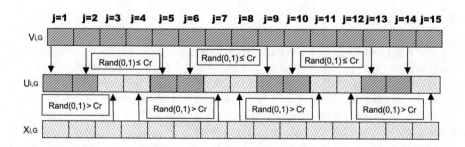

Fig. 2. Crossover operation

4.6 Selection

The selection process decides which vector survives for the next generation, either target vector or trial vector. Both of these vectors are evaluated to find fitness function. The target vector $X_{i,G}$ is compared with the trial vector $U_{i,G+1}$ and one with the lowest function value is selected for the next generation as shown below,

$$X_{i,G+1} = \begin{cases} U_{i,G+1} & if(fitness(U_{i,G+1}) \leq fitness(X_{i,G})) \\ X_{i,G} & otherwise \end{cases} \tag{8}$$

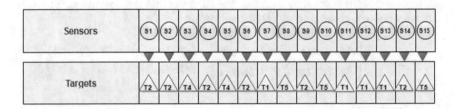

Fig. 3. Random assignment of sensors to targets are in the coverage range

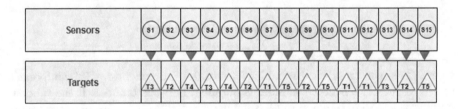

Fig. 4. Optimal assignment of sensors to targets are in the coverage range

5   Illustration

Consider a wireless sensor network with 15 sensors $S = \{s_1, s_2, \ldots, s_{15}\}$ and 5 targets $T = \{t_1, t_2, \ldots, t_5\}$ as shown in Fig. 1b. An edge between target and sensor represents a cover of a target by a sensor. Now, random assignment of sensor to target has been done as shown in Fig. 3 which obtains a vector (chromosome) as shown in Table 1. The integer 1 in the cell indicates i^{th} sensor assigned to j^{th} target; otherwise, this integer is 0. The variable $S_{count}(T_j)$ represents number of sensors assigned to the target T_j. The value of fitness of vector is computed using the Eq. 3.

The Table 1 represents a vector with a random assignment of sensors to targets. The Table 2 represents a vector with an optimal assignment of sensors to targets as shown in Fig. 4. The lowest σ_T value indicates the most fit vector over others. Therefore, vector in Table 2 is better than the vector in Table 1. It is also observed from the Table 1 that the lifetime of global networks is zero because the target T_3 is not in the cover of any sensor.

Algorithm 4.1. The proposed DE based Target Coverage Algorithm for WSN(DETC)

Input: m Targets, n Sensors, and Coverage Matrix of size $m \times n$
Output: Optimal sensor and target assignment matrix $(X_{Best,G})$, best fitness value($Best_{fitness}$), disjoint Target Cover Set $(T_j), \forall j \in m$, Cover Sets C_k, and $Network\_lifetime$.

1. // Generate initial popualtion of size P
2. **for** $i = 1$ **to** P
3.   Initialize each ith individual // Using random function.
4. // Differential algorithm starts
5. **for** $itr = 1$ **to** $Max_{iteration}$ // Generation
6.   **for each** member vector of population $X_{i,G}$
7.   Compute the fitness using Equation (3)
8.   Select best member vector $X_{best,G}$ using best fitness value.
9.   Select two random $X_{a,G}$ and $X_{b,G}$, such that $a, b \in [1, P]$, $a \neq b \neq best$, and set F=1.
10.   Perform mutation operation using Equation (4)
11.   Set crossover probability(C_r=0.5).
12.   Perform crossover operation using Equation (7)
13.   Perform selection operation using Equation (8)
14. Obtain $Best_{fitness}$ and $X_{Best,G}$
15. // Obtain disjoint Target Cover Set T_j from $X_{Best,G}$
16. **for** $j = 1$ **to** m
17. $T = T$ union T_j
18. // Obtain Cover sets C_k , from each Target Cover Set $T_j \in T$.
19. $k = 0$, $C_k = \phi$
20. **repeat**
21. $k = k + 1$ // To create each Cover Set
22.   **for** $j = 1$ **to** m
23.   select random sensor $s_r \in T_j$
24. $C_k = C_k \cup S_r$
25. $T_j = T_j - S_r$ // set difference operation
26. **until** $T \neq \phi$
27. // Calculate life time of the network
28. Initialize lifetime of each sensor S_r to 1
29. set $Network\_lifetime = 0$
30. **for** each cover set C_k
31.   **repeat**
32.   lifetime_S_r= lifetime_S_r - w // w is lifetime granularity [2]
33. $Network\_lifetime = Network\_lifetime + 1$
34.   **if** lifetime_S_r == 0
35.   **then**
36. $C_k = C_k - S_r$
37.   **until** $C_k \neq \phi$ and each target covered by at least one sensor
38. **return** $Network\_lifetime$

Table 1. Vector encoding of random sensors and targets assignment.

Targets	Sensors															$S_{count}(T_j)$	R_j
	S1	S2	S3	S4	S5	S6	S7	S8	S9	S10	S11	S12	S13	S14	S15		
T1	0	0	0	0	0	0	1	0	0	0	1	1	1	0	0	4	0.26
T2	1	1	0	1	0	1	0	0	1	0	0	0	0	1	0	6	0.4
T3	0	0	0	0	0	0	0	0	0	0	0	0	0	0	0	0	0
T4	0	0	1	0	1	0	0	0	0	0	0	0	0	0	0	2	0.13
T5	0	0	0	0	0	0	0	1	0	1	0	0	0	0	1	3	0.2
$\sigma_T = 0.148$																	

Table 2. Vector encoding of optimal sensors and targets assignment.

Targets	Sensors															$S_{count}(T_j)$	R_j
	S1	S2	S3	S4	S5	S6	S7	S8	S9	S10	S11	S12	S13	S14	S15		
T1	0	0	0	0	0	0	1	0	0	0	1	1	0	0	0	3	0.2
T2	0	1	0	0	0	1	0	0	1	0	0	0	0	1	0	4	0.267
T3	1	0	0	1	0	0	0	0	0	0	0	0	1	0	0	3	0.2
T4	0	0	0	1	0	1	0	0	0	0	0	0	0	0	0	2	0.13
T5	0	0	0	0	0	0	0	1	0	1	0	0	0	0	1	3	0.2
$\sigma_T = 0.048$																	

6 Experimental Results

In this section, we discuss simulation results to compare the performance of the DE-based scheme and random scheme. In these results, we have considered parameters like network lifetime versus number of targets, network lifetime versus range of sensors, and network lifetime versus targets. For simulation, we have used MATLAB R2017b. A set of random wireless sensor networks are generated within a field size 100×100. The number of sensors are varied from 500 to 2500 in steps of 500; The sensor range and the number of targets are varied from 10 to 50 in steps of 10. The network is assumed to be homogeneous and initial energy of each sensor is $1J$. The sensor lifetime granularity is assumed as $w \in (0, 1]$ [2]. For our proposed algorithm DETC, we considered population of 100 vectors and 100 generations. We have taken crossover rate (C_r) and scaling factor(F) as 0.5 and 1 respectively. The Fig. 5 illustrates performance improvement of DE-based scheme over Random scheme when the number of sensors increases from 500 to 2500 with fixed targets 10 and each sensor range value as 10. The Fig. 6 illustrates performance improvement of DE-based scheme over Random scheme when range of each sensor increases from 10 to 50 with 1500 sensors and 10 targets. The Fig. 7 illustrates performance improvement of DE-based scheme over Random scheme when number of targets increases from 10 to 50 with 1500 sensors and each sensor range value as 10. It can be observed from Fig. 7 that number of network lifetime decreases when targets increases. This is due to the fact that when the number of targets increase, less number of sensors cover each target. Figure 8 drawn by increasing sensors from 500 to 2500 and targets from 10 to

50 with a constant ratio of 50 between them and each sensor range value as 10. Finally, the graph depicted by taking sensors to targets ratio Vs networks lifetime. This graph shows scale-up of sensors and targets never degrades the performance. However, it has been observed from experiments that scaling up of network size increases the convergent time.

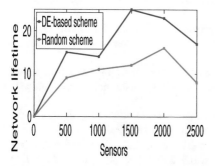

Fig. 5. Performance comparison in terms of the number of sensors

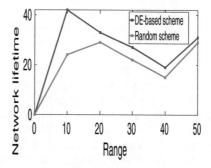

Fig. 6. Performance comparison in terms of range

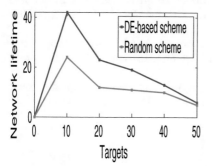

Fig. 7. Performance comparison in terms of the number of targets.

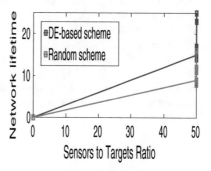

Fig. 8. Scalability test using a constant sensors to targets ratio of 50.

7 Conclusion

In this paper, a differential evolution based meta-heuristic technique is proposed for the target coverage problem in the wireless sensor network. The differential evolution based scheme assigns an optimal disjoint set of sensors to targets. Our algorithm performs better than the random scheme. The simulation has been carried out with various parameter settings. The research work might be extended by adding more parameters like distance between sensors and targets, Q-coverage, and K-connected coverage.

Acknowledgment. The authors would like to acknowledge the support by National Institute of Technology Karnataka, India to carry out research.

References

1. Akyildiz, I.F., Su, W., Sankarasubramaniam, Y., Cayirci, E.: Wireless sensor networks: a survey. Comput. Netw. **38**(4), 393–422 (2002)
2. Cardei, M., Thai, M.T., Li, Y., Wu, W.: Energy-efficient target coverage in wireless sensor networks. In: Proceedings of the 24th Annual Joint Conference of the IEEE Computer and Communications Societies, INFOCOM 2005, vol. 3, pp. 1976–1984. IEEE (2005)
3. Gupta, G.P., Jha, S.: Biogeography-based optimization scheme for solving the coverage and connected node placement problem for wireless sensor networks. Wirel. Netw. 1–11 (2018)
4. Gupta, S.K., Kuila, P., Jana, P.K.: Genetic algorithm approach for k-coverage and m-connected node placement in target based wireless sensor networks. Comput. Electr. Eng. **56**, 544–556 (2016)
5. Kuila, P., Jana, P.K.: A novel differential evolution based clustering algorithm for wireless sensor networks. Appl. Soft Comput. **25**, 414–425 (2014)
6. Mini, S., Udgata, S.K., Sabat, S.L.: Sensor deployment and scheduling for target coverage problem in wireless sensor networks. IEEE Sens. J. **14**(3), 636–644 (2014)
7. Panda, B.S., Bhatta, B.K., Mishra, S.K.: Improved energy-efficient target coverage in wireless sensor networks. In: International Conference on Computational Science and its Applications, pp. 350–362. Springer (2017)
8. Singh, D., Chand, S., Kumar, B., et al.: Target coverage heuristics in wireless sensor networks. In: Advanced Computing and Communication Technologies, pp. 265–273. Springer (2018)
9. Slijepcevic, S., Potkonjak, M.: Power efficient organization of wireless sensor networks. In: IEEE International Conference on Communications, ICC 2001, vol. 2, pp. 472–476. IEEE (2001)
10. Storn, R., Price, K.: Differential evolution–a simple and efficient heuristic for global optimization over continuous spaces. J. Glob. Optim. **11**(4), 341–359 (1997)

Implementation of Threshold Comparator Using Cartesian Genetic Programming on Embryonic Fabric

Gayatri Malhotra[✉], V. Lekshmi, S. Sudhakar, and S. Udupa

U R Rao Satellite Centre, Bangalore, India
gayatri@isac.gov.in
https://www.isro.gov.in

Abstract. Recent research in the area of evolvable design clearly indicates its advantage in the electronics system domain. This biologically inspired approach for design automation and reconfiguration is required for the hardware that needs online adaptation. The Cartesian Genetic Programming (CGP) represents a circuit genotype in the form of grid of nodes. In satellite for safe mode detection, a threshold comparator circuit is used. The comparator circuit can be evolved using CGP architecture by evolutionary algorithm. An evolutionary algorithm (EA) is designed and applied on the CGP pattern of comparator. The evolved comparator in the cascaded form can be further implemented on embryonic fabric. The embryonic fabric has cellular structure that makes it suitable for self-healing and self-replication. The CGP approach to generate configuration data for embryonic array is better than the LUT based data generation approach. In this paper a 2-bit comparator is evolved using customized evolutionary algorithm. The implementation of cascaded 8-bit threshold comparator on the embryonic array is also demonstrated. This comparator design needs configuration data for 2-bit and the data can be cloned to next cells for scalable design on embryonic fabric.

Keywords: CGP · Evolutionary algorithm · Comparator ·
Evolvable hardware · Embryonics

1 Introduction

Embryonic architecture is multicellular in nature so enables cellular division to achieve fault tolerance [1,2]. The ability to grow, self-repair and self-replication is due to the kind of architecture. The CGP configuration data represents a circuit genotype, while the corresponding circuit implementation is called its phenotype. The CGP phenotype can be implemented on the embryonic fabric to enable it for self-repair and self-replication [3]. Any digital circuit mapped onto embryonic cellular array implements fault tolerance by self-repairing [4]. The configuration data required for 8-bit comparator using LUT is 2^{16}, while in

© Springer Nature Switzerland AG 2019
A. Abraham et al. (Eds.): IBICA 2018, AISC 939, pp. 93–102, 2019.
https://doi.org/10.1007/978-3-030-16681-6_10

Table 1. CGP parameters description

Parameter	Description
G	Genotype
n_n	Node's inputs
n_o	Program outputs
F	Function set
n_f	Functions in F set
n_r	Row nodes
n_c	Column nodes
L	Level back parameter

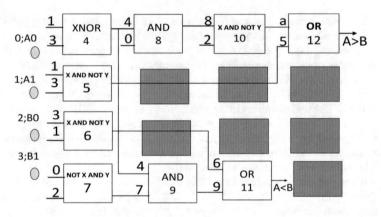

Fig. 1. Configuration of 2-bit comparator in CGP

CGP 2-bit comparator data of 120 bits can be sufficient that can be cloned four times to acheive 8-bit.

The safe mode detection function in sun pointing satellite detects the loss of sun presence. It is based on the 4Pi sun sensor data computation and compare it with the threshold value. The threshold value of sun sensor depends on the field of view of the sensor used. The error between the measured and threshold value is used to initiate detection logic. For 4Pi sun sensor the implemented magnitude comparator is 8 bits and the no. of gates used are very high. As the gates optimization can be achieved through evolutionary algorithm for a digital circuit [5], comparator circuit is got implemented using EA. Due to the modularity of comparator design a 2-bit comparator configuration data can be cloned to achieve 8-bit comparator. The evolutionary algorithm is customized for comparator application and simulated for it. The evolved configuration data can be decoded using defined node functions. The fitness is calculated separately for greater and smaller functions. The total fitness is defined as together of 'smaller' $(A < B)$ and 'larger' $(A > B)$ function fitness. The digital circuit mapping to

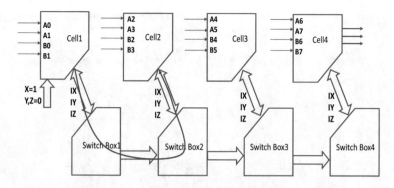

Fig. 2. 8-bit Comparator mapped on embryonic fabric

Table 2. Set1: Comparator logic functions

Bit pattern	Logic function	Bit pattern	Logic function
0	NAND	1	NOR
2	XNOR	3	AND
4	OR	5	XNOR
6	XNOR	7	AND
8	AND	9	OR
10	NOTX AND Y	11	NOTX AND Y
12	AND	13	NAND
14	NOR	15	X AND NOTY

novel embryonic fabric using switch box is already demonstrated [7]. The fabric was simulated for regular circuits of adder and multiplier [8]. The K-map generated comparator circuit is modeled into CGP form and the corresponding genotype is used as best configuration data. This data along with random population is used for evolving the optimized comparator circuit. Section 2 describes the CGP representation of magnitude comparator. Section 3 is about customized evolutionary algorithm and the percentage of genetic operators used. Section 4 is the comparator mapping to embryonic fabric for 8-bit implementation. Section 5 describes and discusses simulation results. Conclusion discusses the limitations and future scope.

2 CGP Representation of Magnitude Comparator

A digital circuit can be represented in CGP as $n_c X n_r$ nodes. A program set is defined as $[G, (n_i), (n_o), (n_n), F, (n_f), (n_r), (n_c), L]$.

The parameters are defined in the Table 1. For 2-bit comparator, $n_i = 4$; (A0, A1, B0, B1), $n_o = 2 or 3$; (equal, large, small), $n_f = 16$; (sixteen functions), $L = 3$; (three previous column connectivity). The best configuration pattern for

Fig. 3. Fitness results for different mutation rate

comparator is 132; 13f; 31f; 02b; 048; 47c; 28b; 699; a59; b; c. The configuration of each node is represented as $in1, in2, function$. Thus node1: 132 means, A1 (1) and B1 (3) are in1 and in2. The function is XNOR (2). Thus first node implements - A1 XNOR B1. The first node output is numbered as '4' as inputs are 0 to 3.

The last two nibbles in the configuration pattern 'b' and 'c' are for two outputs. Here each node is represented into 4 bits (nibble) as the best configuration pattern uses 9 gates for larger and smaller logic outputs. The comparator in cgp form is shown in the Fig. 1. The set of functions applied on comparator CGP are depicted in the Table 2. The best configuration uses function as 2, 8, 9, b, c and f. Other functions are arbitrary selected and the EA run with different functions shows different convergence rate.

3   Implementation of Evolutionary Algorithm

The various developed Genetic Algorithm (GA) approaches relevant to hardware implementation are discussed in [9]. One of the type is Hsclone GA that is 'half-sibling-and-a-clone' crossover technique. The customized evolutionary algorithm (EA) applied for comparator circuit is based on Hsclone GA. The genetic operators are selected/appplied based on the fitness evaluation. If fitness is near to best fit, the mutation is applied, else crossover is applied. It has utilized both crossover and mutation operators to attain convergence faster. The crossover performed is on 2-point at most, while single-point crossover is also applied. The crossover rate is customized based on the fitness calculated. The mutated random pattern is used for crossover with best configuration data. The mutation is performed at most of 42% where almost every alternate bit is flipped. The random population is continuously added to broaden the search space. The pseudo code of EA is shown as-

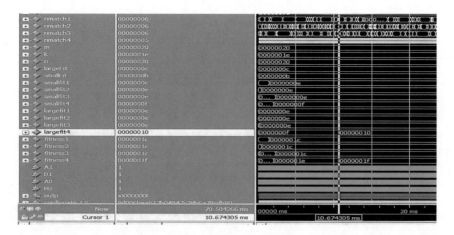

Fig. 4. Maximum fitness '10' for fourth pattern-largefit4

- Randomly generate four initial populations
- Perform constraints and crossover/mutation on initial population
- Evaluate fitness of population
- If fitness is equal to best fit, new circuit pattern is evolved else perform genetic operators
- if fitness is less than 70% crossover with random pattern
- if fitness is more than 70% mutation of current pattern
- Fitness evaluation
- Repeat from step 1 till new pattern has evolved.

The configuration pattern is 120 bits wide as 3×4 bits x 9 nodes = 108 bits, additional one node is added as spare node. Each node represents a logical function block depicted in CGP form. The mutation rate is also varied based on fitness and to satisfy the constraints of CGP. The applied mutation rates are 16bits/120bits - 15% and 50bits/120bits - 42%. The CGP does not allow feedback connection (for combinational circuit) so the output of a gate can not be applied as input of previous gate. This is taken as one of constraint of CGP and the pattern generated is checked as per this. The constraint criteria is defined as-

1. nodefn(0) and nodefn(1) < 4
   (inputs of first node $\neq$ output of first node)
2. nodefn(3) and nodefn(4) < 5
   (inputs of second node $\neq$ output of second node)
3. nodefn(6) and nodefn(7) < 6 and so on.

As per the steps of EA mentioned above the simulation results obtained were fitness = 15 out of 16. The algorithm was further modified for fitness more/less than 60% for selecting the genetic operators. Accordingly the mutation percent is changed and the crossover point is changed. the convergence rate was changed with the modified EA. Finally the convergence is achieved for fitness = 16.

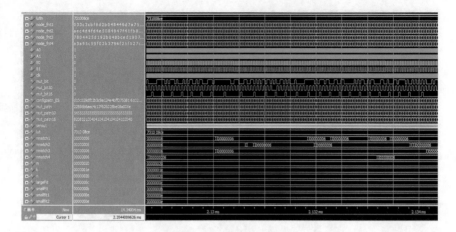

Fig. 5. Node functions and fitness match parameters

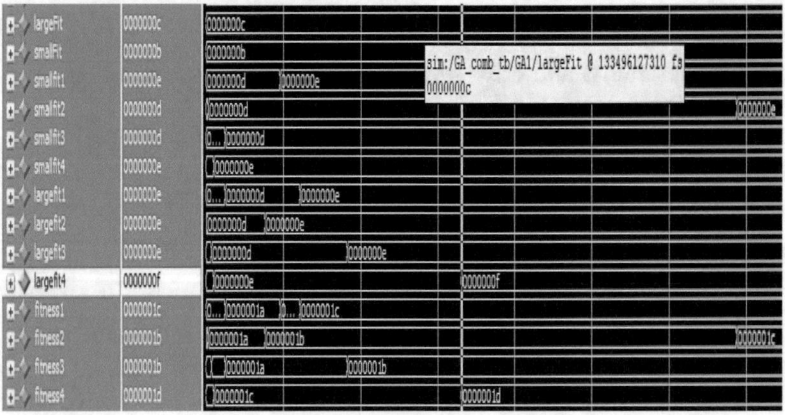

Fig. 6. Fitness results for changed mutation rate and crossover point

4   CGP Mapping on to Embryonic Fabric

Embryonics is defined as multi-cellular structure where mother cell has complete genome data and it can be further divided to create complete organism [10]. The electronic circuit in the similar way can be evolved from a single genome. It is based on the process of ontogenesis that is the development of an organism from a single cell. Basic cell genome of 2-bit comparator can be replicated to develop higher bit comparator [6]. The circuit implemented on the embryonic fabric can be easily adopted for self-replication. The 2-bit comparator can be mapped into a embryonic cell and then further cloned to create 8-bit or more. The 2-bit comparator equations are-

Table 3. Set2: comparator logic functions

Bit pattern	Logic function	Bit pattern	Logic function
0	XNOR	1	XNOR
2	XNOR	3	AND
4	OR	5	NOTX AND Y
6	OR	7	X AND NOTY
8	AND	9	OR
10	OR	11	NOTX AND Y
12	AND	13	X AND NOTY
14	AND	15	X AND NOTY

$$X = x_1.x_0[EQUAL] \tag{1}$$

$$x_1 = A_1.B_1 + not(A_1).not(B_1) \tag{2}$$

$$x_0 = A_0.B_0 + not(A_0).not(B_0) \tag{3}$$

$$Y = A_1.not(B_1) + x_1.A_0.not(B_0) \tag{4}$$

$$Y = 1 \, for \, A > B \tag{5}$$

$$Z = not(A_1).B_1 + x_1.not(A_0).B_0 \tag{6}$$

$$Z = 1 \, for \, A < B \tag{7}$$

The magnitude comparator for 8-bit can be constructed by cascading 2-bit comparators. The 8-bit implementation to embryonic fabric is shown in Fig. 2. Each 2-bit comparator outputs for $A = B(X), A > B(Y) \, and \, A < B(Z)$ is routed through switch box to next cell. This way four cells and four switch boxes can make the 8-bit comparator. The configuration data is 120 bits for each of cell, thus the data can be cloned to each cell. The cloning procedure to perform on embryonic fabric is discussed in [8]. Cloning is initiated with the genome data and the number of times cloning is required. Here the data is 120 bits and the three clones are to be created. Thus embryonic fabric is initialized with one cell and then cloning creates three more cells. After that all inputs (two sets of 8 bits to compare) are applied to the fabric. The three outputs: equal, large and small are generated.

5   Simulation Results

As per the fitness criteria of 70% the simulation results obtained are shown in Figs. 3, 4 and 5. In Fig. 3 different mutation rates are shown and the fitness calculation for smaller and larger function is shown as 'smallfit' and 'largefit'. The best pattern output is shown as 'smallfit' = 'b' or '11' and 'largefit' = 'c' or '12'. this is as per the CGP representation in the Fig. 1. In Fig. 4 the evolved

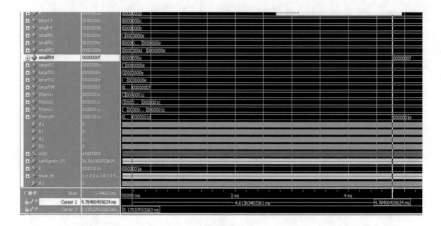

Fig. 7. Fitness results for smallfit4 and largefit4

fitness for largefit4 is shown as '10'. The largefit4 is converged to '10' at 10.67 ms. Thus the algorithm has converged to required fitness as '16'. The set of functions applied on comparator CGP are depicted in the Table 3.

Further in Fig. 5 the four node functions are shown and the corresponding fitness match parameter as 'nmatch' is shown. The nmatch is checked for the 70% or 60% and accordingly the genetic operators are applied.

When the algorithm is modified with changed mutation and crossover operation, the results are shown in Figs. 6 and 7. The first fitness as 'f' is converged earlier than the previous case. The 'largefit4' is converged first and at 0.17 ms. The smallfit4 is converged at 4.78 ms for fitness 'f'. In Fig. 8 the largefit4 is converged to 'f' at 0.04 ms and smallfit4 is converged at 2.18 ms.

The converged results for largefit4 generates new pattern for comparator. The new pattern data is 310; 5c7; 14d; d2b; 048; 47c; 28b; 6a6; fef; 972. This is shown in Fig. 9. The 'largest' in the Fig. 9 denotes the output node. It is '0000000b', thus output from eighth node gives the output. In this 'one' spare node is added thus total = 10 nodes are used.

The new pattern result shows many nodes like 2, 4, 6, 9 an 10 are invalid as per constraints. As nodes 2, 4 and 6 are invalid, the largefit is using nodes 1, 3, 5, 7 and 8 only. Thus five gates are used to generate the output. It is same number of gates as used by initial circuit in Fig. 1.

To achieve the convergence at earliest the constraints are not applied for the data. Once the new pattern is achieved, the invalid nodes can be removed to achieve optimized circuit. The disadvantage is that the single pattern is not generating both smallest and largest output as in initial pattern. To achieve both fitness in the same pattern, the configuration data length needs to be increased. The processing speed will reduce for the increased data length.

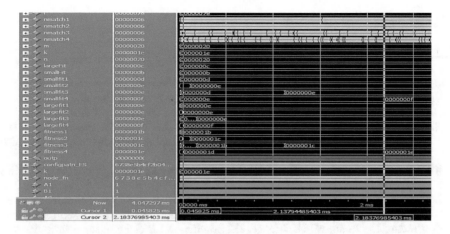

Fig. 8. Fitness results for smallfit4 and largefit4

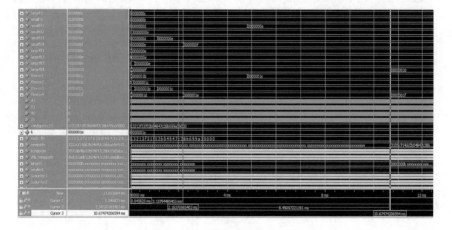

Fig. 9. New configuration data pattern

6   Conclusion and Future Scope

The 8-bit comparator is got implemented on generic embryonic array. The EA is run on the CGP generated data and found its convergence with different genetic operators. The EA is found to be converged for fitness as 16 as required for largefit only. The smallfit is not converged to '10'.

This implies that the additional spare nodes are to be included as part of CGP data. As shown in the CGP block some spare nodes (6) are to be utilized. This will increase the genotype data length. As the spare gates are still required to further do any self-repair, the genotype data length can be increased. The same approach is to be explored to be implemented for other satellite applications like thruster reconfiguration. The self fault detection also is to made automatic to initiate the reconfiguration automatically.

References

1. Tempesti, G., Mange, D., Stauffer, A.: Toward robust integrated circuits: the embryonics approach. Proc. IEEE **88**, 516–543 (2000)
2. Mange, D., Sipper, M., Stauffer, A., Tempesti, G.: Bio-inspired computing architectures: the embryonics approach. In: Proceedings of the Seventh International Workshop on Computer Architecture for Machine Perception (CAMP) (2005)
3. Miller, J.F., Thomson, P.: Cartesian genetic programming. In: Proceedings of the Third European Conference on Genetic Programming (EuroGP2000), vol. 1802, pp. 121–132 (2000)
4. Yang, S., Wang, Y.: A new self-repairing digital circuit based on embryonic cellular array. In: IEEE (2006)
5. Chong, K.H., Aris, I.B., Sinan, M.A., Hamiruce, B.M.: Digital circuit structure design via evolutionary algorithm method. J. Appl. Sci. **7**(3), 380–385 (2007)
6. Prodan, L., Tempesti, G., Mange, D., Stauffer, A.: Biology meets electronics: the path to a bio-inspired FPGA. In: ICES 2000. LNCS, vol. 1801, pp. 187–196 (2000)
7. Malhotra, G., Nagalakshmi, A.M., Sudhakar, S., Udupa, S.: Switch box configuration for generic embryonic cells routing. In: Proceedings of the World Congress on Engineering and Computer Science (WCECS 2015), pp. 414–417 (2015)
8. Malhotra, G., Becker, J., Ortmanns, M.: Novel field programmable embryonic cell for adder and multiplier. In: 9th Conference on Ph.D. Research in Microelectronics and Electronics (PRIME-2013), pp. 153–156 (2013)
9. Zhu, Z., Mulvaney, D., Chouliaras, V.: A novel genetic algorithm designed for hardware implementation. Int. J. Comput. Intell. **3**(4), 281–288 (2007)
10. Zhuo, Q., Qian, Y., Li, Y., Wang, N., Li, T.: Embryonic electronics: state of the art and future perspective. In: The 11th IEEE International Conference on Electronic Measurement and Instruments, ICEMI (2013)

Design and Implementation of an IoT-Based Háptical Interface Implemented by Memetic Algorithms to Improve Competitiveness in an Industry 4.0 Model for the Manufacturing Sector

Roberto Contreras[1]([⊠]), Alberto Ochoa[1], Edgar Cossío[2],
Vicente García[1], Diego Oliva[3], and Raúl Torres[4]

[1] Doctorado en Tecnología, UACJ, Ciudad Juárez, Mexico
rcontreras@itcj.edu.mx
[2] Universidad Enrique Díaz de León, Guadalajara, Mexico
[3] CUCEI, Universidad de Guadalajara, Guadalajara, Mexico
[4] Universidad de Cuenca, Cuenca, Ecuador

Abstract. In the manufacturing industry, priority is given to quality models in the product, in order to make companies competitive, that is why it has been proposed to implement haptic interfaces capable of detecting anomalies in the operation of the electronic equipment of the companies. Cars in the manufacturing sector, that is why through the implementation of Artificial Intelligence and Smart Manufacturing Models, we can get to build intelligent systems for the correct decision making in the auto parts sector in Mexico. Intelligent manufacturing is a subset that employs various techniques of artificial intelligence and emerging technologies coupled with computer control and high levels of adaptability to adapt to changes in product improvement. Intelligent manufacturing is focused on taking advantage of advanced information technologies and even intelligent analysis of data and manufacturing via the Internet of things to allow flexibility in physical processes to address a dynamic market in each society and from a global perspective. There is more training related to the implementation of artificial intelligence of the workforce for such flexibility of adaptability of products and use of emerging technology instead of specific tasks as is usual in traditional manufacturing, and which requires a larger group of individuals for it.

Keywords: Haptical interface · Internet of Things · Smart Manufacturing · Memetic Algorithms and Smart City

1 Introduction

Industry 4.0

The industrial revolution was not an isolated event of the nineteenth century, but it has evolved. Historians agree that the first industrial revolution was characterized by mechanical manufacture. The second was for the introduction of the assembly line and

A. Abraham et al. (Eds.): IBICA 2018, AISC 939, pp. 103–117, 2019.
https://doi.org/10.1007/978-3-030-16681-6_11

serial production. The third revolution was to introduce flexible manufacturing, robotics and quality control and optimization by 1970. Each of these revolutions has named it by number: Industry X.0, where X is the ordinal number of the revolution (Govindarajan et al. 2018). During the fair in Hannover, Germany in 2011, the term Industrie 4.0 (original of the German language) was coined, which is the fourth industrial revolution (Urquhart and Mcauley 2017; Xu et al. 2018) and was later announced in official form in 2013 as a strategic initiative of Germany. Smart Manufacturing is a term coined in the United States of America by the Smart Manufacturing Leadership Coalition initiative in 2014, which coincides with the German term Industrie 4.0, which is the one used in Europe. In the literature both terms are found interchangeably, and are based on cyberphysical systems, the internet of things and cloud computing. Industry 4.0 (hereinafter referred to as I4.0 and including Smart Manufacturing) has key elements that have been accepted by the community dedicated to this topic. The Boston Consulting Group (Rüßmann et al. 2015) has identified nine pillars of I4.0, which are (i) Big Data and Analytics, (ii) Autonomous Robots, (iii) Simulation, (iv) Vertical and Horizontal Integration of Systems, (v) Industrial Internet of Things (IoT), (vi) Cybersecurity, (vii) Cloud or Cloud, (viii) Additive Manufacturing including 3D printing, and (ix) Augmented Reality. These pillars can all be implemented in factories, or take some depending on the case you want to improve.

Internet of Things
The Internet of Things (IoT) is a network of physical devices, vehicles, appliances, and other devices with embedded electronics, software, sensors, actuators, and connectivity that enable these objects to connect and exchange data. The applications of IoT were mainly observed in logistics and transport, health care, intelligent environments, and personal and social applications, proposing ubiquity as a concept that could materialize in advance (Angelini et al. 2018; Atzori et al. 2010). In Industry 4.0, the IoT is a fundamental component and its penetration in the market is growing. Companies dedicated to telecommunications such as Cisco Inc. expects that by 2020 there will be 50 billion (trillion in English) of connected devices (Khan and Salah 2018) and Ericsson Inc. estimates 18 billion (Scott 2018) and other firms such as Gartner estimates that in 2020 there will be 20.4 billion (Gartner 2017). These estimated quantities of connected devices will be due to the increase in technological development, development in telecommunications and adoption of digital devices, and this will invariably lead to the increase in the generation of data and digital transactions, which leads to the mandatory increase in regulations, for security, privacy and informed consent in the integration of these diverse entities that will be connected and interacting with each other and with the users. The increase of these aspects is presented as one of the factors for the increase of cyber-attacks to users (Raymond Choo et al. 2018). IoT is also a complex environment due to the amount of components and layers that make it up. These layers contain a large number of sensors, actors and processing devices with heterogeneous software and different manufacturers (Cheng et al. 2018; Harbers et al. 2018). The aforementioned growth requires and depends on connectivity, forming favorable environments for IoT. The main environment is the computational cloud, which IBM (2009) defines as "the delivery of computing resources on demand". However, small connectivity ecosystems that use the same Wi-Fi or Bluetooth

techniques to connect and form two new levels of environments have been developed at the same time. The first is called Fog Computing and was introduced by Cisco in 2012, and is a layer model to enable ubiquitous access to shared and scalable computing resources, performing data analysis by applications running on the device instead of the cloud (Iorga et al. 2018; Roman et al. 2018; Yannuzzi and Milito 2014). Fog Computing (the fog) is located between the intelligent devices and the centralized services offered by the cloud. The second environment is Mist Computing (in Spanish it can be called the dew) which is a rudimentary and lightweight form of Fog Computing (Iorga et al. 2018) and allows communication with smart devices to be distributed and faster, thanks to its power of computation with microprocessors and microcontrollers.

Augmented Reality/Virtual Reality
Augmented reality (AR) and virtual reality (VR) are not something new in the world of technology. (Milgram et al. 1995) define the continuity Reality-Virtuality and its parallel relationship with the continuity known world extension (EWK) which is known as Milgram Continuum or Mixed Reality Continuum. Nintendo Virtual Boy was launched in 1995 and failed. Simply the computing power was not prepared for this. The power of computation and current processing is much higher and this enables the RA and VR are possible, developing what is known as immersive technologies and when applied to the industrial field is known as Industrial Immersive Technologies (IIT). In the IIT ontology proposed by (Govindarajan et al. 2018) it mentions three types of technology and one application. The technologies included are (i) Brain-Machine Interface (BMI), (ii) Virtual Reality (VR) and (iii) Augmented Reality (AR). AR and VR technologies have several keywords, among which are: Haptic, Interface, Sensor, Mobile and Wearable. In the application focused on industrial engineering, the key words stand out: Agile, Manufacturing, Quality Assurance, Safety. This gives indications that in industrial engineering a haptic interface with sensors can be applied to improve or facilitate a manufacturing process, which can be to minimize the downtime of a machine or to optimize the quality process of a product.

Memetic Algorithms to Complete the Hybrid Operation of Our Proposed Solution
MAs are population-based metaheuristics. This means that they maintain a set of candidate solutions for the problem considered. According to the jargon used, in EAs, each of these tentative solutions is called an individual. As anticipated previously, the nature of the MAs suggests that the agent term is nevertheless more appropriate. The basic reason is the fact that "individual" denotes a passive entity that is subject to evolutionary processes and rules, while the term "agent" implies the existence of an active behavior, directed to the resolution of a certain problem. Saying active behavior is reflected in different typical constituents of the algorithm, such as for example, local search techniques.

Figure 1 shows the general outline of an MA. As in the EAs, the population of agents it is subject to the processes of competition and mutual cooperation. The first is achieved through of the well-known selection procedures (line 6) and replacement (line 12): from the information that provides an ad hoc guide function determines the goodness of the agents in pop; then, a part of them is selected to move to the repro-ductive phase attending to such kindness. Subsequently, this information is used again to determine which agents will be eliminated from the population to make room for the new agents. In both cases - selection and replacement - any of the typical strategies of the EAs, e.g., tournament, ranking, elitism, among other heuristic operators. In terms of cooperation, this is achieved through reproduction. In this phase, create new agents from existing ones through the use of a series of operators of reproduction. As shown in Fig. 1, lines 7–11, can be considered an arbitrary number #op of such operators, which are applied sequentially to the population of segmented way, giving rise to several intermediate populations auxpop [i], 0 6 i 6 #op.

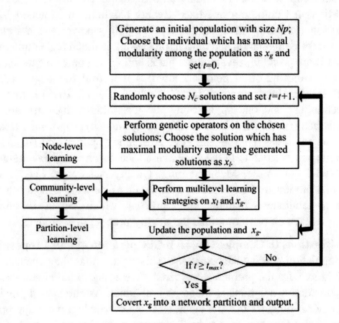

Fig. 1. General template of an MA

Memetic Algorithm

Input: an instance I of a problem P.
Output: a sol solution.
// generate initial population
1: for j ← 1: popsize to do
2: be ind ← GenerateHeuristicSolution (I)
3: be pop [j] ← ImproveLocal (ind, I)
4: end
5: repeat // generational loop
// Selection
6: be breeders ← SelectDePoblaci'on (pop)
// Segmented playback
7: be auxpop [0] ← pop
8: for j ← 1: #op do
9: be auxpop [j] ← ApplyOperator (op [j], auxpop [j - 1], I)
10: end
11: be newpop ← auxpop [#op]
// Replacement
12: be pop ← UpdateProb (pop, newpop)
// Check convergence
13: yes Convergence (pop) then
14: be pop ← RefrescarPoblaci'on (pop, I)
15: finsi
16: up CriteriaTermination (pop, I)
17: Return Better (pop, I)

Where auxpop [0] is initialized to pop, and auxpop [#op] is the final descent. In practice, the most typical situation is to simply use three operators: recombination, mutation, and local improvement. Approve on line 9 of the pseudo code that these operators receive not only the solutions on which they act, but also the instance I that you want to solve. With this it illustrates the fact that the operators of an MA are aware of the problem, and base their functioning in the knowledge they incorporate about it (unlike the models more classic EA). One of the reproductive processes that best encapsulates the cooperation between agents (two, or more [16]) is recombination. This is achieved through the construction of new solutions to from the relevant information contained in the cooperating agents. By "relevant" is meant that the information elements considered are important when determining (in one sense or another) the quality of the solutions. This is undoubtedly an interesting notion' that moves away from the most classic synthetic manipulations, typical of simple AEs. We will be back to this later, in the next section. The other classic operator - the mutation - fulfills the role of "keeping the fire alive" by injecting new information in the population continuously (but at a low rate, since otherwise the algorithm would degrade to a pure random search). Of course, this interpretation is which comes from the 'area of genetic algorithms, and does not necessarily coincide with the other researchers (those in the area of evolutionary programming without going any further). From in fact, it has sometimes been argued that recombination is not more than a macro-mutation, and certainly that may be the case in numerous applications of EAs in which this operator of recombination simply performs a random mix of information. However, no a similar assessment should be made in the field of MAs, since in these 'recombination' is typically done through the use of astute strategies, and therefore contribute essential way to the search. Finally, one of the most distinctive characteristics of MAs is the use of strategies of local search (LS). These (note that different LS strategies can be used in different points of the algorithm) constitute one of the essential reasons why it is appropriate to use the term "agent" in this context: its operation is local, and sometimes even self-employed. In this way, an MA can be viewed as a collection of agents who perform Autonomous exploration of the search space, sometimes cooperating through recombination, and competing for computational resources through the mechanisms of selection/replacement. The pseudo-code in Fig. 1 shows a component that also deserves attention: The Refresh Population procedure (lines 13–15). This procedure is very important with a view to the use of computational resources: if in a certain moment of the execution all the agents have a similar state (that is, convergence has taken place), the progress of the search becomes very complex. This type of circumstance can be detected at through the use of measures such as the entropy of Shannon [18], setting a minimum threshold below which it is considered that the population has degenerated. Obviously, said threshold depends on the representation of the problem that is being used, and it must be decided therefore particular way in each case.

2   Proposed Solution

Haptic Interface
The haptic term is related or based on the sense of touch, from the Greek word haptestahai; it can also be characterized by a predilection for the sense of touch, e.g. a haptic person (definition taken from the Merriam-Webster dictionary). The research firm Gartner defines haptic as the use of tactile interfaces to provide touch feedback or force as part of its user interface. (Vafadar 2013) also defines haptics as the "feedback generation of touch and strength information." This can be applied in the automotive industry to the user interface in automobiles to alert the driver of a pedestrian willing to cross the street by vibrating the seat. Haptic technology has the potential to add new forms of communication with the user, improve usability and user experience (UX, User Experience in English), and improve information applications. In technology, everything related to haptic receives the term "haptics" or "haptics" in English. Haptics can be studied in three major areas: (i) Human Haptics, which is relative to the touch perceived by humans, (ii) Computational Haptics, which is the software for touching and feeling virtual objects, and (iii) Haptic Machines, which refers to the design and use of machinery that can increase or replace human touch. The haptic feedback

channels can be tactile sensations, such as pressure, texture, temperature, etc., can be tactile vibro-stimulating Meissner's corpuscles that detect 5–50 Hz vibrations and Pacinian's corpuscles that detect vibrations of 40–400 Hz, or kinetic perception, which detects the state of the body. Haptic devices can be classified according to their interaction: (i) Take or grasp, (ii) carry, and (iii) touch (Culbertson et al. 2018; Halunen et al. 2017). The inclusion of IoT with haptic interfaces is a topic where few works have explored the principles of tangible interaction that can guide the design of interfaces for IoT located in the physical world (Angelini et al. 2018). On the other hand, the combination of VR and AR in three-dimensional scenarios with haptic feedback has also been carried out in other industries, such as the health industry, where virtual surgeries, rehabilitation systems, video games for training, etc. have been explored including studying the brain-computer interfaces that receive haptic information among other stimuli (Albert et al. 2017; Wu et al. 2017). The development by (Choi et al. 2018) called CLAW, is a device that uses force feedback and index finger movement information to take a virtual object, press virtual surfaces and shoot an object. Finally, haptic interfaces can allow machinery to operate remotely, but the user experience should be similar to interacting directly with the machine. (Aijaz et al. 2017; Van Den Berg et al. 2017) mention the challenge to transmit sensations through the network, called the "1 ms challenge of the Touch Internet" and have an "ultrareliable" and "ultra-responsive" network. In addition to this, the security of the transmitted data is necessary to avoid situations such as those mentioned in the Cybersecurity section. Another of its great challenges is to make the device feel exactly like the original artifact (Tiwari 2016).

The Human Role in Industry 4.0 and Decision Making

Within the manufacturing is also the role of the human and how it works and interacts with I4.0. It is important to underline the four cognitive functions of the human being within I4.0:

(i) Perception and Consciousness, where infrastructure sensors and portable sensors (wearables) now take part that capture activities, behavior, context and attention (e.g. cameras, microphones, tactile gloves);

(ii) Modeling and Understanding, which allows the interpretation and progress of the work flow, the identification of the level of experience of the operator and the interpretation of the context of the environment;

(iii) Reasoning and Decision Making, to select the best auxiliary and assistance measures depending on progress in the work flow and experience level; (iv) execution of the activities, which now become autonomous acts in I4.0 with actuators of visual, auditory and haptic infrastructures, as well as portable visual, auditory and haptic actuators. The I4.0 seeks to automate the manufacturing process as much as possible and move decision-making and monitoring of the human to the machines, but the human factor in the manufacturing processes is still necessary in manufacturing and therefore takes into account the process of decision making that the individual will make based on the modeling and recognition of the workflow and its tasks. Decision making is necessary for complex systems and processes that exist in the manufacturing industry and that have not been automated or cannot be automated. (Chong et al. 2018; Haslgrübler et al. 2017; Whitmore et al. 2015).

3   Resolution Problem

The technological development reported in the literature shows the progress in IoT, I4.0 and haptic interfaces. However, these are independent studies, which do not integrate their capabilities to increase their efficiency and benefit. Also, there are still important challenges such as evaluation, latency and processing of the transmission of information. There are proposals for industrial IoT architectures, but they do not interact with the human directly or do so with a computer interface, minimizing or not using haptics. Finally, there are already works developed to take advantage of AR for the process of maintenance of products and machinery, which provide detailed information to identify parts and review the work done, as in a welding process of a car (Halim 2018; Ni et al. 2017; Yelamarthi et al. 2017).

The manufacturing industry is using IoT and has pre-established programs for its implementation. Bosch, General Electric, AT & T, are some examples of companies that have these strategic plans and have a certain degree of progress. These advances and plans are focusing on sensors and actuators, but not on haptic interfaces. Also, there is machinery in the manufacturing industry (and in others such as health care) that requires precision and delicacy - necessary force, to operate it. Qualified technicians are the only ones who can correct a problem in this delicate equipment. When a team of these is broken down, they usually need to transfer the skilled and qualified technician to the plant to carry out the maintenance processes. The associated cost is high. To this it is added that if the machinery is in two different plants, in different geographies it needs attention, the human limitations cause that there is a delay in the attention and composing of the machinery giving rise to important losses for the factory. On the other hand, in quality processes it is common that the error can be measured in variables that are quantifiable, e.g. distance, thickness, volume increases, cracks, resistance, etc., but it is very complicated to have a quality control where the sense of touch is involved. This is where the use of haptic interfaces can be useful and if we integrate them into a process where IoT intervenes; it could result in more efficient and effective processes, obtaining improvements in the quality of production. Also, when there are processes that involve physical risk to the operator and these processes have to be done in situations that could favor the drowsiness of the operator, it would be very useful to have sensors of the movements, awareness and coordination of the operators and through a haptic interface of vibration alert the operator to wake up or pay attention to reduce accidents and safeguard their physical integrity. Another example where smart devices are used to safeguard the integrity of the person is described in the intelligent traffic management model to help people with some type of color blindness. However, a haptic interface that alerts the driver when the light is red or green is not considered (Ochoa-zezzatti and Hernández 2018). The problem that is observed is that there is not yet -or is very limited research work, the use of haptic interfaces in combination with IoT and that seeks to make decisions, and this makes the manufacturing industry is not taking advantage of the technology and the nine pillars that support I4.0. Then it is valid to investigate to design and implement a human interface that uses sensors and haptic actuators, taking advantage of the sensors and actuators of things to obtain data and execute tasks, which improve the way of production in factories and decision making.

General Purpose

Develop an integrative model of IoT and a haptic interface to increase the capacity of decision making in manufacturing processes in Industry 4.0. Figure 2 describes the different actors in the cognitive process within I4.0 that will be used to jointly improve the competitiveness of the manufacturing industry through efficient decision making and supported by IoT and haptic interfaces.

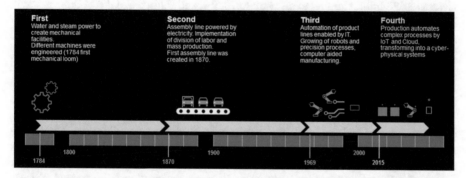

Fig. 2. Integrative model for decision making supported by IoT and haptic interfaces in Industry 4.0

Specific Objectives

- Structure and detail the mental maps of Industry 4.0 or Smart Manufacturing to have a complete model.
- Relate the key indicators of Industry 4.0 and generate the industrial trends that are based on IoT to be visualized in haptic interface.
- Create a data repository to perform the analysis required for decision making.
- Build the intelligent tool based on IoT and haptic interfaces that improve decision making and improve competitiveness.
- Define the appropriate plans and tests to corroborate the proposed solution
- Determine the feasibility and feasibility of the project.

Justification

Industry 4.0 is supported by technology that is now viable thanks to the power of computer processing and a scientific area that is developing a lot that is data science - the basis of machine learning. IoT produces a very large amount of data resulting from physically deployed sensors. However, when introducing haptic interfaces to Industry 4.0 and this is combined to help make decisions effectively and efficiently, then this research generates a very important contribution to technology that is not only theoretical knowledge, but can be applied and help to obtain benefit to the manufacturing industry or any other. Also, the impact of Industry 4.0 on the economic level for countries is important. Taking Germany as an example (as the country that coins the term of Industrie 4.0), they estimate that in the next five years the industry will grow from € 90 to € 150 billion thanks to the adoption of Industry 4.0. This increase is estimated by the improvements in productivity in conversion costs, which exclude the cost of the material, and which should be improved between 15 and 25%. Industry 4.0 also brings growth in revenue (revenue in English) that only in Germany can reach up to € 30 billion per year, which represents almost 1% of the gross domestic product (GDP) of this country (Rüßmann et al. 2015). Worldwide, this estimated trend is similar. IDC (2018) estimates that the Asia/Pacific region will increase its IoT investment by $ 500 billion with an expected growth of 13.1%; America will invest close to $ 328 billion hoping to grow up to 14.5%; and Europe and the Middle East will invest about $ 260 billion, estimating to grow 17%. In Mexico, the importance of Industria 4.0 as an alternative to increase competitiveness is also being considered. However, Mexico lags behind in this area because it allocates about 0.6% of GDP while the average of the countries of the Organization for Economic Cooperation and Development (OECD) is 2.4% (Industry Forum 4.0: Challenges for Mexico, 2018). So, this research is important because of the growth factors that the industry will have worldwide and to increase the competitiveness of Mexico. Likewise, the social impact of this research considers a very important aspect for the country. Contrary to what one might think, when introducing Industry 4.0 to manufacturing, machines replace man and put people's jobs at risk, it is observed that it is the opposite. When the machines become intelligent entities, the production lines are enriched and humanized. Simple tasks tend to disappear, transforming workers into coordinators, placing them in action where help is required, in a flexible and timely manner, improving the worker's work-life balance. In Germany it is estimated that the introduction of these technologies and paradigms will serve to increase the employment rate by 6% over the next 10 years. Also, the demand for employees with special skills (skilled worker is the term used in the market) as mechanical engineers, mechatronic engineers and information technology professionals (IT) will grow even more, probably reaching 10% increase in demand. In Mexico, the Ministry of Economy promotes the adoption of Industry 4.0 technology tools through the generation of Industrial Innovation Centers (CII) in Aguascalientes, Baja California, Chihuahua, Mexico City, Colima, Jalisco, Nuevo Leon, Queretaro and Tamaulipas. This is a historic opportunity for young people where up to 10,000 graduates could be served to work with Industry 4.0 (Castañón 2018). This is why IoT research with haptic interface to improve the competitiveness of the manufacturing industry serves the national development project and addresses issues that are necessary to generate opportunities for young graduates to have a field of action to work on what they studied Last but not least, Industry 4.0 also helps the sustainability of companies and therefore the environment. For example, by finding ways to deal with energy, resource and environmental constraints, Industry 4.0 can find solutions to these challenges. The production can reduce the consumption of energy and helps the sustainability of the business. Inclusive, the way of working, the habilitation of remote work, the sensors and actuators, can allow the reduction of transfer of people and this helps to reduce the emission of contaminants by transportation. Inclusive, the way of working, the habilitation of remote work, the sensors and actuators, can allow the reduction of transfer of people and this helps to reduce the emission of contaminants by transportation. The application of efficiency in the decision making of a production process makes time and use of resources optimized, and this translates into the aforementioned benefits to the environment. Although the research to be conducted does not directly seek the benefit to the environment, it is reasonable to intuit that indirectly the environment will benefit by providing a way to streamline a process.

Scope and Limitations

The model will allow analyzing and forecasting the effectiveness of decision making for a task within a manufacturing process by adding IoT and haptics. It is expected that this defined model is feasible and feasible by calculating the effectiveness index offered by the automatic learning analysis and forecasting techniques. The practice indicates that a model with an effectiveness index higher than 80% is feasible and feasible. Also, this model limited to a task can serve as a base for the manufacturing industry to replicate it and implement it in other tasks of different manufacturing processes, not limiting its application to this industry exclusively, but it is expected that it can be used for other areas, such as maintenance for example, and in other industries such as transportation, health care, energy, to name a few.

4   Implementation Solution

Taking as reference Fig. 3, the development of the project is proposed as follows:

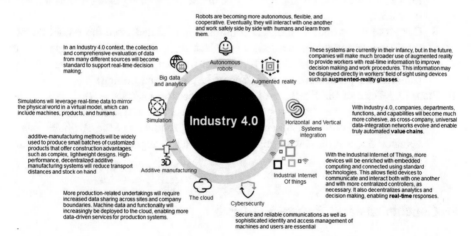

Fig. 3. Proposed methodology for the implementation of the project design and implementation of an IoT-based haptic interface to improve the competitiveness of an Industry 4.0 model in the manufacturing sector

A. Structure Industry 4.0 modeling
  1. Detail each of the mental maps of the Industry 4.0 model.
  2. Create an index file that allows to relate each of the Industry 4.0 indicators that have been registered. (Technical support with Excel and SQL Server Express)

B. Produce the first indicators from the modeling and using the haptic interface

   3. Generate industrial trends based on IoT to be visualized through the use of a characterized Háptic Interface for the Manufacturing Industry

   4. Store the information in a database repository in order to have enough information to properly apply the IoT in the Haptic Interface to improve decision making and industrial competitiveness (SQL Server Express Technical Support)

C. Build the intelligent tool based on IoT to improve competitiveness

   5. Create specific trends to the problem to be solved, this will allow building an adequate tool for decision making, considering the changes in the Manufacturing Industry and determining both a tactical plan and a strategic plan applicable to the conditions of the problem.

   6. Build an Industry 4.0 model according to the existing problems and improve the performance of the Haptic Interface for the correct Decision Making.

D. Verify the model with manufacturing data

E. Structure the consultation and visualization windows of the tool for decision making

   7. Apply the results to the previously constructed trend maps in order that the resolution of the problem can interact with the Industry 4.0 model, and determine if it is feasible and feasible to make paradigmatic changes with respect to the Decision Making in the Industry of the Manufacturing using haptic and IoT interfaces.

   8. Carry out a corroboration of the solution implemented from the model developed and applying multicriteria analysis techniques in order to support the project to be carried out through a correct Industrial Decision Making. (Weka technical support, Tableau, SQL Server Express, data mining)

F. Plans and Tests of the Intelligent System for decision making.

   9. Determine the effectiveness index of the project and assess its validation with other projects based on Smart Manufacturing models. (In the Artificial Intelligence area, a project with an effectiveness index higher than 80% is feasible and feasible).

5 Conclusions and Future Research

Using two different perspectives for solving a problem related with Industry 4.0 we were able to verify its model with different technics, such a Metaheuristic method as Memetic Algorithms, additionally we could discover that the use of Data Mining to determine previous successful models of aid distribution would be very useful for posterior evaluation of the results of the portfolio financed. With this at the future we could verify if the target population would agree on how to access this type of social support. Also, as future research, social networks could be considered in the evaluation of social experiences. Data Mining and a specific Topsis Model based on agents can improve the understanding of change for the better substantial paradigm that ranks communities' agents appropriately in terms of their relationship attribute approach. Goal programming and Ant Colony Optimization provide a powerful alternative to optimization problems. For this reason, is that it provides a comprehensive overview of the cultural phenomenon. This technology leads to the possibility of the generation of experimental knowledge, created by the community of agents for a given application domain. For the most part, the extent of this knowledge is cognitively community of agents that is a topic for future work, as is proposed in Fig. 4.

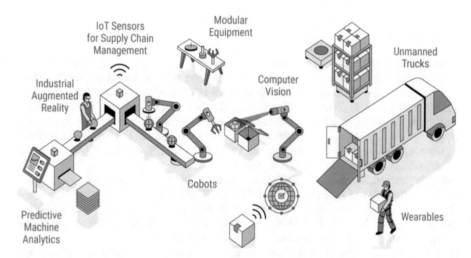

Fig. 4. Our proposal model of Industry 4.0 to a model of Smart Manufacturing in a smart city.

The specification of each artificial intelligent environment would be a contradiction to traditional systems that don't consider all the factors related with the domain, that's why at the end, the last one generate unfair and inefficient distribution of resources, as is shown in Fig. 5.

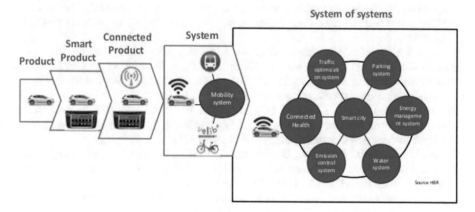

Fig. 5. Incremental model of the manufacture of an intelligent product in a smart city.

A new artificial intelligence could be in charge of these systems, but it is still far in the horizon, in the same way that we still lack methods to understand the original and peculiar aspects of each society. In future work will attempt to analyze the discussion of priority values for each different necessity of people and what could be the reason for being prioritized over others necessities.

References

Aijaz, A., Dohler, M., Hamid Aghvami, A., Friderikos, V., Frodigh, M.: Realizing the tactile internet: haptic communications over next generation 5G cellular networks. IEEE Wirel. Commun. **24**(2), 82–89 (2017). https://doi.org/10.1109/MWC.2016.1500157RP

Albert, B., Maire, J.-L., Pillet, M., Zanni-Merk, C., De Bertrand De Beuvron, F., Charrier, J., Knecht, C.: A haptic quality control method based on the human somatosensory system (n.d.)

Angelini, L., Mugellini, E., Abou Khaled, O., Couture, N.: Internet of tangible things (IoTT): challenges and opportunities for tangible interaction with IoT. Informatics **5**(1), 7 (2018). https://doi.org/10.3390/informatics5010007

Atzori, L., Iera, A., Morabito, G.: The internet of things: a survey. Comput. Netw. **54**(15), 2787–2805 (2010). https://doi.org/10.1016/J.COMNET.2010.05.010

Castañón, J.P.: Retos para méxico: industria 4.0. In Industria 4.0 Retos para México (2018)

Cheng, J., Chen, W., Tao, F., Lin, C.-L.: Industrial IoT in 5G environment towards smart manufacturing (2018). https://doi.org/10.1016/j.jii.2018.04.001

Choi, I., Ofek, E., Benko, H., Sinclair, M., Holz, C.: CLAW: a multifunctional handheld haptic controller for grasping, touching, and triggering in virtual reality. In: Chi 2018 (2018). https://doi.org/10.1145/3173574.3174228

Chong, L., Ramakrishna, S., Singh, S.: A review of digital manufacturing-based hybrid additive manufacturing processes. Int. J. Adv. Manuf. Technol. (2018). https://doi.org/10.1007/s00170-017-1345-3

Culbertson, H., Schorr, S.B., Okamura, A.M.: Haptics: the present and future of artificial touch sensations. Annu. Rev. Control Robot. Auton. Syst. **11225**(1), 1–12 (2018). https://doi.org/10.1146/annurev-control-060117

Govindarajan, U.H., Trappey, A.J.C., Trappey, C.V.: Immersive technology for human-centric cyberphysical systems in complex manufacturing processes: a comprehensive overview of the global patent profile using collective intelligence. Complexity (2018). https://doi.org/10.1155/2018/4283634

Halim, A.Z.A.: Special Issue Maintenance Process in Automotive Industry (2018)

Halunen, K., Latvala, O.-M., Karvonen, H., Häikiö, J., Valli, S., Federley, M., Peltola, J.: Human verifiable computing in augmented and virtual realities (2017). http://www.vtt.fi/inf/julkaisut/muut/2017/human-verifiable-computing-white-paper.pdf

Harbers, M., Van Berkel, J., Bargh, M.S., Van Den Braak, S., Pool, R., Choenni, S.: A conceptual framework for addressing IoT threats: challenges in meeting challenges. In: 51st Hawaii International Conference on System Sciences (2018)

Haslgrübler, M., Fritz, P., Gollan, B., Ferscha, A.: Getting through. In: Proceedings of the Seventh International Conference on the Internet of Things - IoT 2017, October, pp. 1–8 (2017). https://doi.org/10.1145/3131542.3131561

https://www.gartner.com/en/newsroom/press-releases/2017-02-07-gartner-says-8-billion-connected-things-will-be-in-use-in-2017-up-31-percent-from-2016

Iorga, M., Feldman, L., Barton, R., Martin, M.J., Goren, N., Mahmoudi, C.: Fog Computing Conceptual Model. NIST Special Publication, 500–325 (2018). https://doi.org/10.6028/NIST. SP.500-325

Khan, M.A., Salah, K.: IoT security: review, blockchain solutions, and open challenges. Future Gen. Comput. Syst. (2018). https://doi.org/10.1016/j.future.2017.11.022

Milgram, P., Takemura, H., Utsumi, A., Kishino, F.: Augmented reality: a class of displays on the reality-virtuality continuum, vol. 2351, pp. 282–292 (1995). https://doi.org/10.1117/12. 197321

Ni, D., Yew, A.W.W., Ong, S.K., Nee, A.Y.C.: Haptic and visual augmented reality interface for programming welding robots. Adv. Manuf. 5(3), 191–198 (2017). https://doi.org/10.1007/ s40436-017-0184-7

Ochoa-zezzatti, A., Hernández, A.: Intelligent traffic management to support people with color blindness in a Smart City (n.d.)

Raymond Choo, K.-K., Bishop, M., Glisson, W., Nance, K.: Internet- and cloud-of-things cybersecurity research challenges and advances (2018). https://doi.org/10.1016/j.cose.2018. 02.008

Roman, R., Lopez, J., Mambo, M.: Mobile edge computing, Fog et al.: a survey and analysis of security threats and challenges. Future Gen. Comput. Syst. (2018). https://doi.org/10.1016/j. future.2016.11.009

Rüßmann, M., Lorenz, M., Gerbert, P., Waldner, M., Justus, J., Engel, P., Harnisch, M.: Industry 4.0. The Boston Consulting Group, 1–20 (2015). https://doi.org/10.1007/s12599-014-0334-4

Scott, R.: Comparative Study of Open-Source IoT Middleware Platforms. KTH Estocolmo, Suecia (2018)

Tiwari, A.: A review on haptic science technology and its applications. In: Proceeding of International Conference on Emerging Technologies in Engineering, Biomedical, Manage-ment and Science, ETEBMS 2016, March, pp. 78–83 (2016). http://www.sdtechnocrates. com/ETEBMS2016/html/papers/ETEBMS-2016_ENG-EC3.pdf

Urquhart, L., Mcauley, D.: Avoiding the internet of insecure industrial things (2017). https://doi. org/10.1016/j.clsr.2017.12.004

Vafadar, M.: Virtual reality: opportunities and challenges. Int. Res. J. Eng. Technol. (IRJET) 3 (2), 1139–1145 (2013). https://www.irjet.net/archives/V5/i1/IRJET-V5I1103.pdf

Van Den Berg, D., Glans, R., De Koning, D., Kuipers, F.A., Lugtenburg, J., Polachan, K., Venkata, P.T., Singh, C., Turkovic, B., Van Wijk, B.: Challenges in haptic communications over the tactile internet. IEEE Access 5, 23502–23518 (2017). https://doi.org/10.1109/ ACCESS.2017.2764181

Whitmore, A., Agarwal, A., Da Xu, L.: The internet of things—a survey of topics and trends. Inf. Syst. Front. 17(2), 261–274. https://doi.org/10.1007/s10796-014-9489-2

Wu, H., Liang, S., Hang, W., Liu, X., Wang, Q., Choi, K.S., Qin, J.: Evaluation of motor training performance in 3D virtual environment via combining brain-computer interface and haptic feedback. Procedia Comput. Sci. 107(Icict), 256–261 (2017). https://doi.org/10.1016/j.procs. 2017.03.096

Xu, L.D., Xu, E.L., Li, L.: Industry 4.0: state of the art and future trends. Int. J. Prod. Res. (2018). https://doi.org/10.1080/00207543.2018.1444806

Yannuzzi, M., Milito, R.: Key ingredients in an IoT recipe Fog Computing. In: Computer Aided Modeling and Design of Communication Links and Networks (CAMAD), pp. 325–329 (2014)

Yelamarthi, K., Aman, M.S., Abdelgawad, A.: An application-driven modular IoT architecture. Wirel. Commun. Mob. Comput. (2017). https://doi.org/10.1155/2017/1350929

An Effective Approach for Party Recommendation in Voting Advice Application

Sharanya Nagarjan$^{(\boxtimes)}$ and Anuj Mohamed

School of Computer Sciences,
Mahatma Gandhi University, Kottayam, Kerala, India
sharu01@gmail.com

Abstract. When it comes to the current political scenario the large number of political parties put voters, especially first time voters, women, youngsters etc., in confusing state regarding whom to vote for. To solve this issue of, Voting Advice Applications (VAAs) which are questionnaire based recommender systems are used. VAAs are online tools that suggest the user with the most suitable party based on answers to the policy based questions. Even though this is an active area of research in the western political scenario, the performance of the existing algorithms are not very appreciable. Also the existing works have not imparted the human decision making behavior in developing algorithms. This research work aims in proposing novel approaches based on the human decision making which can efficiently suggest suitable parties for the voters. Soft set and Fuzzy Soft Set are techniques that are found to be good in modeling human decision making as it supports parameterization and vagueness. The proposed work uses these techniques to develop algorithms that can effectively suggest the voter with a suitable party. The research is carried out in the domain of political scenario in Kerala, where this is the first research in the area of VAAs. The developed algorithms were evaluated on a data set collected from various parts of the state and found promising.

Keywords: Voting Advice Application · Soft set · Fuzzy soft set

1 Introduction

The exponential growth of digital information in the World Wide Web and the number of visitors to the Internet has made a dramatic increase in the size and complexity of websites, which made searching for information cumbersome and time consuming [1]. Researchers are nowadays keen on developing techniques that brings to the user the information that is in accordance with the user's interest and preference. Thus Recommender systems were developed in order to suggest the users with items of their interest. Recommender system can simply be defined as software tools and techniques providing suggestions for user in the Internet. They ensure that the websites are more personalized with service and content of his preference and interest.

In the current political scenario, many developed nations are being threatened by the reduced voter turnout [2]. Also the increasing number of political parties along with massive amount of online information puts people in dilemma on whom to vote for.

© Springer Nature Switzerland AG 2019
A. Abraham et al. (Eds.): IBICA 2018, AISC 939, pp. 118–128, 2019.
https://doi.org/10.1007/978-3-030-16681-6_12

Voting Advice Applications (VAAs) are Party Recommender Systems, which are online tools that assist voters by suggesting them the most suitable party. Studies have shown that VAAs play a substantial role in shaping voters' perception of the position of the political parties. VAAs are web applications that provide voters with information about which political party comes closest to their own political preference [3]. The theoretical background on which VAAs work is the model of issue voting; according to which voters choose the party that is closest to their own preference on a set of political issues. VAAs produce the result as a ranked list of parties according to the proximity with the voter. VAAs have become very popular in many European countries and have a large impact on increasing the voter turnout and increase the vote switching [2]. A typical VAA works as follows: A set of questions based on policies are given to experts or politicians. The answers corresponding to each of the party are recorded. When a user answers the same set of policy based questions, the answers are compared and the party with the most similar answers is suggested to the user/voter.

In the proposed work, experts were asked to answer the questions, based on the policy of different political parties. When the users marked the answers they were compared with the experts' answers for various parties and the party with the most similar or close answers was recommended to the voter. Then the concept of soft set and fuzzy soft set were used to find the most suitable party for the user, based on his answers to the policy statements.

The closeness is measured based on the distance between the user's answers and the average answers of political parties. The lesser the distance more close is the party's policy to the user's policy. So the party with the least distance is the most suitable party and is the one suggested.

The proposed methods were compared with the existing methods discussed in [2] and the classifications methods used in [4], where a variant of VAA called Social Voting Advice Application (SVAA) is explained. SVAA eliminates the need for getting party responses either from candidates or experts.

2   Related Work

2.1   Recommender System in Politics

We are all familiar with the fact that a large number of political parties are upcoming these days and choosing the right candidate also becomes a bothersome matter, and users prefer to have suggestions on whom to vote for. In recent electoral politics, one of the most striking internet-related developments is the increasingly widespread use of VAAs. VAAs are online Decision support systems that try to match voters with political parties or candidates in elections, based on how each responds to a number of policy issue statements. A VAA can be a website, app or of any ot1her online format. It acts as a vote recommendation system that helps users in deciding which party/candidate to vote for. The first VAAs were websites, based on policy views and recommended parties' positions. Such systems were developed in Finland and the Netherlands in the mid-1990s, and the concept has since spread to most European countries, USA and Canada. In many countries proliferation of VAAs has made a rapid increase in the number of people who vote. VAAs influences decision making process the citizens who are undecided, women, youth or first time voters. VAAs does not limit being a digital product but has a great influence on journalism, predicting elections and also help parties in analyzing their winning strategies. VAAs make users and political parties to mention their views (usually Likert scale) as on a set of policy based statements and based on the similarity between the user's interest and party's policy, it recommends the user the best party. For this users and parties are asked to fill a questionnaire that contains a number of policy statements, created according to the current issues of that nation or state and concerned about political, economic and social issues [4–6].

Another key benefit of VAAs is that, in traditional media large and powerful parties get more coverage than smaller parties. VAAs collect opinions from all candidates and thus ensure a fair coverage to all candidates. VAAs take political opinion into account and find an unfiltered answer based on the closeness between party and one's opinion. VAAs also try to exclude parties based on stricter criteria and also promises in providing a balanced way of finding suitable parties. But how much each variable is valuable for a user varies between different users, influence of media and prominences of candidates are some of the key criterions to be taken into consideration. There are various stakeholders when considering a VAA, the voter, the candidate and the media/website. VAAs work in complex intersection of many interests like political, financial and informational. The policy distance is defined as a scaled metric, where the Euclidean distance between the candidate's policy and voter's policy is scaled by maximum distance between them [3].

2.2 VAA Methods

The conventional way of voting recommendation in VAAs was spatial model used along with the paper-pencil model, which was abandoned later. The key component in designing a VAA is an effective method which ensures the match-making between the voter and the party or the candidate. This is done by mapping the user's preference to that of party's involvement in various issues [7]. To find the suitable party which was close to the user's preference various distance measures were used [8]. VAAs are based on the idea that sequence of answer pattern characterizes the typical voters of a party. The method suggested in [2] uses Hidden Markov Models (HMM) as classifiers for party-user similarity estimation. This is because HMM classifiers very well denoted the path users follow while answering and enables Machine Learning to data which is structured as correlated sequential data.

2.3 Applications of Soft Sets and Fuzzy Soft Sets

The real world is full of uncertainty, imprecision and vagueness and most of the concepts we are meeting in everyday life are vague rather than precise. In recent years vague concepts are common in different areas like medical applications, pharmacology, economics and engineering since the classical mathematical methods for modeling and decision making are inadequate to solve many complex problems in these areas [9]. But classical mathematics, on the other hand, requires that all mathematical notions must be exact, otherwise precise reasoning would be impossible [10]. It is obvious, that the soft set theory can be applied to a wide range of problems in economics, engineering, physics, and so on [11].

In [9] Soft set theory was used for developing knowledge-based system in medicine and devises a prediction system named soft expert system (SES) by using the prostate specific agent (PSA), prostate volume (PV) and age data of patients based on fuzzy sets and soft sets and calculate the patient's prostate cancer risk. A rule-based system that works according to the rules is developed so as to determine the risk of prostate cancer and help the doctor to determine whether the patient needs biopsy or not. This work involves various steps like fuzzification of data, transforming the data into soft set, reduction of parameters, defining the soft rules, analysis using the soft rules and calculation of risk. The system had a promising accuracy.

Classification is one of the key tasks in research field of Data mining and analysis. The classification tasks in medical area are to classify medical dataset based on the result of medical diagnosis or description of medical treatment by the medical specialist. The large amount of information available and data warehouse in medical databases needs sophisticated tools for storing, retrieving, analysis and usage of stored knowledge and data. Intelligent methods such as neural networks, fuzzy sets, decision trees, and expert systems are being applied for classification task in medical domain. Recently, Bijective soft set theory has been proposed as a new intelligent technique for the discovery of data dependencies, data reduction, classification and rule generation from databases. A novel approach based on Bijective soft sets for the generation of classification rules from the data set and investigational results from applying the Bijective soft set analysis to the set of data samples are given and evaluated and found to be promising [12].

Even though a number of mathematical theories such as probability theory, fuzzy set theory, rough set theory, vague set theory, etc. have been devised to solve such problems and have been found to be only partially successful. Among these, the fuzzy set theory approach has been found to be the most appropriate for dealing with uncertainties [11]. Fuzzy soft set allows both parameterization and vagueness, which very well mimics human decision making methods.

A theoretical approach for decision making using fuzzy soft sets is defined in [11] which explains an application of fuzzy soft set theory in object recognition problem. The recognition strategy is based on multi-observer input parameter data set. The scope of using fuzzy soft set for classification is explained in [13] which use the fuzzy approach in the pre-processing stage to obtain features, and similarity concept in the process of classification. This method gave a better performance than the soft set approach. Fuzzy logic has been extensively used in the design of a recommender system to handle the uncertainty, impreciseness and vagueness in item features and user's behavior.

Use of Fuzzy logic in recommender systems has made the following possible [14]:

1. Efficient modeling of vagueness in user input
2. Improves determination of similar users
3. Easily find similar products
4. Easy handling of demographic data

Fuzzy systems and variants of fuzzy systems are also used in recommender systems for medical diagnosis. Fusion of Fuzzy clustering and intuitionistic fuzzy recommender systems called HIFCF (Hybrid Intuitionistic Fuzzy Collaborative Filtering) is used for medical diagnosis and the results are found to be promising [15]. In the current marketing scenario the high end consumer products are highly dynamic in nature and changes over a short span of time, which makes old data useless for product recommendation because of the changing customer interest. Also the number of diverse product for different market segments makes it difficult for the consumers to find their preferred products. To address the above challenges [10] proposes the Fuzzy Cognitive Pairwise Comparison for Ranking and Grade Clustering (FCPC-RGC) for comparison ranking and clustering. This method effectively handles the recommendation of trending products.

3 Proposed Work

The proposed methods in this paper model the problem of VAA as a human decision making problem. Two approaches are proposed in which the decision, here recommending the suitable party, is made based on the 30 policy based questions, which act as the criteria for making the decision. The user will be presented with these 30 questions in the form of a questionnaire as shown in Fig. 1. Figure 1 shows the initial page of the questionnaire.

The two proposed methods use soft set and fuzzy soft approach along with distance measure to find the party which is close to user preference.

3.1 Soft Set Approach

One of the approaches we have developed is to use the concept of soft sets for finding the most suitable party for the user. The data consists of answers to 30 policy based question marked in Likert scale ranging 0 to 4 (0 - Completely Disagree, 1 - Disagree, 2 - Neutral/No opinion, 3 - Agree, 4 - Completely Agree).

The set of 30 policy based questions are divided into 5 categories:

i. **Human Development:** This category consist of policy questions that are related to the development of the citizens and their better life pattern
ii. **Financial Development:** This category deals with questions that correspond to the financial development of the citizen and the state.
iii. **Community Based Development:** This category corresponds to the questions that deal with the development of minority groups, women, and other religious policies.
iv. **National Merit:** This category corresponds to the policies related to the development of the entire nation and other questions of national level interest.
v. **Political based Development:** The certain policies related to the political views and political developments of the nation fall under this category.

The proposed algorithm consists of two phase, in the initial phase experts are asked to mark the policy based questionnaire based on the policy of four different political

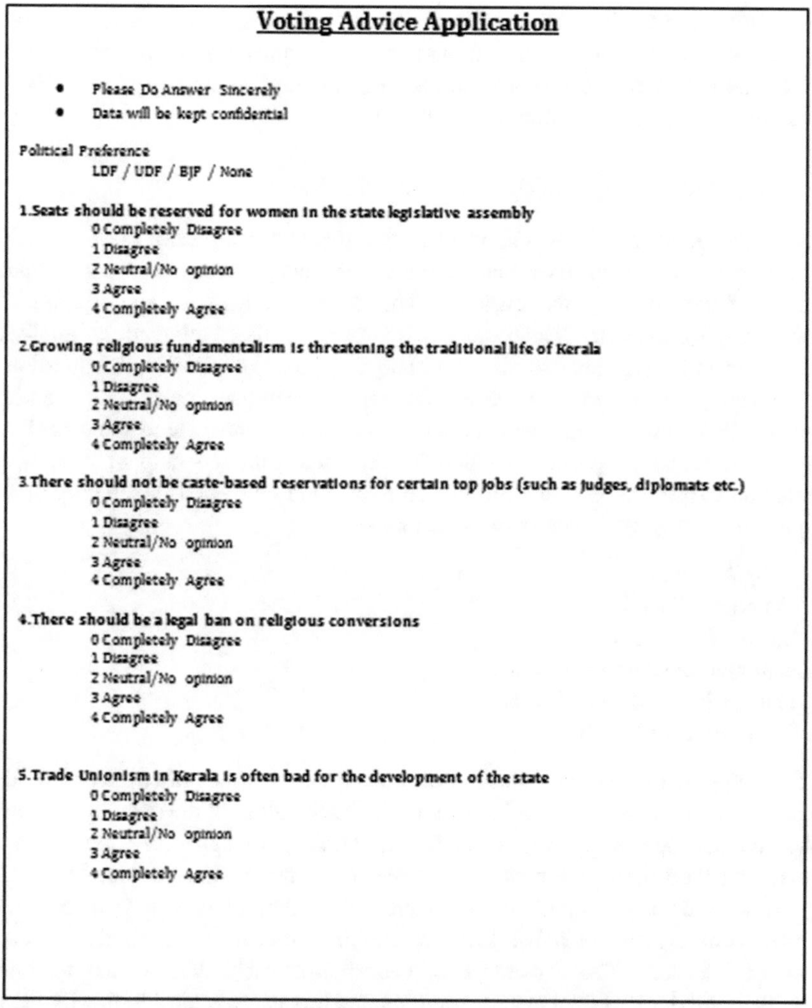

Voting Advice Application

- Please Do Answer Sincerely
- Data will be kept confidential

Political Preference
   LDF / UDF / BJP / None

1.Seats should be reserved for women in the state legislative assembly
   0 Completely Disagree
   1 Disagree
   2 Neutral/No opinion
   3 Agree
   4 Completely Agree

2.Growing religious fundamentalism is threatening the traditional life of Kerala
   0 Completely Disagree
   1 Disagree
   2 Neutral/No opinion
   3 Agree
   4 Completely Agree

3.There should not be caste-based reservations for certain top jobs (such as judges, diplomats etc.)
   0 Completely Disagree
   1 Disagree
   2 Neutral/No opinion
   3 Agree
   4 Completely Agree

4.There should be a legal ban on religious conversions
   0 Completely Disagree
   1 Disagree
   2 Neutral/No opinion
   3 Agree
   4 Completely Agree

5.Trade Unionism in Kerala is often bad for the development of the state
   0 Completely Disagree
   1 Disagree
   2 Neutral/No opinion
   3 Agree
   4 Completely Agree

Fig. 1. First page of the questionnaire

parties in Kerala. The answers in the range 0 to 4 are mapped to 0 and 1. The values 0, 1 and 2 (0 - Completely Disagree, 1 - Disagree, 2 - Neutral/No opinion) are mapped to 0 and the values 3 and 4 (3 - Agree, 4 - Completely Agree) are mapped to 1. Then the questions are categorized into the above 5 categories. Then the average scores for each of the categories are identified. This is done for each of the political parties, by collecting opinions from political science experts. At the end of this step we will be left with 5 scores, corresponding to each category, for each of the party. Now in the second phase, i.e. when a voter marks the questionnaire his scores are compared with that of the different parties and the party with closest scores is suggested to the user. The answers on Likert scales are mapped to 0 and 1 as discussed before and then categorized. Then the average score for each category is found which is then compared with each of the party's average score. Then the party that lies close to the scores of user is suggested. The closeness with parties is determined by computing the distance between the user's category score and each of the party's category scores. The party with smallest distance is recommended to the user.

3.2 Fuzzy Soft Set Approach

Human decision always deals with vagueness and parameters, hence there is a need for a concept that handles fuzziness and parameters is always a concern among researchers and hence fuzzy soft set was developed. This concept is used for identifying the best suitable party for the voter. The developed approach builds a voting advice application. The steps to build are same as discussed above. Instead of dividing the question into different categories, we identify various factors that each question is comprised of and fuzzify the data based on the factors. Each question may contribute to some or all of the factors in different proportions, which is estimated with the help of experts. This information is used to calculate how much a question belongs to a category and thus the question is fuzzified. The various factors are:

i. Religious Factor
ii. Economic Factor
iii. Sustainable Factor
iv. National Developmental Factor
v. Human Development Factor
vi. Governmental Factor

The fuzzy soft set based algorithm developed also consists of two steps similar to that of the above method. The difference lies in fuzzification of the data. In the previous method the data was mapped to 0 and 1, here the data is mapped to a range 0 to 1. In the mapping the Likert scale value is divided by 4, the maximum value in the Likert scale. So the values 0, 1, 2, 3, 4 in the Likert scale will be mapped to 0, 0.25, 0.5, 0.75 and 1 respectively. Then it is fuzzified based on the statement's contributions to each of the above 7 factors. The percentage of contribution each question has to different factors is found out based on experts' opinion and this value is used to fuzzify the data. This method also consists of 2 phases as the above discussed approach. The initial phase finds the average scores for each of the factors for all the parties. This phase is performed only once and at the end of the initial phase there will 7 values for each of the party. The 7 values depict the average factor score for each of the 7 factors mentioned above. In the second phase when a user marks the answers, they are initially mapped and the fuzzified as discussed above and the average score corresponding to each factor is found. A distance measure is used to find the distance between the users' factor scores and the party's factor scores for each party and the party close to the user's policy is found. The party with the minimum distance is the party said to be close and this party is suggested.

4 Results and Discussion

Expert answers for each of the party were collected from Research scholars in Political Science, this answers were used to find the average scores of each parties. Data for testing the performance of the proposed methods were collected from around 200 voters of different age groups from different parts of Kerala. They were given a questionnaire with 30 questions related to the current social and political issues in the state of Kerala and were asked to mark their answers in Likert Scale. The questions were collected from PreferenceMatcher, PreferenceMatcher is an academic consortium involving political scientists, social psychologists, computer scientists, and communication specialists from the University of Zurich, University of Twente, Cyprus University of Technology and Oxford Brookes University who collaborate in developing e-literacy tools designed to enhance voter education [16]. Also they were asked to mention their political preference under an assurance that the data will be kept confidential. These answers marked by the voters were used to find the suitable party using both the proposed algorithms. To find the closeness of the user to party different distance measures were tried and the results are given in Tables 1 and 2. The precision, recall and f-score for the soft set approach and fuzzy soft set approach is as given in Tables 1 and 2 respectively.

Table 1. Performance of different distance measure in soft set approach

Sl. no	Distance measure	Precision	Recall	F-score	Accuracy
1	Euclidean distance	70	65	66	65.21
2	City block distance	77	45	54	44.92
3	Relative euclidean distance	77	45	54	44.92
4	Chi square distance	100	13	23	13.04

Table 2. Performance of different distance measure in fuzzy soft set approach

Sl. no	Distance measure	Precision	Recall	F-score	Accuracy
1	Euclidean distance	82	72	73	72.09
2	City block distance	67	60	61	60.04
3	Relative euclidean distance	92	26	28	25.59
4	Chi square distance	67	60	61	60.04

It was found that Euclidean Distance measure outperformed all the other measures and gave and accuracy of 65.21% was obtained for the soft set approach and an accuracy of 72.09% was obtained for the fuzzy soft set approach. So the final algorithm was devised using Euclidean Measure.

Also the performance of using classifiers was tested on the data to analyze and compared the performance with the proposed methods. Seven superior classifiers including those explained in [4] were used. The classifiers used were Decision Tree,

Logistic Regression, Support Vector Machine, Naïve Bayesian Classifier, Artificial Neural Network, Random Forest and AdaBoost. For this scikit-learn package from python was used. It was found that the proposed method using soft set and fuzzy soft set outperformed the performance of exiting techniques. Precisely the fuzzy soft set method was found to have excellent and satisfying performance in terms of precision, recall and f-score.

The precision, recall and f-score of both the proposed methods and of other classifications techniques are as shown in Fig. 2. It is clearly understood that the performance of the proposed algorithms using soft sets and fuzzy soft set is very satisfying when compared with the existing techniques. The fuzzy soft set technique outperforms all the other methods in terms of precision, recall and f-score.

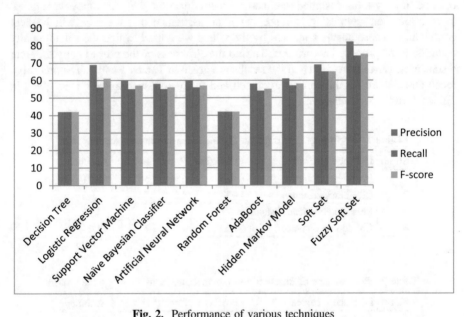

Fig. 2. Performance of various techniques

5 Conclusion and Future Enhancement

Since Voting Advice Applications have to deal with human decision making, the data may not be always crisp and precise, they are usually vague. So the conventional modeling techniques are not very well suited. Uses of fuzzy sets are also not suitable due to lack of parameterization. So the scope of using Soft Set was analyzed as it allows parameterization. The proposed method using soft sets gave a satisfying result in terms of recall, precision and f-score. But unfortunately soft set uses crisp values 0 and 1 which caused a distortion and does not accommodate fuzzy values. Therefore the concept of fuzzy soft sets was exploited and a new algorithm was developed. The method using fuzzy soft sets performed outstandingly well as it can handle details in minute and gave better results than all the other existing methods. The proposed soft set approach gave a promising accuracy of 65.21% and an f-score of 66% and the fuzzy soft set method gave an accuracy of 72.09% and an f-score of 73%. These values show that the proposed algorithms work efficiently than the existing methods.

In this work, all the parameters used to make the decision are treated equally; fuzzy parameterization can be used to prioritize the parameters or policies. Also the methods or techniques to alleviate the errors caused due to imbalance in data can be identified, so as to make the system more robust. Another key area of enhancement is to extract political related data from social media like twitter and Facebook and then use them to suggest the party for the user, by analyzing sentiments in the social media data. Also the preference of friends or followers on Facebook or twitter can be found out and used to suggest the most suitable party. Also faceted approach can be incorporated to VAAs so that the application becomes more user friendly, flexible and time saving, which from a recommender system point of view increases user satisfaction.

Acknowledgement. The authors acknowledge the support extended by DST-PURSE (Phase II), Government of India.

References

1. Isinkaye, F.O., Folajimi, Y.O., Ojokoh, B.A.: Recommendation systems: principles, methods and evaluation. Egypt. Inform. J. **3**(16), 261–273 (2015)
2. Agathokleous, M., Tsapatsoulis, N.: Applying hidden Markov models to voting advice applications. EPJ Data Sci. **5**, 34 (2016)
3. Pajala, T., Korhonen, P., Malo, P., Sinha, A., Wallenius, J., Dehnokhalaji, A.: Accounting for political opinions, power, and influence: a voting advice application. Eur. J. Oper. Res. **266**(2), 702–715 (2018)
4. Katakis, I., Tsapatsoulis, N., Mendez, F., Triga, V., Djouvas, C.: Social voting advice applications -definitions, challenges, datasets. IEEE Trans. Cybern. **44**, 1039–1052 (2014)
5. Gemenis, K., Rosema, M.: Voting advice applications and electoral turnout. Electoral. Stud. **36**, 281–289 (2014)
6. Germann, M., Mendez, F., Wheatley, J., Serdült, U.: Spatial maps in voting advice applications: the case for dynamic scale validation. Acta Polit. **50**, 214–238 (2015)
7. Thomas, F., Joel, A.: What's the point of voting advice applications? Competing perspectives on democracy and citizenship. Electoral. Stud. **36**, 244–251 (2014)
8. Louwerse, T., Rosema, M.: The design effects of voting advice applications: comparing methods of calculating matches. Acta Politcia **2**(50), 214–238 (2015)
9. Yuksel, S., Dizmanr, T., Yildizdan, G., Sert, U.: Application of soft sets to diagnose the prostate cancer risk. J. Inequalities Appl. **2013**, 229 (2013)
10. Yuen, K.K.F.: The fuzzy cognitive pairwise comparisons for ranking and grade clustering to build a recommender system. An application of smartphone recommendation. Eng. Appl. Artif. Intell. **61**, 136–151 (2017)
11. Maji, P.K., Biswas, R., Roy, A.R.: An application of soft sets in a decision making problem. Comput. Math Appl. **44**(8–9), 1077–1083 (2002)
12. Kumar, S.U., Inbarani, H.H., Kumar, S.S.: Bijective soft set based classification of medical data. In: 2013 International Conference on Pattern Recognition Informatics and Medical Engineering (PRIME), pp. 517–521 (2013)

13. Handaga, B., Herawan, T., Deris, M.M.: FSSC: an algorithm for classifying numerical data using fuzzy soft set theory. Int. J. Fuzzy Syst. Appl. **2**, 29–46 (2012)
14. Zenebe, A., Norcio, A.F.: Fuzzy modeling for item recommender systems or a fuzzy theoretic method for recommender systems (2006)
15. Thong, N., Son, L.: HIFCF: an effective hybrid model between picture fuzzy clustering and intuitionistic fuzzy recommender systems for medical diagnosis. Expert Syst. Appl. **42**, 3682–3701 (2015)
16. http://www.preferencematcher.org/?page_id=18

Fuzzy Logic Based Dynamic Plotting of Mood Swings from Tweets

Srishti Vashishtha$^{(\boxtimes)}$ and Seba Susan

Delhi Technological University, Delhi 110042, India
srishtidtu@gmail.com

Abstract. Twitter is one the most popular social media platforms. Users express their feelings easily on social media regarding any trending event. In this paper, we propose a fuzzy logic based approach for dynamic plotting of mood swings from tweets. The novelty of the paper is use of linguistic hedges with fuzzy logic to compute the sentiment of tweet. Comparison of our approach with existing methods, on real-time tweets extracted from online website confirms the superiority and efficiency of our method. The tweets used in our experiments are extracted from the timeline of the India Vs Pakistan final ICC world-cup match in June 2017. They reflect the moods of the twitter users as the match progresses. Using our fuzzy logic based approach, we successfully plot the dynamic mood vs time and compute the polarity of the sentiment at each time instant.

Keywords: Dynamic plotting · Mood swing · Twitter · Sentiment analysis · Fuzzy · Text mining

1 Introduction

Twitter is one of the most trending and mainstream platforms for microblogging services and online social media. Huge amount of data is created each day on the internet, as of April 2017 a total of 3.8 billion internet users are there. Social media is enormousreports reveal that social media gains 840 new users each minute [1]. Twitter has some prominent features compared with traditional media like: Timeliness, broad coverage of themes and diverse channels for broadcasting information. This makes twitter a popular and easy to use social media sensor for tracking various fascinating and thought-provoking incidents, events, themes, etc. [2]. Tweets usually express a user's viewpoint and opinion on a topic of interest, and research has shown they provide worthy insights on issues or trending events.

Twitter has created tremendous interest for sentiment analysis researchers. Sentiment analysis or opinion mining is the process of analyzing the mindset of people related to specific topic, incident, event, etc. Public opinion is a complex collection of assumptions of different people about specific subjects. The evolution of the social web has granted researchers to observe and analyze how people feel and react regarding a large variety of topics, events, etc. currently happening in the world. The attractive fact is that the users are sharing information freely and willingly. Social web provides an environment where users can express, post and exchange messages, with limit on the size of message, using any language [3].

© Springer Nature Switzerland AG 2019
A. Abraham et al. (Eds.): IBICA 2018, AISC 939, pp. 129–139, 2019.
https://doi.org/10.1007/978-3-030-16681-6_13

Text mining can be applied to tweets to extract logical and emotional patterns. It is the integration of information retrieval, artificial intelligence, machine learning, computational linguistics, natural language processing, statistics, database systems and others [4]. Our application of text mining is related to human thinking, the way human expresses their opinion about a specific event currently going on. Their opinions are transformed to sentiments; further these sentiments are interpreted and analysed over a period of time. Text mining along with natural language processing (NLP) is used to study different relations existing between the human emotions, feelings and opinions. The researchers collect information and knowledge on how humans understand, and applies language to create models, tools and techniques via NLP [5].

In 1965, Zadeh [6] introduced fuzzy set theory and fuzzy logic to deal with uncertain reasoning. Real-world problems are complex so they require smart and intelligent systems to integrate knowledge, techniques and methodologies from various sources. To deal with such real-world problems the concept of fuzziness is extremely useful. A fuzzy set consists of such elements which don't have any crisp boundary [7]. Each element in the fuzzy set is usually associated with a membership function, suppose the information is not in numeric form or the information in numeric form is co-related to words in natural language; or in natural language form then fuzzy set theory cannot directly handle the linguistic elements. To deal with such linguistic elements, Zadeh introduced linguistic variables. Linguistic hedges are modifiers, they are adverb or adjective like very, slightly, more or less, fairly, etc. These linguistic hedges have the effect of modifying the membership function for a basic atomic term [8]. This paper carries out dynamic analysis of tweets using fuzzy sets and linguistic hedge-based approach.

The rest of the paper is organized as follows. Section 2 discusses different papers on dynamics of twitter data. Section 3 describes the proposed fuzzy approach for dynamic analysis of tweets. In Sect. 4, experimental setup of our approach is discussed. The results are presented in Sect. 5. The overall conclusions are drawn in Sect. 6.

2   Related Work

Twitter sentiment analysis is a particular problem and of great interest to researchers in the field of computational linguistics. Twitter is one of the most popular social media platforms where live public opinions are easily accessible. These public opinions are available in continuous time streams thereby helps researchers to interpret and analyze the opinions dynamically. Mining logical, emotional and useful patterns from tweets with accuracy and efficiency from such unstructured form is treated as a critical task. One of the tasks is to identify which topic or theme is trending on twitter currently. Zhao *et al.* [2] has proposed a dynamic query expansion (DQE) model for theme tracking in Twitter. Specifically, DQE characterizes the theme consistency among heterogeneous entities through semantic and social relationships, including co-occurrence, replying, authorship, and friendship. The type of trending topics in twitter can be detected based on certain features. Further technical analysis and classification of trends into positive, negative or neutral trends classes can be done [9]. Social network analysis researchers have inspected the influence of brand related tweets to gain attention by users in the twitter network. In order to perform efficient sentiment classification related to brands, Ghiassi *et al.* [10] proposed a twitter-specific-lexicon with brand-specific terms for brand-related tweets. Another approach to brand-related twitter sentiment analysis address challenges associated with the unique properties of the twitter language, and mild sentiment expressions that are of interest to brand management practitioners. This approach has less dimensions but greater feature density [11]. Twitter has also been used by researchers as a source of information to explain the occurrence of various major events and demonstrating the value of these communications. Some researchers have explained and predicted the outcomes of major political elections through the analysis of tweets on the candidates and political issues. Marquez *et al.* analyzed time series obtained from twitter messages related to the 2008 U.S. elections using ARMA/ARIMA and GARCH models [3]. Another paper proposes a system that analyzes the sentiment from the whole twitter traffic related to the election, declares the results instantly and continuously. While conventional content analysis takes days or weeks to complete, the sentiment in the entire twitter traffic [12]. Some researchers got interest to study the evolution of hashtags to know how some groups in twitter retain their members, attract new ones and grow over time. Certain measures can be used for figuring out the factors that influences a user to attract other users to a certain hashtag [13]. A linguistic-inspired study of how these tags are created, used and disseminated by the members of information networks was carried out in [14]. Interesting and useful facts about hashtag and its characteristics were discovered. Liu *et al.* proposed a semi-supervised topic-adaptive sentiment classification (TASC) model to train and classify topics available on twitter. It updates topic-adaptive features based on the collaborative selection of unlabeled data [15]. Cheng *et al.* proposed a content-based opinion influence framework for predicting sentiment in opinion dynamics. It uses the content information along with historical exchanged textual information. The expression style of users also reflects user's influential power, thereby providing valuable information for companies [16].

3 Proposed Approach for Dynamic Tweet Analysis

Our proposed approach for dynamic analysis of the sentiment of tweets is based on fuzzy logic. It uses SentiWordNet (SWN) [17], a lexicon resource for opinion mining. Following is the description of the approach followed.

3.1 Tokenization and Lemmatization of Tweets

Tweets are stored in a document/paragraph form along with their corresponding timestamp. We need to first extract tweets from the document. Splitting up of paragraphs into sentences is termed as sentence tokenization [18]. Tokenizing a sentence is a process of splitting a sentence into a list of words. In other words, a tokenizer parses a sentence into a list of tokens (words). The output of sentence tokenization process will be stored in a dynamic list. The tokenized words are lemmatized to get the root word.

3.2   Bag of Words

In sentiment analysis, Bag of words means those important words which are essential for mining reviews, opinions, etc. It is a model that transforms documents into numerical form, where each word in the document is assigned some value in the range of 0 to 1. Next, we discuss about computing fuzzy scores for the bag of words. A fuzzy set A can be represented as $A = \{(x, \mu_A(x))\}, x \in U$ where x is the element from the universal set U and $\mu_A(x)$ is the membership of element x. In our approach, Universe of discourse is the set of all the words in each tweet. Each tweet contains a bag of word list. Each word has positive and negative score (Eqs. (3) and (4)) obtained from SWN [17]. These scores can be interpreted as fuzzy memberships pertaining to the fuzzy sets *Pos* and *Neg*. In our proposed model, we have used NLTK POS Tagger [18] to extract words which are nouns, adjectives, verbs or adverbs. The fuzzy sets *Pos* and *Neg* are represented as:

$$Pos = \{(a, \mu_{Pos}(a))\}, a \in X_i \tag{1}$$

$$Neg = \{(a, \mu_{Neg}(a))\}, a \in X_i \tag{2}$$

where a is the word and it belongs to the ith Bag of words X_i, where i = 1, 2,..., *total* corresponding to the *total* tweets. Each tweet has its own unique bag of words. The membership functions $\mu_{Pos}(a)$ and $\mu_{Neg}(a)$ associated with the tokenized word *a* are defined below.

$$\mu_{Pos}(a) = \frac{\sum\limits_{synsets} [syn.pos\_score()]}{length(synsets)} \tag{3}$$

$$\mu_{Neg}(a) = \frac{\sum\limits_{synsets} [syn.neg\_score()]}{length(synsets)} \tag{4}$$

where syn.pos_score() and syn.neg_score() are the scores obtained from SWN [17]; synsets is the set of synonyms of each word present in SWN.

3.3   Formulation of Fuzzy Sets for Plotting Mood Swings

First of all, Union of fuzzy sets is carried out to check the presence of emotion flowing through the tweets. Union fuzzy set is Pos ∪ Neg and the membership function is:

$$\mu_{Union}(a) = \mu_{Pos}(a) \vee \mu_{Neg}(a), a \in X_i \tag{5}$$

Linguistic hedges are modifiers, they are adverbs or adjectives such as *Very, slightly, More or Less, fairly*, etc. These linguistic hedges have the effect of modifying the membership function for a basic atomic term [8, 19, 20]. In our approach we have used the linguistic hedges *Very*, *More or Less* and *Not* in Eqs. (6) and (7).

$$\mu_{Verypos}(a) = [\mu_{Pos}(a)]^2, \ \mu_{NotVerypos}(a) = 1 - [\mu_{Pos}(a)]^2, \ \mu_{MoreLesspos}(a) = [\mu_{Pos}(a)]^{0.5} \quad (6)$$

$$\mu_{Veryneg}(a) = [\mu_{Neg}(a)]^2, \ \mu_{NotVeryneg}(a) = 1 - [\mu_{Neg}(a)]^2, \ \mu_{MoreLessneg}(a) = [\mu_{Neg}(a)]^{0.5} \quad (7)$$

We have proposed *FuzzyPos* and *FuzzyNeg* sets for all tweets for capturing the sentiment even if it is present in moderate amount. The formulation is explained below:

$$FuzzyPos = \{(a, \mu_{FuzzyPos}(a))\}, a \in X_i \quad (8)$$

$$FuzzyNeg = \{(a, \mu_{FuzzyNeg}(a))\}, a \in X_i \quad (9)$$

where X_i is the *i*th BOW and a is the word in BOW. The membership functions $\mu_{FuzzyPos}$ and $\mu_{FuzzyNeg}$ are defined as:

$$\mu_{FuzzyPos}(a) = \mu_{Morelesspos}(a) \wedge \mu_{NotVeryneg}(a) \quad (10)$$

$$\mu_{FuzzyNeg}(a) = \mu_{Morelessneg}(a) \wedge \mu_{NotVerypos}(a) \quad (11)$$

Now let W be the set of words $\{w\}$ in tweet at timestamp t, m is the number of words in set W and $p = 1, 2 \dots total$, where *total* is the total number of tweets. Next, we calculate cardinality of *FuzzyPos* and *FuzzyNeg* for each tweet p, as shown below:

$$Card\_FuzzyPos(p) = \sum_{w=1}^{m} \mu_{FuzzyPos}(w), w \in W \quad (12)$$

$$Card\_FuzzyNeg(p) = \sum_{w=1}^{m} \mu_{FuzzyNeg}(w), w \in W \quad (13)$$

The fuzzy processing for single tweet using our proposed fuzzy approach is shown in Fig. 1. Next for dynamic plotting of the mood swings, we fix time windows containing 3 tweets each. We sum up the cardinalities for all 3 tweets in each time window as shown below:

$$FuzzyPos\_value(n) = \sum_{p=3n+1}^{3n+3} Card\_FuzzyPos(p) \quad (14)$$

$$FuzzyNeg\_value(n) = \sum_{p=3n+1}^{3n+3} Card\_FuzzyNeg(p) \quad (15)$$

where n is the number of time windows $n = 0, 1 \dots (total/3) - 1$.

Finally, *Sentiment* value is calculated for each time window by de-fuzzifying of *FuzzyPos* and *FuzzyNeg* values to Positive (P), Neutral (Neu) and Negative (N) respectively. This is done to get crisp result, i.e. the sentiment of tweets.

$$Sentiment(n) = \begin{cases} P, FuzzyPos\_value > FuzzyNeg\_value \\ Neu, FuzzyPos\_value = FuzzyNeg\_value \\ N, otherwise \end{cases} \quad (16)$$

where *Sentiment* is the sentiment at particular time window n.

3.4 Overall Methodology

Following are the steps of overall methodology applied in this paper:

(1) Procure single tweet with timestamp t.
(2) Do the fuzzy processing of tweet as per Fig. 1.
(3) Repeat (1) and (2) for 3 consecutive tweets.
(4) Sum up the respective cardinality for all 3 tweets as per Eqs. (14) and (15).
(5) Decode the final sentiment for current time window using Eq. (16).
(6) Repeat steps1–5 for all consecutive non-overlapping time windows.
(7) Plot the *FuzzyPos_value* and *FuzzyNeg_value* with respect to time to show the dynamic mood swings.

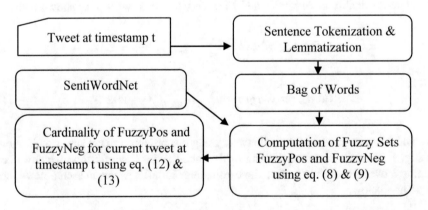

Fig. 1. Fuzzy processing of a single tweet by our proposed fuzzy approach

4 Experimental Setup and Implementation

Our proposed approach is based on the social media: Twitter. We have collected 30 English plaintext tweets from online websites [23, 24]. These tweets have been posted by cricket fans who were watching India Vs Pakistan ICC Champions Trophy Final World Cup match on 18th June, 2017. The identity of fans is kept anonymous. They are expressing their feelings, mood. These 30 tweets available at [25] are used as a test case while experimenting the proposed fuzzy approach. We have used these tweets as input to fuzzy processing method. The dynamic graph can be plotted in real-time. We have compared our method with two methods, first given by Cavalcanti *et al.* [21] and second given by Ortega *et al.* [22]. All methods are unsupervised techniques. Same set of tweets are processed by these methods. For dynamic plotting of mood swings by the proposed approach, we add up scores of all words for every 3 tweets in each time window in the same procedure as mentioned in the above section.

5   Results and Discussion

In this section we discuss the results of our proposed fuzzy logic based approach. We implemented the methodology, given in Sect. 3, for processing tweets posted by cricket fans. Following are the results.

5.1   Union

The first operation we have applied is the union of positive and negative scores of each tweet as per Eq. (5). It shows that some kind of emotion (either positive or negative) and sentiment is running. Overall reading of graph is high. Figure 2(a) shows the union values for all 30 tweets individually and Fig. 2(b) aggregates 3 consecutive tweets for 10 time windows.

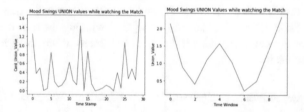

Fig. 2.   Union values for (a) 30 tweets plotted individually (b) cumulative plot-10 time windows

5.2   FuzzyPos

The next implementation is our proposed FuzzyPos set in Eq. (8) which states that there is more or less positivity but not very negativity in tweets. Figure 3(a) shows the values for all 30 tweets individually and Fig. 3(b) aggregates 3 consecutive tweets for 10 time windows. We can observe initially that the tweets are positive about the match, but as match proceeds there is no positivity in tweets 10, 11, 12 hence the values are 0 and we can see a dip. Later on, due to the fall of wickets of Indian team, the fans get angry and post negative tweets like tweet 5: "SharmaJi ka beta fail ho Gaya!! #INDvPak#CT17Final", tweet 23: "Hurts #IndvsPak #PakistanMurdabad", tweet 25: "TV sets broken on the streets of Ahmedabad after India's defeat by Pakistan in the #CT2017Final #IndVsPak". But at the end some kind of positivity came up to support

Indian team like tweet 30: "Do not become sad ur team (India) is one of the best team i think they loss becoz of overconfidence."

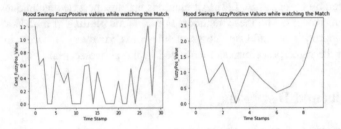

Fig. 3. Fuzzy positive values for (a) 30 tweets individually (b) cumulative plot-10 time windows

5.3   FuzzyNeg

Another implementation proposed is FuzzyNeg set in Eq. (9) which states that there is more or less negativity but not very positivity in tweets. The graph is shown in Fig. 4. Figure 4(a) shows the values for all 30 tweets individually and Fig. 4(b) aggregates 3 consecutive for 10 time windows. We can observe initially the tweets are less negative about the match but as match proceeds India starts losing the match the negativity starts rising and reaches its peak, this implies the fans are sad and angry like in tweet 17: "We lost the match bulleya, They played so well bulleya, We need a slap bulleya, Cup unka, cup unka #INDvPAK". The peak is achieved due to words "massacre" in tweet 10, "not" in tweet 11, "down" in tweet 12 and highest at tweet 14 due to the double occurrence of "not". Some fans are still in hope of winning the match and supporting the team hence the negativity falls in between but finally when India loses the match negativity rises again.

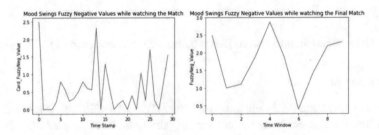

Fig. 4. Fuzzy negative values for (a) 30 tweets individually (b) cumulative plot-10 time windows

5.4   Comparison with Other Methods

We have calculated the sentiment of each tweet at each and particular time window, using our method, Cavalcanti *et al.* [21] method and Ortega *et al.* [22] method and displayed in Table 1.

Table 1. Sentiment of tweets.

Time window	FuzzyPos	FuzzyNeg	Proposed method sentiment		Cavalcanti et al. [21] sentiment		Ortega et al. [22] sentiment
TW1	**2.523**	2.4826	2.523	P	0.0084	P	P
TW2	0.6599	**1.0039**	1.0039	N	−0.0061	N	N
TW3	**1.3005**	1.1038	1.3005	P	−0.0011	N	N
TW4	0	**1.8912**	1.8912	N	−0.0125	N	N
TW5	1.1975	**2.8758**	2.8758	N	−0.0091	N	N
TW6	0.7265	**1.8899**	1.8899	N	−0.0053	N	Neu
TW7	0.3536	**0.4007**	0.4007	N	−0.0002	N	Neu
TW8	0.5445	**1.4324**	1.4324	N	−0.003	N	Neu
TW9	1.2115	**2.1987**	2.1987	N	−0.0057	N	P
TW10	**2.6014**	2.3113	2.6014	P	−0.0018	N	P

Table 1 corresponds to time graph for dynamic plotting of mood swings. It shows the computed sentiment values for all three methods. We can see initially the tweets are positive implying fans are positive for match. As the match progresses and India starts losing the match, people become sad, angry and thus the tweets start getting negative. In the last time window, sentiment become positive again because people are supporting team India and portray their love for team India. Our fuzzy logic based approach catches the optimism at the end and computes the sentiment at each time window correctly. Comparison with Cavalcanti *et al.* method's [21] sentiment values show that it is not able to detect positive sentiment at 3rd and 10th time window and Ortega *et al.* method's [22] sentiment values show that it is not able to detect the negative emotion in the 6th–9th time windows. Our approach is sensitive to dynamic mood swings. Analysis and interpretation of our proposed fuzzy logic based method demonstrates that processing of plaintext tweets for decoding the mood swings of cricket fans while watching India Vs Pakistan ICC Champions final world cup match is superior than Cavalcanti *et al.* method [21] and Ortega *et al.* method [22].

6   Conclusion

In this paper, we have proposed a fuzzy logic based approach for dynamic plotting of mood swings of tweets. These tweets were posted by cricket fans while watching India Vs Pakistan final ICC world-cup match in June 2017. The novelty of our approach is use of linguistic hedges: *Very, More or Less* and *Not*, with fuzzy logic. These hedges describe the mood of user in tweet and captures the sentiment even when it is present in moderate amount. The fuzzy membership values of these hedges are used to plot the fuzzy positive and fuzzy negative sentiment of tweet with respect to time. Finally, we apply defuzzification to get the sentiment of each tweet. We have compared our method with existing methods. The analysis demonstrates that our approach is more sensitive to mood swings and decodes the correct sentiment at each time window.

References

1. Microfocus. https://blog.microfocus.com/how-much-data-is-created-on-the-internet-each-day/. Accessed 12 Sept 2018
2. Zhao, L., Chen, F., Lu, C.T., Ramakrishnan, N.: Dynamic theme tracking in Twitter. In: 2015 IEEE International Conference on Big Data (Big Data), pp. 561–570. IEEE (2015)
3. Marquez, F.B., Avello, D.G., Mendoza, M., Poblete, B.: Opinion dynamics of elections in Twitter. In: 2012 Eighth Latin American Web Congress (LA-WEB), pp. 32–39. IEEE (2012)
4. Irfan, R., King, C.K., Grages, D., Ewen, S., Khan, S.U., Madani, S.A., Kolodziej, J.: A survey on text mining in social networks. Knowl. Eng. Rev. **30**(2), 157–170 (2015)
5. Chowdhury, G.G.: Natural language processing. Ann. Rev. Inf. Sci. Technol. **37**(1), 51–89 (2003)
6. Zadeh, L.A.: Fuzzy sets. Inf. Control **8**(3), 338–353 (1965)
7. Zadeh, L.A.: Calculus of fuzzy restrictions. In: Fuzzy Sets and Their Applications to Cognitive and Decision Processes, pp. 1–40 (1975)
8. Ross, T.J.: Fuzzy Logic with Engineering Applications. Wiley, Hoboken (2009)
9. Saquib, S., Ali, R.: Understanding dynamics of trending topics in Twitter. In: 2017 International Conference on Computing, Communication and Automation (ICCCA), pp. 98–103. IEEE (2017)
10. Ghiassi, M., Skinner, J., Zimbra, D.: Twitter brand sentiment analysis: a hybrid system using n-gram analysis and dynamic artificial neural network. Expert Syst. Appl. **40**(16), 6266–6282 (2013)
11. Zimbra, D., Ghiassi, M., Lee, S.: Brand-related twitter sentiment analysis using feature engineering and the dynamic architecture for artificial neural networks. In: 2016 49th Hawaii International Conference on System Sciences (HICSS), pp. 1930–1938. IEEE (2016)
12. Wang, H., Can, D., Kazemzadeh, A., Bar, F., Narayanan, S.: A system for real-time twitter sentiment analysis of 2012 us presidential election cycle. In: Proceedings of the ACL 2012 System Demonstrations, pp. 115–120. Association for Computational Linguistics (2012)
13. Daher, L.A., Zantout, R., Elkabani, I.: Dynamic evolution of hashtags on Twitter: a case study from career opportunities groups. In: 2017 International Conference on New Trends in Computing Sciences (ICTCS), pp. 314–319. IEEE (2017)
14. Cunha, E., Magno, G., Comarela, G., Almeida, V., Gonçalves, M.A., Benevenuto, F.: Analyzing the dynamic evolution of hashtags on Twitter: a language-based approach. In: Proceedings of the Workshop on Languages in Social Media, pp. 58–65. Association for Computational Linguistics (2011)
15. Liu, S., Cheng, X., Li, F., Li, F.: TASC: topic-adaptive sentiment classification on dynamic tweets. IEEE Trans. Knowl. Data Eng. **27**(6), 1696–1709 (2015)
16. Chen, C., Wang, Z., Li, W.: Tracking dynamics of opinion behaviors with a content-based sequential opinion influence model. IEEE Trans. Affect. Comput. **1**, 1 (2018)
17. SentiWordNet. http://sentiwordnet.isti.cnr.it/. Accessed 12 Sept 2018
18. Natural Language Toolkit. http://www.nltk.org/#natural-language-toolkit. Accessed 12 Sept 2018
19. Hameed, I.A.: Enhanced fuzzy system for student's academic evaluation using linguistic hedges. In: 2017 IEEE International Conference on Fuzzy Systems (FUZZ-IEEE), pp. 1–6. IEEE (2017)
20. Zamali, T., Lazim, M.A., Osman, M.T.A.: Sensitivity analysis using fuzzy linguistic hedges. In: 2012 IEEE Symposium on Humanities, Science and Engineering Research (SHUSER), pp. 669–672. IEEE (2012)

21. Cavalcanti, D.C., Prudêncio, R.B.C., Pradhan, S.S., Shah, J.Y., Pietrobon, R.S.: Good to be bad? Distinguishing between positive and negative citations in scientific impact. In: 2011 23rd IEEE International Conference on Tools with Artificial Intelligence (ICTAI), pp. 156–162. IEEE (2011)
22. Ortega, R., Fonseca, A., Gutierrez, Y., Montoyo, A.: SSA-UO: unsupervised twitter sentiment analysis. In: Second Joint Conference on Lexical and Computational Semantics (* SEM), vol. 2, pp. 501–507 (2013)
23. Scoopwhoop. https://www.scoopwhoop.com/48-Tweets-Which-Define-How-Embarrassing-Indias-Champions-Trophy-Defeat-To-Pakistan-Was/#.l1uoc3uu5. Accessed 12 Sept 2018
24. Twitter. https://twitter.com/. Accessed 12 Sept 2018
25. File. https://drive.google.com/file/d/1Gz9Toz7GcfRYPfMw_S3zFd6wRhLDVmgC/view?usp=sharing. Accessed 31 Oct 2018

Application of Parallel Genetic Algorithm for Model-Based Gaussian Cluster Analysis

Peter Laurinec, Tomáš Jarábek[✉], and Mária Lucká

Faculty of Informatics and Information Technologies,
Slovak University of Technology in Bratislava, Ilkovičova 2,
842 16 Bratislava, Slovak Republic
{peter.laurinec,tomas.jarabek,maria.lucka}@stuba.sk
https://petolau.github.io/

Abstract. Proposed paper presents a new model-based Gaussian clustering method and defines new optimization criteria for model-based clustering, which are used as fitness functions in genetic algorithm. These optimization criteria are based on different properties of covariance matrices. The proposed model-based Gaussian clustering method is compared with the well-known K-Means method that is solved by genetic algorithm or by Particle Swarm Optimization method. Our method achieves higher similarity between real classification and computed clustering results on all six presented real-world datasets. Because of the high computational requirements of the used methods we have focused on their parallelization. Due to the chosen parallel computer architecture we have combined both MPI and OpenMP programing interfaces. We show that parallelization of the proposed method is very effective and scalable on many execution units.

Keywords: Model-based Gaussian cluster analysis ·
Genetic algorithm · Parallel programming · K-Means

1 Introduction

Clustering is a process of linking objects to groups called clusters, where each group consists of objects with similar properties and different groups have each to other more different characteristics [1]. It is used mainly in the context of data mining and statistical or exploratory analysis of large volumes of data for finding patterns in the data, for natural classification and data compression.

The algorithms of cluster analysis can be divided into a few categories [5]. In our work we use model-based methods which divide the data so that a predefined mathematical model is met. A model-based Gaussian clustering method requires a statistical model so we present several optimization criteria based on different properties of covariance matrices. The model-based clustering compared to centroid-based clustering methods like K-Means has some advantages,

© Springer Nature Switzerland AG 2019
A. Abraham et al. (Eds.): IBICA 2018, AISC 939, pp. 140–150, 2019.
https://doi.org/10.1007/978-3-030-16681-6_14

because can find better solutions for overlapping clusters or clusters of elliptic shapes, that is in general a difficult task. The optimization problem of objects clustering by model-based Gaussian clustering method and centroid-based methods are NP-hard problems. In our work, the model-based Gaussian clustering is solved by a genetic algorithm [24]. The proposed method is compared with centroid-based K-Means solved by genetic algorithm or the biologically inspired algorithm Particle Swarm Optimization (PSO) [10].

Papers dealing with model-based cluster analysis explain the probability model, and the process of finding a solution by minimizing a model's criterion. Recently, model-based clustering has been applied to many practical problems including identifying children's profile fact development [22], improving energy use in crop production [11] or discovering novel molecular subtypes in DNA [12]. Paper [19] shows how to search over the vast solution space by means of genetic algorithm, with a new probability model based on subset clustering. With the enormous increase of volumes of data, many scientific datasets become high-dimensional. Clustering of such data is a problem for model-based methods [23]. Authors in [23] solve this problem by partition clustering of subspaces. Biologically inspired algorithms help in avoiding a local optimum and in finding a global optimum [7]. Biologically inspired clustering methods also accelerate finding a global optimum by using more individuals searching for the solution [7]. One of these algorithms is PSO with many applications in cluster analysis [3].

Together with increasing amount of data the need for fast and effective processing has brought wide recognition of parallel processing. Its role in providing scalable performance and lower computational costs is reflected in a wide variety of applications [8]. Because we use parallel computing for speeding up the used methods, we briefly introduce basic frameworks for parallel computing. We use parallel master-slave paradigm and show its advantages and effectivity. Besides the standard parallelization techniques such as OpenMP and Message Passing Interface (MPI) [20], we also exploit its combination.

This paper is organized as follows: in Sect. 2 multivariate Gaussian distribution and model-based cluster analysis are described. Section 3 presents genetic algorithm as optimization method for model-based clustering. In Sect. 4 K-Means and Particle Swarm Optimization methods are shortly presented. Section 5 discuss parallelization models. Section 6 describes experiments and results and Sect. 7 concludes the paper.

2   Methodology

In this section, we describe concepts and principles of multivariate Gaussian distribution and model-based cluster analysis.

2.1   Multivariate Gaussian Distribution

Multivariate Gaussian distribution [13] is very important in model-based clustering and can be defined as follows:

Consider a random vector $\boldsymbol{X} = (X_1, X_2, \ldots, X_p)^T$, given vector $\boldsymbol{\mu} = (\mu_1, \mu_2, \ldots, \mu_p)^T$ and a positive symmetric definite matrix $\boldsymbol{\Sigma}$ of type $p \times p$. Consider a vector of observations $\boldsymbol{x} = (x_1, \ldots, x_p)$ characterizing objects that are realizations of the random vector $\boldsymbol{X}$. Suppose, that the p-dimensional random vector $\boldsymbol{X}$ has regular p-dimensional normal distribution with parameters $\boldsymbol{\mu}$ and $\boldsymbol{\Sigma}$, if $\boldsymbol{X}$ is absolute continuous with density

$$f(\boldsymbol{x}) = f(x_1, \ldots, x_p) = \frac{1}{(2\pi)^{\frac{p}{2}} \det(\boldsymbol{\Sigma})^{\frac{1}{2}}} \exp\left\{ -\frac{1}{2}(\boldsymbol{x} - \boldsymbol{\mu})^T \boldsymbol{\Sigma}^{-1}(\boldsymbol{x} - \boldsymbol{\mu}) \right\},$$

where $\boldsymbol{\mu}$ is a vector of expected values, a $\boldsymbol{\Sigma}$ is covariance matrix, that is a positive symmetric defined matrix of type $p \times p$ of vector $\boldsymbol{X}$. This fact is denoted as $\boldsymbol{X} \sim N_p(\boldsymbol{\mu}, \boldsymbol{\Sigma})$.

With the help of matrix calculus, we can interpret multivariate Gaussian distribution geometrically. Let $f(\boldsymbol{X}, \boldsymbol{\mu}, \boldsymbol{\Sigma})$ be a density function of random vector $\boldsymbol{X}$ with distribution $N_p(\boldsymbol{\mu}, \boldsymbol{\Sigma})$. If the level set (a set where the function achieves a given constant value c) is non-empty and is defined as

$$L_c(f) = \{\boldsymbol{x} \in \mathbb{R}^p : f(\boldsymbol{x}, \boldsymbol{\mu}, \boldsymbol{\Sigma}) \geq c\},$$

then this set is an ellipsoid. The direction of principal axes of ellipsoids are given by eigenvalues of vectors of the covariance matrix $\boldsymbol{\Sigma}$. The relative length of the major axis is given by square roots of eigenvalues of the covariance matrix $\boldsymbol{\Sigma}$.

If $\boldsymbol{\Sigma} = \boldsymbol{U}\boldsymbol{\Lambda}\boldsymbol{U}^T = (\boldsymbol{U}\boldsymbol{\Lambda}^{\frac{1}{2}})(\boldsymbol{U}\boldsymbol{\Lambda}^{\frac{1}{2}})^T$ is a decomposition based on eigenvalues, where $\boldsymbol{U}$ is an orthogonal matrix of eigenvectors and $\boldsymbol{\Lambda}$ is a diagonal matrix of eigenvalues, then the distribution $N_p(\boldsymbol{\mu}, \boldsymbol{\Sigma})$ can be understood as the "spheric" distribution $N_p(0, \boldsymbol{I})$, which is scaled by $\boldsymbol{\Lambda}^{\frac{1}{2}}$, rotated by matrix $\boldsymbol{U}$ and translated by $\boldsymbol{\mu}$.

2.2   Model-Based Gaussian Cluster Analysis

Generally, cluster analysis means to group a set of objects (observations) in such a way that objects in the same group (called cluster) are more similar (in one sense or another) to each other than to those in other groups (clusters). At first, we define notations, that we use further in the cluster analysis.

Let N be a number of objects and K denotes the number of clusters. Let $\Gamma_{N,K}$ be a set of vectors of length N, whose elements belong to the set $\{1, \ldots, K\}$ and its permutations. By classification of N objects to K clusters, we denote each vector $\gamma \in \Gamma_{N,K}$. For $j \in \{1, \ldots, K\}$ and $i \in \{1, \ldots, N\}$ the equation $\gamma_i = j$ is interpreted as classification γ assigning i-th observation to j-th cluster.

The set $K_j(\gamma) = \{i : \gamma_i = j, i \in \{1, \ldots, N\}\}$ is denoted as j-th cluster for classification γ. The number of objects in a set $K_j(\gamma)$ is denoted as $N_j(\gamma)$, where $j \in \{1, 2, \ldots K\}$. The optimal classification to clusters will be denoted as $\hat{\gamma}$. Covariance matrix of a cluster $K_j(\gamma)$ is denoted as

$$\boldsymbol{S}_j(\gamma) = N_j^{-1}(\gamma) \sum_{i \, \in \, K_j(\gamma)} (\boldsymbol{x}^{(i)} - \overline{\boldsymbol{x}}_j(\gamma))(\boldsymbol{x}^{(i)} - \overline{\boldsymbol{x}}_j(\gamma))^T,$$

where $\bar{x}_j(\gamma)$ is the mean vector of objects in j-th cluster. The overall covariance matrix depending on the classification of clustering is the matrix $S(\gamma) = \frac{1}{N} \sum_{j=1}^{K} N_j(\gamma) S_j(\gamma)$, which is the weighted average of covariance matrices $S_j(\gamma)$.

The main assumption by the model-based normal cluster analysis is that the vectors of observations $\{x^{(i)}\}_{i=1}^{N}$ characterizing objects are realizations of random vectors $\{X^{(i)}\}_{i=1}^{N}$, and these have p-dimensional Gaussian distribution $X^{(i)} \sim N_p(\mu_{\gamma_i}, \Sigma_{\gamma_i})$, where μ_{γ_i} is a p-dimensional vector and Σ_{γ_i} is a $p \times p$ covariance matrix. This assumption allows us to use known statistical methods. The goal is to find optimal classification to clusters $\hat{\gamma}$ by maximizing the likelihood

$$L(\gamma, \mu_1, \ldots, \mu_c, \Sigma_1, \ldots, \Sigma_c | x^{(1)}, \ldots, x^{(n)}) = \prod_{i=1}^{n} f(x^{(i)}, \mu_{\gamma_i}, \Sigma_{\gamma_i}).$$

A big advantage of this method is the possibility to change (vary) all features of distribution (orientation, shape and volume) between clusters. The key to these changes is re-parametrization possibility of covariance matrices Σ_j in terms of decomposition based on eigenvalues [2] as mentioned above. The orientation of principal components of Σ_j is given by matrix U_j, while the matrix Λ_j determines the volume and shape of a cluster j.

Based on the previous descriptions we have derived four optimization criteria based on four conditions for covariance matrices Σ_j which are summarized in Table 1, whereby each criterion is an objective function that is minimized.

Table 1. Summarization of the optimization criteria.

#	Optimization criterion	Condition for Σ_j	Distribution	Orientation	Shape	Volume
1.	$\sum_{j=1}^{K} N_j(\gamma) \ln \det(S_j(\gamma))$	None	Ellipsoidal	Variable	Variable	Variable
2.	$\ln \det(S(\gamma))$	Σ	Ellipsoidal	Equal	Equal	Equal
3.	$\sum_{j=1}^{K} N_j(\gamma) \ln \operatorname{tr}(S_j(\gamma))$	$\lambda_j I$	Spherical	NA	Equal	Variable
4.	$\ln \operatorname{tr}(S(\gamma))$	λI	Spherical	NA	Equal	Equal

3 Genetic Algorithm

To solve the described optimization problem of clustering objects into groups, we have chosen a genetic algorithm. Genetic algorithms are modern heuristic optimization techniques that belong to the class of evolutionary algorithms and apply techniques from biology – heredity, mutation, natural selection and crossbreeding [17]. Genetic algorithms are iterative, typically each iteration of the algorithm moves a solution closer to the global optimum.

A genetic algorithm consists of the following steps:

1. Initialization of population.
2. Computation of fitness and ordering of population.

3. Natural selection and mating.
4. Selection of mates.
5. Mutating.
6. Convergence check (number of iterations). If the stopping rule is not satisfied, go to step 2.

Next, we explain how genetic algorithm solves clustering problem with the proposed optimization criteria based on the Gauss model. We want to assign N objects into K clusters optimally. A population consists of $Npop$ chromosomes in each iteration. Each chromosome is represented by N integers, in the range $\{1, \ldots, K\}$ representing its classification to a cluster of observations. In the next part, we consider the fitness function as an argument of minimum of optimization criterion.

According to the chosen fitness f (functions from Table 1), the value of fitness of all chromosomes is calculated. Then chromosomes are ordered (from the best to the worst fitness) and a combination of natural selection with a rank selection method is applied. The process is repeated ($Npop/2$ - $elite$) times, where $elite$ is a number of elite chromosomes. Mates are selected with a probability of:

$$prob[i] = \frac{(\frac{Npop}{2i})^5}{\sum_{j=1}^{Npop/2} j^5}, \quad i = 0, 1, 2, \ldots, \frac{Npop}{2} - 1.$$

To preserve the best chromosomes, the first few chromosomes are chosen as elite and used in the next population. As a mating method, the two-point crossover is used. The two-point crossover selects two random numbers in the range 1 to N and cut chromosomes in these places. Then everything between the two points is swapped, thus creating children. The children are chromosomes in the new population joined with elites.

The last step before the convergence check is mutation. Mutation is randomly changed classification of an observation, it can even change to a previous classification. Mutation is not applied to elite chromosomes. The count of mutations to a chromosome is constant and it is defined at the beginning of the algorithm.

In each iteration the fitness of all chromosomes are calculated, population is ordered and the best-rated chromosome is marked as best. In our experiments as convergence check (stopping rule), the number of iterations $tmax$ is chosen.

4   K-Means and PSO

K-Means [9] is a centroid-based clustering method. The aim of the centroid-based clustering is to divide objects into clusters so that each cluster is represented by a centroid. An object belongs to its nearest centroid. K-Means minimizes a fitness function, which is the sum of the distances between the centroid and its associated observations. We propose two variants of solving K-Means method; one with the described genetic algorithm and second with Particle Swarm Optimization (PSO) algorithm [10].

PSO is a stochastic optimization method simulating the movement of particles in flocks searching for food. Each particle moves alone but is affected by the movement of other particles. The position of the particles is used as centroids of clusters in fitness computation and clustering result. Before computation, the size of the population and number of clusters must be determined. More details on PSO can be found in [10] but we can summarize the algorithm in the following steps:

1. Initialize the population.
2. Compute fitness.
3. Update local best solutions of all particles.
4. Update the global best solution.
5. Update the position and speed of particles.
6. Convergence check (number of iterations). If the stopping rule is not satisfied, go to step 2.

5   Parallelization and Performance Results

Parallelization is one possibility for speeding up the calculation. Our calculations have been parallelized by using OpenMP [15] and Message Passing Interface (MPI) [14]. The tasks were running in parallel across multiple computers whereby OpenMP paradigm employed all available cores within each computer. With the help of MPI we used multiple computers whereby every computer used OpenMP.

In the proposed method, the computation of the fitness function by genetic algorithm took 99.6% runtime, so we have focused on its parallelization. In each iteration (before fitness calculation), the population is divided into groups that are distributed to available nodes by means of MPI. For the simultaneous calculation of fitness of multiple chromosomes on multiple cores OpenMP is used. The implementation was written in the C programming language [18] and is available online[1]. All calculations were realized on the cluster at the Centre for High Performance Computing of Slovak University of Technology in Bratislava, equipped with Intel Xeon X5670 2.93 GHz processors. Different configurations with varying number of cores across different number of computers were used.

In our experiments, we were not able to run the code at this Centre using a single core. For testing purposes, the program was always running with identical parameters but on a different number of computational units. Table 2 presents runtime results and shows that the runtime with six cores is approximately one-half of the runtime with three cores, with a difference of five hours.

As can be seen the proposed method is not suitable for large datasets. This is caused by the slow convergence rate of the proposed genetic algorithm.

---

[1] https://github.com/PetoLau/ParallelGenClust.

Table 2. Runtimes and executions units for model-based Gaussian clustering solved by genetic algorithm on SDSS dataset with 9000 observations. Here the second optimization criterion was used (see Table 1). The genetic algorithm executed 600,000 iterations with population size of 200 chromosomes.

Processors	Cores	Runtime [sec]	Runtime [h]
1	3	258332	71,76
1	4	180163	50,05
1	5	160621	44,62
1	6	137704	38,25
2	3	136595	37,94
2	6	75066	20,85
4	3	75118	20,87
4	6	45784	12,72
8	3	45540	12,65
8	6	30885	8,58

6   Data Analysis and Results

The proposed methods were evaluated on six datasets: the IRIS Flower dataset[2], the SEEDS X-ray attributes dataset[3], the BREAST Cancer Wisconsin (Diagnostic) dataset[4], MHOUT [21] Multi-Hop Labelled Readings dataset, SHOUT [21] Single-Hop Labelled Readings dataset and the SDSS point sources [6] dataset.

In the IRIS dataset there are three species of iris, in SEEDS dataset there are three varieties of wheat, in BREAST dataset there are two types of tumor, in MHOUT and SHOUT there are data of two types (normal and abnormal) and in SDSS one can find three types of astronomical objects (Table 3).

Table 3. Datasets used in experiments.

Dataset	# Observations	# Dimensions	# Classes
IRIS	150	4	3
SEEDS	210	7	3
BREAST	569	30	2
MHOUT	4690	2	2
SHOUT	5041	2	2
SDSS	9000	4	3

---

[2] Source: https://archive.ics.uci.edu/ml/datasets/Iris.

[3] Source: https://archive.ics.uci.edu/ml/datasets/seeds.

[4] Source: https://archive.ics.uci.edu/ml/datasets/Breast+Cancer+Wisconsin+(Diagnostic).

As external validation methods, we have used Rand index [16] and Fowlkes-Mallows index [4], which are measures of similarity between the real classification and results of clustering. The Rand index is a standard measure used in clustering results evaluation, but in the case of unrelated clusterings, the value of the Rand index approaches one as the number of observations increases.In contrast, the value of Fowlkes-Mallows index quickly approaches zero [4].

The results of experiments are presented in Table 4. The numbers 1 to 4 in the second row reference corresponding criteria introduced in Table 1. Table 4 show the best results (in the meaning of fitness value) from several runs. The proposed model-based Gaussian clustering performed better in all datasets compared to K-Means and PSO. The second criterion performed best in three cases with datasets IRIS, SEEDS and SDSS. The third criterion performed best in two cases with datasets MHOUT and SHOUT, but the results of K-Means were similar. The fourth criterion performed best in one case with dataset BREAST.

Table 4. Values of the Rand index and the Fowlkes-Mallows index. The numbers 1 to 4 in the second row reference corresponding model-based criteria introduced in Table 1. The KM-GA notation refers to K-Means solved by genetic algorithm and KM-PSO refers to K-Means solved by PSO. The first part corresponds to the value of the Rand Index and the second part corresponds to the value of the Fowlkes-Mallows index.

Dataset	Rand index						Fowlkes-Mallows index					
	1	2	3	4	KM-GA	KM-PSO	1	2	3	4	KM-GA	KM-PSO
IRIS	0.8121	**0.9575**	0.8859	0.8859	0.8797	0.8859	0.7168	**0.9355**	0.8321	0.8294	0.8208	0.8306
SEEDS	0.7619	**0.9573**	0.8744	0.8744	0.8744	0.8704	0.6409	**0.8443**	0.8106	0.8106	0.8106	0.8052
BREAST	0.8194	0.8279	0.5326	**0.8423**	0.8279	0.7308	0.8269	0.8410	0.7286	**0.8535**	0.8410	0.7419
MHOUT	0.5051	0.5105	**0.5150**	0.5105	0.5107	0.5112	0.7021	0.7062	**0.7094**	0.7061	0.7063	0.7066
SHOUT	0.5000	0.6021	**0.7472**	0.6134	0.6147	0.7146	0.7026	0.7727	**0.8626**	0.7800	0.7809	0.8433
SDSS	0.8269	**0.8271**	0.7426	0.7485	0.7504	0.7472	0.7875	**0.7911**	0.6699	0.6767	0.6821	0.6911

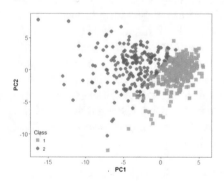

Fig. 1. BREAST - real

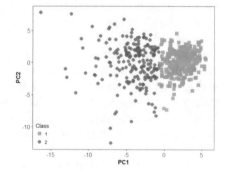

Fig. 2. BREAST - clustered

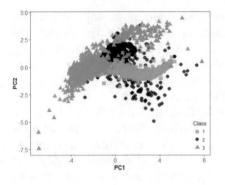
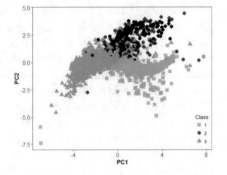

Fig. 3. SDSS - real   **Fig. 4.** SDSS - clustered

For better illustration the best clustering results (according to Tables 4) and real classification to groups for two of the datasets are shown on plots. For these multidimensional datasets, the first two principal components (using Principal Component Analysis) were used for visualization in two dimensions. The BREAST dataset and its best clustering result are shown in Figs. 1 and 2. The SDSS dataset and its best clustering result are shown in Figs. 3 and 4.

7   Conclusion

In this paper, we described four variations of model-based Gaussian cluster analysis based on different conditions of covariance matrices. We defined four optimization criteria and precisely explained the application of genetic algorithm used in solving optimization problem of clustering. The proposed model-based Gaussian clustering method solved by genetic algorithm was compared with K-Means solved by genetic algorithm or PSO. Experiments on six real-world datasets proved that quality of clustering, according to accuracy of classification to clusters, of our model-based method performed generally better than K-Means. To show that the proposed method suitable all mentioned methods were parallelized using both OpenMP and MPI interfaces. We proved that model-based Gaussian clustering solved by genetic algorithm method is efficiently parallelizable and reduces the runtime significantly. For achieving high quality clustering results the proposed model-based Gaussian method needs to execute more iterations in comparison with the K-Means method solved by genetic algorithm or PSO. The proposed method is also not suitable for large datasets because of the slow convergence rate of the genetic algorithm.

Acknowledgment. We would like to thank Lukáš Csóka for his assistance with programing the method in the C programming language during his studies at the Faculty of Informatics and Information Technologies of the Slovak University of Technology in Bratislava. We would also like to thank Radoslav Harman for his supervising while this method was being created.

This work was partially supported by the Scientific Grant Agency of The Slovak Republic, Grant No. VG 1/0458/18 and APVV-16-0484.

References

1. Andrews, J.L., Mcnicholas, P.D.: Using evolutionary algorithms for model-based clustering. Pattern Recogn. Lett. **34**(9), 987–992 (2013)
2. Banfield, J.D., Raftery, A.E.: Model-based Gaussian and non-Gaussian clustering. Biometrics **49**(3), 803–821 (1993)
3. Chen, C.Y., Ye, F.: Particle swarm optimization algorithm and its application to clustering analysis. In: 2004 IEEE International Conference on Networking, Sensing and Control, vol. 2, pp. 789–794 (2004)
4. Fowlkes, E.B., Mallows, C.L.: A method for comparing two hierarchical clusterings. J. Am. Stat. Assoc. **78**(383), 553–569 (1983)
5. Fahad, A., et al.: A survey of clustering algorithms for big data: taxonomy and empirical analysis. IEEE Trans. Emerg. Top. Comput. **2**(3), 267–279 (2014)
6. Feigelson, E., Babu, G.: Modern Statistical Methods for Astronomy: With R Applications. Cambridge University Press, Cambridge (2012)
7. Fong, S., Deb, S., Yang, X.S., Zhuang, Y.: Towards enhancement of performance of k-means clustering using nature-inspired optimization algorithms. Sci. World J. **2014**, 16 p. (2014). https://doi.org/10.1155/2014/564829. Article ID 564829
8. Grama, A.: Introduction to Parallel Computing. Pearson Education. Addison-Wesley (2003)
9. Hartigan, J.A., Wong, M.A.: Algorithm as 136: a k-means clustering algorithm. J. R. Stat. Soc. Ser. C (Appl. Stat.) **28**(1), 100–108 (1979)
10. Kennedy, J., Eberhart, R.: Particle swarm optimization. In: Proceedings of the IEEE International Conference on Neural Networks, vol. 4, pp. 1942–1948, November 1995
11. Khoshnevisan, B., et al.: A clustering model based on an evolutionary algorithm for better energy use in crop production. Stochast. Environ. Res. Risk Assess. **29**(8), 1921–1935 (2015)
12. Koestler, D.C., Houseman, E.A.: Model-based clustering of DNA methylation array data, pp. 91–123. Springer, Dordrecht (2015)
13. Lamoš, F., Potocký, R.: Pravdepodobnosť a matematická štatistika: Štatistické analýzy. Alfa (1989)
14. Message Passing Interface Forum: A message-passing interface standard version 2.1 (2008)
15. OpenMP Architecture Review Board: OpenMP application program interface version 3.1 (2011)
16. Rand, W.M.: Objective criteria for the evaluation of clustering methods. J. Am. Stat. Assoc. **66**(336), 846–850 (1971)
17. Raposo, C., et al.: Automatic clustering using a genetic algorithm with new solution encoding and operators. In: Computational Science and Its Applications – ICCSA 2014, Proceedings, Part II, Cham, pp. 92–103 (2014)
18. Schildt, H.: The Annotated ANSI C Standard American National Standard for Programming Languages–C: ANSI/ISO 9899–1990. Osborne/McGraw-Hill, Berkeley (1990)
19. Scrucca, L.: Genetic algorithms for subset selection in model-based clustering, pp. 55–70. Springer, Cham (2016)

20. Si, M., et al.: MT-MPI: multithreaded MPI for many-core environments. In: Proceedings of the 28th ACM International Conference on Supercomputing, ICS 2014, pp. 125–134. ACM, New York (2014)
21. Suthaharan, S., et al.: Labelled data collection for anomaly detection in wireless sensor networks. In: 2010 Sixth International Conference on Intelligent Sensors, Sensor Networks and Information Processing (ISSNIP), pp. 269–274, December 2010
22. Vanbinst, K., Ceulemans, E., Ghesquière, P., Smedt, B.D.: Profiles of children's arithmetic fact development: a model-based clustering approach. J. Exp. Child Psychol. **133**, 29–46 (2015)
23. von Borries, G., Wang, H.: Partition clustering of high dimensional low sample size data based on values. Comput. Stat. Data Anal. **53**(12), 3987–3998 (2009)
24. Whitley, D.: A genetic algorithm tutorial. Stat. Comput. **4**(2), 65–85 (1994)

EEG Based Feature Extraction
and Classification for Driver Status Detection

P. C. Nissimagoudar$^{(\boxtimes)}$, Anilkumar V. Nandi, and H. M. Gireesha

B.V.B. College of Engineering and Technology, Hubballi, Karnataka, India
pcngoudar@gmail.com

Abstract. Driver status determination is one of the important features present in today's automotive. EEG analysis based driver status indication is one of the effective ways to measure driver status. This paper discusses about EEG analysis using wavelet transforms for separating EEG signal frequencies and extracting time and frequency domain features for further classification. EEG is a non-stationary signal and analysis only in time or frequency domain is not preferred. Wavelet transforms analyze the signals in both time and frequency domain. EEG rhythms consisting of different frequency bands are separated using Daubitius DB8 wavelet transform. Sleep data sets from Physionet were used for the proposed study. The drowsy status is indicated with alpha and theta frequency rhythms. The features representing alpha and theta activity were extracted and can be used to classify the driver status. The statistical features in time and frequency domain are used classify alert and drowsy state of the driver. Variants of SVM models were used to classify the signals and cubic SVM is found to give highest classification accuracy of 93.9%. The proposed method can be used to analyze driver status and further to analyze different sleep stages.

Keywords: EEG analysis · Wavelet transforms · Feature extraction ·
Classification · Driver drowsiness

1 Introduction

There are wide range of biomedical signals Electrocardiogram (ECG), Electroen-cephalogram (EEG), Electromyogram (EMG), Electroneurogram (ENG), Electro-Occulogram, (EOG) Vibroarthogram (VAG), Phonocardiogram (PCG) and Vibromyogram (VMG) signals. These signals represent the collective electric activity happening inside the human organ. The information is in the frequency, amplitude, and phase and also as a function of time. The techniques to analyze biomedical signals such as electrocardiogram, electroencephalogram are significantly contributing in health care systems. Amongst these biomedical signals, EEG is one the most exploited signal used to analyze human brain status. Examination of EEG signals is found to be important in diagnosing brain health disorders and for cognitive analysis of brain. In spite of EEG signals providing great deal of information about brain status, research in analysis and evaluation of EEG signals is limited. Therefore there is a need for automatic analysis and evaluation of EEG signals. Classification techniques can be used to differentiate

© Springer Nature Switzerland AG 2019
A. Abraham et al. (Eds.): IBICA 2018, AISC 939, pp. 151–161, 2019.
https://doi.org/10.1007/978-3-030-16681-6_15

EEG segments and to decide whether a person is healthy. A challenge is for extracting representative features and selecting significant features [1].

The electroencephalography (EEG) recording is an electrical activity in the brain generated by nerve cells. EEG parameters and patterns indicate the health or state of the brain. The electrical activities within are of mainly two types. Based on the reason that arouses potential change, the electrical activity can be classified as spontaneous brain electrical response and evoked potential response. The potential change taking place in the brain without any external stimulus is called spontaneous brain electrical response and potential change happening with external stimulus like sound, light and electricity is called evoked potential response. Both potentials have information related to brain state and can be analyzed for various conditions of brain [2].

1.1   EEG Signals Characteristics and Behavior

EEG signals are mainly classified based on different frequency bands as alpha, beta, theta, gamma and delta waves. Each band is characterized by its shape, head distribution and symmetry property and it is specific certain age and brain state [9]. Figure 1 represents different frequency bands of EEG signals [3].

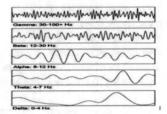

Fig. 1. EEG frequency bands

1.2   EEG Analysis Stages

An EEG analysis can be broadly divided into four stages. The first stage brain signal acquisition, second stage is preprocessing, third stage is about feature extraction and fourth stage is about classification of EEG information. The brain activity measurement or signal acquisition phase captures the brain signals using EEG electrodes. After acquiring the EEG signals, the information which is contaminated with various artifacts has to be preprocessed without losing the EEG information. During the next stage of feature extraction, EEG signals extracted after removing the noise. In the last stage which is EEG signal classification involves classifying EEG signals according to user's intention. In this stage very good classifying techniques have to be used to interpret the information [8].

In EEG the information of interest is in both time and space domain and is usually overlapped. The extraction of information is highly challenging and requires efficient techniques. Various techniques are discussed in the literature such as Principal Component Analysis (PCA), Independent Component Analysis (ICA), Fast Fourier Transforms (FFT), Auto Regression (AR), Wavelet Transforms (WT), Wavelet Packet

Decomposition (WPD) and also some simple techniques like band pass filters [20]. Feature extraction stage of EEG analysis requires the data of interest to be separated efficiently involving less computations and less time. Once the desired features are extracted the classification of information is the next step. Various training and machine learning techniques can be used for this process. The most common techniques are Artificial Neural Network (ANN), k-Nearest Neighbor (k-NN), Linear Discriminant Analysis (LDA) and Support Vector Machine (SVM). The performance of the classification technique depends on the type and size of the data sets so, these techniques cannot be comparable. The selection of the technique depends on the type of analysis we are performing and on data sets.

The paper discusses about EEG analysis and feature extraction methods with various significant features used to classification EEG signals for driver state monitoring.

2 Methodology

This section discusses about the methodology used for analysis of EEG signals for feature extraction. The analysis of EEG signals is demonstrated with the following schematic diagram, EEG data from Physionet website is collected and stored in the required format for further analysis. The schematic in Fig. 2, shows EEG analysis stages.

Fig. 2. Schematic of the EEG signal analysis

Filtering and Noise Removal: The raw EEG data is expected to be noisy and non-stationary, so the data is filtered using band pass filter with the frequency band between 0.5 Hz to 90 Hz to remove DC shifts and other artifacts. This process also helps in minimizing the epoch limits. A 50 Hz notch filter is applied to remove 50 Hz line noise [4].

Epoch Extraction (Splitting the Data): Each wakeup and sleep data is separated in to different time regions as epochs for processing. Epochs considered can be continuous, separated or overlapping [5].

Rhythm Isolation: EEG data after filtering and epoch selection has to be separated in to different frequency regions namely delta (0.5–4 Hz), theta (4–8 Hz), alpha (8–12 Hz), beta (12–30 Hz), and gamma (1–40 Hz). As EEG being non stationary signal, FFT is not suited. So, wavelet transforms (Daubechies db8) which translates the information into both frequency and time domain is chosen [20].

Feature Vector Construction: Features were computed both in frequency and time domain for theta and alpha bands, as these two bands represent the drowsy condition.

Spectral activity of gamma, theta, alpha, theta and delta were calculated using Fourier transforms. The features like absolute powers, relative powers and feature obtained from wavelet coefficients represent the frequency domain features.

Classification: Parameter estimation of algorithm is to be done to measure the optimization of algorithm in terms of computational ability, speed, size of the code and power consumed. Fusion of EEG information with other related information like heart rate, driver movement to confirm the driver status; which can be realized using other existing sensors in the vehicle [17].

2.1 Data Sets

The EEG data, in particular drowsy or sleep data is obtained in the form of European Data Format (EDF) from Physionet database. The data is recorded at, Bob Kemp, Sleep Centre, MCH-Westeinde Hospital, Den Haag, The Netherlands. The data consists of 61 polysomnograms (PSGs) in the form of *PSG.edf files and its related hypnograms (annotations of sleep stages by experts) in the form of *Hypnogram.edf files. *PSG.edf files are whole-night recordings of polysomnographic sleep recordings consisting of EEG from Fpz-Cz and Pz-Oz locations of electrodes, the eye movement recordings Electrooculogram (EOG), submental chin movement Electromyogram (EMG) and also contain oro-nasal respiration and rectal body temperatures. *Hypnogram.edf files contain annotations of sleep stages corresponding to PSGs. These files contain sleep stages W, R, 1, 2, 3, 4, REM, M (Movement time) and? (Not scored). Well-trained technicians have manually scored these data according to according to the 1968 Rechtschaffen and Kales manual [5], but based on Fpz-Cz/Pz-Oz EEGs instead of C4-A1/C3-A2. The EEG and EOG signals are sampled at 100 Hz [6]. The downloaded .edf files have the data for entire sleep stage, for classifying between drowsy or initial sleep stages and wake-up stage the data required is of W stage and S1 stage and is to be separated. The data is separated with annotations through Polyman software and stored in separate files. The *PSG.edf and *Hypnogram.edf files for wake-up stage W and drowsy stage S1 are shown in Fig. 3.

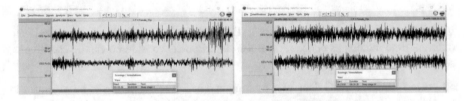

Fig. 3. EEG signal for sleep stage S1 and wakeup stage W

The PSG database from Physionet contains the EEG recordings from Fpz-Cz, and Pz-Oz electrodes, only Fpz-cz electrode information is used in this experimentation. This is because EEG electrode positioned at prefrontal region gives ease placement of electrode for driver.

Preprocessing: EEG data is usually contaminated with various artifacts and is non stationary in nature. So the data needs to be filtered appropriately. Basically two types of filters are used in preprocessing stage, band pass filter of 0.5 Hz to 60 Hz, which is FIR filter. A band reject notch filter is used to remove 50 Hz power line nose [19].

2.2   EEG Frequency Band Isolation

After removal of noise signals, the EEG information needs to be separated in to five different frequency bands, namely δ, θ, α, β and γ. From the literature [7, 18, 19] wavelets are an alternative and efficient methods used for analyzing EEG signals apart from conventional FFT analysis methods. Wavelets provide time-frequency domain analysis and are more suited for non-stationary signals like EEG. In the current work the discrete wavelet transform is used to recognize different brain rhythms.

2.3   Feature Extraction

Feature extraction and selection is the most important step in EEG signal analysis and classification as it determines the performance of the classifier. The features were extracted from EEG signals divided into smaller segments. The sleep stage 1 and wake up data obtained from physionet is further divided in to segments of two minutes to extract the features.

Frequency domain features: Spectral activity of gamma, theta, alpha, theta and delta were calculated using Fourier transforms. The features like absolute powers, relative powers and feature obtained from wavelet coefficients represent the frequency domain features [15].

The absolute power was computed for all frequency bands of EEG signal ranging from delta (0.5–4 Hz), theta (4–8 Hz), alpha (8–12 Hz), beta (12–30 Hz), and gamma (1–40 Hz) as Delta power (Pδ_abs), Theta power (Pθ_abs), Alpha power (c), Beta power (Pβ_abs) and Gamma power (Pγ_abs). From absolute powers relative powers were estimated using following expressions,

$$P\delta\_rel = P\delta\_abs/Ptotal\_power$$

Relative power is one of the important features as it represents contribution of individual frequency band to the overall EEG power [12].

Time domain features: The statistical features like mean, variance, complexity measures (Sample entropy, kurtosis and skewness) are determined from time domain analysis.

Mean: It determines the mean of selected band EEG amplitude. It is one of the common features in time domain and is represented as,

$$mean(\mu) = \frac{1}{N}\sum_{n=1}^{N} x_n$$

Variance: It is another common time domain statistical feature represented as

$$var = \frac{1}{N} - 1\sum_{n=1}^{N}(x_n - \mu)^2$$

Standard Deviation: It is represented as $std(\sigma) = \sqrt{\frac{1}{N-1}\sum_{n=1}^{N}(x_n - \mu)^2}$

Skewness: It's a measure of asymmetry of the probability distribution of a random variable (real valued) about its mean.

$$skew = \frac{\frac{1}{N}\sum_{n=1}^{N}(X_n - \mu)^3}{\sigma^3}$$

Kurtosis: Kurtosis is measure of fourth order cumulative or peakness of probability distribution and is represented as, $kurt = \frac{\frac{1}{N}\sum_{n=1}^{N}(X_n-\mu)^4}{\sigma^4}$

Sample entropy: It is one of the very important features used to assess the complexity of time domain physiological signals.

Features were computed both in frequency and time domain for theta and alpha bands as these two bands represent the drowsy condition.

2.4   Classification

Once the desired features are extracted the classification of information is the next step. Various training and machine learning techniques can be used for this process. The most common techniques are Artificial Neural Network (ANN), k-Nearest Neighbor (k-NN), Linear Discriminant Analysis (LDA) and Support Vector Machine (SVM). For our experimentation we used different variants SVM to classify the signal as drowsy or awake state and the performance is evaluated in terms of sensitivity and confusion matrix [13].

3   Experimental Results

A set of experiments were conducted on the data sets mentioned in previous section to extract and select features. Variants of SVM were tested for classification performance. It is observed that cubic SVM gives the best accuracy of 93.9%. The experiments were conducted on MATLAB (version R2018a on a computer with 3.40 GHz Intel core i7 CPU processor machine with 8.00 GB RAM.

3.1   EEG Frequency Band Isolation Using Daubechies Db8 Wavelet Transform

To separate EEG signals in to five frequency bands, db8 Daubechies wavelet is used. The procedure is as follows,

- Data is divided into smaller segments of two minutes for both S1 and wakeup state.
- Data is sampled at 500 Hz, Wavelet function of type Daubechies db8 with 8 level decomposition is applied.

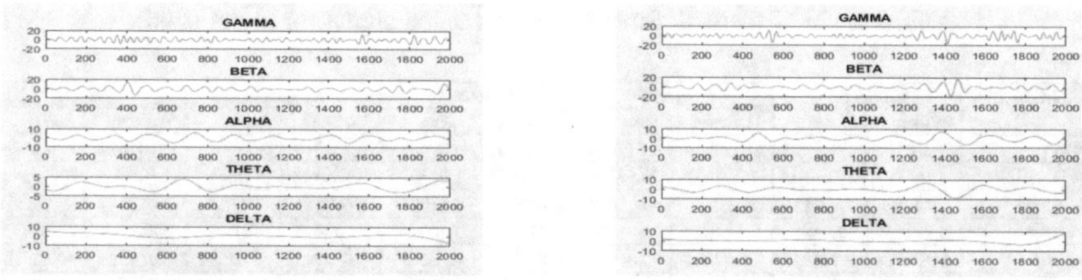

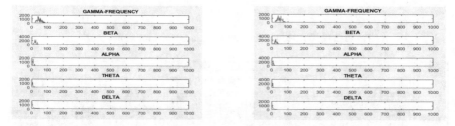

Fig. 4. (a) Time domain wake up EEG rhythms; (b) Time domain sleep S1 EEG rhythms

Fig. 5. (a) Frequency domain wake up EEG rhythms; (b) Frequency domain sleep S1 EEG rhythms

- Level 4 to 8 decomposition represents gamma, beta, alpha, and theta and delta frequency bands and time domain; shown in Fig. 4. FFT is applied to represent EEG rhythms in frequency domain; shown in Fig. 5.

EEG rhythm extraction is an important phase in EEG signal analysis. All the EEG frequency bands (gamma to delta) are extracted using db8 wavelet function. Wavelet gives the signal representation in both frequency and time domain.

3.2 Time and Frequency Domain Feature Extraction

Time domain and frequency domain features are computed for delta, theta, alpha, beta and gamma frequency bands. For drowsiness estimation features related alpha and theta are found to be significant from the literature.

The result of all above mentioned features in previous section were computed for wakeup signal and sleep S1 data of alpha band as shown in Table 1. The columns represent the five subjects under test. The rows represent the features in the order, power, rms, skewness, kurtosis, mean, variance, standard deviation and sample entropy.

Table 1. EEG features for wakeup and sleep signal for alpha band

EEG Features	Subject-1		Subject-2		Subject-3		Subject-4		Subject-5	
	Wakeup	Sleep	Wakeup	Sleep	Wakeup	Sleep	Wakeup	Sleep	Wakeup	Sleep
Power	0.026111	0.026772	0.020024	0.077216	0.042382	0.074054	0.111316	0.167313	0.102479	0.168996
RMS	1.908477	1.565503	1.671277	2.658682	2.431468	2.603679	3.940552	3.913613	3.780894	3.933244
Skewness	-0.6195	1.840115	0.25942	-0.31432	0.002596	0.110231	0.096461	0.0089	0.057031	-0.15066
Kurtosis	4.838732	12.73815	2.644618	4.168931	2.328073	3.449427	4.074994	6.795037	2.841757	4.736413
Mean	0.473318	0.345502	0.024087	-0.07439	0.048139	-0.08667	0.062993	-0.02086	0.018573	-0.00813
Variance	3.419964	2.332594	2.793984	7.066587	5.912677	6.775021	15.53175	15.32359	14.30196	15.47808
Std. Deviation	1.849314	1.527283	1.671521	2.658305	2.4316	2.602887	3.941034	3.914536	3.781794	3.934219
Samp. Entropy	-10.6774	-12.0549	-10.2649	-9.49959	-10.8913	-11.6681	-12.1937	-11.6889	-11.8077	-10.7179

Table 2. EEG features of four second epochs for alpha band

EEG Features	Alpha Band											
	Wakeup						Sleep					
	0-4 seconds	4-8 seconds	8-12 seconds	12-16 seconds	16-20 seconds	Std. Deviation	0-4 seconds	4-8 seconds	8-12 seconds	12-16 seconds	16-20 seconds	Std. Deviation
Power	0.020024	0.029988	0.060115	0.035492	0.050758	0.014404	0.077216	0.058137	0.069539	0.065704	0.062326	0.006512
RMS	1.671277	2.109561	2.898867	2.818324	2.38688	0.455363	2.658682	2.290725	2.506014	2.122087	2.460573	0.184776
Skewness	0.25942	0.000545	-0.06797	0.193364	0.180285	0.124755	-0.31432	-0.17951	-0.51793	0.207035	-0.13894	0.238049
Kurtosis	2.644618	2.347049	4.351925	2.626961	2.193178	0.778536	4.168931	3.055496	3.983001	2.72716	2.345284	0.708696
Mean	0.024087	0.088386	-0.04451	0.080443	-0.02487	0.053685	-0.07439	-0.01886	0.014422	-0.01464	0.117758	0.063341
Variance	2.793984	4.444655	8.405649	7.940447	5.699428	2.108932	7.066587	5.24969	6.28304	4.505288	6.043571	0.879958
Std. Dev.	1.671521	2.108235	2.89925	2.81788	2.387347	0.455447	2.658305	2.29122	2.506599	2.122566	2.458368	0.184402
Samp.Ent.	-10.2649	-9.3148	-9.22256	-9.30548	-10.1374	0.453748	-9.49959	-8.24406	-8.38183	-8.38269	-9.35312	0.538443

EEG features for different epochs representing four seconds of data is computed as shown in Table 2 for alpha and theta bands for both sleep and wake up data. From the results it is observed that there is comparative variation in sleep and wakeup data and similarity in features for different epochs over the range the same is shown Fig. 6(a) and (b) for alpha band.

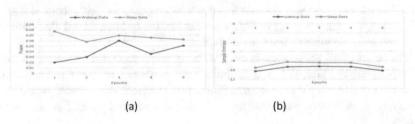

(a)   (b)

Fig. 6. Plot of EEG features wakeup Vs. Sleep S1 (a) Relative power (b) Sample entropy for different epochs

The features of wake up and sleep S1 are plotted against each other to compare the behavior. Frequency domain features like relative power and relative power and time domain features like skewness, kurtosis and sample entropy show the distinguishing behavior for wakeup and sleep stages and found to be important for classification.

3.3 Classification Study with Different Variants of SVM

Support Vector Machine (SVM) is a supervised learning technique used to analyze the data for classification and or regression. For a set of data the algorithm categorizes the data whether it belongs to one or the other among two categories. Technically SVM uses hyper plane to separate the data. Every data that is sampled is matched with two categories, conscious and drowsy before being reflected into a high dimensional space using kernel function. The model is created to assign a new sample to one of the two categories. The hyper plane in the high dimensional space divides the samples and accordingly hyper planes are located. The SVM maximizes the distance between these two parallel hyper planes. A greater distance or difference between parallel hyper planes indicates a smaller total SVM error rate. SVM is very commonly used in EEG analysis as it is simpler to implement and robust against high dimensionality. In our study we experimented with different models of SVM which included linear, quadratic, cubic, medium Gaussian and Course Gaussian SVM techniques. Cubic SVM is found to give highest accuracy of 93.9%. Figure 7 shows accuracy levels of different SVM technique and confusion matrix for the cubic SVM.

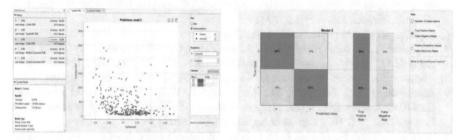

Fig. 7. Classification using different models of SVM and confusion matrix for the model cubic SVM

The performance of the classification technique depends on the type and size of the data sets so, these techniques cannot be comparable. The selection of the technique depends on the type of analysis we are performing and on data sets. Two main factors can be considered for selecting classifier are curse of dimensionality and bias variance trade off. Curse of dimensionality refers to the size of training data needed to give good results increases exponentially with the dimension of the feature vectors, and bias-variance tradeoff refers to high bias with low variance.

4 Conclusion

In this paper, the statistical features in time and frequency domain are used to classify the drowsy i.e. S1 and awake signals. The EEG signals are separated using wavelet transforms into different frequency bands and the significant bands alpha and theta are used to analyze the driver state. Wavelet technique offers the advantage of interpreting signals both in time and frequency domain. The statistical features are used to classify signals using different SVM model; from the study it is observed that cubic SVM gives the classification accuracy level of 93.9% compared to other SVMs. This method can help to determine the drowsy driver condition in automotive. The method also can be applied to analyze different sleep stages.

References

1. Rangayyan, R.M.: Biomedical Signal Analysis. Wiley, Hoboken (2002)
2. Tompkins, W.J.: Biomedical Digital Signal Processing. Prentice-Hall, Upper Saddle River (1995)
3. Sun, Y., Ye, N., Wang, X., Xu, X.: The research of EEG analysis methods based on sounds of different frequency. In: IEEE/ICME International Conference on Complex Medical Engineering, pp. 1746–1751 (2007)
4. Kumar, J.S., Bhuvaneshwari, P.: Analysis of electroencephalography (EEG) signals and its categorization-a study. In: International Conference on Modeling, Optimization and Computing (ICMOC 2012). Elsevier Publications (2012)
5. Rechtschaffen, A., Kales, A.E.: A manual of standardized terminology, techniques and scoring systems for sleep stages of human subjects. UCLA Brain Information Service. Brain Research Institute, Los Angeles, 10 (1968)
6. https://physionet.org/physiobank/database/sleep-edfx/
7. da Silveira, T., de Jesus Kozakevicius, A., Rodrigues, C.R.: Drowsiness detection for single channel EEG by DWT best m-term approximation. Res. Biomed. Eng. 31(2), 107–115 (2015)
8. Blinowska, K., Durka, P.: Electroencephalography (EEG). Wiley, New York (2006)
9. Aboalayon, K.A.I., Faezipour, M., Almuhammadi, W.S., Moslehpour, S.: Sleep stage classification using EEG signal analysis: a comprehensive survey and new investigation. Entropy 18, 272 (2016). https://doi.org/10.3390/e18090272
10. Ilyas, M.Z., Saad, P., Ahmad, M.I.: A survey of analysis and classification of EEG signals for brain-computer interfaces. In: 2nd International Conference on Biomedical Engineering (ICoBE), 30–31 March 2015, Penang (2015)
11. Chun-Lin, L.: A tutorial of the wavelet transforms, February 2010
12. Awais, M., Badruddin, N., Drieberg, M.: A hybrid approach to detect driver drowsiness utilizing physiological signals to improve system performance and wearability. Sensors 17, 1991 (2017). https://doi.org/10.3390/s17091991
13. Mandal, B., Li, L., Wang, G.S., Lin, J.: Towards detection of bus driver fatigue based on robust visual analysis of eye state. IEEE Trans. Intell. Transp. Syst. 18(3), 245–557 (2017)
14. Hossan, A., Kashem, F.B., Hasan, Md.M., Naher, S., Rahman, Md.I.: A smart system for driver's fatigue detection, remote notification and semi-automatic parking of vehicles to prevent road accidents. 978-1-5090-5421-3/16/©2016 IEEE (2016)
15. Huang, Y.-P., Sari, N.N., Lee, T.-T.: Early detection of driver drowsiness by WPT and FLFNN models. In: 2016 IEEE International Conference on Systems, Man, and Cybernetics, SMC 2016, 9–12 October 2016, Budapest, Hungary (2016)
16. Putra, A.E., Atmaji, C., Utami, T.G.: EEG-based microsleep detector using microcontroller. In: 2016 8th International Conference on Information Technology and Electrical Engineering (ICITEE), Yogyakarta, Indonesia (2018)
17. Khunpisuth, O., Chotchinasri, T., Koschakosai, V., Hnoohom, N.: Driver drowsiness detection using eye-closeness detection. In: 2016 12th International Conference on Signal-Image Technology & Internet-Based Systems (2016)

18. Kim, D., Han, H., Cho, S., Chong, U.: Detection of drowsiness with eyes open using EEG based power spectrum analysis. 978-1-4673-1773-3/12/©2013 IEEE (2012)
19. Wang, R., Wang, Y., Luo, C.: EEG-based real-time drowsiness detection using Hilbert-Huang transform. In: 2015 7th International Conference on Intelligent Human-Machine Systems and Cybernetics (2015)
20. Yu, Y.-H., Lai, P.-C., Ko, L.-W., Chuang, C.-H., Kuo, B.-C., Lin, C.-T.: An EEG-based classification system of passenger's motion sickness level by using feature extraction/selection technologies. 978-1-4244-8126-2/10/ ©2010 IEEE (2010)

Security Challenges in Semantic Web of Things

Sanju Mishra[1(✉)], Sarika Jain[2], Chhiteesh Rai[3], and Niketa Gandhi[4]

[1] Department of Computer Applications, Teerthanker Mahaveer University,
Moradabad, India
sanju.tiwari.2007@gmail.com
[2] Department of Computer Applications, National Institute of Technology,
Kurukshetra, Haryana, India
jasarika@nitkkr.ac.in
[3] Department of Computer Science and Engineering,
Institute of Technology and Management, Gorakhpur, India
chhiteesh.rai@gmail.com
[4] Machine Intelligence Research Labs (MIR Labs), Auburn, WA, USA
niketa@gmail.com

Abstract. Internet of Things (IoT) is expressed by heterogeneous technologies at the system level in various application domains. It proposes the ability to connect billions of resources, devices and things with each other on the Internet. The formed data and devices are mainly applied to design area-centric Internet of Things (IoT) applications. These applications are not successfully interoperable with each other from a data-centric perspective. The major challenges are to reuse, exploit, combine and interpret sensor data to guide the different user or different machines in designing inter-domain IoT applications. 'Semantic Web of Things' (SWoT) is a current exploration area targeting to assimilate Semantic Web-based technologies with the Internet of Things. It can also be considered as a transformation of the Web of Things (WoT) by incorporating semantics. Semantic Web of Things (SWoT) targets the ability to exchange and use information among data and ontologies. In this situation, the achievement of security and privacy needs plays an elementary role. The high number of interconnected devices arises scalability, heterogeneity and several interoperability issues; hence an adjustable infrastructure is required, which can handle security threats in such type of environment. This paper represents the main research challenges and security incidents of SWoT.

Keywords: Security · Privacy · Internet of Things · Semantic Web of Things · Ontologies

1 Introduction

Semantic Web (SW) technologies have been represented as tools to improve the ability to exchange and use information and information processing in everywhere computing. These technologies are largely employed to design interoperable IoT applications. Semantic approaches provide efficient searches for data from complex and voluminous data sets. The Semantic Web technologies and Internet of Things (IoT) prototype are combining more and more towards the so-termed Semantic Web of Things (SWoT)

© Springer Nature Switzerland AG 2019
A. Abraham et al. (Eds.): IBICA 2018, AISC 939, pp. 162–169, 2019.
https://doi.org/10.1007/978-3-030-16681-6_16

[1, 2]. It empowers semantic-based universal computing by enclosing intelligence into general objects and environments by a huge number of independent micro-devices, each transmitting a little volume of information. Heterogeneity, diversity, and dynamicity of devices, data and networks are among the biggest challenges hampering the worldwide acceptance of IoT technologies. Semantic technologies have been effectively applied in several domains, specifically, to indicate the heterogeneity challenge to (i) ease the association of such data, (ii) infer new knowledge to design smart applications and (iii) provide interoperability at data storage, processing, and management. To achieve these challenges, privacy and security play a significant role. These requirements incorporate data integrity, confidentiality, trustworthiness, access control and authentication within the network of the Internet of Things, privacy and security between things and users, and the enforcement of privacy and security policies. Such security requirements can be accomplished by assimilating semantics and reasoning engine in insecure devices. For integrating semantics and reasoning in IoTs, Semantic Web languages (RDF, OWL, and SPARQL) play a significant role. Ontologies are considered as a backbone of the semantic web. Various security ontologies have been developed to improve the security of SWoT. The semantic web and its network need devices to be able to understand each other: it should be semantically interoperable. Ontology technology can facilitate a formalized and unified description to resolve the issues of semantic heterogeneity, scalability, flexibility, and interoperability in the IoT security domain.

The remaining paper is constructed as follows. Section 2 represents the existing work. After studying the literature, Sect. 3 discusses and identifies several technical challenges and interoperabilities of SWoT. Section 4 focuses the security challenges of the semantic web faces to be compliant with the IoT constraints. Finally, in Sect. 5 concludes the paper and proposes some perspectives for the future of SWoT.

2 Literature Review

OWASP IoT25 [3] describes the concern to enhance security in IoT by categorizing the topmost 10 vulnerabilities viz. insufficient authorization/authentication, privacy concerns, insecure network services, insecure cloud interface, insecure mobile interface, insecure web interface, lack of transport encryption, insufficient security configure ability weak physical security and insecure software/firmware.

OneM2M [4] can be observed as security attacks that can occur in the OneM2M architecture and the associated panacea. In OneM2M, the authors describe the significance of securing network communications and M2M applications.

Linked Open Vocabularies for IoT (LOV4IoT) is considered as a dataset assigning almost 300 ontology-based projects in several domains significant for IoT applications such as weather, healthcare, smart home, agriculture, etc. [5]. Further, these domain ontologies are categorized according to their status, such as not available, online and following best practices.

Borgohain et al. represent the fundamental security problems specific to RFID and wireless sensor networks [6]. Authors do not propose security problems for other technologies associated with M2M or IoT such as web applications, cellular technologies etc.

Cheng et al. describe privacy prevention process to secure M2M devices, processing, and storage of M2M data, M2M communications [7].

Bandyopadhyay et al. suggested the urgency to secure IoT architectures at the development phase [8]. Alam et al. outline the requirement of security reasoning for the Internet of Things through semantic rules and ontologies [9]. Authors frame security needs for it such as authentication, integrity, trustworthiness, confidentiality, access control, authorization etc., but do not guide which security processes should be combined in the secure Internet of Things (IoT) applications.

SPITFIRE represented the SWoT concept to incorporate Semantic Web (SW) technologies to the Web of Things [1]. SPITFIRE linked the Web with the real world of things by introducing concepts, software infrastructure, and methods. SPITFIRE implemented tools such as LD4Sensors for connecting sensor data, smart-service-proxy, Web-based Task Assignment and execution (WebTAsX), gateway connection mapper and visualizer server.

OpenIoT is termed as an open-source IoT platform implementing the semantic interoperability of IoT services in the cloud [10]. OpenIoT represented the "Sensing-as-a-Service" principle by combining cloud frameworks with the Internet of Things applications providing IoT service formation and deployment.

IntellegO is considered as a semantic concept approach to reason and interprets on sensor data [11]. It employs an abductive logic infrastructure and Parsimonious Covering Theory (PCT) to translate data based on an "ontology of perception". The creation and reuse of the background knowledge acquired for translating data is a critical process and it is not incorporated in this work.

CityPulse is primarily focused on real-time intelligence and large-scale analysis to obtain meaningful perceptions and knowledge from heterogeneous data streams [12]. It works on bridging the gap between the real world data streams and application technologies on the Web of Thing.

VITAL is considered as a platform that concentrates on lessening the cost of deployment and development of smart city applications [13, 14] by employing the functionality to interrelate the IoT sensor collected data with legacy data by using semantics.

An overview [15] is provided for the evolution from the IoT to the WoT and SWoT. This paper is mainly focused on the role standards in interoperability and the integration of semantic web technologies in standards.

A study of the semantic web technology roles in the IoT and challenges is presented in [16]. This work evaluates the Semantic Web of Things by presenting its unique levels to attempt an Internet of Things convergence in terms of device abstraction, heterogeneous device integration, and various semantic descriptions, existing for the IoT.

3   Technical Challenges and Interoperability in SWoT

One of the challenging issues with the existing IoT applications is that used devices are little interoperable or not interoperable with each other since the data of these devices is based on predefined formats and do not employ general terms or a vocabulary to explain interoperable data of Internet of Things. Semantic web technologies have already presented their importance in other domains than IoT. IoT interoperability is still in its infancy phase. Semantic web technologies can be employed in IoT to overcome the challenge in dealing with interoperability of data produced by devices and already employed in real-life. Sindice15 and Swoogle17 are the Semantic search engines but not enough prepared for obtaining relevant domain ontologies for the Internet of Things [1].

Authors of [17] propose a mapping of LDP to CoAP30, based on UDP31. CoAP is a protocol, specially designed for constrained applications, with reduced headers and limiter packet body. Such initiative allows IoT nodes to be connected to the LOD, and therefore to extend the WoT, while respecting the constraints of IoT nodes. In an attempt to resolve interoperability concern, various IoT networks are being associated with the WoT, and semantic web technologies and principles are applied to raise the WoT into the SWoT. In order to be able to analyze the existing IoT architectures introduced by the SWoT society, researchers discussed LMU-N, a unifying IoT architectural platform issued from a bottom-up analysis.

Interoperability is considered as the ability of two or more systems to interchange data and user information. This feature provides several challenges on how to obtain the information, use the information and exchange data in understanding it and being able to process it [18].

Information is available everywhere on the web. Information is typically data that make sense. Several interoperability challenges are shown in Fig. 1.

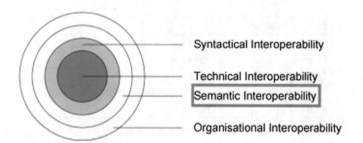

Fig. 1. Types of interoperability [18]

Technical interoperability that employs heterogeneous hardware and software (e.g., heterogeneity of communication protocol).

Syntactical interoperability that employs data formats (e.g., XML or JSON). It is also an issue for reusing and combining and semantic datasets or ontologies designed with unique software handling with unique syntaxes (e.g., XML/RDF, N3).

Semantic interoperability that employs (1) heterogeneity of ontology, (2) terms used to express data (3) the meaning of content interchanged according to the context. This is significant to later translate Internet of Things data and constructs interoperable semantic-based Internet of Things applications.

Organizational interoperability employs heterogeneity of the various infrastructures.

Primary challenges that are notified in Semantic Interoperability and needs research in future include but not limited to:

• Data Exchange and Data Modelling and
• Ontology matching/merging & alignment
• Semantic Annotation for Data/Event
• Knowledge Representation and associating ontologies
• Sharing of Knowledge
• Knowledge Consistency Revision
• Reasoning & Analysis

The more accurate model can be designed by the semantic analysis of data in the SWoT for diagnosing the privacy and security issues. These issues are significant in different areas like Information Processing, Gathering, Invasion and Dissemination. This section disparages the distinct security challenges represented in a SWoT environment and simplifies issues from a security aspect that are connected with computing applications, devices, and networks. The concepts of security and privacy are identified and certain security requirements are noted. In IoT applications, there are several categories for incidents and types of incidents. These incidents are discussed in Table 1.

Table 1. Listing of SWoT security incidents

Incident category	Incident type
Information security	Unauthorized deletion/access/updation
Fraud	Wastes of resources
Information collection	Phishing, sniffing
Malware	Infection, undetermined
Availability	DoS/DDoS

Information Security (Unauthorised Access) consists the majority of security problems for access control in SWoT applications. Various access control frameworks are there that are formed upon SW languages (RDF, OWL, SPARQL) for preventing their resources from unauthorized access. SW languages support enforcement policies and designs access control policies. For restricting access to RDF data, the IS category also provides encryption techniques. As encryption is considered as the cornerstone of information security, hence authorized encryption has been getting enough attention in the SWoT context to promote integrity and confidentiality. Lightweight and fast encryption plays a significant role for acquiring the adapted security and privacy for IoT. Ontologies [19–21] are considered as an interesting area for IS as a basis for detecting and analyzing security issues. Ontologies are generally applied as representation schema, knowledge bases, or an annotation vocabulary. Infection analysis or malware detection considered as knowledge-intensive security issues. Fraud,

Information Collection, Malware and availability is another common security area such as phishing or SPAM.

4 Security Challenges of Semantic Web with IoT Constraints

Internet of Thing has several security issues, it may arise as a variety of damage and threaten human lives and properties. To enhance the abilities to provide emergency response, monitoring, and predicting the designing trends of IoT security. This paper highlights some security challenges that need to be fixed such as:

Confidentiality
SWoT services can have sensitive data; so for protecting the data, IoT connected objects data should be stored confidentially. By the encryption process, one can achieve confidentiality. For ensuring confidentiality there are several existing asymmetric and symmetric encryption schemes. However, choosing of a specific type of encryption is extremely device and application ability dependent. For example, consider a healthcare environment that maintains the information about the patient's activity at the hospital. The doctor will never wish that anyone in the hospital will access the data by monitoring devices of patient's activity.

Integrity
IoT services interchange crucial data with other services and also with the third parties that keeps forward rigorous demand that stored, sensed and transferred data should not be damaged either accidentally or maliciously. To protect the integrity of sensor data is critical for designing dependable and reliable SWoT applications. It is assured, with message authentication codes (MAC) applying one-way hash functions. The choice of MAC technique also depends on device and application capabilities. For example, a smart home that is linked with the smart grid. For producing electricity bill, this smart grid provides a monitoring of electricity consumption. The provider never welcomes that the consumed data can be damaged during the transmission period.

Availability
Our anticipated SWoT environment may consist of sensor node hosted services. Hence, it is highly significant that these services exist from everywhere at any time in order to produce semantic-based information. To satisfy this property, there is no security protocol. However, various pragmatic measures may be accepted to assure the availability.

Authentication
It assigns to the means applied for the identity verification. In SWoT term, mutual authentication is needed because SWoT data are applied inactuating processes and decision making. Therefore, the service consumer and the service provider need to be ensured that the service is approached by authentic service and the user is provided by an authentic source. Applying any authentication process needs to register the identities of user and resource limitation of SWoT objects pose restrict constraints to empower the techniques of authentication.

Authorization

It assigns to the means of describing the access policies that usually appoint certain privileges to subjects. The SWoT environment requires facilitating reuseable, fine-grained, dynamic, updating, and easy to use policies describing the mechanism. Hence, it is important to externalize the definition of policy and enforcement process of SWoT services.

Access Control

This is considered as an enforcement process that permits access to the resources for only authorized users. The enforcement is generally based on the outcomes of the access control. It is highly significant to reveal users' data only to the authorized parties.

Trustworthiness

Several applications that are delicate in nature, such as health care services, safety critical services, need to analyze trustworthiness of various entities indulged. From a SWoT application perspective, analyzing trustworthiness of sensors data and sensor is significant. Non-trustworthy sensor data or malicious sensor nodes can lead to a disaster in a safety critical place. Untrusted sensor data can enter from a trusted sensor node. Non-trustworthy nature can have two reasons: unintentional errors and intentional misbehavior. It can be easier to assure trustworthiness of SWoT by incorporating trustworthiness analysis.

5   Conclusion

This paper discusses semantic web technologies with IoT systems. The evolution of web of thing is combined with semantic web technologies for making a secure semantic web of things (SWoT). A deep survey is represented in this paper and has highlighted the gaps in the domain. Security is an important feature of all applications. Many several security challenges such as confidentiality, availability, integrity, trustworthiness, authentication, and authorization are discussed. Several SWoT security incidents are represented in a table. The focus was also on interoperability of IoT applications that will be incorporated in future work. The integration of semantic technologies with adapted machine learning algorithms may provide a smarter IoT in the future.

References

1. Pfisterer, D., Romer, K., Bimschas, D., Kleine, O., Mietz, R., Truong, C., Hasemann, H., Kroller, A., Pagel, M., Hauswirth, M.: SPITFIRE: toward a semantic web of things. IEEE Commun. Mag. **49**(11), 40–48 (2011)
2. Scioscia, F., Ruta, M.: Building a Semantic Web of Things: issues and perspectives in information compression. In: Proceedings of the 3rd IEEE International Conference on Semantic Computing Semantic Web Information Management (SWIM 2009), pp. 589–594. IEEE Computer Society (2009)

3. https://www.owasp.org/index.php/OWASP_Internet_of_Things_Project
4. OneM2M TR-0007-V2.3.0: Study of Abstraction and Semantics Enablements. OneM2M, MAS Working Group 5 (2015)
5. Gyrard, A., Serrano, M., Atemezing, G.: Semantic web methodologies, best practices and ontology engineering applied to Internet of Things. In: Proceedings of the IEEE World Forum on Internet of Things (WF-IoT), 14–16 December 2015, Milan, Italy (2015)
6. Borgohain, T., Kumar, U., Sanyal, S.: Survey of security and privacy issues of internet of things. arXiv preprint arXiv:1501.02211 (2015)
7. Cheng, Y., Naslund, M., Selander, E., Fogelstrom, G.: Privacy in machine-to-machine communications a state-of-the-art survey. In: IEEE International Conference on Communication Systems (ICCS), pp. 75–79 (2012)
8. Bandyopadhyay, D., Sen, J.: Internet of things: applications and challenges in technology and standardization. Wirel. Pers. Commun. 58(1), 49–69 (2011)
9. Alam, S., Chowdhury, M., Noll, J.: Interoperability of security-enabled internet of things. Wirel. Pers. Commun. 61(3), 567–586 (2011). https://doi.org/10.1007/s11277-011-0384-6
10. Soldatos, J., Kefalakis, N., Hauswirth, M., Serrano, M., Calbimonte, J.P., Riahi, M., Aberer, K., Jayaraman, P., Zaslavsky, A., Zarko, I.: OpenIoT: open source internet-of-things in the cloud. In: Interoperability and Open-Source Solutions for the Internet of Things, pp. 13–25. Springer (2015)
11. Henson, C.: A semantics-based approach to machine perception. Wright State University (2015)
12. Barnaghi, P., Tonjes, R., Holler, J., Hauswirth, M., Sheth, A., Anantharam,P.: CityPulse: real-time IoT stream processing and large-scale data analytics for smart city applications. In: Europen Semantic Web Conference (ESWC). Springer (2014)
13. Petrolo, R., Loscri, V., Mitton, N., Soldatos, J., Hauswirth, M., Schiele, G.: Integrating wireless sensor networks within a city cloud. In: Eleventh Annual IEEE International Conference on Sensing, Communication, and Networking Workshops (SECON Workshops), pp. 24–27 (2014)
14. Petrolo, R., Loscri, V., Mitton, N.: Towards a smart city based on cloud of things, a survey on the smart city vision and paradigms. Trans. Emerg. Telecommun. Technol. 28(1) (2015)
15. Barnaghi, P., Wang, W., Henson, C., Taylor, K.: Semantics for the Internet of Things: early progress and back to the future. Int. J. Semant. Web Inf. Syst. 8(1), 1–21 (2012)
16. Jara, A.J., Olivieri, A.C., Bocchi, Y., Jung, M., Kastner, W., Skarmeta, A.F.: Semantic Web of Things: an analysis of the application semantics for the IoT moving towards the IoT convergence. Int. J. Web Grid Serv. 10(2–3), 244–272 (2014)
17. Seydoux, N., Drira, K., Hernandez, N., Monteil, T.: Capturing the contributions of the semantic web to the IoT: a unifying vision. In: 16th Semantic Web Technologies for the Internet of Things Workshop Colocated with ISWC-2017 (2017)
18. Patel, K.K., Patel, S.M.: Internet of things-IOT: definition, characteristics, architecture, enabling technologies, application & future challenges. Int. J. Eng. Sci. Comput. 6(5), 6122–6131 (2016)
19. Mishra, S., Jain, S.: Ontologies as a semantic model in IoT. Int. J. Comput. Appl. 40, 1–11 (2018)
20. Mishra, S., Jain, S.: A unified approach for OWL ontologies. IJCSIS 14(11), 747–754 (2016)
21. Mishra, S., Malik, S., Jain, N.K., Jain, S.: A realist framework for ontologies and the semantic Web. Procedia Comput. Sci. 70, 483–490 (2015)

A Theoretical Approach Towards Optimizing the Movement of Catom Clusters in Micro Robotics Based on the Foraging Behaviour of Honey Bees

K. C. Jithin and Syam Sankar[(✉)]

Department of Computer Science and Engineering, NSS College of Engineering,
Palakkad, Kerala, India
jithinkc22@gmail.com, syam.sankar8@gmail.com

Abstract. Claytronics is a future technology in the field of artificial intelligence and modular robotics to create programmable nanoscale robots (catoms). Each catom acts like a computer. It has the ability to adopt different shapes. Many methods are being used for the cluster formation of catoms and are mainly based on the message passing techniques. The energy consumed and the time delays taken in cluster formation are depending on the number of messages sent between catoms and the topology factor. In this paper, we report a new optimisation algorithm for cluster formation in claytronics. The proposed methodology is based on the foraging behaviour of the honey bee swarm. The algorithm introduces many user-defined functions for optimising the operations in cluster formation. It also minimises the number of message transfers among catoms and chooses a maximum topology factor value to form a particular shape. The paper theoretically analyses the operation of cluster formation and makes a comparison with previous methods. The analysis shows that our work optimises the energy and message transfer in the catom-cluster formation.

Keywords: Catom clusters · Claytronics · Rejoin algorithm ·
Micro robotics · Honeybee swarm · DPRSim

1 Introduction

In Claytronics, the programmable matter may organize itself into the form of an object and renders its outer surface to match with the visual look of the same item [3]. Claytronics deals with the design of 3D objects from individual parts referred to as catoms. The movement of catoms in a particular direction generates an object of the desired shape. Creation and updating of a physical replica of an arbitrary moving 3D object in real time involve many challenges [4]. The distributed system containing millions of computers (catoms) synchronizes state information continuously to attain an actual shape gradually [2]. Catoms can

© Springer Nature Switzerland AG 2019
A. Abraham et al. (Eds.): IBICA 2018, AISC 939, pp. 170–181, 2019.
https://doi.org/10.1007/978-3-030-16681-6_17

form an object through proper communication by sending messages in order to determine their relative location and orientation [7]. Based on this information, they will form a distributed network and organizes themself into a structure [3].

2 Related Works

2.1 Dynamic Simulation

DPRSim [14] is a software program developed by a research team assembled by Carnegie Mellon University and Intel. It enables researchers to create models of claytronic ensembles. The simulator allows us to program and control the performance of individual catoms. The simulated world of DPRSim manifests an environment of catoms and its characteristics. In this environment, every catom owns a mailbox manager with many mailboxes received from every other catoms [8]. DPRSim message unit is designed in adherence with the feature of catoms and it does not have a catom recipient address [8]. Each catoms broadcasts its mailbox which contains the actual message. Once the mailbox is received by some other catoms, it will be treated as low-level feature interface and also as a low-level message. However, at the same time, they are directly processed by the corresponding code modules. The following steps describe the whole processing of DPRSim [8].

1. Obtaining the present position of all catom from the Open Dynamic Engine (ODE) [8] library and computing the neighbouring lists.
2. The individual processing of every catoms.
3. Change the states of magnets according to the data received from Step 2.
4. Advancing physical simulation- here catoms move in particular direction an forms a shape.

Multiple physical simulations are possible for ODE. Once Step 4 is started, it remains uninterruptible as long as it finishes.

2.2 Rejoin Algorithm

The microrobots estimate their position for supporting rejoin operations [9]. The catom area unit is used for map building functions (structural part formation). The sensor data from catom walkers facilitates the formation of structures. Rejoin algorithm [10] will execute after the catom walkers finish the map building task and thereby enabling central ensemble to get the knowledge about position and map and finally algorithm reconnect all catoms to form a desired object [5].

The Fig. 1 shows [10] three catom clusters, each with some number of catoms, placed at an equal distance. In order to form a bigger cluster, they need to have some communication by sending messages. Several kinds of messages are exchanged between catoms for the transmission of relevant information like position, neighbouring data, cluster data etc. [6]. The messages are listed below [10]

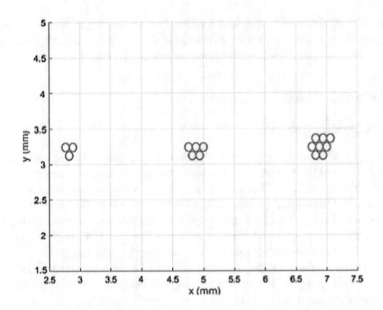

Fig. 1. Three catom clusters at equal distance

- *helloMsg()* is the basic message sent by catoms to notify their presence to other catoms in the surroundings.
- *reqMsg()* is sent by a router which has already received a *helloMsg()*. This is for requesting the movement of catoms belonging to smaller clusters to bigger clusters.
- *perMsg()* is a permission message to move and join when a reqMsg is received. When a router receives a *perMsg()* it starts to move toward the position of the bigger cluster.
- The functions like *moveToPosition()* and *connect feature()* are primitives to connect two clusters through their features.
- If the number of catoms is smaller than 3, *helpMsg()* is sent. When a router receives *helpMsg()*, it designates one of them as catom walker which then joins with the nearby catom.

3   Proposed Catom Cluster Join Method

Our proposed algorithm works based on the communication model followed in the foraging behavior of honeybees [1]. In analogous to selecting a good food source by the bees, our work finds the best possible catom structure by having proper cluster formation. The structural information is obtained from a tool called Linear Distributed Predicate (LDP) [2]. It continuously checks different structural information based on the distributed pattern matching algorithm and distributed watchpoint algorithm [6]. The input to our optimization algorithm is the output of the LDP. The total number of catoms can be initially divided equally into two, namely, employed and onlooker bees (or employed and onlooker catoms), and one of the employed bees acts as the scout bee (or scout catom) [1]. The scout catom plays an important role in the cluster formation. It collects the structural information from the LDP and then passes it to the employed catoms. The employed catoms then update its structural information and select the best structure based on the topology factor (a measure of connectedness). The best structure information is passed to onlooker catoms. Then onlooker catoms choose the nearest employed catoms to join according to the structural information. The ultimate role of an onlooker catom is to join with the employed catom to form a catom cluster. And after joining, all the catoms in the cluster act as employed catoms. The optimization in join operations can be done by considering two important parameters like topology factor and number of messages sent between catoms [13]. The optimization efficiency is assessed based on how much we reduce the total time delay and the energy consumption rate of the catoms [11]. The proposed algorithm works based on some conditions:

- If N is even, $V = N/2$ and $U = (N/2 - 1)$
- If N is odd, $V = (N - 1)/2$ and $U = (N - 1)/2$
- The number of structures will be same as number of employed catoms.
- Scout bees collect structural information from LDP.

Here, N is the total number of catoms or catom clusters, U is the total number of onlooker catoms and V is the total number of employed catoms.

3.1   Join Algorithm

In the algorithm, the variables a_i and a_j indicate the addresses of the ith and j^{th} catoms, p_i and p_j shows the positions of ith and j^{th} catoms, α_i, and α_j are the topology factors of corresponding structures, n_i is the number of catoms in the cluster and t_i is the communication time delay experienced by the catom [10]. The algorithm has the efficiency to select a structure with highest topology factor from all possible structures received from LDP. Our algorithm uses the terminologies (variables and functions) suggested by an existing rejoin algorithm [10]. We focus on improving its performance in terms of number of messages sent between catoms and energy consumption.

Algorithm 1: Catom Cluster join

Procedure: Catom_Cluster_join( )
initialization;
N=total no: of catoms/cluster;
if *N is odd* **then**
| V=(N-1)/2;
| U=(N-1)/2 ;
end
if *N is even* **then**
| V=N/2;
| U= (N/2 -1);
end
Random_selection(1);
Random_selection(V);
Random_selection(U);
if *N≤2* **then**
| $p_j \leftarrow$ selectCatom();
| rescuer←selectClosestCatom(p_j);
| enableRouting(rescuer);
| sendCatomWalker (rescuer, p_j);
else
| scout_catom();
end
End

Algorithm 2: scout_catom

Procedure:scout_catom ()
if *V > 1 and $n_i \neq$ N-1* **then**
| **while** *i = 1 to V* **do**
| | $(a_i, p_i, t_i, \alpha_i, n_i) \leftarrow$ Collect_structure ();
| | employed_catom $(a_i, p_i, t_i, \alpha_i, n_i)$;
| **end**
else
| $p_j \leftarrow$ selectCatom();
| rescuer←selectClosestCatom(p_j);
| enableRouting(rescuer);
| sendCatomWalker (rescuer, p_j);
end
End

Algorithm 3: employed_catom

Procedure: employed_catom$(a_i, p_i, t_i, \alpha_i, n_i)$
if $U > 1$ and $V > 1$ **then**
  | nearest(a_i, p_i);
  | onlooker_catom $(a_i, p_i, t_i, \alpha_i, n_i)$;
end
if $U = 0$ and $V > 1$ **then**
  | **while** $V \neq 1$ **do**
  | | $(a_i, p_i, t_i, \alpha_i, n_i) \leftarrow$ Collect_structure ();
  | | Topology(α_i, a_i, p_i);
  | | nearest(a_i, p_i);
  | | V=V-1;
  | | employed_catom$(a_i, p_i, t_i, \alpha_i, n_i)$;
  | **end**
end
if $U == 0$ and $V == 1$ **then**
  | $p_j \leftarrow$ selectscoutCatom();
  | re$\leftarrow$selectClosestCatom(p_j);
  | moveToPosition(re);
  | connectFeatures();
else

end
End

Algorithm 4: onlooker_catom

Procedure: onlooker_catom $(a_i, p_i, t_i, \alpha_i, n_i)$
Topology(α_i, a_i, p_i);
SelectCatom(a_j);
moveToPosition(p_i);
connectFeatures();
U=U-1;
End

The algorithm introduces some new user defined functions as listed in the Table 1.

Our optimization algorithm to perform rejoin operations among catoms or catom clusters comprises mainly four phases, detailed as follows.

Initial Phase: In this phase, we need to calculate the total no of catoms/clusters that have to join and the variable N is assigned with that number. The algorithm calculates the values of U and V and it also chooses one of the employed catoms as scout catom (*Algorithm 1*). If the value of N is less or equal to 2, we have only two (or less than two) catoms or catom clusters to join.

Table 1. User defined functions of join algorithm.

Function	Meaning
Random selection ()	Classify all the available catoms into three (random selection) groups: Employed catoms, Onlooker Catoms and Scout catom. It also updates the group information like address and count of catoms
Collect structure ()	It will collect different structural information from LDP and updates the values of ai, pi, ti, ai and ni
Topology ()	It will collect the optimal level of topology factor from different possible structures. It outputs maximum topology factor and updates the corresponding structure information
Nearest ()	This function will select the nearest catoms information with respect to ai and pi

Scout Bee Phase: After identifying the employed and onlooker catoms, scout catom passes the information to corresponding employed catoms (*Algorithm 2*). At the end of whole join operations, the scout catom joins with the nearest catom cluster and forms a single catom cluster. When there is a failure in the scout catom, one of the employed catoms or clusters will be randomly selected as scout immediately and the values of U and V are re-calculated.

Employed Bee Phase: The employed catoms or clusters collect both the structural information and distance measurement (of onlooker catoms) and passes this data to onlookers (*Algorithm 3*).

Onlooker Bee Phase: The onlooker catoms or clusters check the topology factor of the corresponding structure with nearest employed catoms (*Algorithm 4*). They select the structure with the highest topology factor and simply join with the employed catoms or clusters. After this phase, we have only employed catoms or clusters. They start communication by sending messages to perform join operations further.

4   Results and Discussion

In this part, we analyze the performance of both the re-join [10] and the proposed catom cluster join algorithm. Consider the joining of three clusters, each having three, five and eight catoms, initially positioned at an equal distance of 2 mm as shown in Fig. 1. Cluster A comprises three catoms at coordinates (3, 3.5), cluster B consists of 5 catoms at coordinates (5, 3.5) and Cluster C comprises eight catoms at coordinates (7, 3.5) [10]. The function Random selection () is used to select the employed, onlooker and scout cluster in this case. The scout cluster gets the structural information with topology factors. One of the important parameters that we take into account is topology factor. It is measure of maximum connectedness among catoms in a cluster or structural arrangement of catoms in a cluster. The topology factor can be calculated as [10],

$$\alpha = \left\{ \begin{array}{ll} 1 & \varphi \leq n+1 \\ 0.8 & \varphi > n \\ 0.5 & \varphi < n-1 \end{array} \right\} \tag{1}$$

The most preferred values of topology factor are 1 and 0.8. The value of 0.5 indicates linear structure in most cases. The variable n indicates number of catoms.

$$No: of\ active\ features(\varphi) = No: of\ triangles(sides) - No: of\ shared\ sides \tag{2}$$

The Eq. (2) gives the number of active feature for further calculation [10]. The value of topology factor varies in accordance with the number of active features. The Fig. 2 shows one of the best structures obtained after joining the three catom clusters (A, B and C) and it consists of 33 active features. In this scenario, our algorithm finds the value of topology factor as $1(\alpha = 1)$ in order to perform join operations. The topology factor (α) affects the time delay in movement of catoms or clusters. The time delay in movement is calculated as [10],

$$t = (2.n.t_\varphi.n_d)/\pi.\alpha \tag{3}$$

Where t_φ is the time delay for feature activation [10]. Here, the distance (D) value is taken as 4 mm.

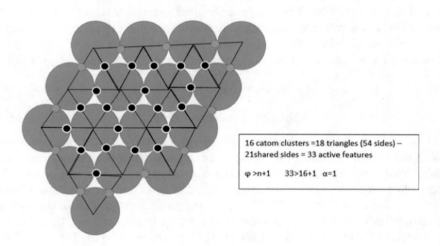

16 catom clusters =18 triangles (54 sides) –
21shared sides = 33 active features

φ >n+1   33>16+1 α=1

Fig. 2. Joining of three catom clusters

$$n_d = D/2r \tag{4}$$

$$t_\varphi = 60/\omega \tag{5}$$

$$\omega = v/r \tag{6}$$

Where r is the catom radius, ω is the angular velocity and v is the constant velocity (0.01 mm/ms). The total time delay, T, can be calculated as,

$$The\ total\ time,\ T = N * t \tag{7}$$

Where N is the number of clusters (here it is 3). The re-join algorithm and the proposed catom cluster join algorithm have the same topology factor ($\alpha = 1$). Hence the total time delay in movement is obtained as indicated in Table 2.

Table 2. Total time delay

	n_d	ω	t_φ	n	α	t	T
Re-join algorithm [10]	33.3	0.16666	360	16	1	61144.05 fs	183432.1562 fs
Catom cluster join algorithm (proposed)	33.3	0.16666	360	16	1	61144.05 fs	183432.1562 fs

Unlike the existing rejoin algorithm [10], the proposed algorithm has the efficiency to choose most perfect topology factor (preferably 1 or 0.8) based on the structural analysis. It is accomplished with the function Topology () as mentioned in the algorithm. The energy is the next parameter to be considered for performance analysis. Our algorithm attempts to reduce the energy consumption during cluster formation and thereby increasing the lifetime of catoms [12]. The energy used to send ($Epacket_{tx}$) or receive a message ($Epacket_{rx}$) is [10],

$$Epacket_{tx} = Nbits * W * Epul_{tx} \tag{8}$$

$$Epacket_{rx} = Nbits * Epul_{rx} \tag{9}$$

Where Nbits is the number of bits and its value is 116(constant message size), W is the weight of coding (W = 0.5) and the pulse transmission and receiving rate, $Epul_{tx} = 1pj$ and $Epul_{rx} = 0.0948pj$ respectively. The total energy is depending on the number of messages sent. The total number of messages sent by re-joining algorithm is 9 [10]. The only way to optimize or reduce the energy consumption is to decrease the number of message transmissions. The following Fig. 3 shows the communication between three catom clusters (A, B and C). The three catom clusters are randomly categorized as scout, employed and onlooker clusters [1]. The number of message transmissions between clusters based on our optimization algorithm is 5, as indicated in the Fig. 3. The description of each message transfer is given in the Table 3. The Main Ensemble Energy (MEE) can be calculated as follows [10]:

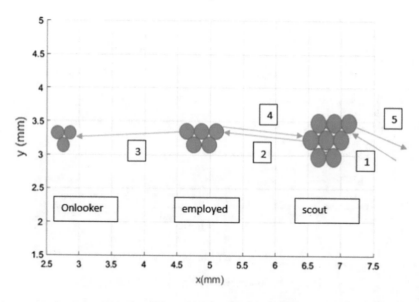

Fig. 3. communication between 3 clusters based on our optimization algorithm

Table 3. Messages and its description

Msg. no	Description
1	The scout cluster gets the structural information from LDP
2	Communicating the structural information to all employed clusters
3	Informing the nearest onlooker clusters about the different structural information and its corresponding topology factors
4	The onlooker cluster selects the structure with the highest topology factor and joins with the employed clusters The new cluster information is passed to the scout cluster
5	The employed cluster join with the scout cluster and the total cluster information is given back to DPRSim

$$MEE = (Epacket_{tx} + Epacket_{rx}) * Avg.msg \qquad (10)$$

Hence the total energy calculated as [10].

$$Totalenergy = MEE * Avg.msg \qquad (11)$$

The Fig. 4 shows the single cluster obtained after the joining of the three catom clusters. The amount of energy consumed during joining operation is affected by the number of message transmissions. The calculated value of total energy required is shown in Table 4.

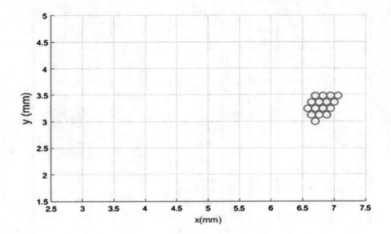

Fig. 4. The single cluster after joining

Table 4. Comparison with the total energy

	Total no: of msg	Total no: of clusters	Avg. msg	$Epacket_{tx}$	$Epacket_{rx}$	MEE	Total energy
Rejoin algorithm [10]	9	3	3	58	11	207pj	621pj
Catom cluster join algorithm (proposed)	5	3	1	58	11	68pj	68pj

The energy consumption rate is obtained as 621pj when we apply the existing rejoin method [10] to a group of three clusters, each having three, five and eight catoms. The same with the proposed one is calculated as 68pj. So the energy consumption is reduced by 89.05%.

5   Conclusion

Our work aims at optimizing the re-join operations in claytronics. The energy consumption rate and the time delay taken in re-join operations of catoms are high when applied with existing algorithms. The proposed algorithm optimizes the re-join algorithm by using the concept of foraging behavior of honey bees. The proposed approach decreases the number of messages sent between catoms for communication and thereby reducing the rate of energy consumption. As per the analysis, it is proved that the energy is reduced by 89.05% when the proposed join algorithm is applied to a group of three clusters, each having three, five and eight catoms.

Acknowledgment. We would like to thank Carnegie Mellon University, Intel, and all Claytronics groups that shared valuable information regarding the various concepts in Claytronics and also thankful to the Department of Computer Science and Engineering, NSS College of Engineering, Palakkad, for providing all the required facilities for doing the work in a systematic way.

References

1. Chen, H., et al.: Artificial bee colony optimizer based on bee life-cycle for stationary and dynamic optimization. IEEE Trans. Syst. Man Cybern.: Syst. **47**(2), 327–346 (2017)
2. De Rosa, M., Goldstein, S., Lee, P., Pillai, P., Campbell, J.: Programming modular robots with locally distributed predicates. In: 2008 IEEE International Conference on Robotics and Automation, Pasadena, CA, pp. 3156–3162 (2008)
3. Kalyani, V.L.: Claytronics is an unimaginable shape shifting future tech. J. Manag. Eng. Inf. Technol. (JMEIT) **2**(4), 21–29 (2015)
4. Lyke, J.C., Christodoulou, C.G., Vera, G.A., Edwards, A.H.: An introduction to reconfigurable systems. Proc. IEEE **103**(3), 291–317 (2015)
5. Jahanshahi, M.R., Shen, W., Mondal, T.G., Abdelbarr, M., Masri, S.F., Qidwai, U.A.: A brief review. Int. J. Intell. Robot. Appl. **1**, 287–305 (2017)
6. Piranda, B., Bourgeois, J.: A distributed algorithm for reconfiguration of lattice-based modular self-reconfigurable robots. In: 24th Euromicro International Conference on Parallel, Distributed, and Network-Based Processing (PDP), Heraklion, pp. 1-9 (2016)
7. Bourgeois, J., Goldstein, S.C.: Distributed intelligent MEMS: progresses and perspectives. IEEE Syst. J. (2013). https://doi.org/10.1109/JSYST.2013.2281124
8. Boillot, N., Dhoutaut, D., Bourgeois, J.: New applications for MEMS modular robots using wireless communications. IEEE Syst. J. **11**(2), 1094–1106 (2017)
9. Lakhlef, H., Bourgeois, J.: Fast and robust self-organization for micro-electro-mechanical robotic systems. Comput. Netw. (Part 1) **93**, 141–152 (2015)
10. Ferranti, L., Cuomo, F.: Nano-wireless communications for microrobotics: an algorithm to connect networks of microrobots. Nano Commun. Netw. **12**, 53–62 (2017)
11. Ackerman, E.: A thousand kilobots self-assemble into complex shapes. IEEESpectrum (2014)
12. Bouchard, S.: LineScout robot climbs on live power lines to inspect them. IEEE Spectrum Online (2010)
13. Bourgeois, J., Cao, J., Raynal, M., Dhoutaut, D., Piranda, B., Dedu, E., Mostefaoui, A., Mabed, H.: Coordination and computation in distributed intelligent MEMS. In: Proceeding of the 27th IEEE International Conference on Advanced Information Networking and Applications, Spain. AINA, pp. 118–123 (2013)
14. Intel Research Pittsburgh, Dprsim. http://www.pittsburgh.intel-research.net/dprweb/. 4.1, 5, 5.3

Sensor Data Cleaning Using Particle Swarm Optimization

Parag Narkhede$^{(\boxtimes)}$, Shripad Deshpande, and Rahee Walambe

Symbiosis Institute of Technology, Symbiosis International (Deemed University),
Pune 412115, India
parag.narkhede@sitpune.edu.in

Abstract. Sensors play an important role in the monitoring and decision making of autonomous systems. However, the sensor measurements
suffer from the addition of random noise due to various reasons. These
noisy observations affect the decisions taken by the sensory system.
Removal of noise is a critical aspect of any such system. This paper
presents an optimization based novel procedure for cleaning the noisy
sensor measurements. The proposed algorithm is designed with the help
of a well-known particle swarm optimization technique. The developed
algorithm is tested with the simulated standard data having different
noise intensities and validated with the help of real-time measurements
taken using the gyroscope sensor. Results show the feasibility of using
the optimization based algorithm for cleaning the sensory data.

Keywords: Data cleaning · Noise cancellation · Sensors ·
Particle swarm optimization

1 Introduction

In this technological era, autonomous systems are replacing the human-operated
systems. These systems require accurate knowledge of the work environment to
take proper decision. The important part of autonomous systems is their sensing
unit, usually known as a sensor. Different kinds of sensors are being employed
in the systems depending upon their use cases and applications. Sensors have
wide applications in the fields ranging from medical to military and agricultural
to industrial etc.

The sensor data usually suffers from errors, which can either be systematic
or random. The systematic errors are generally known as bias. Changes in the
operating conditions (e.g. temperature, moisture etc.) are the major reasons
for this type of error. Bias can be addressed by the process of calibration [1,
2]. However, bias estimation is out of the focus for this paper and hence not
discussed in details.

The random error observed in the measurement is known as a noise. Usually,
due to the noise, sensor observations are highly corrupted and distorted. The randomness in the measurements can be low quality of sensing units, environmental

© Springer Nature Switzerland AG 2019
A. Abraham et al. (Eds.): IBICA 2018, AISC 939, pp. 182–191, 2019.
https://doi.org/10.1007/978-3-030-16681-6_18

conditions or due to any arbitrary effect of external sources [3]. The unwanted voltage and current fluctuations in the measurement system, analog signal processing system, analog to digital conversion unit, or even the processing unit are the culprits for the presence of noise in measurement system [4]. The reasons for noise additions in the measurements are random and hence it may not be possible to model the noise accurately. Inaccurate noise model may lead to performance degradations of the measurement system. The performance of a sensor varies significantly depending upon its individual precision and accuracy, fault tolerance to the hardware faults etc. [3]. The noisy Global Positioning System (GPS) may lead to a decrease in the position estimation accuracy. Noisy heartbeat monitoring system may indicate the continuously fluctuated beats count, by using which a physician may not take corrective decisions. Disturbed vehicle orientation estimation system may lead to the wrong direction selection. The noisy temperature monitoring system may generate false alarms. These are some of the cases where noisy measurements may yield inappropriate decisions. Even, due to high fluctuations in the observations, the decision making may lead to misleading decision. Therefore, it can be stated that for efficient operation and decision making, error free sensor data is essential. Using the errorless data a query raised by the system can be addressed efficiently and effectively [5,6].

The removal of random noise or reduce the effect of noise on the system performance is a key solution for appropriate decision making [7], query handling, event detection, outlier detection etc. The process of noise reduction i.e. cleaning the sensory data is usually known sensor data cleaning or just a data cleaning [8].

Use of regression modeling, a well known technique of system modeling is used in the process of data cleaning. It is an offline process where the prior knowledge of the environment and sensor measurements is used to form a system model. The model is then used for cleaning the sensor data [8,9]. Moving average technique is a widely used and well-known procedure of data cleaning [10–12]. In this method, at any time instant t the moving average algorithm finds the average of last k samples to get the noiseless measurement $\hat{x}_t$ as $\hat{x}_t = \sum_{i=t-k}^{k} \frac{x_i}{k}$. When multiple sensors (let's say N) are monitoring the same environment, average over the neighbouring sensors (set $S(i)$) is computed to get more accurate measurement data $\hat{x}_t = \sum_{i=1}^{N} \frac{\hat{x}_i}{N}$. However, when deep valleys and/or heights are present in the data, moving average filter fails to estimate them accurately. To overcome this, a weighted moving average filter is used in [5,13]. In this method, each sample of the measurement is assigned with a weight such that the sum of all weights is unity. The weighted average is then computed to predict the noiseless measurement value.

Kenda et al. used the Kalman Filter (KF) based approach for the process of data cleaning [14,15]. KF is a two step procedure with steps performing prediction and correction of data in respective steps [16]. The use of the spatial and temporal relationship between the data has been studied in [17]. By using the machine learning technique, the missing values in the measurement sample are then predicted. The attribute correlation along with spatial-correlation theory has been applied for the data cleaning operation in [18]. Zhang et al. proposed an iterative method for repairing the sensor data which follows the minimum change principle [19]. However, most of these discussed techniques are applicable when there exist multiple sensors sensing the same environment.

In this paper, the use of an optimization technique namely Particle Swarm Optimization (PSO) is proposed to solve the problem of sensor data cleaning. PSO is evolved from the nature and architecture of bird searching for food. The algorithm is designed such that it computes a value from which the sum of the distances of all the sample points under consideration is minimum. The detailed algorithm is discussed in the further sections. The proposed algorithm is tested on the standard test signals and then validated with the real world data logged using the gyroscope sensor.

The complete paper is organized as follows: Sect. 2 discusses the PSO algorithm in detail. The proposed PSO based data cleaning procedure is elaborated in the methodology, Sect. 3. Section 4 describes the experimentation design along with the results description and Sect. 5 concludes the paper.

2   Particle Swarm Optimization

Kennedy and Eberhart [20] in 1995 came up with the new optimization technique which follows the structure of birds searching for their food in the multi-dimensional space. In this technique, each bird is considered as a particle and its motion/movement is monitored by its own best position achieved in the past as well as positions of its neighbors. The swarm of particles is generated using all the particles in which, each particle represents the solution for the optimization problem under consideration. It is an iterative process and in each iteration, the particles' characteristics (position and velocity) are updated. The movement of an individual swarm is control by taking into consideration the cost function, which is generally to be minimized. Since the invention, PSO has been successfully applied to solve multiple problems in the varied functional and multi-dimensional optimization domains. Each particle in the PSO is governed by its position vector pos, velocity vector vel, the best position achieved by the particle in the past $pBest$. Other than these local parameters, a global parameter, i.e. global best position $gBest$ which signifies the best position achieved amongst all the particles in the swarm. In each iteration, the velocity and position of the particles are updated using the Eqs. 1 and 2 respectively.

$$vel = w * vel + c_1 * r_1 * (pBest - pos) + c_2 * r_2 * (gBest - pos) \qquad (1)$$

$$pos = pos + vel \qquad (2)$$

Where w denotes the inertial weight influencing the search space. c_1 and c_2 are the acceleration parameters generally assigned with a constant value 2, whereas r_1 and r_2 are the random parameters distributed in the range of $[0, 1]$. When each particle reaches to global best position or the initially set value of the cost function, then PSO is said to be converged. The complete procedure of the PSO algorithm is as discussed Algorithm 1.

Algorithm 1. Particle Swarm Optimization

    Input: *swarm_size, max_iterations*, r_1, r_2, c_1, c_2
    Output: *gBest*
1 Initialize *pos* and *vel* for all particles
2 **repeat**
3   Compute *fitness_cost* of all particles
4   *gBest* = particle with minimum *fitness_cost*
5   **if** $fitness\_cost_{current} > fitness\_cost_{previous}$ **then** *pBest* $\leftarrow$ *pos*
6
7   compute *vel* and *pos* using the Eqs. 1 and 2
8 **until** *max_iterations*

---

3   Proposed Methodology

In this work, PSO is applied over the series of data to reduce the fluctuations observed in the sensor measurements. Cumulative difference between the single observation and the remaining observations in the considered window is considered as a cost function while applying the PSO. For example, in certain measurement the observations are $[b, c, d, e]$ then the value $|a-b|+|a-c|+|a-d|+|a-e|$ is minimized using PSO to obtain the appropriate $'a'$ as shown in Fig. 1.

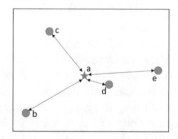

Fig. 1. Sample measurements optimization example

Here $'a'$ may or may not belong to the set of observations. However, PSO will assure that the obtained $'a'$ will be at minimum distance from all other observations available in the considered window. The proposed PSO based Sensor Data Cleaning (PSO-SDC) algorithm can be stated as in Algorithm 2.

The final *gBest* will give the expected point/value which is equidistant from all the observation samples under consideration. This single value will now be used to represent the window of noisy/fluctuated observations. The convergence property of PSO will make sure that the obtained *gBest* value is at the centre of all observations under considerations, maintaining a minimum possible distance from all observations in the window.

Algorithm 2. PSO-SDC Algorithm

Input: *swarm_size* ← number of observations to be considered for cleaning, N
    max_iterations
    r_1 and r_2 ← random number between [0,1],
    c_1 and c_2 ← constant value 2
Output: *gBest*
1 Initialize *pos* = *pBest* = ← observations to be considerd for cleaning
    vel ← *zero*
2 **repeat**
3   Compute $pBest\_cost(i) = \sum_{j=1}^{N} |pBest(i) - pBest(j)|$
4   **if** $pBest\_cost(i)_{current} < pBest\_cost(i)_{previous}$ **then** $pBest(i) \leftarrow pos(i)$
5
6   **if** $min(pBest\_cost) < gBest\_cost$ **then** $gBest = \underset{pBest}{\mathrm{argmin}}(pBest\_cost)$
7
8   compute *vel* and *pos* using the Eqs. 1 and 2
9 **until** *max_iterations*

---

4   Experimentation, Results and Discussion

Feasibility of the proposed PSO-SDC algorithm for data cleaning is checked in two steps. In the first step standard sinusoidal signals are generated and used for feasibility analysis. In the second phase, data is logged using the actual sensor-based system and analysis is carried out.

The ideal sinusoidal signal is added with a random noise having zero mean. In different experimental trials, the intensity of noise is varied and the PSO-SDC algorithm is applied. The comparison of the proposed estimates is made with the Kalman Filter estimates. A Kalman Filter model is designed and implemented for data cleaning application. Using the generated noisy data, offline simulations are carried out. Simulation results are shown in the below figures. In all the figures describing the results of the PSO-SDC algorithm, a certain color code is followed. A line in red color is used to show the noise-corrupted version of the ideal data and the cleaned data using Kalman Filter is indicated by green color. The data obtained using the PSO-SDC algorithm is represented by black color (dash line). For the quantitative validation of the algorithm, root mean square error (RMSE) is used. RMSE can be calculated mathematically using the equation in 3.

$$RMSE = \sqrt{\frac{1}{n} \sum_{1}^{n} (x_{reference} - x_{measured})^2} \qquad (3)$$

Where, n represents the total number of data points in the given sample of data, $x_{reference}$ is the reference data and $x_{measured}$ denotes the measured/estimated data.

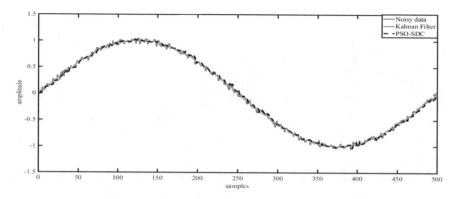

Fig. 2. Sinusoidal signal with low intensity noise

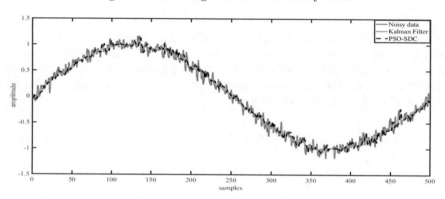

Fig. 3. Sinusoidal signal with medium intensity noise

The sinusoidal wave in Fig. 2 is corrupted with a low-intensity noise and PSO-SDC algorithm is applied. In this experimentation, RMSE (with respect to the ideal data) for cleaned data and that for noisy data are found to be 0.025 and 0.029 respectively. After the successful performance of the PSO-SDC algorithm on the low noise data, a medium intensity noise is added to the sinusoidal wave. On this medium noise data, data cleaning is performed using the proposed PSO-SDC algorithm and results are shown in Fig. 3. For the signal corrupted with medium noise, the RMSE obtained for cleaned data and noisy data are 0.055 and 0.086 respectively. When the reference sinusoidal signal is corrupted with a high-intensity noise, the RMSE values obtained for cleaned data and noisy data are found to be 0.283 and 0.492 respectively. The results are graphically shown in Fig. 4. The comparison of RMSEs with respect to the ideal data obtained for (i) generated noisy data, (ii) cleaned data using Kalman Filter and (iii) PSO-SDC cleaned data in all the three experimentation trials are shown in Fig. 5. It can be observed that in all the three experimentation trials, the proposed PSO-SDC algorithm outperformed the noisy data. At the same time, with a comparison

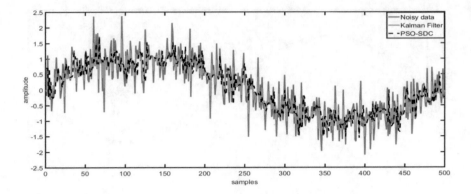

Fig. 4. Sinusoidal signal with high intensity noise

to Kalman Filter, the results are comparable. The Kalman Filter is designed and implemented for data cleaning applications. The filter parameters tuning is a tedious task and requires adaptive tuning in the dynamic environment. However, here for the comparison, parameters are tuned manually by trial and error method for the given set of data.

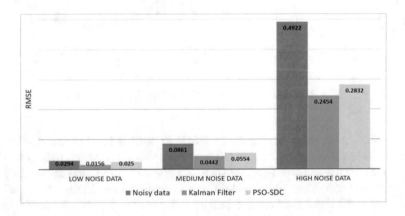

Fig. 5. RMSE comparison

In the second phase of the experimentation, the proposed PSO-SDC algorithm is validated with the help of real-time experimentation setup. A sensor-based system is developed using an Arduino Uno board. Arduino Uno is open source microcontroller-based platform widely used for fast prototyping. It is designed with the help of Atmel's ATmega328 microcontroller. In this trial, MPU6050 module is used to record the real-time data for offline simulation. It is an inertial measurement unit consisting of a triaxial gyroscope and triaxial accelerometer. However, in this validation process, only the gyroscope measurements along z-axis are considered. The gyroscope measures the angular velocity

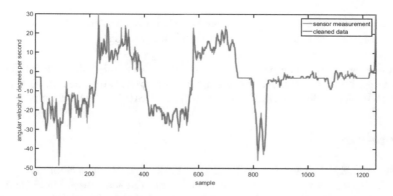

Fig. 6. Angular velocity measured using gyroscope

of the moving object it is attached to. The selection of the gyroscope sensor is done at random. Here, the angular velocity measured by gyroscope along the z-axis is logged and used for offline validation of the proposed PSO-SDC algorithm. While logging the data, random motion with random varying velocity is provided to the sensor module. Figure 6 shows the obtained results in graphical format.

From all these experimental results, it can be seen that the proposed algorithm is a feasible solution to clean the noisy sensor data and is also a useful algorithm for noise reduction from the data. Even though the comparison shown in Fig. 5 shows the results of KF better then PSO-SDC algorithm, the KF considerd here is properly tuned. After multiple trial and errors the best obtained results are considerd for comparison. The proposed algorithm does not require any pre-knowledge of the work environment nor the major initial configurations.

5 Conclusion

The paper addresses a critical issue of any sensory system. The sensors, whose measurements generally suffer from the addition of noise, are the most important source of information in autonomous systems. The issue of noise cancellation is addressed in this paper. The process of noise removal and predicting the smooth data is termed as data cleaning. A novel particle swarm optimization based procedure for cleaning the sensor observations is developed in this paper. The proposed algorithm is designed in a way that it tries to predict a value such that the sum of differences of other measurements from that value is minimum in the given window of observations. In this regards, the sum of differences is minimized using PSO. The proposed PSO-SDC algorithm is developed in MAT-LAB. Simulations are carried out for testing and validation of the algorithm for the desired objective of sensor data cleaning. Although the results are not optimal, they show the usefulness of the proposed algorithm for noise reduction. The forum is open for the researchers for further modifications and improvements. The accurate sensor measurements will lead to improving the efficiency of autonomous systems by appropriate decision making.

Acknowledgement. This work is sponsored by Symbiosis International (Deemed University), Pune, India under the research project grant sanctioned under the Minor Research Project Grant 2017-18 scheme vide number SIU/SCRI/MRP Approval/2018.

References

1. Bychkovskiy, V., Megerian, S., Estrin, D., Potkonjak, M.: A collaborative approach to in-place sensor calibration. In: Proceedings of the 2nd International Conference on Information Processing in Sensor Networks, IPSN 2003, pp. 301–316. Springer, Berlin (2003)
2. Rajan, R.T., van Schaijk, R., Das, A., Romme, J., Pasveer, F.: Reference-free calibration in sensor networks. arXiv preprint arXiv:1805.11999 (2018)
3. Elnahrawy, E., Nath, B.: Cleaning and querying noisy sensors. In: Proceedings of the 2nd ACM International Conference on Wireless Sensor Networks and Applications, pp. 78–87. ACM (2003)
4. Bordoni, F., D'Amico, A.: Noise in sensors. Sens. Actuators A: Phys. **21**(3), 17–24 (1990)
5. Tasnim, S., Pissinou, N., Iyengar, S.: A novel cleaning approach of environmental sensing data streams. In: 2017 14th IEEE Annual Consumer Communications & Networking Conference (CCNC), pp. 632–633. IEEE (2017)
6. Cheng, H., Feng, D., Shi, X., Chen, C.: Data quality analysis and cleaning strategy for wireless sensor networks. EURASIP J. Wireless Commun. Netw. **2018**(1), 61 (2018)
7. Abbasi, A.Z., Islam, N., Shaikh, Z.A., et al.: A review of wireless sensors and networks' applications in agriculture. Comput. Stand. Interfaces **36**(2), 263–270 (2014)
8. Aggarwal, C.C.: Managing and Mining Sensor Data. Springer, Heidelberg (2013)
9. Sathe, S., Papaioannou, T.G., Jeung, H., Aberer, K.: A survey of model-based sensor data acquisition and management. In: Managing and Mining Sensor Data, pp. 9–50. Springer, Heidelberg (2013)
10. Ali, A., Li, W., He, X.: Simple moving voltage average incremental conductance MPPT technique with direct control method under nonuniform solar irradiance conditions. Int. J. Photoenergy **2015**, 12 (2015)
11. Pires, I.M., Garcia, N.M., Pombo, N., Flórez-Revuelta, F., Spinsante, S.: Approach for the development of a framework for the identification of activities of daily living using sensors in mobile devices. Sensors **18**(2), 640 (2018)
12. Bassey, E., Whalley, J., Sallis, P.: An evaluation of smoothing filters for gas sensor signal cleaning. In: Proceedings of the Fourth International Conference on Advanced Communications and Computation, Paris, France, pp. 20–24 (2014)
13. Zhuang, Y., Chen, L., Wang, X.S., Lian, J.: A weighted moving average-based approach for cleaning sensor data. In: 27th International Conference on Distributed Computing Systems, ICDCS 2007, p. 38. IEEE (2007)
14. Kenda, K., Škrbec, J., Škrjanc, M.: Usage of the kalman filter for data cleaning of sensor data (2013)
15. Kenda, K., Mladenić, D.: Autonomous sensor data cleaning in stream mining setting. Bus. Syst. Res. J. **9**(2), 69–79 (2018)

16. Kalman, R.: A new approach to linear filter and prediction theory. Basic. Engr. D **82**, 35–45 (1960)
17. Kurasawa, H., Sato, H., Yamamoto, A., Kawasaki, H., Nakamura, M., Fujii, Y., Matsumura, H.: Missing sensor value estimation method for participatory sensing environment. In: 2014 IEEE International Conference on Pervasive Computing and Communications (PerCom), pp. 103–111. IEEE (2014)
18. Lei, J., Bi, H., Xia, Y., Huang, J., Bae, H.: An in-network data cleaning approach for wireless sensor networks. Intel. Autom. Soft Comput. **22**(4), 599–604 (2016)
19. Zhang, A., Song, S., Wang, J., Yu, P.S.: Time series data cleaning: from anomaly detection to anomaly repairing. Proc. VLDB Endowment **10**(10), 1046–1057 (2017)
20. Kennedy, J., Eberhart, R.: Particle swarm optimization. In: Proceedings of ICNN 1995 - International Conference on Neural Networks, vol. 4, pp. 1942–1948, November 1995

Comparative Analysis of Chaotic Variant of Firefly Algorithm, Flower Pollination Algorithm and Dragonfly Algorithm for High Dimension Non-linear Test Functions

Amrit Pal Singh[1][(✉)] and Arvinder Kaur[2]

[1] Bharati Vidyapeeth's College of Engineering, GGSIPU, New Delhi, India
amritpal1986@gmail.com
[2] University School of Information and Communication Technology, GGSIPU,
New Delhi, India

Abstract. Non-linear test functions are NP-Class problems. To solve them, Swarm Algorithms (SA) have been used in last two decades very effectively. In this work, three swarm based algorithms (i.e. Firefly Algorithm (FFA); Flower Pollination Algorithm (FPA) and Dragonfly Algorithm (DA)) have been used. Chaos is familiarized with swarm algorithm to improve their performance. As per our knowledge, most of the studies have applied chaos on one standard SA and compared it with other standard algorithm(s). No comparison has been shown among the chaotic variant of different algorithms. Comparison of Chaotic variants of FFA, FPA & DA with their standard algorithms has been performed using four high dimensions non-linear test functions on the basis of Mean fitness (i.e. $P1$) and convergence rate (i.e. $P2$). The results indicate that chaotic variant has performed better than standard and FFA evaluates best fitness for multi-modal function (i.e. $f3$ and $f4$).

Keywords: Swarm Algorithms · Firefly Algorithm ·
Flower Pollination Algorithm · Dragonfly Algorithm · Chaos theory ·
Non-linear test functions

1 Introduction

Various Swarm Algorithms (SA) have been applied to problems classified as NP-class and aim to find the optimal solution for these problems [1,2]. Some well known Swarm Algorithms are Particle Swarm Optimization (PSO) [3], Firefly Algorithm (FFA) [4], Bat Algorithm (BA), Krill Heard (KH) [5], Flower Pollination Algorithm (FPA) [6], Dragonfly Algorithm (DA) [7] etc. In this paper, out of well known algorithms three algorithms (i.e. Firefly algorithm (FFA); Flower Pollination Algorithm (FPA) and Dragonfly Algorithm (DA)) have been chosen on the basis of their performance in previous studies [2,7–11,26]. FFA has been

© Springer Nature Switzerland AG 2019
A. Abraham et al. (Eds.): IBICA 2018, AISC 939, pp. 192–201, 2019.
https://doi.org/10.1007/978-3-030-16681-6_19

proposed by Yang [4]. It is inspired by the flashing behavior of fireflies. FPA uses pollination concept of flower's and proposed by Yang [25]. However, DA has been proposed by Mirjalili [7]. It is based on collective movement of dragonflies, which has basically two functional domains i.e. hunting and migrating.

Chaos optimization is familiarized with swarm algorithms to free them from converging to a local optimum [12]. Chaos is a basic characteristic of nonlinear systems having series of specific features such as regularity, ergodicity and randomness [13]. In most of the current research the random sequence is substituted with the chaos sequence for effective results [13,14].

Non-linear test functions have been used by various studies [2,9,10,15–18] to test the performance of optimization algorithms. The study by Jamil et al. [19] has collected various unconstrained test functions on the basis of modality (i.e. uni-modal & multi-modal) and separability (i.e. separable & no-separable). Multimodal functions (a function with more than one local optima) are amongst the most difficult problems for numerous algorithms. Another group of test problems is drafted by separable and nonseparable functions. For a function of p variables to be a separable, it is required that it should be written as the sum of p functions of just one variable. While if its variables show interrelations among themselves or are not independent it is called inseparable [19].

This paper opts for FFA, FPA & DA as they have not compared among themselves with their chaotic variants. These algorithms have considered because of the following reasons: FPA can search in both local and global search; DA has a set of equations and variables which are useful when the population of the swarm is massive and collisions are unacceptable; and the study [8] has shown that FFA has performed better than PSO. In this work, FFA, FPA, DA, Chaotic FFA, Chaotic FPA and Chaotic DA have been compared using four non-linear test functions on the basis of two performance measure (i.e. Mean fitness $(P1)$ and Convergence rate $(P2)$). The four test functions are considered on the basis of their four categories: non-separable & uni-modal $(f1)$; separable & uni-modal $(f2)$; non-separable & multi-modal $(f3)$; and separable & multi-modal $(f4)$ as described in Sect. 3.3.

2   Related Work

Table 1 demonstrates previous work based on the three swarm algorithms (i.e. FFA, FPA & DA) along with chaotic variants. [S1]–[S7] show that standard algorithm's chaotic variant has been compared with standard algorithm. However, [S6] shows that chaotic variant has been compared with other standard algorithms. In [S1] & [S2] PSO has been used. Whereas [S3] and [S5] have worked on FFA. FPA and its chaotic variant is implemented in [S4] & [S7]. However, [S6] study focused on DA. [S5] & [S7] have used flavors of chaotic maps and analyze which chaotic map performed well for particular algorithm.

Previously comparisons have been performed among the standard algorithms and chaotic variant proposed in study (i.e. [S1]–[S7]). As per our knowledge, different algorithms chaotic variant have not compared among themselves. This

Table 1. Related work of chaotic variant of FFA, FPA & DA

Sr. No.	Authors and year	Algorithms used
S1	Song et al. [20]	Tent map chaotic PSO
S2	Hongwu [21]	PSO; Chaotic PSO
S3	Yang [22]	FFA; Chaotic FFA
S4	El-henawy et al. [23]	FPA; Chaotic FPA
S5	Gandomi et al. [11]	FFA; Twelve Chaotic FFA
S6	Sayed et al. [26]	DA; Chaotic DA; PSO; ABC; GWO
S7	Kaur et al. [9]	FPA; Eleven Chaotic FPA

work focuses to compare the chaotic variants of FFA, FPA & DA with their standard algorithms.

3 Research Methodology

Various methods to conduct the research have been presented here: Sect. 3.1 is used to explain chaos theory; Swarm algorithms and their learning equation has been presented in Sect. 3.2; Non-linear test functions, performance measure and experimental setup have been presented in Sects. 3.3, 3.4 and 3.5 respectively.

3.1 Chaos Theory

Chaos theory has three characteristics: non-asymptotically periodic; bounded; and positive Lyapunov exponent [24]. In this paper, chaos has been used with FFA, FPA & DA to bind the swarm agent's value in a search space. Let us consider logistic chaotic map as defined in Eq. 1. Where z_t is chaotic sequence at t time stamp.

$$g(z_t) = \mu z_t (1 - z_t) \qquad \text{where } 0 < \mu \leq 4 \tag{1}$$

3.2 Swarm Algorithms Used

As per discussion three Swarm Algorithms are used i.e Firefly Algorithm (FFA); Flower Pollination Algorithm (FPA); and Dragonfly Algorithm (DA) in this paper. Their detailed description can be seen from their bases papers. However, this section discuss characteristics of base algorithms with its learning equations and our updated equations using logistic chaotic map. Table 2 describe the learning equations of standard algorithms and their respective chaotic variant. Updated chaotic equations (as shown in Table 2) used learning equation to introduce the chaotic disturbance in x_i^t.

Table 2. Learning equations of algorithms

Sr. No.	Algorithm	Learning equation	Chaotic equation[1]
1	FFA [4]	$x_i^{t+1} = x_i^t + \beta o e^{-\gamma r_{i,j}^2}(x_j^t - x_i^t) + \alpha\epsilon_i$ Where, $r_{i,j}$ is Cartesian distance; second term (i.e. $\beta o e^{-\gamma r_{i,j}^2}(x_j^t - x_i^t)$) is due to the attraction; third term (i.e. $\alpha\epsilon_i$) is randomization with α being the randomization parameter	$x_i^{t+1} = g(x_i^{t+1})$
2	FPA [25]	$x_i^{t+1} = x_i^t + L*(x_i^t - g^*)$ (GLOBAL POLLINATION) $x_i^{t+1} = x_i^t + \epsilon\left(x_j^t - x_k^t\right)$ (LOCAL POLLINATION) Where, L is levy's distribution and switch probability $p = 0.8$ is used to switch between global and local pollination	$x_i^{t+1} = g(x_i^{t+1})$
3	DA [7]	$\Delta X_{t+1} = (sS_i + aA_i + cC_i + fF_i + eE_i) + w\Delta X_t$ Where, s = separation weight; S_i = separation of ith fly; a = alignment weight; A_i = alignment of ith fly; c = cohesion weight; C_i = cohesion weight of ith fly; f = food factor; F_i = food source of ith fly; e = enemy factor; E_i = enemy position for i^{th} fly; w = inertia weight; t = current iteration. $X_{t+1} = X_t + \Delta X_{t+1}$	$x_i^{t+1} = g(x_i^{t+1})$

[1] Update the learning equation to introduce the chaotic behavior.

Firefly Algorithm. Flashing characteristics of fireflies is used to develop firefly-inspired algorithm. Firefly Algorithm (FA or FFA) [4] developed by Xin-She Yang at Cambridge University in 2007, use the following three idealized rules [2,4]:

Rule1: All the fireflies are unisex so it means that one firefly is attracted to other fireflies irrespective of their sex.

Rule2: Attractiveness and brightness are proportional to each other, so for any two flashing fireflies, the less bright one will move towards the one which is brighter. Attractiveness and brightness both decrease as their distance increases. If there is no one brighter than other firefly, it will move randomly.

Rule3: The brightness of a firefly is determined by the view of the objective function and brightness is simply proportional to the value of the objective function.

Flower Pollination Algorithm. The pollination concept has been used in optimization problems by Yang [25] to enhance the capability of optimization algorithm to search in global space [6,25]. The objective of flower pollination in nature is the survival of the fittest and optimal reproduction of plants. FPA can be idealized using the natural phenomena as: Cross pollination is considered as global pollination and self-pollination is considered as local pollination. Global pollination is performed with pollen carrying pollinators which used Levy flights and local pollination is used where the distance between the pollen is less [6,25].

Dragonfly Algorithm. Dragonflies are known for their unique swarming behavior. They swarm for the purpose of hunting and migrating. While hunting refers to searching for food exhaustively in a confined area where dragonflies abruptly change their velocities, migrating means the mass movement of large number of dragonflies from one place to another. Dragonfly algorithm (DA) is inspired from these special swarming patterns of dragonflies. The hunting swarming pattern is called static swarm and migratory swarming pattern is called dynamic swarm. Dynamic swarm and static swarm reflects exploration and exploitation abilities of DA respectively [7]. The behavior of dragonflies in the swarm is controlled by three prime parameters which are separation, cohesion and alignment (Reynolds 1987). Separation (S) is the static collision avoidance among individuals in a neighborhood. Cohesion (C) defines the aptness of the particles of swarm towards the center of mass of the neighborhood. Alignment (A) is the matching of velocity i.e. direction and magnitude of the dragonflies in a particular neighborhood. Distraction from enemy is computed by adding an individual dragonfly's position to the position of the enemy, and attraction to food source is computed by subtracting individual's current position from the position of the food [7].

3.3   Non-linear Test Functions Used

The following four D-dimension non-linear test functions have been considered on the basis of modality and separability of the function:

$f1$: Stepint function (Uni-modal & Separable)

$$f1(x) = 25 + \sum_{i=1}^{D} \lfloor |x_i| \rfloor \tag{2}$$

where $x_i \in [-5.12, 5.12]$ and $f(x^*) = 25$.

$f2$: Zakharov's function (Uni-modal & Non-separable)

$$f2(x) = \sum_{i=1}^{D} x_i^2 + \left(\frac{1}{2} \sum_{i=1}^{D} i x_i \right)^2 + \left(\frac{1}{2} \sum_{i=1}^{D} i x_i \right)^4 ; \quad -5 \leq x_i \leq 10 \tag{3}$$

where $x_i \in [-5, 5]$ and $f(x^*) = 0$.

$f3$: Styblinski-Tang function (Multi-modal & Separable)

$$f3(x) = \frac{1}{2} \sum_{i=1}^{D} (x_i^4 - 16x_i^2 + 5x_i) \tag{4}$$

where $x_i \in [-5, 5]$ and $f(x^*) = 0$.

$f4$: Levy 8 function (Multi-modal & Non-separable)

$$f4(x) = sin^2(\pi y_1) + \sum_{i=2}^{D-2} \left\{ (y_i - 1)^2 \left[1 + 10sin^2(\pi y_i + 1) \right] \right\}$$
$$+ (y_{D-1} - 1)^2 \left[1 + sin^2(2\pi y_{D-1}) \right] \tag{5}$$

where $x_i \in [-10, 10]$; $f(x^*) = -39.1659 * D$; and

$$y_i = 1 + \frac{x_i - 1}{4}; \quad i = 1, 2 \ldots \ldots D, -10 \leq x_i \leq 10 \tag{6}$$

3.4   Performance Measure

The following two performance measures are considered to evaluate three algorithms and their variants:

$P1$: *Mean fitness*; Fitness value has been evaluated by an algorithm for thirty runs. The mean of fitness values has been calculated has defined in Eq. 7

$$P_2 = \frac{\sum_{r=1}^{30} f(x_r)}{30} \tag{7}$$

where $f(x_r)$ is the evaluated $f(.)$ at r^{th} run by an algorithm and P_2 is the mean $f(x_r)$ for thirty runs.

$P2$: *Convergence rate*; If an algorithm will find function's optimum value with minimum number of iteration then it has high convergence rate.

3.5   Experimental Setup

The experiment has been conducted in MATLAB R2017a simulation environment. The system configuration used in this study is CPU @1.6 GHz Intel processor with 6 GB RAM. Implementation of FFA by [2], FPA by [25] and DA by [7] have been used and their chaotic variants are implemented but the equations as defined in Sect. 3.2. Every algorithm was run for thirty times to reduce biasness and increase the stability of algorithms. For the optimal results experiment is performed on optimal parameters are described in Table 3. All algorithms uses five hundred iterations and twenty agents to get accurate fitness value and precise convergence rate. Chaotic variants have been implemented by logistic map using any random $\mu \in [0, 4]$. The lower and upper bound of search space of x_i for every function are defined as given in Sect. 3.3.

Our previous study [9] has focused to evaluate complexity over the dimensions of function. The study conclude that function has high complexity with more number of dimensions. So in this paper, $D = 100$ is set to evaluate the performance of algorithms.

Table 3. Parameter settings

Sr. No.	Algorithm	Parameter's values
1	FFA	Alpha$(\alpha) = 0.6$; Gamma $(\gamma) = 0.1$
2	FPA	Probability$(p) = 0.8$; Lambda$(\lambda) = 1.5$ Step length$(s) = 0.1$
3	DA	Inertia weight $w = 0.9$ Adaptive variable $my_c = 0.1$

4   Experimental Result

As per experimental setup defined in Sect. 3.5, three algorithms and their chaotic variant using logistic map has been compared in this section. Two performance measure has been used for comparison i.e. Mean fitness $(P1)$ and Convergence rate $(P2)$. P1 results has been shown in Table 4 and P2 result's can be viewed in Fig. 1. The following results have been drawn on the basis of performance measures:

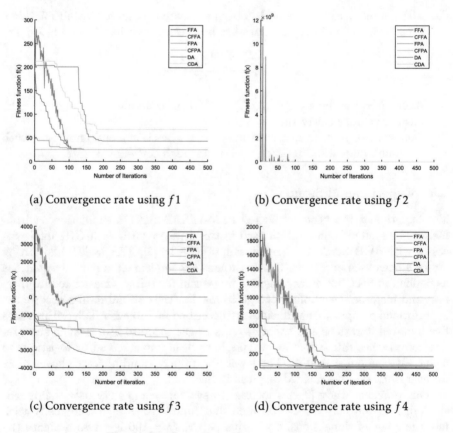

(a) Convergence rate using $f1$

(b) Convergence rate using $f2$

(c) Convergence rate using $f3$

(d) Convergence rate using $f4$

Fig. 1. Graphs to show the convergence rate.

Table 4. Results on the basis of fitness value

	$f1$		$f2$		$f3$		$f4$	
	Min	Mean	Min	Mean	Min	Mean	Min	Mean
FFA	25	29.5	260.2	378.8	−3417.6	−3310.7	3.51	15.52
CFFA	25	25	0	0	0	0	44.01	44.01
FPA	25	38.5	48	117.7	−2324.4	−2129.6	46.71	67.02
CFPA	25	33.2	54.89	116.4	−12464.8	−941.5	2.25	67.47
DA	52	67.3	256.49	541.4	−2868.7	−2520.6	42.01	96.3
CDA	52	67.3	0	259.3	−1298.19	−1060.97	965.8	112.8

On the basis of mean fitness value (P1): Clearly, the best mean $f1$ has been found only by CFFA and minimum $f1$ has been found by FFA, CFFA, FPA, CFPA. Further, minimum $f2$ and mean $f2$ have been evaluated by CFFA and CDA. Similarly, FFA calculates best $f3$ & $f4$ mean and minimum fitness.

On the basis of convergence rate (P2): Figure 1a demonstrate the results of $f1$. CFPA converges earlier than others with optimum value. Further, $f2$ results has been shown in Fig. 1b. The results indicate that all algorithms perform approximately equal. Similarly, FFA and CFPA converges earlier than other algorithms on the basis of $f3$ and $f4$ respectively.

Chaotic map has played a profound factor of swarm algorithm. As seen through the results there are some colossal effects. However, In some cases the effects are not excellent. So, researcher need to select swarm algorithm and its corresponding chaotic variant cautiously.

5   Conclusion and Future Work

This work compared three swarm algorithms (i.e. Firefly Algorithm (FFA), Flower Pollination Algorithm (FPA) & Dragonfly Algorithm (DA)) and their chaotic variants (i.e. CFFA, CFPA & CDA) using four high dimension nonlinear unconstrained test functions on the basis of fitness function (i.e. P1) and convergence rate (i.e P2). Experimental results discussed in this paper conclude the following points:

- CFFA evaluates best fitness for $f1$ and $f2$. However, CDA calculates best min $f2$. Overall, FFA and its chaotic variant evaluates best fitness.
- FFA evaluates best fitness values for multi-modal functions (i.e. $f3$ and $f4$)
- CFPA convergences earlier for $f1$ and $f4$. Whereas FFA converges earlier for $f3$.
- Overall, CFPA has high convergence rate for both uni-modal and multi-modal.

In future following areas can be considered: (a) Other algorithms such as Krill herd, Spider monkey etc. can be choose to evaluate their chaotic performance with these algorithms; (b) the convergence rate has been calculated using random run, in future algorithms can be run for thirty times and its mean fitness will be calculated over every iteration to have stable convergence rate.

References

1. Van den Bergh, F., Engelbrecht, A.P.: A study of particle swarm optimization particle trajectories. Inf. Sci. **176**(8), 937–971 (2006)
2. Yang, X.-S.: Nature-Inspired Metaheuristic Algorithms. Luniver Press, Cambridge (2010)
3. Eberhart, R., Kennedy, J.: A new optimizer using particle swarm theory. In: Proceedings of the Sixth International Symposium on Micro Machine and Human Science, MHS 1995, pp. 39–43. IEEE (1995)
4. Yang, X.-S.: Firefly algorithm. In: Nature-Inspired Metaheuristic Algorithms, vol. 20, pp. 79–90 (2008)
5. Gandomi, A.H., Alavi, A.H.: Krill herd: a new bio-inspired optimization algorithm. Commun. Nonlinear Sci. Numer. Simul. **17**(12), 4831–4845 (2012)
6. Yang, X.-S., Karamanoglu, M., He, X.: Multi-objective flower algorithm for optimization. Procedia Comput. Sci. **18**, 861–868 (2013)
7. Mirjalili, S.: Dragonfly algorithm: a new meta-heuristic optimization technique for solving single-objective, discrete, and multi-objective problems. Neural Comput. Appl. **27**(4), 1053–1073 (2016)
8. Pal, S.K., Rai, C., Singh, A.P.: Comparative study of firefly algorithm and particle swarm optimization for noisy non-linear optimization problems. Int. J. Intell. Syst. Appl. **4**(10), 50 (2012)
9. Kaur, A., Pal, S.K., Singh, A.P.: New chaotic flower pollination algorithm for unconstrained non-linear optimization functions. Int. J. Syst. Assurance Eng. Manag. **9**(4), 853–865 (2018)
10. Abdel-Raouf, O., El-Henawy, I., Abdel-Baset, M., et al.: A novel hybrid flower pollination algorithm with chaotic harmony search for solving sudoku puzzles. Int. J. Modern Educ. Comput. Sci. **6**(3), 38 (2014)
11. Gandomi, A.H., Yang, X.-S., Talatahari, S., Alavi, A.H.: Firefly algorithm with chaos. Commun. Nonlinear Sci. Numer. Simul. **18**(1), 89–98 (2013)
12. Liu, H., Abraham, A., Clerc, M.: Chaotic dynamic characteristics in swarm intelligence. Appl. Soft Comput. **7**(3), 1019–1026 (2007)
13. Ouyang, A., Pan, G., Yue, G., Du, J.: Chaotic cuckoo search algorithm for high-dimensional functions. JCP **9**(5), 1282–1290 (2014)
14. He, X., Huang, J., Rao, Y., Gao, L.: Chaotic teaching-learning-based optimization with lévy flight for global numerical optimization. Comput. Intell. Neurosci. **2016**, 43 (2016)
15. Nabil, E.: A modified flower pollination algorithm for global optimization. Expert Syst. Appl. **57**, 192–203 (2016)
16. Lukasik, S., Kowalski, P.A.: Study of flower pollination algorithm for continuous optimization. In: Intelligent Systems 2014, pp. 451–459. Springer, Heidelberg (2015)
17. Bansal, J.C., Sharma, H., Jadon, S.S., Clerc, M.: Spider monkey optimization algorithm for numerical optimization. Memetic Comput. **6**(1), 31–47 (2014). https://doi.org/10.1007/s12293-013-0128-0

18. Jadon, S.S., Bansal, J.C., Tiwari, R., Sharma, H.: Artificial bee colony algorithm with global and local neighborhoods. Int. J. Syst. Assurance Eng. Manag. **9**(3), 589–601 (2018)
19. Jamil, M., Yang, X.-S., Zepernick, H.-J.: Test functions for global optimization: a comprehensive survey. In: Swarm Intelligence and Bio-Inspired Computation, pp. 193–222. Elsevier (2013)
20. Song, Y., Chen, Z., Yuan, Z.: New chaotic PSO-based neural network predictive control for nonlinear process. IEEE Trans. Neural Netw. **18**(2), 595–601 (2007)
21. Hongwu, L.: An adaptive chaotic particle swarm optimization. In: ISECS International Colloquium on Computing, Communication, Control, and Management, CCCM 2009, vol. 2, pp. 324–327. IEEE (2009)
22. Yang, X.-S.: Chaos-enhanced firefly algorithm with automatic parameter tuning. In: Recent Algorithms and Applications in Swarm Intelligence Research, pp. 125–136. IGI Global (2013)
23. El-henawy, I., Ismail, M.: An improved chaotic flower pollination algorithm for solving large integer programming problems. Int. J. Digit. Content Technol. Appl. **8**(3), 72 (2014)
24. Ashwin, P.: Cycles homoclinic to chaotic sets; robustness and resonance. Chaos: Interdisciplinary J. Nonlinear Sci. **7**(2), 207– 220 (1997)
25. Yang, X.-S.: Flower pollination algorithm for global optimization. In: International Conference on Unconventional Computing and Natural Computation, pp. 240–249. Springer, Heidelberg (2012)
26. Sayed, G.I., Tharwat, A., Hassanien, A.E.: Chaotic dragonfly algorithm: an improved metaheuristic algorithm for feature selection. Appl. Intell., 1–18 (2018)

Analysis of Optimised LQR Controller Using Genetic Algorithm for Isolated Power System

Anju G. Pillai[✉], Elizabeth Rita Samuel, and A. Unnikrishnan

Department of Electrical and Electronics Engineering,
Rajagiri School of Engineering & Technology, Ernakulam, Kerala, India
anjugpillai@gmail.com, elizabethrs@rajagiritech.edu.in,
unnikrishnan_a@live.com

Abstract. Automatic Generation Control (AGC) is an important tool to ensure the stability and reliability of power systems. For stable operation of power systems, the frequency of the system should be reserved within the nominal value. Towards this, the estimation of states is of supreme implication. In this paper, a comparison is made on the estimation of the states using Kalman filter method and optimal control approach to the Automatic Generation Control (AGC) of an isolated power system. The performance of optimised Linear Quadratic Regulator (LQR) in pole placement is compared with Kalman filter estimating the states for pole placement. Genetic Algorithm is used to optimise the weighting matrices Q and R of an LQR controller. Kalman filter based controller estimates the states of the system by measuring only one output signal i.e. the frequency output of the system considered. The comparison is made on the basis of the mean of the variances of frequency estimate of both the approaches under different noise levels from independent Monte Carlo simulations. Modeling of an isolated power system is done using Simulink/MATLAB.

Keywords: Automatic Generation Control (AGC) ·
Genetic Algorithm (GA) · Linear Quadratic Regulator (LQR) ·
Kalman filter · Single area power system

1 Introduction

Automatic Generation Control (AGC) in a power system has a function of controlling the reliable operation of the system. It is well known that any load change in the system results in frequency deviation [1]. The primary responsibility of AGC is to attain the equilibrium state by sustaining the nominal frequency of the system. Large frequency deviations can cause damages to the equipment, overload transmission lines, degrade load performance and negatively affect the performance of system protection schemes. Deviation in frequency can also affect the stability of the system. In order to maintain the stability of the

© Springer Nature Switzerland AG 2019
A. Abraham et al. (Eds.): IBICA 2018, AISC 939, pp. 202–211, 2019.
https://doi.org/10.1007/978-3-030-16681-6_20

system, imbalances between the load and generation has to be corrected well in time to prevent frequency deviation beyond limits. The problem of controlling the frequency, is resolved by adjusting the production of generating units in response to the load demand, is termed as Automatic Generation Control. AGC is also stated as Load Frequency Control (LFC) [1].

In power system operation and control, AGC is a very important issue for supplying sufficient and reliable electric power with good quality. Electrical power system is also becoming more complex, with the inclusion of randomly varying load, resulting in the rapid variation of the frequency related to that area. In normal cases, the generators working in corresponding area will often adjust their power generation to compensate for the frequency variation [2].

To minimize the frequency deviation of the system, AGC functions by varying the power generation. If the power generation is greater than the demand then the frequency of the system increases and vice versa. An effective way is to bring down frequency deviation to zero, by adjusting the speed of the generator. This process of adjusting the generated power and frequency of the system is called Automatic Generation Control [2].

In a single area power system, the frequency deviation gains significance, as the system inertia is very small. For stabilizing a system, a controller is required which provides higher performance in terms of settling time [3]. This paper focuses on the implementation of two techniques (i) GA based LQR controller, which optimizes a cost function based on the state variables sensed and (ii) a Kalman Estimator estimating the states, for pole placement only from the frequency measurement. Then the controllers are compared based on the mean of the variances of the frequency response using Monte Carlo simulations.

2 System Description

In a complex power network, the imbalance in generated and consumed active power determines the range of frequency deviation, along with other state variables [2]. Thus in order to maintain the balance, state of the systems must be estimated accurately. The states of the system is described as (1).

$$x(t) = [\Delta P_v, \Delta P_m, \Delta f] \tag{1}$$

ΔP_v is the real power command
ΔP_m is the change in mechanical power
Δf is the change in frequency.

AGC is accomplished in two control loops; primary control loop and supplementary control loop. The change in frequency is sensed by a speed governor in the primary control loop [2]. Figure 1 shows AGC for an isolated power system.

3 Brief Overview of LQR and KF

An analysis of optimal controller, LQR and Kalman filter based controller is made on the single area system considered. In the first section, LQR controller

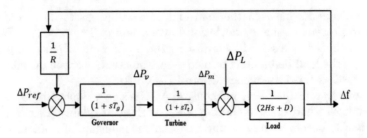

Fig. 1. Single area power system

is designed for a linearized single area power system assuming all system states are available [4]. This designed regulator creates a control vector based on the measured states. In the later section, a Kalman filter based controller is designed with control input and measured output of the system. In this case with only one measured output, three states of the system are estimated.

3.1   Optimal Controller – LQR

Linear Quadratic Regulator (LQR) is defined as a state space based optimal control technique for linear systems [5]. The performance of which is controlled by minimizing a quadratic performance index [6]. In this paper LQR is used to obtain the state feedback gain, K for a single area power system. Consider a plant described by the linear state equation

$$\dot{x}(t) = Ax(t) + Bu(t) \tag{2}$$

The problem is to find K of the Control law (3)

$$u(t) = -Kx(t) \tag{3}$$

Which minimizes the quadratic performance index, J

$$J = \int_{0}^{t} (x^T Q x + u^T R u) dt \tag{4}$$

where the two matrices Q and R are positive semi definite matrix and a real symmetric matrix respectively. For obtaining the solution in terms of the states, the method of Lagrange Multipliers is used [6]. The problem can be reduced to the form of minimizing the unconstrained equation

$$L[x, \lambda, u, t] = [x^T Q x + u^T R u] + \lambda^T [Ax + Bu - \dot{x}] \tag{5}$$

After solving, value of K is obtained as

$$K = R^{-1} B^T P \tag{6}$$

where P is the solution of continuous time algebraic matrix Riccati equation given as:

$$PA + A^T P - PBR^{-1} B^T P + Q = 0 \tag{7}$$

3.2 Kalman Filter

For fully obtaining the advantage of state feedback, all the states of the system should be fed back. A state observer then suggests itself as a solution that gives the estimate of internal state of the system from the input and output measurements [7]. If a system is observable then it is possible to obtain the designs from its output measurements, for which the Kalman filter is used, which can cater for measurement errors. The state estimation problem is given as

$$\dot{x}(t) = Ax(t) + Bu(t) + \omega \tag{8}$$

$$Y(t) = Cx(t) + Du(t) + v \tag{9}$$

where A, B, C and D are plant state matrices. ω and v are the plant noise and measurement noise respectively. In a two-step predictor corrector approach, the Kalman Filter estimates the states, corrected by the measurements.

4 Genetic Algorithm

Genetic Algorithm is an optimisation algorithm based on natural selection and genetics. It was invented by Holland in 1975. GA starts without having any prior knowledge about the solution and it depends on the genetic operators to arrive at the best solution. It starts at independent initial points and search is progressed in a parallel fashion so that this technique avoids local minima and converges to the optimal solution. Hence GA shows high capability in solving problems related to complex domains avoiding the difficulties associated with dimensional complexities [4].

GA starts with initialising a random population. This population is usually represented as real number or a binary string called a chromosome. The algorithm passes through genetic phenomena similar to those in natural evolution in order to attain a minimised cost function. The objective function evaluates each individual in terms of their fitness values. Selection is performed on the population by keeping the individuals with best fitness value. This optimisation algorithm is an iterative one. Different genetic operators such as crossover operator and mutation operator are used for the evolution of new generation. Individuals having better fitness values are allowed to survive in the next generation. Algorithm is repeated until termination criterion or maximum number of generations are reached [8].

In LQR problem, the weighting matrices Q and R have significant effect on the performance of the controller. Finding the best values for Q and R are highly time consuming by implementing computer simulations or trial and error methods. For effective solution some intelligent optimisation techniques can be applied. The main objective of the genetic algorithm implemented here is to determine the weighting matrices Q and R in order to have better performance for AGC [8]. Q and R are usually represented as diagonal matrices.

$$Q^{GA} = \begin{bmatrix} q_{11} & 0 & 0 \\ 0 & q_{22} & 0 \\ 0 & 0 & q_{33} \end{bmatrix} \qquad R^{GA} = q_{44} \tag{10}$$

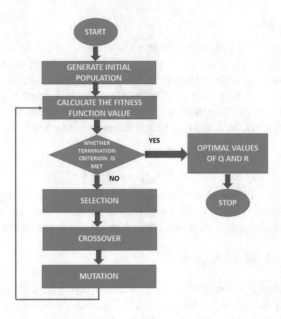

Fig. 2. Flowchart for optimising LQR using GA

Figure 2 represents the flowchart for optimisation of Q and R matrices for the LQR controller.

4.1 Genetic Algorithm Parameters

Population: 50, Generations: 400, Selection: Stochastic Uniform, Mutation: 0.5, Crossover fraction: 0.8, Crossover: Single point.

The objective function [8] to find the optimal value of Q and R matrices is

$$F = \frac{1}{q_{11}x_1^2 + q_{22}x_2^2 + q_{33}x_3^2 + q_{44}u^2} \qquad (11)$$

MATLAB program is used to obtain the optimal values of Q and R. After 130 generations the optimal values are obtained as

$$Q^{GA} = diag(20, 10, 0.003) \qquad (12)$$

$$R^{GA} = 0.18 \qquad (13)$$

Figure 3 shows the MATLAB simulation window for Genetic Algorithm solution for optimisation of LQR.

5 Simulation Study

The performance of the isolated power system is analyzed in this section. The study is done using MATLAB/Simulink. These techniques are compared based

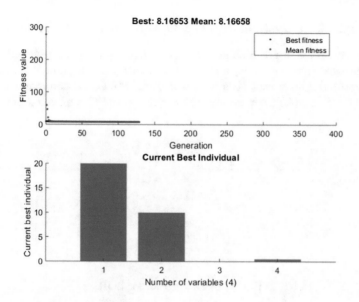

Fig. 3. MATLAB simulation of Genetic Algorithm for optimal values Q and R

on the mean of the variances of frequency estimated, after 100 independent Monte Carlo simulations, for a given noise variance. The states of the system are x(t) = [ΔP_v, ΔP_m, Δf] as described in (1). The system level variables of the considered example are mentioned in Table 1.

Table 1. System level variables

Variable	Value
Turbine time constant (Tt), sec	0.5
Governor time constant (Tg), sec	0.2
Inertia constant (H)	5
Speed regulation (R), p.u.	0.05
Frequency sensitive load coefficient (D)	0.8
Nominal frequency (f), Hz	60
Base power (S), MVA	1000

The parameters of state space model for the system in Table 1 is as shown below [2].

$$A = \begin{bmatrix} -5 & 0 & -100 \\ 2 & -2 & 0 \\ 0 & 0.1 & -0.08 \end{bmatrix} \qquad B = \begin{bmatrix} 0 \\ 0 \\ -0.1 \end{bmatrix} \tag{14}$$

5.1   Single Area System with GA Optimised LQR Controller

An LQR controller for the single area system considered is designed and the state feedback gain vector is obtained [7]. The block diagram is as shown in Fig. 4. Optimal gain vector K is obtained using MATLAB command lqr which satisfies (6). The MATLAB command for gain vector K is as given below.

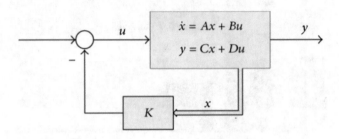

Fig. 4. Block diagram for LQR

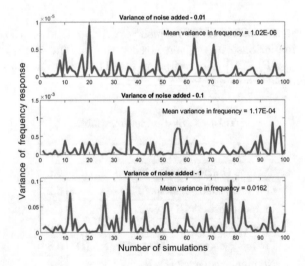

Fig. 5. Variance of frequency response with GA based LQR controller

$$[K, P] = lqr(A, B, Q, R) \tag{15}$$

The optimal gain K for LQR is found using MATLAB function and obtained as

$$K = \begin{bmatrix} 6.415 & 1.101 & -112.475 \end{bmatrix} \tag{16}$$

Variance of output of the system is shown in Fig. 5. GA based LQR controller is incorporated in the system. With 0.01, 0.1 and 1 as noise variances of the Gaussian Noise added, the mean variance of the frequency, with 100 Monte Carlo simulations are observed as 1.02E−06, 1.17E−04 and 0.0162 respectively.

5.2   Single Area System with Kalman Filter

A single area power system is controlled by Kalman filter. The frequency measured from the system is given to the Kalman filter to correct the estimated states. Estimated states are the output of Kalman filter [7]. Block diagram of Kalman filter for AGC is shown in Fig. 6, shows the Kalman filter based controller implemented for an AGC in a single area power system. $\hat{x}$ and $\hat{y}$ indicates estimates of states and output of the system respectively.

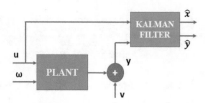

Fig. 6. Block diagram for Kalman filter for AGC

The observer gain, L for Kalman filter is given by the MATLAB command (17).

$$[L, S] = lqe(A, B, C, \omega, v) \tag{17}$$

where L is the Kalman filter optimal gain and S is the solution to the Algebraic Riccati Equation. Optimal gain for Kalman filter is obtained as

$$L = \begin{pmatrix} -1.0538 \\ -0.4251 \\ 0.0662 \end{pmatrix} \tag{18}$$

Figure 7 shows the variance of frequency response of the system with Kalman Filter. With 0.01, 0.1 and 1 as noise variances of the Gaussian Noise added, the mean variance of the frequency, with 100 Monte Carlo simulations are observed as 1.33e−07, 1.4904e−07 and 6.1936e−07 respectively.

Comparison of mean of variance values of the frequency response for single area system with GA based LQR and Kalman filter based controller are shown in Table 2. From Table 2 it is clear that the Kalman estimator is able to produce better results as compared with the output variance of LQR.

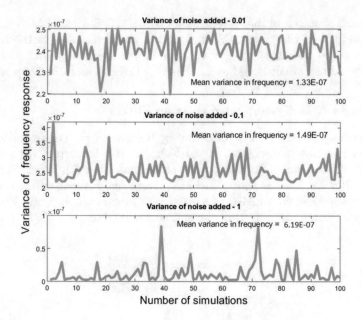

Fig. 7. Variance of frequency response with Kalman filter based controller

Table 2. Comparison of mean of variance values for frequency response

Noise added	Mean of variance	
	Kalman filter	GA based LQR
0.01	1.33E−07	1.02E−06
0.1	1.49E−07	1.17E−04
1	6.19E−07	0.0162

6   Conclusion

This paper analyses the performance of an optimal controller, GA optimised LQR and Kalman filter for the control of a single area power system. For this analysis, a linearized single area power system is used. In LQR, it is assumed that all the states are available for which a regulator based controller is designed. While for a Kalman estimator, a controller is designed based on the states estimated from the limited output measured. For the comparative analysis of both the techniques, the variances of frequency response of the system under different noise levels are evaluated. The results are validated using Monte Carlo simulations. It has been observed that the Kalman estimator is able to produce better output variance in comparison to LQR.

References

1. Kothari, D.P., Nagrath, I.J.: Modern Power System Analysis. Tata McGraw-Hill Education, New York City (2011)
2. Saadat, H.: Power System Analysis. McGraw-Hill Series in Electrical Computer Engineering (1999)
3. Kumar, P., Kothari, D.P.: Recent philosophies of automatic generation control strategies in power systems. IEEE Trans. Power Syst. **20**(1), 346–357 (2005)
4. Ali, M., Zahra, S.T., Jalal, K., Saddiqa, A., Hayat, M.F.: Design of Optimal Linear Quadratic Gaussian (LQG) Controller for Load Frequency Control (LFC) using Genetic Algorithm (GA) in Power System. Int. J. Eng. Works **5**(3), 40–49 (2018)
5. Cavin, R.K., Budge, M.C., Rasmussen, P.: An optimal linear systems approach to load-frequency control. IEEE Trans. Power Appar. Syst. **90**(6), 2472–2482 (1971)
6. Kirk, D.E.: Optimal Control Theory: An Introduction. Courier Corporation, Chelmsford (2012)
7. Shahalami, S.H., Farsi, D.: Analysis of Load Frequency Control in a restructured multi-area power system with the Kalman filter and the LQR controller. AEU-Int. J. Electron. Commun. **31**(86), 25–46 (2018)
8. Jalili-Kharaajoo, M., Moshiri, B., Shabani, K., Ebrahimirad, H.: Genetic algorithm based parameter tuning of adaptive LQR-repetitive controllers with application to uninterruptible power supply systems. In: International Conference on Industrial, Engineering and Other Applications of Applied Intelligent Systems, pp. 583–593. Springer, Heidelberg (2004)

Learning Based Image Super Resolution Using Sparse Online Greedy Support Vector Regression

Jesna Anver[1,2]([✉]) [iD] and P. Abdulla[1] [iD]

[1] Division of Electronics, School of Engineering,
Cochin University of Science and Technology, Kochi, Kerala, India
`jeznaizmail@gmail.com`
[2] Department of Computer Science, Toc H Institute of Science,
Arakunnam, Kerala, India

Abstract. Kernel filters with sparse solutions has become highly advantageous because of the reduced computation. Sparse online greedy support vector regression algorithm, computes the coefficients only after the generation of sparse dictionary. This paper super resolves a low resolution image to high resolution image, with the model generated from the training set using sparse online greedy support vector regression. The method is evaluated with super resolution using support vector regression. Comparisons are done on the PSNR, time and memory scales. The sparse online greedy support vector regression shows good improvement in these scales.

Keywords: Image Super Resolution · Single Image Super Resolution ·
Support Vector Regression · Sparse Online Greedy Support Vector Regression ·
Clustering

1 Introduction

Image super resolution (SR) is an ill posed problem. It is a procedure of creation of High-resolution (HR) image from low-resolution (LR) image. Spatial SR involves enhancing the visual clarity by increasing the number of pixels in the resultant HR image. Availability of HR image is restricted due to technological and economic constraints. SR involves creation of HR image using some algorithms rather than using hardware, hence lowers the expenses. SR can be generated by using single LR image hence called single image super resolution (SISR). Creation of SR by using multiple LR images is called multi-image super resolution (MISR) [1]. The SR task gets better HR image from the LR image based on prior assumptions on the observation model which maps the HR image to the LR ones. HR image contains lager information when compared to LR image. HR image is favorable to attain a better classification to region of interest in certain applications. Recently, there is a growing interest in SISR. Many single-image based SR algorithms make use of some learning algorithms to predict the missing information of the super-resolved images using the relationship among LR and HR images from a training set.

© Springer Nature Switzerland AG 2019
A. Abraham et al. (Eds.): IBICA 2018, AISC 939, pp. 212–218, 2019.
https://doi.org/10.1007/978-3-030-16681-6_21

Image super resolution is an inverse problem for generation of low resolution by down sampling the high resolution image. There are several methods for generating HR images. It includes iterative back projection algorithm [2–4], example-based approach [5–7] and learning-based super-resolution [8–10]. In IBP an error is estimated between the original LR image and the stimulated LR images. It is then back-projected on to the coordinates of HR image. The process is iterated till there is no change. This iteration would converge to one of the solutions in some cases it may oscillate between the solutions. Example-based SR would estimate the HR images by utilizing the examples to LR and HR images. The examples are identified from the existing database. The method removes noise but is computationally very complex. The learning algorithms are expected to have more potential in generalization. Learning-based SR methods start with a training set that contains both LR and HR image patches. These pairs are used to generate coefficients to the regression model. This model further used to predict the HR image. The generated regressions can be linear or non-linear. The non-linear model is capable of providing more details compared to the other. In the above papers the HR samples are created from the LR samples. The SR algorithm in [9] is a classification based method. For classification SR technique in [10] uses kernel k-means algorithm and therefore consumes large amount of time. To avoid this problem spectral clustering [11] is used here. For producing sparse solution SOG-SVR is applied.

There exists wide variety of stochastic algorithms that can be used for pattern recognition applications. For prediction, the regression analytic of support vector machine algorithm called support vector regression [12] is used. SVR has properties like duality, sparseness, convexity and the most appreciable is the generalization ability and good convergence. Even though SVR provides nonlinear sparse solution to the given problem, the sparseness is achieved only after evaluating the quadratic optimization of SVR. Hence the complexity is high and consumes large amount of time for training. Sparse Online Greedy Support Vector Regression (SOG-SVR) [13] is an online greedy algorithm. Initially it finds a subset of linearly independent input samples from the entire training samples. This step leads to an extremely sparse solution. Hence the time complexity is very less compared to SVR. In this paper, the image super resolution using SOG-SVR and SVR [9] is compared. High-resolution (HR) images are desired in many realistic areas such as Satellite imaging [14], medical imaging [15], ultra sound imaging [16] etc. These applications strongly suggest the need of less complex SR algorithm with HR image better quality.

2 Sparse Online Greedy Support Vector Regression

Non linear SVR builds regressions on data that is transformed from the input space to a higher dimension called Hilbert space (or feature space F), i.e. x is mapped to $\phi(x_i)$ in F. The kernel functions reduce the complexity of mapping by computing the inner products of feature vectors. The kernel function is represented as $k(x_i, x_j) = \phi(x_i)^T \phi(x_j)$. where, x_i, x_j are the input vectors and ϕ is the mapping to the F space. The common kernel functions used are the Gaussian kernel with variance σ^2, $k(x_i, x_j) = \exp\{-\|x_i - x_j\|^2 / 2\sigma^2\}$ and the polynomial kernel of degree p, $k(x_i, x_j) =$

$(a\langle x_i, x_j\rangle + b)^p$. The Representer theorem states the solution attained by the kernel methods is represented as

$$\hat{f}(\mathbf{x}) = \sum_{i=1}^{t} \alpha_i k(\mathbf{x}_i, \mathbf{x}) \qquad (1)$$

where f is a linear predictor in Hilbert space.

For the given training samples $\lambda_t = \{x_i, y_i\}_{i=1}^{t}$, where (x_i, y_i) are the input, output pairs. The SVR uses following model to predict the output.

$$\min_{w,b,\xi,\xi^*} \frac{1}{2}||\mathbf{w}||^2 + C\sum_{i=1}^{n}(\xi_i + \xi_i^*)$$

$$\text{s.t.} \begin{cases} (\mathbf{w}^T\phi(\mathbf{x}_i)) + b - y_i \leq \varepsilon + \xi_i \\ y_i - (\mathbf{w}^T\phi(\mathbf{x}_i) + b) \leq \varepsilon + \xi_i^* \\ \xi_i, \xi_i^* \geq 0, i = 1, \ldots, n \end{cases} \qquad (2)$$

where $y \in Y, \forall i$, the two inequalities are upper and lower boundaries of the output of training samples, ξ_i and ξ_i^* are the slack variables.

SVR optimization achieves sparsity by first taking in all training samples as potential contributors. After solution, the coefficients of some samples are found to zero and hence those are eliminated. Since the elimination is after evaluation, the complexity is higher. The sparse online greedy support vector algorithm (SOG-SVR) is fast and online method. SOG-SVR takes in those training samples which can span the sub space that are linearly independent in the entire training set. The SOG SVR starts with an empty dictionary, D and expands the D using the relation

$$\delta_t = \left\|\sum_{i=1}^{t-1} a_{i,t} \cdot \phi(x_i) - \phi(x_t)\right\|^2 < v. \qquad (3)$$

The dictionary size m increases if $\phi(x_t)$ is linearly independent of the existing dictionary samples. $\phi(x_t)$ is then added to the dictionary D. If $\phi(x_t)$ is not linearly independent, the sample can be linearly expanded with the existing dictionary vectors using the coefficients $a_t = (a_{t,1}, \cdots a_{t,m})^T$ hence those samples which are expandable are eliminated.

3   Image Super Resolution Algorithm

The image super resolution starts with the preprocessing step to image patch of size 3×3, to make the data consistent. The neighboring pixels of the input are queued and normalized to generate each input vector. There are two phases, the first phase is training. In training phase, the coefficients are learned using the training samples. For super resolving by a factor of two spatially, four regressions have to be generated. In training, the HR image is down sampled so that, training input p_i is paired to each output pixel $y_{i,j}$. The input is then clustered in to groups using spectral clustering. The spectral clustering performs clustering on the spectrum or the kernel matrix. Each cluster group is then associated with its corresponding HR sample to train with SOG-SVR, to generate the coefficients of the regressions for each cluster. The Gaussian kernel is applied to get better solution. The SOG-SVR formulations for each class 'c' then would become

$$\delta_{t,c} = \left\| \sum_{i=1}^{t-1} a_{i,t,c} \cdot \phi(x_{i,c}) - \phi(x_{t,c}) \right\|^2 < v. \tag{4}$$

The testing phase would utilize the coefficients learned in testing to predict the HR image. The algorithm is given in Fig. 1.

Input: *LR and HR image samples, SOG-SVR parameters, p_i, $x_i=[]$, $l,=-1$, $m=-1$, $c=-2$*
Output: *Super resolved image*

1. *Preprocess each pixel p_i*

    a. *for k = 1 to 9*
       if k!=5
       $x_i = [x_i ; p_{i+m,j+l} - p_{i,j}]$
       if k==3
    m++;l=c;
       else if k==6
       m++;l=c
       l++;

    b. *end*

2. *classify the training input x_i $\forall$ i*
3. *learn the dictionary D for each group using*

   $$\delta_{t,c} = \left\| \sum_{i=1}^{t-1} a_{i,t,c} \cdot \phi(x_{i,c}) - \phi(x_{t,c}) \right\|^2 < v.$$

4. *compute the coefficients.*
5. *find the solution to test set x using*

   $$\hat{f}(x) = \sum_{i=1}^{t} \alpha_i k(x_i, x)$$

6. *generate the super resolved HR image*

Fig. 1. Image super resolution algorithm.

4 Results

The performance of the image super resolution using SOG-SVR is evaluated. The Guassian kernel is used here. The parameters used here are the percentage of support vectors (SVs), time and PSNR. SVs are the non-zero coefficient which defines the regression. Since the algorithm is classified to 'c' clusters, there are $4c$ models used to super resolve the LR image. By applying clustering the training time for finding the coefficients can be educed. Initially the comparison is made with [9]. The optimal parameters are opted using various trials and those which give best results are taken. The images taken for the study are standard images from internet. Table 1 shows the analysis. It shows that the support vectors and time to predict the HR are very less for SOG-SVR compared to SVR. There is also a considerable increase in PSNR. For visual comparison, few images are given in Fig. 2.

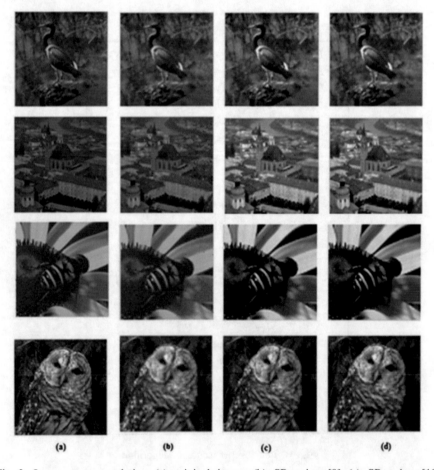

(a)   (b)   (c)   (d)

Fig. 2. Image super resolution (a) original image (b) SR using [9] (c) SR using [10] (d) proposed SR.

Next the comparison with [10] is done. This comparison is mainly focused on behavior of clustering algorithm. The results show that there is a decrease in time. The comparison is given in Table 2. The time taken by image super resolution using spectral clustering is less compared to kernel k-means clustering.

Table 1. Performance comparison of proposed method and [9].

Image	[9]	Proposed
bird		
percentage of SVs	96.76	81.36
time (sec)	3143	138.68
PSNR (dB)	29.07	31.92
Building		
percentage of SVs	95.84	80.13
time (sec)	5974	39.66
PSNR (dB)	26.29	30.46
house		
percentage of SVs	95.31	56.21
time (sec)	8463	28.43
PSNR (dB)	21.51	22.82
fruit		
percentage of SVs	95.28	80.4
time (sec)	3487	54.98
PSNR (dB)	22.05	23.93
bee		
percentage of SVs	96.3	78.72
time (sec)	3257	60.3
PSNR (dB)	26.91	31.35
owl		
percentage of SVs	94.61	83.71
time (sec)	2264	64.21
PSNR (dB)	25.2	27.38

Table 2. Performance analysis of proposed method and [10].

Image	[9]	Proposed
bird time (sec)	159	138.68
building time (sec)	125	39.66
house time (sec)	118	28.43
fruit time (sec)	127.24	54.98
bee time (sec)	28.34	60.30
owl time (sec)	140.16	64.21

5 Conclusion

In SOG-SVR time required for training phase is highly decremented by eliminating linear dependencies in the input space. The metrics like time, support vectors gives an indirect measure to the memory requirement. Also to compare the image quality, the PSNR measure is taken. The SOG-SVR due to its sparse nature in its formulation, gives significant betterment in these metrics compared to super resolution using SVR.

References

1. Nasrollahi, K., Moeslund, T.B.: Super-resolution: a comprehensive survey. Mach. Vis. Appl. **25**, 1423–1468 (2014)
2. Szydzik, T., Callico, G.M., Nunez, A.: Efficient FPGA implementation of a high-quality super-resolution algorithm with realtime performance. IEEE Trans. Consum. Electron. **57**(2), 664–672 (2011)
3. Patel, V., Modi, C.K., Paunwala, C.N., Patnaik, S.: Hybrid approach for single image super resolution using ISEF and IBP. In: Proceedings of International Conference on Communication Systems and Network Technologies, pp. 495–499 (2011)
4. Bengtsson, T., Gu, I.-H., Viberg, M., Lindstrom, K.: Regularized optimization for joint super-resolution and high dynamic range image reconstruction in a perceptually uniform domain. In: Proceedings of IEEE International Conference on Acoustics, Speech and Signal Processing, Japan, pp. 1097–1100 (2013)
5. Elad, M., Datsenko, D.: Example-based regularization deployed to super-resolution reconstruction of a single image. Comput. J. **18**(2), 103–121 (2007)
6. Li, X., Lam, K.M., Qiu, G., Shen, L., Wang, S.: Example-based image super-resolution with class-specific predictors. J. Vis. Commun. Image Represent. **20**(5), 312–322 (2009)
7. Li, Y., Xue, T., Sun, L., Liu, J.: Joint example-based depth map super-resolution. In: Proceedings of IEEE International Conference on Multimedia, Expo, Australia, pp. 9–13 (2012)
8. Wu, W., Liu, Z., He, X.: Learning-based super resolution using kernel partial least squares. Signal Process. Image Commun. **29**, 394–406 (2011)
9. Ni, K.S., Nguyen, T.Q.: Image super resolution using support vector regression. IEEE Trans. Image Process. **16**(6), 1596–1610 (2007)
10. Jesna, A., Abdulla, P.: Single-image super-resolution using kernel recursive least squares. Springer - Signal image video process. **10**(8), 1551–1558 (2016)
11. Girolami, M.: Mercer kernel-based clustering in feature space. IEEE Trans. Neural Netw. **13**(3), 780–784 (2002)
12. Ding, S., Zhang, L., Zhang, Y.: Research on spectral clustering algorithms, prospects. In: Proceedings of International Conference on Computer Engineering, Technology, pp. 16–18 (2010)
13. Engel, Y., Mannor, S., Meir, R.: Sparse online greedy support vector regression. In: Proceedings of the 13th European Conference on Machine Learning, pp. 84–96 (2002)
14. Liebel, L., Körner, M.: Single-image super resolution for multispectral remote sensing data using convolutional neural networks. In: Proceedings of the ISPRS Congress, pp. 883–890 (2016)
15. Rueda, A., Malpica, N., Romero, E.: Single-image super-resolution of brain MR images using overcomplete dictionaries. Med. Image Anal. **17**, 113–132 (2013)
16. O'Reilly, M.A., Hynynen, K.: A super-resolution ultrasound method for brain vascular mapping. Med. Phys. **40**(11), 1–7 (2013)

A Deep Learning Approach for Molecular Crystallinity Prediction

Akash Sharma[⊠] and Bharti Khungar

Birla Institute of Technology Science, Pilani, Pilani, India
{f2012771,bkhungar}@pilani.bits-pilani.ac.in

Abstract. With the success of Convolutional Neural Networks (CNN) in computer vision domain, cheminformatics is slowly moving away from feature Engineering towards Network Engineering. New deep networks and approaches are being proposed to explore the chemical behavior and their properties. In this paper, we propose a deep learning approach using Convolutional Neural Network for predicting the crystallization propensity of an organic molecule. The work is inspired from Chemception and architecture is based on the Inception-Resnet v2 model. The proposed approach only requires a 2D molecular drawing to predict if the molecule has a good probability of forming crystals, without the need of any molecular descriptor, any advanced chemistry knowledge or any study of crystal growth mechanisms. We have evaluated our approach on the Cambridge Structural Database (CSD) and the ZINC datasets. Compared with the machine learning approach of generating molecular descriptors plus SVM classification, our proposed approach gives a better classification accuracy.

Keywords: Computational chemistry · Convolutional Neural Networks · Deep learning

1 Introduction

Crystallization propensity determines the ability of a molecule to form good quality crystals. One of the major challenges in the pharmaceutical industry is determining the propensity of drug molecules and understand the structural features and experimental conditions required for it to crystallize [1]. This prediction helps scientists modify the structures using synthetic alterations required to control the crystallinity according to their needs. They have to study the crystal growth mechanisms and other conditions manually, making this process a bit complex and time consuming.

Machine Learning algorithms have been commonly used in chemistry to model structure-property relationships (QSPR's) [2–4]. Some applications include the prediction of Melting points [5], solubilities [6] and other molecular properties [7]. With the success of deep learning algorithms in computer vision domain, cheminformatics also started incorporating these neural network architectures to solve the domain specific problems [8–10]. Prediction of reactions [11] and toxicities [12] are some common problems in chemistry where Neural networks have been successfully used. These architectures have performed either better than or at par with most of the traditional Machine learning algorithms.

© Springer Nature Switzerland AG 2019
A. Abraham et al. (Eds.): IBICA 2018, AISC 939, pp. 219–225, 2019.
https://doi.org/10.1007/978-3-030-16681-6_22

A lot of work has been done on the Protein crystallinity prediction. Sequence of amino acids are used as molecular descriptors and classification of molecular crystallinity is performed using machine learning algorithms. For organic molecular compounds Wicker et al. [13] calculated the molecular descriptors using RD cheminformatics toolkit [14], did feature selection and finally performed classification using SVM Rbf kernel [15], which showed an error rate of around 10% on unseen molecules. In 2017, Szegedy et al. [16] proposed a deep neural network, called the Inception-ResNet v2 which has one of the highest image classification accuracy and also won the 2015 ILSVRC challenge. It's a hybrid network comprising residual connections in the traditional Inception architecture. The skip connections between the input and output help accelerate the training of Inception networks.

In this work, we propose a novel crystallinity propensity prediction deep learning approach based on CNN. It is inspired from the Chemception [17], which is a general-purpose CNN and has proved to be successful in toxicity and activity prediction. The architecture is based on Inception-ResNet v2 model. The proposed approach only requires a 2D molecular drawing to predict if the molecule has a good probability of forming crystals, without the need of any molecular descriptor generation, any advanced chemistry knowledge or any study of crystal growth mechanisms.

2 Proposed Network Architecture

Table 1 describes the proposed crystallinity prediction architecture. It contains a stem layer, which has a 4 × 4 convolution filter, the Inception-Resnet and Reduction Layers (5 segments in total), and classification layers containing a fully connected layer followed by a Softmax, to get the crystallinity propensity probability.

Table 1. The proposed crystallinity prediction architecture

Segment	Process	Times
Segment 1 (Stem)	4 × 4, stride 2	×1
Segment 2	Inception-Resnet-A	×1
Segment 3	Reduction-A	×1
Segment 4	Inception-Resnet-B	×1
Segment 5	Reduction-B	×1
Segment 6	Inception-Resnet-C	×1
Classification layer	Global Average Pool 1000-d Fc, Softmax	×1

The Fig. 1 above shows the different Inception-Resnet and reduction blocks which form a part of the different segments. (i), (ii), (iii) represent the Inception-Resnet A, B and C respectively, while (iv) and (v) represent the Reduction Blocks A and B. To reduce the number of parameters and for optimization, we have just 1 Inception block in each Inception-Resnet segment. The reduction segments have respective single Reduction blocks. Each layer is assumed to have a RELU activation layer after the convolutional Layer, with a stride of 1 and a similar padding. The different number of convolutional filters are shown in each Layer. Each of the branch in the Inception block has a regularity in the number of convolutional filters. The residual or the skip connections are known to improve the training of Inception networks and are known to preserve the features from previous layers. This network automatically learns the relevant feature representations from the 2D images and predict the chemical property of interest.

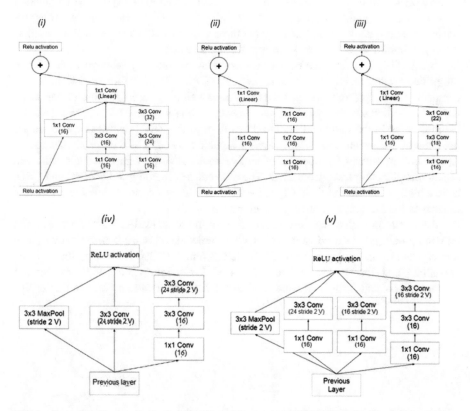

Fig. 1. Inception-Resnet A, B, C (i, ii, iii) and Reduction Blocks A, B (iv, v) used in different segments of the architecture. The number of convolutional filters is specified in each block.

	Smiles	Crystalline
0	CC(=O)Nc1cc(C(=O)NCC(=O)N2CCCc3ccccc32)ccc1C	0
1	CCCCNC(=O)C1CCCN(C(=O)CCC(C)C)C1	0

Fig. 2. Some examples of non-crystalline molecules from ZINC database in SMILES format.

3   Experiments and Analysis

3.1   Dataset

To train the network and evaluate the classification accuracy of our network, we obtained the crystalline and the non-crystalline databases. Cambridge Structural Database (CSD) [18] is a highly curated database of small-molecule metal-organic and organic crystal structures. The molecules have been tested to be crystalline by creating crystals of them, and then using the x-ray diffractions. These molecules form appropriate size and quality crystals to carry out the crystal structure analysis and are used as the crystalline or crystallizable molecule for the analysis.

Zinc28 [19] is a free database which contains biologically relevant molecules in Simplified molecular-input line-entry system (SMILES) [20] format. SMILES is a compact string representation in terms of line notation to describe a molecular structure. Zinc contains both crystalline and non-crystalline molecules. There are around 2 lac molecules in the CSD and around 35 million in the ZINC database. Since we are predicting the crystallization propensity for organic molecules as their properties can be represented in terms of properties of similar molecules, we eliminated the salts and organometallic complexes from CSD. The molecules common to both ZINC and CSD, have a high probability to form Crystal structures and are chosen as reliable crystalline molecules for training and testing our approach.

The common molecules were removed from the downloaded ZINC database. The remaining dataset was used as non-crystalline molecules. Figure 2 above shows some non-crystalline molecules in SMILES format. Around 15,800 molecules for both the crystalline and the non-crystalline were used for training and validation. Similar number of molecules for both were used in the study to avoid any bias in the results.

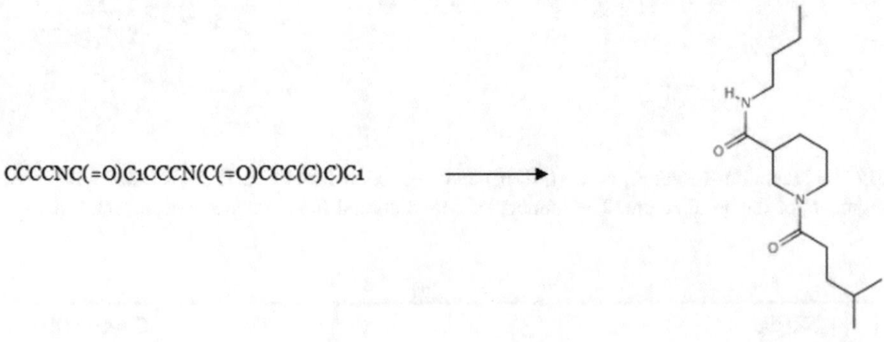

CCCCNC(=O)C₁CCCN(C(=O)CCC(C)C)C₁   ⟶

Fig. 3. SMILES to 2D structure conversion using OpenBabel

3.2   Data Preprocessing

The molecules in the CSD were stored in Crystallographic Information File (CIF) format which were converted into the 2D molecular structures using OpenBabel [21] cheminformatics software. In the ZINC database, the molecules were stored in

Smiles format and were also converted to the structural drawing using the same library. Figure 3 above shows the SMILES to 2D structure conversion. Each of the molecules were then converted to an 80 × 80 greyscale image which were feed to the proposed architecture.

Around 6/7 of the database was chosen as the training data and performance was evaluated using remaining molecules. The number of crystalline and non-crystalline molecules were ensured to be almost equal in the training dataset. Each of the entries was labelled either a 1 or a 0 depending on the number whether it is crystalline or non-crystalline. The Table 2 below shows the summary of the dataset breakdown.

Table 2. Dataset summary and breakdown of testing and test molecules.

Dataset	Property	Size	Training	Testing
CSD	Crystalline	15,835	13,570	2265
ZINC	Non-crystalline	15,880	13,660	2220

3.3 Hyper-parameters

We used Mini batch Stochastic Gradient descent (SGD) to train our network. The size of mini batch was set to 10 while the weight decay and momentum was set to 0.001 and 0.9 respectively. Zero mean Gaussian distribution was used to initialize the convolutional kernels, with the standard deviation fixed at 0.01. To initialize the fully connected layers Xavier initialization [22] was used. The weights follow a Gaussian distribution and are chosen so that the variance for both input and output among each layer remains the same. The bias was disabled.

3.4 Classification Results

We have compared our network architecture classification results with the state of the art Crystallinity prediction approach based on machine learning. The machine learning approach first extracts the molecular descriptors using RDKit Cheminformatics toolkit and then applies SVM algorithm with Rbf kernel. To get the accurate results, both the methods were tested on same molecules. The Table 3 is the confusion matrix for the proposed crystallinity prediction model and the molecular descriptors plus SVM. T(Non-Crystalline) is the percentage of non-crystalline molecules that are correctly predicted, F(Non-Crystallinity) are the non-crystalline molecules predicted to be crystalline, T(Crystalline) are the correctly predicted crystalline molecules while the F(Crystalline) are the wrongly predicted crystalline molecules. The proposed architecture almost outperforms the machine learning based method and has an overall classification accuracy of around 90.3% compared to the 89.15% of SVM.

Table 3. Confusion matrix for the proposed architecture and the molecular descriptors plus SVM

	T(Non-Crystalline)	F(Non-Crystalline)	T(Crystalline)	F(Crystalline)
Proposed	89.4%	10.5%	91.2%	8.8%
Molecular descriptors plus SVM	88.1%	11.85%	89.24%	10.7%

The Fig. 4 below displays the Area under the curve (AUC) for proposed model and the molecular descriptors plus SVM respectively.

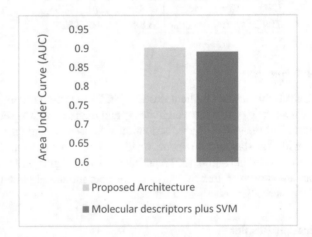

Fig. 4. Area under curve for proposed model and the molecular descriptors plus SVM.

4   Conclusion

This paper proposes a novel approach for organic molecular crystallinity prediction using deep learning based on Convolutional Neural Network (CNN). It is inspired from the Chemception model which is based on the Inception-Resnet v2. With just the 2D images of the chemical structure, the network classifies whether a new molecule has a good probability of forming crystals, without the need of any advanced chemistry knowledge or generation of any relevant molecular descriptors. The proposed method has been compared with the state of the art machine learning approach, that is, generation of molecular descriptors plus SVM classification. Our architecture produces a better classification accuracy and can be a reliable choice when the speed is of utmost priority. Future work can be extending this work to predict the synthetic modifications that can enhance or decrease the crystallization propensity.

References

1. Chen, J., Sarma, B., Evans, J.M.B., Myerson, A.S.: Cryst. Growth Des. **11**, 887–895 (2011)
2. Modarressi, H., Dearden, J.C., Modarress, I.: QSPR correlation of melting point for drug compounds based on different sources of molecular descriptors. J. Chem. Inf. Model. **46**, 930–936 (2006)
3. Le, T., Epa, V.C., Burden, F.R., Winkler, D.A.: Chem. Rev. **112**, 2889–2919 (2012)
4. Mitchell, J.B.O.: Machine learning methods in chemoinformatics. WIREs Comput. Mol. Sci. **4**, 468–481 (2014)
5. Bhat, A.U., Merchant, S.S., Bhagwat, S.S.: Prediction of melting points of organic compounds using extreme learning machines. Ind. Eng. Chem. Res. **47**, 920–925 (2008)
6. Palmer, D.S., O'Boyle, N.M., Glen, R.C., Mitchell, J.B.O.: Random forest models to predict aqueous solubility. J. Chem. Inf. Model. **47**, 150 (2007)
7. Varnek, A., Baskin, I.: Machine learning methods for property prediction in chemoinformatics: quo vadis? J. Chem. Inf. Model. **52**, 1413–1437 (2012)
8. Erić, S., Kalinić, M., Popović, A., Zloh, M., Kuzmanovski, I.: Prediction of aqueous solubility of drug-like molecules using a novel algorithm for automatic adjustment of relative importance of descriptors implemented in counter-propagation artificial neural networks. Int. J. Pharm. **437**, 232–241 (2012)
9. Gawehn, E., Hiss, J.A., Schneider, G.: Deep learning in drug discovery. Mol. Inform. **35**(1), 3–14 (2016)
10. Goh, G.B., Hodas, N.O., Vishnu, A.: Deep learning for computational chemistry. J. Comput. Chem. **38**(16), 1291–1307 (2017)
11. Fooshee, D., Mood, A., Gutman, E., Tavakoli, M., Urban, G., Liu, F., Huynh, N., Van Vranken, D., Baldi, P.: Deep learning for chemical reaction prediction. Mol. Syst. Des. Eng. **3**, 442–452 (2018)
12. Mayr, A., Klambauer, G., Unterthiner, T., Hochreiter, S. DeepTox: toxicity prediction using deep learning. Front. Environ. Sci. **3**(80) (2015). https://doi.org/10.3389/fenvs.2015.00080
13. Wicker, J.G.P., Cooper, R.I.: Will it crystallise? Predicting crystallinity of molecular materials. CrystEngComm **17**, 1927–1934 (2015)
14. Landrum, G.: RDKit: Open-source cheminformatics. http://www.rdkit.org/
15. Suykens, J.A.K., Vanderwalle, J.: Least squares support vector machines. Neural Process. Lett. **9**, 293–300 (1999)
16. Szegedy, C., Ioffe, S., Vanhoucke, V.: Inception-v4, inception-resnet and the impact of residual connections on learning. arXiv:1602.07261 (2016). Inception Resnet v2
17. Goh, G.B., Siegel, C., Vishnu, A., Hodas, N.O., Baker, N.: Chemception: a deep neural network with minimal chemistry knowledge matches the performance of expert-developed QSAR/QSPR models. arXiv preprint arXiv:1706.06689 (2017)
18. Allen, F.H.: Acta Crystallogr. B **58**, 380–388 (2002)
19. Irwin, J.J., Shoichet, B.K.: J. Chem. Inf. Model. **45**, 177–182 (2005)
20. Weininger, D.: SMILES, a chemical language and information-system. 1. Introduction to methodology and encoding rules. J. Chem. Inf. Comp. Sci. **28**(1), 31–36 (1988)
21. O'Boyle, N.M., Banck, M., James, C.A., Morley, C., Vandermeersch, T., Hutchison, G.R.: J. Cheminf. **3**, 33 (2011)
22. Glorot, X., Bengio, Y.: Understanding the difficulty of training deep feedforward neural networks. In: Proceedings of the Thirteenth International Conference on Artificial Intelligence and Statistics, AISTATS2010, Proceedings of Machine Learning-Research, Chia Laguna Resort, Sardinia, Italy, May 2010, vol. 9, pp. 249–256 (2010)

Malayalam to English Translation: A Statistical Approach

Blessy B. John[1](✉), N. V. Sobhana[1], L. Sobha[2], and T. Rajkumar[3]

[1] Computer Science and Engineering, RIT, Kottayam, India
blessybjohn94@gmail.com, sobhananv.rit@gmail.com
[2] AU-KBC Research Centre, Anna University, Chennai, India
sobha@au-kbc.org
[3] College of Engineering Kallooppara, Pathanamthitta, India
rajcek@gmail.com

Abstract. The English language is international language and Malayalam is a morphologically rich language. Malayalam sentence consist of root or stem word with extra suffixes. The Malayalam language uses diverse order of word as distinguish to the language English. Due to the morphological richness of Malayalam leads very few approaches for developing translation system. These approaches are divided into rule based approach and corpus based approach. Statistical machine translation model takes large set of bilingual corpus and monolingual corpus for training. The bilingual corpus contains source sentence and corresponding target sentence. The statistical approach contains translation model, language model and decoder. The translation model assign the probability of Malayalam-English sentence pair based on the maximum likelihood estimation. The decoder is the main step of statistical machine translation. It's find the best target sentence by choosing the highest probability from the product of the translation model and language model.

Keywords: Decoder · Bilingual corpus · Monolingual corpus · Machine translation

1 Introduction

The machine translation is the method of translating the sentence from one natural language into the other language. It is the hardest feature of the natural language processing. There are many difficulties in between the input and output languages. The English language is international language and Malayalam is an agglutinative language and morphologically rich language. Malayalam sentence consist of root or stem word with extra suffixes. The structure of language is difficult to understand. The Malayalam language uses diverse order of word as distinguish to the language English. Due to the morphological richness of Malayalam leads very few approaches for developing translation system. Several approaches are developed for machine translation. These approaches are divided into rule based approach and corpus based approach. The rule based approach of machine translation are divided into Direct, Transfer based MT, Interlingua MT. The corpus based approach of machine translation are divided into

© Springer Nature Switzerland AG 2019
A. Abraham et al. (Eds.): IBICA 2018, AISC 939, pp. 226–234, 2019.
https://doi.org/10.1007/978-3-030-16681-6_23

example based approach and statistical approach. The RBMT contain collection of rules is used for translate the text from input language into the output language. RBMT needs effort of human to generate rules, part-of-speech taggers, parsers, morphological generator, dictionaries etc. The direct approach in the RBMT system is directly translating the text from input language into output language without using any intermediate form. The transfer based approach analyses the input sentence and produce the parse tree of input and target language using transfer rules. The Interlingua approach produces the intermediate representation and generates the target text from this intermediate representation. The corpus based system is fully automatic system which solves the problems of the RBMT system. It's one of the classifier is example based approach which contain already translated example set in the parallel corpus. Then the system check the input text is matched to the stored example set in the bilingual corpus. It produces better translation based on good quality of parallel corpus. The statistical approach uses statistical models extract from the bilingual and monolingual corpus. It takes an assumption that every word in the output language is a translation of the input language words using probability of translation. The best translation has the maximum probability.

Statistical machine translation model takes large set of bilingual corpus and monolingual corpus for training. The bilingual corpus contains source sentence and corresponding target sentence. The monolingual corpus contains target sentences. The statistical model estimates the parameters from the corpus and perform translation. The statistical approach contains translation model, language model and decoder. The translation model assign the probability of Malayalam-English sentence pair based on the maximum likelihood estimation. The language model gives the fluency of the output English sentence. The decoder is the main step of statistical machine translation. It's find the best target sentence by choosing the highest probability from the product of the translation model and language model.

This work is a preliminary report of the project on Malayalam to English Translation: A Statistical Approach. Section 1 includes a general introduction of the project. Section 2 is the literature survey on Malayalam to English Translation: A Statistical Approach and a study about the previous techniques. Section 3 gives the methodology of the system. Section 4 gives the implementation of the system. Section 5 gives the evaluation of the translation system and Sect. 6 concludes the work.

2 Related Works

The different translation techniques are used for translating sentence from one language to another. Some of them are described here.

Nithya et al. [9] proposed a Hybrid Approach to English to Malayalam Machine Translation. The system combines statistical approach and translation memory. The statistical approach builds bilingual corpus and monolingual corpus for automating translation. The system uses statistical model which estimate parameters from the corpus for translation. The statistical model contains each sentence in the output language is consider as the translation of sentence in the input language with probabilities. The sentence has highest probability is the best translation. The translation memory acts as cache which stores the recent previous translation and is used to avoiding the requirement for performing redundant conversions. The system firstly checks the input sentence has a match in the translation memory. If the match is found, then corresponding translation is the output otherwise perform statistical approach. The evaluating the performance of the system using BLEU score.

Sebastian et al. [5] proposed English to Malayalam translation: A statistical approach (2010). This work is a method that translates English text to the Malayalam text using the statistical approach. It uses part of speech information into the parallel corpus to improve the alignment model. To ensure the efficiency of training by using the pre-processes like elimination of stop word from the parallel corpus and suffix separation from the Malayalam corpus used. In the training process, contain two corpuses to implement suffix separation, POS tagging, stop word elimination. The main methods in the decoding phase are tagging and order conversions. This statistical approach used to increase the quality of translation with help of large bilingual corpus. Anju et al. [2] proposed Malayalam to English machine translation: an EBMT system (2014). The Example Based Machine Translation (EBMT) system is the method used to translate the Malayalam to English language. The different modules in the system contains Example acquisition, Matching and recombination. In Example acquisition module, contain already translated example set in the form of parallel corpus. In matching module, it matches the input sentence with the stored sentence in the bilingual corpus and produce best matching result corresponding to the input sentences. Recombination step produce target output by combining translated fragments. This system will produce better translation quality.

Aasha et al. [3] proposed Machine translation from English to Malayalam using transfer approach (2015). This system use transfer approach to translate the English sentence to Malayalam sentence. This system contains different modules such as tokenization, parser, morphological generator and reordering. The initial step is tokenizing the input sentence into tokens and labelling the tokens using POS tagger. Then obtain source parse tree by using Stanford parser. Reorder the words in the source parse tree. Apply transfer rules to the input parse tree and produce output parse tree for translation. The words from the parse tree are mapped to the English to Malayalam bilingual dictionary or using phoneme based transliterator for obtains target text. By using phonetic based transliteration module, the named entities in the sentence are transliterated.

George et al. [12] proposed English to Malayalam Statistical Machine Translation System. The system develops translation of English text into the Malayalam text using statistical approach. The system contains Language Model (LM), Translation Model (TM) and a Decoder. Bayesian network model as Hidden Markov Model (HMM) are required to developing the statistical models. The Bayesian network model is used for align parallel corpus. The language model used n-gram model to produces the probability of output sentence. The translation model produces translation probability for parallel corpus which contain both source and target language using Baum Welch algorithm. The decoder maximizes the probability of translated sentence.

3 Methodology

The statistical approach of translation is used to translate the Malayalam text into the English text. It is corpus based machine translation system.

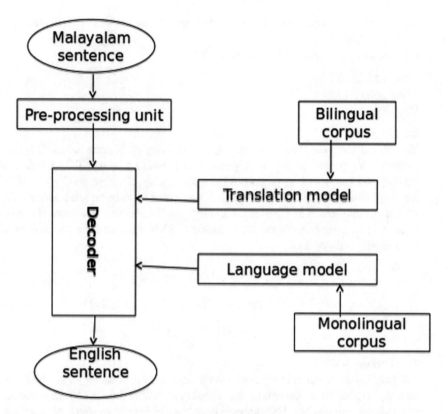

Fig. 1. System model

Statistical approach takes huge collection of parallel corpus and monolingual corpus for training. The bilingual corpus contains source sentence and corresponding target sentence. The monolingual corpus contains target sentences. The statistical approach estimates the parameters from the data set and perform translation. The statistical approach of machine translation is explained as each sentence in the output language has the translation of the sentence in the input language with probability. The best translation of text has the maximum probability. The system model of SMT is shown in the Fig. 1. The main objective of statistical approach is to translate the text from source language into output language. Consider the input sentence i from the

Malayalam language, target English sentence t, the translation probability of translating the Malayalam sentence into English sentence is p(i|t) and probability of target sentence is p(t). Using Bayes Theorem, the best translation is computed by,

$$b = \arg\max_t p(t)\, p(i\mid t) \tag{1}$$

Where P(t) is the language model and P(i|t) is the translation model. This method is known as decoding.

The translation system consists of three units:

(1) Language Model
(2) Translation Model
(3) Decoder

(1) Language Model
     This model gives the fluency of the output text using the n-gram model. It is used to determine the probability of output text. It is built by using SRILM tool. It is trained over the monolingual corpus. The probability of single word given all of the word that precedes it in a sentence is computed by using n-gram model. The size of n-gram model is 1 is termed as unigram, 2 is bigram and soon. By using chain rule, the probability of the sentence p(A) is spilt into the probability of individual words p(v) is

$$P(A) = P(v1, v2, v3 \ldots vn) \tag{2}$$

$$P(v) = P(v1)\, P(v2|v1)\, P(v3|v1v2)\, P(v4|v1v2v3) \tag{3}$$

$$P(vn|v1\ v2 \ldots vn1) \tag{4}$$

(2) Translation Model
     The translation model is trained over the bilingual corpus. Using an unsupervised learning algorithm to determine the translation probability called Expectation-Maximization algorithm. Here phrases based model is used in statistical approach of machine translation. In the phrase based model, one word in input language is translated into two or more words in output language. GIZA++ is a word alignment tool is used for translation model. It helps to arrange words in statistical models. This translation probability p(i|t) is high if the Malayalam sentence and English sentence are equivalent.

(3) Decoder
     Decoder is the crucial part of statistical machine translation system. It finds the best translation. It maximizes the output of the product of language model and translation model. The output from this models is passed into the decoder and it find out the maximum probability. Moses tool is used for decoder. It is used to find out the best translation. The best first search method is used in decoder based on heuristics.

4 Implementation

4.1 Data Set

The basic need of the statistical machine translation is corpus preparation. The bilingual and monolingual corpus is used for producing statistical models. The bilingual corpus contains Malayalam and English sentences. The monolingual corpus contains English sentences. There are 10000 sentences in a data set. These data set is used in training, tuning and testing. Data set is based on information related to India. The corpus is initially pre-processed like tokenization, truecasing and cleaning.

Tokenization: Tokenize the input sentence into tokens.
Truecasing: Decrease data sparsity.
Cleaning: Remove the long sentences and empty sentences.

4.2 Language Model Training

The training of language model is done on monolingual corpus which contain English sentences. The corpus is firstly pre-processed like tokenization and truecasing. It is built by using the SRILM tool. The SRILM (Stanford Research Institute Language Model) is a tool which is required for training the language model. To find the probability of output English sentence using the n-gram model. Here 3-gram model is used.

4.3 Training the Translation System

The translation model is trained on bilingual corpus which contain both Malayalam and English sentences. During the training, create word alignment using GIZA++, phrase extraction tables and scoring, lexicalized reordering tables and lastly generate Moses configuration file. GIZA++ is used for training translation model. It is used to word alignment. It's implement a word alignment model based on the hidden markov model.

4.4 Tuning

Tuning is used for speed up the decoder. The small set of parallel corpus which is separated from training data is used to tuning the system. This parallel corpus is initially subjected to preprocessing like tokenizing and truecasing. Its create moses.ini configuration file with train weight using mertmoses.pl script.

4.5 Testing

This is the last step in the statistical approach of the system. Moses tool is used to run the decoder. The Moses is used to determine the translation of the input sentences using heuristic search procedure. It invokes the moses executable file and configuration file moses.ini.

5   Results

The translation system was trained with various size of parallel corpus. The various parallel corpus contains 1k to 13k sentences were used for training. The monolingual corpus contain English sentences were used for tuning. The system performance was done by testing the translation system using bleu score. We have testing the translation system with tested corpus. From this testing we can clear that the size of the parallel corpus increases then the accuracy of translation also increases. Some of the translated sentences are shown in the Table 1.

Table 1. Translation of some sentences

Malayalam Sentences	English Sentences
ഇതിലെ തന്നെ ചില കൊടുമുടികൾ 14000 അടിയോളം ഉയരമുള്ളവയാണ്.	some of these mountain peaks will be more than 14000 feet high
1999 മികച്ച ചലച്ചിത്ര നിർമാതാവ് – വാനപ്രസ്ഥം	1999 best film producer-vanaprastham .
നമ്പ്യാർ കവിതയുടെ വിമർശനം	The criticism of Nambiar Poems .

5.1   BLUE Evaluation

The Bilingual Evaluation Understudy is an automatic evaluation method used in machine translation system. It's compare the n-gram of the candidate translation to the n-gram of the reference translation and compute the total matches. The bleu score is high means better candidate translation. Bleu score denotes the accuracy of the system. Its calculation method is

$$\text{BLEU} = e^{\min(1-rt/ct,0)} \cdot e^{\left(\sum_{n=1}^{N} (1/N) \log P_n\right)} \tag{5}$$

Here ct and rt are the number of words in the candidate and reference translation respectively. The component brevity penalty $e^{\min\left(1-\frac{rt}{ct},0\right)}$ is less than one if ct < rt the candidate translation is shorter than the reference translation. The bleu score evaluation of the various size of corpus as shown in the Table 2.

Table 2. Analysis of corpus

Corpus	Corpus1	Corpus2	Corpus3	Corpus4	Corpus5
Training (No. of sentences)	1.1k	2.6k	6.5k	10	13
Tuning	150	150	150	150	150
BLEU score	5.65	9.16	9.67	16.32	25.73

Based on the analysis the bleu score is plotted against the size of the corpus is graphically represented as shown in the Fig. 2. This bleu score is small due to the size of the corpus. This translated sentences have correct translation, understandable, partially understandable and incorrect translation. The readability problem is caused due to the word alignment. This result is improved by using millions of sentences. The various size of corpus is tested for finding accuracy as shown in the Table 2.

6 Conclusion

The Bilingual Evaluation Understudy is an automatic evaluation method used in machine translation system. It's compare the n-gram of the candidate translation to the n-gram of the reference translation and compute the total matches. We have tested the system with tested corpus for finding accuracy. After testing the system found that 70% sentences are correctly translated, some sentences are partially understandable and 13% sentences are incorrect. The precision is computed as the ratio of number of correctly translated sentence to the number of input sentences. The value of precision in between 0 and 1. The statistical machine translation system was obtained 70.7% of accuracy. Thus the proposed model performs well and provides more accuracy.

References

1. Nair, L.R., David Peter, S., Ravindran, R.P.: Design and development of Malayalam to English translaor a transfer based approach, vol. 3 (2012)
2. Anju, E.S., Manoj Kumar, K.V.: Malayalam to English machine translation: an EBMT system, vol. 2, pp. 18–23 (2014)
3. Aasha, V.C., Ganesh, A.: Machine translation from English to Malayalam using transfer approach. IEEE (2015)
4. Nair, A.T., Idicula, S.M.: Syntactic based machine translation from English to Malayalam, pp. 198–202. IEEE (2012)
5. Sebastian, M.P., Sheena Kurian, K., Santhosh Kumar, G.: English to Malayalam translation: a statistical approach. In: Proceedings of the 1st Amrita ACM-W Celebration on Women in Computing in India, p. 64. ACM (2010)
6. Haroon, R.P., Shaharban, T.A.: Malayalam machine translation using hybrid approach. In: International Conference on Electrical, Electronics, and Optimization Techniques (ICEEOT) (2016)
7. Jayan, J.P., Rajeev, R., Rajendran, S.: Morphological analyser and morphological generator for Malayalam-Tamil machine translation. Int. J. Comput. Appl. (0975 – 8887) **13**(8) (2011)

8. Rajeev, R., Sherly, E.: A suffix Stripping based morph analyser for Malayalam language. In: Proceedings of 20th Kerala Science Congress, pp. 482–484, 28–31 (2008)
9. Nithya, B., Joseph, S.: A hybrid approach to English to Malayalam machine translation. Int. J. Comput. Appl. (0975 – 8887) **81**(8) (2013)
10. Anisree, P.G., Radhika, K.T.: A hybrid translator: from Malayalam to English. Int. Res. J. Eng. Technol. (IRJET) **03**(07) (2016). e-ISSN 2395-0056
11. Koehn, P.: Moses: open source toolkit for statistical machine translation. In: The Proceedings of the ACL, pp. 177–180, June 2007
12. George, A.: English to Malayalam statistical machine translation system. Int. J. Eng. Res. Technol. (IJERT) **2**(7) (2013)
13. Sanchez-Cartagena, V.M., Sanchez-Martınez, F., Perez-Ortiz, J.A.: Integrating shallow-transfer rules into phrase-based statistical machine translation. Transducens Research Group Departament de Llenguatges i Sistemes Informatics Universitat d'Alacant, E-03071
14. Imamura, K., Okuma, H., Watanabe, T., Sumita, E.: Example-based machine translation based on syntactic transfer with statistical models. ATR Spoken Language Translation Research Laboratories 2-2-2 Hikaridai

GABoost: A Clustering Based Undersampling Algorithm for Highly Imbalanced Datasets Using Genetic Algorithm

O. A. Ajilisa[1], V. P. Jagathyraj[2(✉)], and M. K. Sabu[1(✉)]

[1] Department of Computer Applications,
Cochin University of Science and Technology, Kochi, Kerala, India
{ajilisaaliyar,sabumk}@cusat.ac.in
[2] School of Management Studies, Cochin University of Science and Technology,
Kochi, Kerala, India
jagathy@cusat.ac.in

Abstract. Data sets that have imbalanced class distribution is a challenging problem for many application domains. Learning from imbalanced data can't be done efficiently using current data mining and machine learning tasks. Instead of merely using those algorithms we have to consider some other techniques to learn from those data set. One solution is to develop some preprocessing methods to balance the data sets and combine it with some existing algorithm. In this paper, we propose a new hybrid clustering based undersampling technique using genetic algorithm and AdaBoost, which is called GABoost, for learning from imbalanced data. This algorithm is an attractive alternative for SMOTEBoost, RUSBoost, CUSBoost. Based on the experimental results obtained from 44 imbalanced datasets we strongly recommend GABoost as a striking alternative for improving the performance of the learned classification model which is built using highly imbalanced dataset.

Keywords: Imbalanced data sets · Undersampling · Clustering · Genetic algorithm · Classifier ensembles

1 Introduction

Classification is an essential and integral part of machine learning and pattern recognition problems. Most of the canonical machine learning algorithms assume that the data set has balanced class distribution and can't address the issue of imbalanced data sets efficiently. Imbalanced data sets are those data sets which hold unequal distribution between each class involved. Questions regarding class imbalance have intensively researched in the last decade, and several methods to alleviate this problem was proposed [12]. It is complicated to develop a useful classifier model for imbalanced data sets by using current data mining and machine learning algorithms.

ⓒ Springer Nature Switzerland AG 2019
A. Abraham et al. (Eds.): IBICA 2018, AISC 939, pp. 235–246, 2019.
https://doi.org/10.1007/978-3-030-16681-6_24

There are a wide variety of approaches that have been proposed to solve issues regarding imbalanced dataset. Such methods can be grouped into four different types: Data level methods, Algorithmic methods, Cost-sensitive methods, Ensemble of classifiers. In this paper, the main focus is given to those approaches that follow preprocessing the imbalanced dataset before learning a classifier from it. According to Galar et al., who conducted a brief comparative study on well-known procedures, classifiers that combine data preprocessing methods along with classifier ensemble, outperform with other well-known approaches [5]. Data preprocessing methods involve resampling the imbalanced dataset by either using oversampling the minority class instances or undersampling the majority class instances before learning a classifier. Some representative approaches combine resampling technique with well-known ensemble methods such as Bagging, Boosting, etc.; for example SMOTEBoost, RUSBoost, OverBagging, UnderBagging [14]. In all well-known approaches, undersampling techniques have shown to be a better choice than the oversampling technique because oversampling approach may increase the chance of overfitting. Undersampling can be done by either using random undersampling or non-random undersampling. In the case of random undersampling, there is a chance of eliminating useful data because majority class instances are selected randomly. To eliminate this issue, several approaches replace random undersampling with other technique. One of the essential methods in this category is clustering based undersampling [8].

In this paper, we propose a clustering based undersampling method using a genetic algorithm that provides more optimum clusters. By obtaining such clusters, dataset after resampling will be a good representative of the original data set. Classification model obtained from the resultant data set will outperform other approaches.

The remaining part of this paper is organized as follows. Section 2 reviews some widely used representative approaches in the area of imbalanced data classification and explains their scopes and limitations. Section 3 explains our hybrid preprocessing approach for cluster-based undersampling based on genetic algorithm. It also includes the concepts and techniques needed to understand the genetic algorithm. Section 4 deals with the experimental design and Sect. 5 report the obtained results. Section 6 concludes the paper.

2   Conventional Classifiers for Imbalanced Domain

Different techniques that can be used in the imbalanced domain can be categorized into mainly three categories. These solutions are mostly based on data, algorithm, cost sensitivity [7]. Recently hybrid combinations of data level solutions with ensemble technique gained more attention than other techniques [3,5,8]. The main advantage of this approach is to make sampling and classifier training independent. Using this different approach variety of sampling method can be easily combined with different classification technique. Galar et.al. shows sampling solutions can quickly solve issues regarding class imbalance problem and can improve the performance of the classification model [5].

In 2013, Galar et al. demonstrate combinations of sampling with different ensemble approaches can provide a more accurate classification model [6]. This section briefly describes all the categories and gives an idea about some state of art techniques which is used in this paper.

Data level solutions are based on balancing the imbalanced training data by using either undersampling the majority class instances or oversampling minority class instances before applying classification technique. It does not modify existing classification algorithm rather it is considered as a preprocessing technique before applying classification.

Algorithmic level solutions involve algorithms that can directly handle imbalanced dataset. It may use either threshold method or one class learning. In the case of the threshold method, it involves assigning different threshold values for different classes during the classification. But in one class learning, it involves learning the classifier model from the dataset that contains instances of one specific class so that learned classifier can identify the features of the specified class.

Cost sensitive solutions include all the approaches including data level solutions, algorithmic level solutions or combination of both. It tries to increase the classification ability by assigning larger misclassification cost for minority class instances and tries to minimise higher cost errors [9].

2.1   Combinations of Resampling and Ensemble Classifier

We briefly describe ensemble based data preprocessing approaches which are related to our proposal.

SMOTEBoost. SMOTE is considered as one of the most commonly used oversampling approaches for solving class imbalance problem. It generates synthetic training examples by using linear interpolation of minority class instances. Various classification techniques can be clubbed together with SMOTE to improve the performance of imbalanced classification [4,5].

RUSBoost. It is based on random undersampling and AdaBoost algorithm. This approach provides a simpler and faster method to implement undersampling. This method will select majority class instances randomly until the data set is balanced. This will increase the classifier performance and improves run time. But removing majority class instances may lead to loss of useful information [13].

Clustering Based Undersampling Technique. In order to overcome the limitations of random undersampling techniques, Lin et.al proposed a clustering based undersampling technique. In this case, simple K-Means clustering algorithm is used to cluster majority class instances such that inter-cluster similarity is low and intracluster similarity is high. Then each cluster centroid which is based on the mean of the cluster elements is used to represent the whole cluster. Thereby reducing the size of the majority class. This approach reduces the risk of removing useful data from the majority class enables to outperform the classifiers developed using a random undersampling strategy [8]. But the main problem associated with the K-Means algorithm is it may be stuck at a local optimum solution and the result depends on the selection of initial centroid. There are other technique, like graph-theoretic approach and hierarchical approach, to implement clustering process which is not simple as K-Means. But we want an algorithm that should be conceptually simple like K-Means and it will provide good results irrespective of starting point.

3   Proposed Method

3.1   Genetic Algorithm

Genetic Algorithms are randomized search and optimization techniques based on the concept of natural selection and natural genetics. It is motivated by Darwinian theory of evolution. Genetic Algorithm performs the search in complex, large and multi-modal landscapes and provides near-optimal solutions for the fitness function of an optimization problem [10]. In genetic algorithms, the parameters of the search space that we want to optimize through different generations are encoded in a string known as chromosomes. A collection of all chromosome strings are known as population. Initially random population is created which contains a set of strings (initial chromosomes). There is a fitness function based on what we want to optimize. Value of fitness function is computed for each chromosome that represents degree of goodness of the string or how good a solution it is to a particular problem. It carries out a process of chromosome selection based on its fitness value and recombine selected chromosome to produce chromosome for next generation. Parent chromosomes are selected and their genetic materials are recombined to produce child chromosome. This newly generated population of chromosome represents candidates for the next generation. As this process is iterated, successive generations are evolved and their fitness value become better and better. This process is continued until a specific stopping criteria is reached. In this way, the genetic algorithm gives an optimal solution to a problem [10,11].

3.2   GA-Clustering Algorithm

In this work, the optimization capability of the genetic algorithm is used for obtaining K optimum cluster centers such that inter-cluster similarity is low and intra-cluster similarity is high. Then it can be used for optimally clustering n different points. Sum of the Euclidean distances of the points from their respective cluster centers will be taken as a fitness function for the genetic algorithm. GA aims to find k optimum cluster centers such that this clustering metric is minimized.

12.3	34.2	14.6	32.1	15.6	17.9	23.1	43.2

Fig. 1. Representation of a chromosome

String Representation. Each chromosome represents a set of real numbers representing K cluster centers. When we deal with N-dimensional space, it will be N * K real numbers, where first N real values indicate the first cluster center, second N real values indicate the second cluster center and so on [10]. For example, let N = 2 and K = 4 where we are dealing with two-dimensional space and number of clusters we are dealing with are 4. Then the length of the chromosomes will be 8. First two values represent the first cluster center. Next two values represent next cluster center and so on [10]. Figure 1 represents four cluster centers (12.3, 34.2), (14.6, 32.1), (15.6, 17.9), (23.1, 43.2).

Population Initialization. Each chromosome represents k cluster centers in N-dimensional space. So k points in N-dimensional space are selected randomly from the dataset, and it is encoded into a chromosome. This process is repeated P times where P is the number of chromosomes in the population or size or length of the population. Size of the population is determined on a random basis, and it is fixed throughout the experiment.

Fitness Computation. Fitness value for each chromosome will be computed in this step. This phase consists of two different processes. In the first stage data values are clustered according to given cluster centers encoded in the chromosome and clustering metric will be computed in the next stage. This clustering metric will be treated as a fitness value for the particular chromosome. The process is repeated for each chromosome in the entire population range. In first stage every instance in data set x_i where i = 1, 2, 3, 4....n, where n is the total number of instances in the data set will be assigned to one of the clusters C_j with center z_j which is encoded in the chromosome, such that $\| x_i - z_j \| < \| x_i - z_p \|$ where p = 1, 2...k and p ≠ i. After clustering, cluster centers are replaced by mean values of numbers that belongs to particular cluster. In other words for each cluster C_i the new center z_i^* is computed as

$$z_i^* = \frac{\sum_{x_j \epsilon c_i} x_j}{n_i} i = 1, 2...K \tag{1}$$

$$\mu = \sum_{i=1}^{K} \mu_i \tag{2}$$

$$\mu_i = \sum_{x_j \epsilon c_i} \| x_j - z_i \| \tag{3}$$

After clustering, the metric for clustering is calculated for each chromosome. Clustering metric can be calculated by using either of the following two methods. In the first method very instance is simply assigned to one of the cluster centers which is encoded in the chromosome and the sum of the distance between each cluster center and its corresponding elements will be calculated. This value corresponds to the clustering metric or fitness value of the chromosome. In the second method, by taking the cluster center which is encoded in the chromosome as initial cluster center, K-Means clustering will be performed. After completion of K-Means, clustering metric will be computed and it will be taken as a fitness value for the chromosome. We completed implementation with the second method.

Selection. Chromosomes will be arranged according to the sorted order computed fitness value. During the process of selection, chromosomes with the highest fitness value will be selected from the mating pool.

Elitism. Elitism ensures that best chromosomes are allowed to pass their traits to the next generation to reduce genetic drift. Some genes of chromosomes which is having high fitness value remain unchanged and will be allowed to pass to next iteration. Elitism rate is fixed as during the entire process.

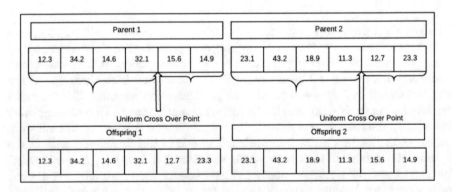

Fig. 2. Cross over operation

Crossover. In the process of crossover, two-parent chromosomes will exchange their segments and generate two new offspring. The process of crossover in the clustering scenario that we considered is given below. Peng et al. experimentally states that single point crossover performs better than the multipoint crossover [1]. So in this proposed method, we use single point crossover to perform crossover between two parent chromosomes and the process of crossover is given in Fig. 2.

Mutation. In order to maintain genetic diversity from one generation of a population to the next generation, the mutation operator is applied to the chromosome. Generally in a binary representation of a chromosome, mutation is applied by simply flipping one of the gene value in a chromosome with a fixed mutation probability μ_m. Here we are considering the fixed-point representation of cluster centres, so mutation operation that we used is given in the following equation. A number δ is generated in the interval $[0, 1]$ with uniform distribution. Let ν is the value present at the gene position, then after mutation value will be

$$\nu = \nu \pm 2 * \delta * \nu, v \neq 0 \tag{4}$$

$$\nu = \nu \pm 2 * \delta, \nu = 0 \tag{5}$$

'+' or '−' will be occur at equal probability. Here the process of fitness computation, selection, crossover, mutation are executed a fixed number of times. We have implemented elitism in each iteration that preserves best chromosomes in the current generation. After all iteration, best chromosome in the final iteration represents optimized cluster centres.

3.3 GA-Boost: Clustering Based Undersampling Based on Genetic Algorithm

The required number of clusters (denoted by K) is determined using imbalance ratio $(K = \frac{M}{N})$. Consider a data set that contains 100 majority class instances and 20 minority class instances. Here the number of clusters required is taken as 5 $(K = \frac{100}{20})$. Then Genetic Algorithm based clustering technique is applied to obtain K optimized cluster centres. By taking this centre as the initial centroid, K-Means clustering technique will be applied to find K optimum cluster. From each cluster, $\frac{N}{K}$ elements will be selected randomly and combine with minority

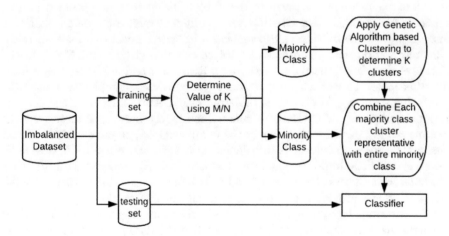

Fig. 3. GABoost model

class instances. In the above example, from each of the 5 clusters, 4 elements will be retrieved and the set of all these elements will be combined with minority class instances. AdaBoost classifier is trained using this newly generated dataset. The entire process will be represented in the Fig. 3.

4   Experimental Framework

To evaluate the performance of our proposed method, we use 33 imbalanced real world binary-data sets from KEEL data set repository which is publicly available on the corresponding web page [2]. To handle multi-class data set, it will be converted to a binary class in such a way that the union of one or more classes will be treated as majority class, and same will be applied for getting minority class instances. Table 1 explains the properties of the data sets such as the number of instances, number of features, the imbalance ratio which indicates the number of instances of each class. In this paper, the classification performance is evaluated using AUC, which is the most commonly used evaluation metric in the imbalanced domain. In this experiment, we compared the performance of the proposed undersampling ensemble approach based on genetic algorithm based clustering (GABoost) with other state of art techniques such as AdaBoost, RUSBoost, SMOTEBoost, CUSBoost. Performance of each method is evaluated using AUC (Area Under ROC Curve). As for base learner, we used c4.5 decision tree induction integrated with Boosting ensemble approach.

5   Results and Discussion

Table 2 shows the classification performance of the various ensemble approaches that we used for conducting the experiment. For better readability, we again divide the results based on the imbalance ratio of data sets and average performance of each classifier is given in the Table 3. From this, we obtained performance of various classifiers with data sets of different value for imbalance ratio. It also shows how classification performance varies with respect to imbalance ratio. The information that we obtained is plotted and it is given in the Fig. 4. In the case of slightly imbalanced data sets (0–5), RUSBoost and GABoost outperform all other methods. But when we consider computational complexity along with predictive accuracy, RUSBoost is preferred over GABoost Because the computational complexity of GABoost is constrained by the number of majority class instances and it will be high due to the complexity of genetic algorithm based clustering process involved. If imbalance ratio is in the range 5–10, SMOTE-Boost shows little improvement in performance when compared with all other methods. But In all other cases of highly imbalanced data sets (10–20, 20–30, >30), GABoost provide better results than all other methods.

From the above experiment, we can see that our proposed hybrid method outperforms other state of art techniques when the data set imbalance ratio is higher. When the imbalance ratio is lower, it shows an average performance. But in all cases, it shows better performance when compared with CUSBoost that

Table 1. Data set characteristics

	Datasets	No. of data samples	No. of features	Imbalance ratio
1	Abalone9-18	731	8	16.68
2	Abalone19	4174	8	128.87
3	Ecoli-0_vs_1	220	7	1.86
4	Ecoli-0-1-3-7_vs_2-6	281	7	39.15
5	Ecoli1	336	7	3.36
6	Ecoli2	336	7	5.46
7	Ecoli3	336	7	8.19
8	Ecoli4	336	7	13.84
9	Glass0	214	9	3.19
10	Glass0123vs456	192	9	10.29
11	Glass016vs2	184	9	19.44
12	Glass016vs5	214	9	1.82
13	Glass1	214	9	10.39
14	Glass2	214	9	15.47
15	Glass4	214	9	22.81
16	Glass5	214	9	22.81
17	Glass6	214	9	6.38
18	Haberman	306	3	2.68
19	Iris0	150	4	2
20	Newthyroid1	215	5	5.14
21	New-thyroid2	215	5	4.92
22	Page-blocks0	5472	10	8.77
23	Page-blocks13vs2	472	10	15.85
24	Pima	768	8	1.9
25	Segment0	2308	19	6.01
26	Shuttle0vs4	1829	9	13.87
27	Shuttle2vs4	129	9	20.5
28	Vehicle0	846	18	3.23
29	Vehicle1	846	18	2.52
30	Vehicle2	846	18	2.52
31	Vehicle3	846	18	2.52
32	Vowel0	988	13	10.1
33	Wisconsin	683	9	1.86
34	Yeast0567vs4	528	8	9.35
35	Yeast1	1484	8	2.46
36	Yeast3	1484	8	8.11
37	Yeast4	1484	8	28.41
38	Yeast5	1484	8	32.78
39	Yeast6	1484	8	39.15
40	Yeast1vs7	459	8	13.87
41	Yeast1289vs7	947	8	30.56
42	Yeast1458vs7	693	8	22.1
43	Yeast2vs4	514	8	9.08
44	Yeast2vs8	482	8	23.1

Table 2. Classification Performance of different approaches

Datasets	IR	AdaBoost	RUSBoost	SMOTEBoost	CUSBoost	GABoost
Abalone9-18	16.4	0.6934	0.7051	0.7195	0.7243	0.7321
Abalone19	128.87	0.5723	0.5923	0.5783	0.6090	0.6220
Ecoli	71.5	0.6354	0.6174	0.6597	0.6589	0.6987
Ecoli-0_vs_1	1.86	0.6934	0.7865	0.7841	0.7543	0.7569
Ecoli-0-1-3-7_vs_2-6	39.15	0.7164	0.7237	0.7642	0.7582	0.7793
Ecoli1	3.36	0.6971	0.7187	0.7154	0.7075	0.7131
Ecoli2	5.46	0.7031	0.7165	0.7174	0.6993	0.7098
Ecoli3	8.19	0.7151	0.7075	0.7158	0.6976	0.7046
Ecoli4	13.84	0.7121	0.7020	0.7213	0.7298	0.7231
Glass0	3.19	0.7545	0.7996	0.7985	0.7995	0.8031
Glass0123vs456	10.29	0.7654	0.7543	0.7677	0.7896	0.7954
Glass016vs2	19.44	0.7543	0.7585	0.7439	0.7675	0.7653
Glass016vs5	1.82	0.6952	0.6994	0.6990	0.6875	0.7032
Glass1	10.39	0.6954	0.6958	0.7143	0.6971	0.7121
Glass2	15.47	0.7831	0.7997	0.7842	0.7932	0.8052
Glass4	22.81	0.7645	0.7656	0.7657	0.7859	0.7995
Glass5	22.81	0.7363	0.7563	0.7905	0.8031	0.8091
Glass6	6.38	0.6998	0.7154	0.7321	0.7332	0.7326
Haberman	2.68	0.6985	0.7054	0.7089	0.7060	0.7096
Iris0	2	0.6998	0.7063	0.7131	0.7056	0.7001
led7digit	10.97	0.8937	0.8879	0.8873	0.8612	0.8721
Newthyroid1	5.14	0.6998	0.6937	0.6932	0.6934	0.6948
New-thyroid2	4.92	0.6457	0.6732	0.6551	0.6751	0.6851
Page-Blocks	164	0.7605	0.7775	0.8123	0.8281	0.8356
Page-blocks13vs2	15.85	0.7564	0.7522	0.7854	0.7786	0.7896
Pima	1.87	0.6223	0.6376	0.6597	0.6629	0.6654
Segment0	6.02	0.9061	0.9570	0.9510	0.9430	0.9487
Shuttle (statlog)	8.53	0.8331	0.8327	0.8445	0.8242	0.8274
Shuttle0vs4	13.87	0.9764	0.9754	0.9675	0.9589	0.9694
Shuttle2vs4	20.5	0.9656	0.9698	0.9629	0.9757	0.9899
Vehicle0	3.23	0.9565	0.9754	0.9643	0.9626	0.9676
Vehicle1	2.52	0.9599	0.9633	0.9432	0.9543	0.9587
Vehicle2	2.52	0.9765	0.9743	0.9526	0.9638	0.9687
Vehicle3	2.52	0.9643	0.9754	0.9698	0.9679	0.9765
Vowel0	10.1	0.6973	0.6899	0.7042	0.7132	0.7155
Wisconsin	1.86	0.6987	0.6876	0.6987	0.6875	0.6853
Yeast0567vs4	9.35	0.9121	0.9262	0.9041	0.9120	0.9152
Yeast1	2.46	0.9043	0.8976	0.9176	0.9089	0.9117
Yeast3	8.11	0.9172	0.8965	0.9120	0.9074	0.9121
Yeast4	28.41	0.8454	0.8665	0.8663	0.8815	0.8975
Yeast5	38.73	0.8731	0.8871	0.9050	0.9134	0.9263
Yeast6	39.15	0.9043	0.8961	0.9176	0.8976	0.9321
Yeast1vs7	13.87	0.8454	0.8487	0.8326	0.8453	0.8573
Yeast1289vs7	30.56	0.8765	0.8854	0.9105	0.9117	0.9176
Yeast1458vs7	22.1	0.6975	0.7154	0.7179	0.7232	0.7385
Yeast2vs4	9.08	0.6975	0.7175	0.7169	0.7187	0.7119
Yeast2vs8	23.15	0.7589	0.7382	0.7410	0.7603	0.7905

Table 3. Average classification performance of different approach based on imbalance ratio

IR	AdaBoost	RUSBoost	SMOTEBoost	CUSBoost	GABoost
0–5	0.7833	0.8002	0.7989	0.7956	0.8003
5–10	0.7871	0.7951	0.7986	0.7921	0.7952
10–20	0.7794	0.7790	0.7844	0.7872	0.7943
20–30	0.7947	0.8020	0.8074	0.8216	0.8375
>30	0.7626	0.7685	0.7925	0.7967	0.8159

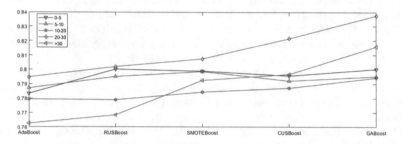

Fig. 4. Graph

uses simple K-Means algorithm. But in the case of slightly imbalanced datasets, both CUSBoost and GABoost failed to show a better performance.

6   Conclusion

In this paper, we examined the following research question: Is it possible to develop an efficient classifier from an imbalanced dataset by using a genetic algorithm based clustering for undersampling. The results obtained from the experiments have shown that the performance of proposed classifier that uses genetic algorithm based clustering for undersampling has improved AUC value when compared with other considered boosting extensions, which are known to be well-performed classifiers for imbalanced learning problems. In the case of a slightly imbalanced classification problem, it does not show a significant improvement in the performance of the classifier. But it shows a substantial improvement in the case of a highly imbalanced dataset. In future work, we intend to extend this work by implementing this proposed methodology in a big-data framework to solve classification problems in highly imbalanced big-data domain. We will continue with an application of the proposed method in multiclass classification problems. We will enhance genetic algorithm based clustering algorithm to obtain more optimized cluster centers which can be used in the domain of imbalanced dataset.

References

1. Peng, P., Addam, O., Elzohbi, M., Özyer, S.T., Elhajj, A., Gao, S., Liu, Y., Özyer, T., Kaya, M., Ridley, M., Rokne, J.: Reporting and analyzing alternative clustering solutions by employing multi-objective genetic algorithm and conducting experiments on cancer data. Knowl.-Based Syst. **56**, 108–122 (2014)
2. Alcala-Fdez, J., Fernández, A., Luengo, J., Derrac, J., Garc'ia, S., Sanchez, L., Herrera, F.: KEEL data-mining software tool: data set repository, integration of algorithms and experimental analysis framework. J. Multiple-Valued Log. Soft Comput. **17**, 255–287 (2010)
3. Beyan, C., Fisher, R.: Classifying imbalanced data sets using similarity based hierarchical decomposition. Pattern Recogn. **48**(5), 1653–1672 (2015)
4. Chawla, N., Lazarevic, A., Hall, L., Bowyer, K.: SMOTEBoost: improving prediction of the minority class in boosting. In: Proceedings of Principles of Knowledge Discovery in Databases, pp. 107–119 (2003). cited By 43
5. Galar, M., Fernandez, A., Barrenechea, E., Bustince, H., Herrera, F.: A review on ensembles for the class imbalance problem: bagging-, boosting-, and hybrid-based approaches. IEEE Trans. Syst. Man Cybern. Part C (Appl. Rev.) **42**(4), 463–484 (2012)
6. Galar, M., Fernández, A., Barrenechea, E., Herrera, F.: EUSBoost: enhancing ensembles for highly imbalanced data-sets by evolutionary undersampling. Pattern Recogn. **46**(12), 3460–3471 (2013)
7. Kotsiantis, S., Kanellopoulos, D., Pintelas, P.: Handling imbalanced datasets: a review. GESTS Int. Trans. Comput. Sc. Eng. **30**(1), 25–36 (2006). Cited By 258
8. Lin, W.C., Tsai, C.F., Hu, Y.H., Jhang, J.S.: Clustering-based undersampling in class-imbalanced data. Inf. Sci. **409–410**(Supplement C), 17–26 (2017)
9. Longadge, R., Dongre, S., Malik, L.: Class imbalance problem in data mining: review. Int. J. Comput. Sci. Netw. **2**(1), 83–87 (2013). Cited By 70
10. Maulik, U., Bandyopadhyay, S.: Genetic algorithm-based clustering technique. Pattern Recogn. **33**(9), 1455–1465 (2000)
11. McCall, J.: Genetic algorithms for modelling and optimisation. J. Comput. Appl. Math. **184**(1), 205–222 (2005). Special Issue on Mathematics Applied to Immunology
12. Napierała, K., Stefanowski, J., Wilk, S.: Learning from imbalanced data in presence of noisy and borderline examples. In: Rough Sets and Current Trends in Computing, pp. 158–167. Springer, Heidelberg (2010)
13. Seiffert, C., Khoshgoftaar, T., Van Hulse, J., Napolitano, A.: RUSBoost: a hybrid approach to alleviating class imbalance. IEEE Trans. Syst. Man Cybern. Part A: Syst. Hum. **40**(1), 185–197 (2010). Cited By 365
14. Yen, S.J., Lee, Y.S.: Cluster-based under-sampling approaches for imbalanced data distributions. Expert Syst. Appl. **36**(3, Part 1), 5718–5727 (2009)

A Modified IEEE 802.11 Protocol for Increasing Confidentiality in WLANs

Abhijath Ande$^{(\boxtimes)}$, Nakul Singh, and B. K. S. P. Kumar Raju

National Institute of Technology, Andhra Pradesh, Tadepalligudem 534101, India
abhijath97@gmail.com, nakuldydx20@gmail.com, pavan0712@gmail.com

Abstract. In this paper, a modified version of the original IEEE 802.11 security protocol is proposed and simulated. The protocol in a C++ program is compared against existing standard IEEE 802.11 protocols. Since their conception, Wireless LANs have explosively risen in popularity and have been extensively deployed in commercial and industrial fields, due to the simplicity, robustness and flexibility of usage they offer. Their widespread usage makes WLANs all the more susceptible to third party interference. Hence, it also becomes increasingly important to secure such WLANs with proper protocols to prevent unwanted snooping, disruption and misuse of the resources the network provides.

Keywords: Computer networks · Network security · Cryptography

1 Introduction

The past decade has seen a rise in the popularity of public wireless LANs and they are being used at almost everywhere. Their ease of usability, simplicity of setup and flexibility in mobility, are the main factors of their popularity. Wireless LANs also known as Wi-Fi networks which can connect multiple devices wirelessly i.e. without the use of any additional hardware components other than the devices themselves. Such wireless networks allow a certain range of mobility of connected devices, while still being connected to the network. With a gateway present, all the devices on the network can also access the Internet.

Their wide range usage makes it necessary to secure them, in order to prevent unauthorized access of such networks. Presently, most WLANs employ the use of WPA2-PSK [1], the most secure security protocol till date. It uses CCM mode with 128-bit AES for confidentiality, making it highly secure for modern day applications.

All WLANs that provide confidentiality with a PSK (pre-shared key), involve a stage called a handshake (4-way handshake in the case of WPA2-PSK), in which the connecting station and the access point, both prove to each other that both hold the same PSK, without actually ever revealing the PSK itself.

The 4 way-handshake in WPA2-PSK protocol, is known to be vulnerable, since the data transmitted during the handshake can be exploited to obtain the

© Springer Nature Switzerland AG 2019
A. Abraham et al. (Eds.): IBICA 2018, AISC 939, pp. 247–256, 2019.
https://doi.org/10.1007/978-3-030-16681-6_25

PSK (Pre-Shared Key). A more powerful attack has recently emerged, namely KRACK (Key Reinstallation Attack) [3], and this has shown the sensitiveness of the 4 way-handshake.

These vulnerabilities call for more secure and robust security protocols.

2   Related Work

There has been a lot of work on constructing WLAN security protocols, out of which most of them use public key cryptography or key exchange systems to realize their protocols [4,8]. Unless the protocol is optimized, hardware which lack in computation power may show degraded performances. Upcoming device in the future show promising performances which may support public key cryptography.

The authors in [4] describes an efficient security protocol which employs public key encryption to provide improved confidentiality and authenticity. But in reality, due to lack of proper hardware in most devices, this protocol may result in less-than-optimal performance.

The work in [5] describes a modified version of the deprecated WPA protocol, employing SHA3 as a message integrity check function, instead of the WPA's Michael. As per date, SHA3 is the strongest cryptographic hash function and hence, greatly improves integrity in WLANs. But their work is yet to be applied to the WPA2 protocol.

The authors in [6] introduced a new approach to improve confidentiality and authenticity. Their WEP-PIV-SDK enhancement involves encrypting the usually transmitted-in-plain-text IV used in cryptographic procedures, and also having a fixed secret key, use it to generate a new key everyday and use this day key as the passphrase for the WLAN for that day. But again, their work is only researched on the deprecated WEP protocol, and not on the present standard's WPA2 protocol.

The work in [7] describes a completely new security protocol namely State Based Key Hop protocol, which has been shown to be more efficient in the amount of computations performed, but also provides a similar level of confidentiality at the same time, when compared against existing standard protocols like WEP, WPA and WPA2.

3   Problem Formulation

Generally, any WLAN security protocol should contain two fundamental aspects

- A Zero Knowledge Password Proof (ZKPP) derived method, as defined in IEEE P1362.2 [2], to prove the correctness of a pre-shared key between an Access Point and a Station.
- A cryptographic algorithm, to provide strong confidentiality while transmitting data.

The problem of introducing a secure, data confidentiality algorithm is glossed over most of the time. The paper aims to introduce a new security protocol, which improves over the standard WPA2-PSK [1] algorithm and improves data confidentiality.

The algorithm basically works by using a pair of independent keys, unique for each station, to encrypt upstream and downstream data using each key respectively, from the connected station to the access point.

4   Proposed Protocol Description

The WPA2-PSK protocol uses a single password, to encrypt all traffic in the WLAN, as shown in Fig. 1. The proposed protocol uses two cryptographically independent passwords, to encrypt upstream and downstream traffic separately, upstream and downstream w.r.t. the Station (STA). The upstream key is pre-shared and the downstream key is configured within the Access Point (AP), as shown in Fig. 2.

For a Station to establish a connection to the Access Point, it must hold the correct upstream key. Upon successful verification of the upstream key, the downstream key is encrypted and sent to the Station from the Access Point.

Hence from a layman's perspective, there is no change in the usage of password protected WLANs. All they need to know to connect to a wireless network is a pre-shared key.

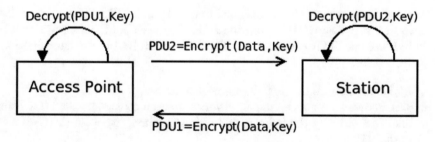

Fig. 1. Data Flow between an Access Point and a Station as per the WPA2 PSK protocol

4.1   Algorithm Description

The improvement in confidentiality produced by the proposed protocol, boils down to how both the upstream and downstream keys are successfully installed on both STA and AP via the 4-way handshake procedure. The following subsections describe the handshake process of both WPA2-PSK and the proposed protocol in detail.

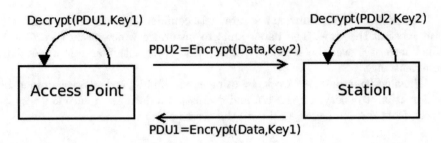

Fig. 2. Data Flow between an Access Point and a Station as per the proposed protocol

WPA2-PSK 4-Way Handshake. The 4-way handshake of the WPA2-PSK has been explained below. It consists of 4 steps, through which an Access Point (AP) can verify if a connecting Station (STA) holds the correct PSK.

Before the 4-way handshake begins, the STA is already authenticated and associated with the AP. Hence, both STA and AP know each other's MAC addresses. The successful completion of a 4-way handshake yields the following at both AP and STA

- Pairwise Transient Key (PTK) unique to a session between the STA and AP. This PTK is used to encrypt all unicast traffic.
- Group Temporal Key (GTK), which is used to encrypt all multicast and broadcast traffic.

At the beginning of the handshake, both AP and STA derive a key called Pairwise Master Key (PMK) from the PSK they each hold. This PMK will be used in PTK and GTK construction. The 4 steps involved in the handshake are explained below

  i. AP sends a nonce to the STA, namely ANonce.
 ii. STA generates a nonce, namely SNonce. It then constructs the PTK using the PMK. Finally the SNonce and a message integrity code (MIC) is sent to the AP.
iii. AP constructs the GTK using the PMK and computes the MIC for the whole frame. Then The GTK and MIC are sent to the STA.
 iv. STA sends an acknowledgement to the AP.

This 4-way handshake of WPA2-PSK is visually depicted in Fig. 3.

Proposed Protocol 4-Way Handshake. The successful completion of a 4-way handshake of the proposed protocol yields the following at both AP and STA.

- Two Pairwise Transient Keys (PTK1 and PTK2) unique to a session between the STA and AP. The PTK1 is used to encrypt all unicast traffic from STA to AP. The PTK2 is used to encrypt all unicast traffic from AP to STA.

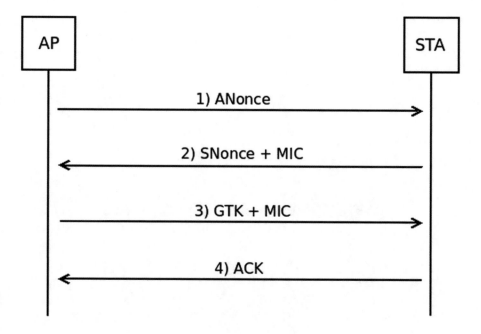

Fig. 3. WPA2-PSK 4-way handshake

– Group Temporal Key (GTK), which is used to encrypt all multicast and broadcast traffic.

At the beginning of the handshake, both AP and STA derive a key called Pairwise Master Key 1 (PMK1) from the upstream key (PSK1) they each hold. This PMK1 will be used in PTK1 and GTK construction. The 4 steps involved in the proposed handshake process are explained below

   i. AP sends a nonce to the STA, namely ANonce.
  ii. STA generates a nonce, namely SNonce. It then constructs the PTK using the PMK1. Finally the SNonce and a message integrity code (MIC) is sent to the AP.
 iii. AP constructs the GTK using the PMK1. It also encrypts the PTK2 using the PTK1 and computes the MIC for the whole frame. Then the GTK, encrypted PTK2 and MIC are sent to the STA.
 iv. STA sends an acknowledgement to the AP.

This 4-way handshake of the proposed protocol is visually depicted in Fig. 4.

5   Experimental Observations

The proposed protocol was benchmarked against two existing WLAN protocols namely, Open Channel (No Password Protection) and WPA2-PSK.

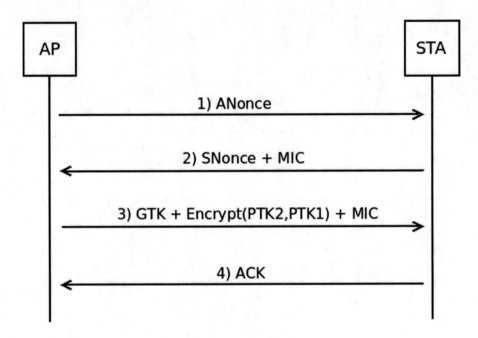

Fig. 4. Proposed protocol 4-way handshake

For a wireless network operating using Open Channel configuration, a STA is said to be connected to an AP if it undergoes successful authenication and association steps in the specified order.

For a wireless network operating using WPA2-PSK configuration or the proposed protocol's configuration, a STA is said to be connected to an AP if it undergoes successful authenication, association and 4-way handshake steps in the specified order.

The topology used in all the simulations, is a star topology with the Access Point as the central vertex and all Stations as surrounding vertices, with an edge representing a link between each Station and the Access Point.

Number of stations in Fig. 5 are for illustration purposes only. Number of stations used in the simulations have been varied to derive multiple observations.

Extensive simulations with varying number of clients, have been run and their results aggregated to observe two properties, namely Link Setup Time and End to End Delay.

5.1 Link Setup Time

Link Setup Time is the time taken to setup a wireless link between the connecting Station and the Access Point. This is a necessary initial overhead at the beginning of any connection, during which all parameters are initialized.

Multiple simulations were run with varying number of Stations, and the results were aggregated. Figure 6 depicts these results.

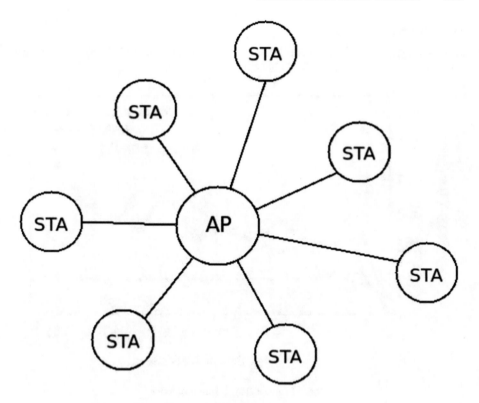

Fig. 5. Illustrative star topology used in the simulations

5.2   End to End Delay

End to End Delay is the time taken for a network layer packet sent from a Station to reach another Station on the same WLAN via the Access Point. ICMP traffic was generated by random number of nodes at different time intervals.

Multiple simulations were run with varying number of Stations and number of ICMP echo requests made, and the results were aggregated. Figure 7 depicts these results.

6   Results

Multiple simulations were run with different number of STAs and different rates of traffic, and the results shown below are consolidated from data on all such simulations.

6.1   Link Setup Time

The average of all Link Setup Time values for each protocol from Fig. 6 are calculated and show below in Table 1.

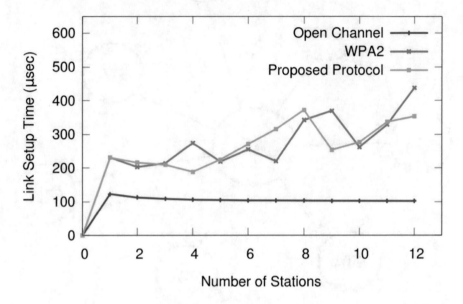

Fig. 6. Link Setup Time statistics

Table 1. Average Link Setup Time of different WLAN protocols (in μsec)

Open Channel	WPA2-PSK	Proposed protocol
98.09	269.11	309.76

6.2   End to End Delay

The average of all End to End Delay values for each protocol from Fig. 7 are calculated and show below in Table 2.

Table 2. Average End to End Delay of different WLAN protocols (in μ*sec*)

Open Channel	WPA2-PSK	Proposed protocol
1152.38	1125.71	1232.36

From the above results, it is evident that our proposed protocol has approximately the same execution time and performance as WPA2-PSK protocol, and simultaneously has increased the confidentiality of the wireless network.

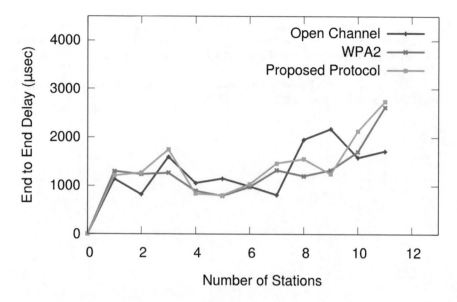

Fig. 7. End to End Delay statistics

7 Conclusion

As the results indicate, our proposed protocol provides better confidentiality than the standard WPA2-PSK protocol, at almost the same cost of time and computation. If chosen to implement, this protocol would mean only a small change to the existing firmware installed in APs and STAs and requires no additional hardware requirements.

References

1. IEEE Std 802.11i. 2004. Amendment 6: Medium Access Control (MAC) Security Enhancements
2. IEEE Standard 1363-2000. Standard Specifications for Public Key Cryptography
3. Vanhoef, M., Piessens, F.: Key reinstallation attacks: forcing nonce reuse in WPA2. In: Proceedings of the 24th ACM Conference on Computer and Communications Security (CCS). ACM (2017)
4. Pisa, C., et al.: WI-FAB: attribute-based WLAN access control, without pre-shared keys and backend infrastructures. In: Proceedings of the 8th ACM International Workshop on Hot Topics in Planet-scale mObile computing and online Social neT-working. ACM (2016)
5. Monga, K., Arora, V., Kumar, A.: Analyzing the behavior of WPA with modification. In: 2015 International Conference on Communication Networks (ICCN). IEEE (2015)
6. Xiao, Y., Bandela, C., Pan, Y.: Vulnerabilities and security enhancements for the IEEE 802.11 WLANs. In: IEEE Global Telecommunications Conference, GLOBE-COM 2005, vol. 3. IEEE (2005)

7. Michell, S., Srinivasan, K.: State based key hop protocol: a lightweight security protocol for wireless networks. In: Proceedings of the 1st ACM International Workshop on Performance Evaluation of Wireless Ad Hoc, Sensor, and Ubiquitous Networks. ACM (2004)
8. Lee, S.-W., et al.: Parallizable simple authenticated key agreement protocol. ACM SIGOPS Oper. Syst. Rev. **37**(2), 13–18 (2003)

Temporal Event Detection Using Supervised Machine Learning Based Algorithm

Rakshita Bansal$^{(\boxtimes)}$, Monika Rani, Harish Kumar, and Sakshi Kaushal

Department of Computer Science and Technology,
University Institute of Engineering and Technology (UIET),
Panjab University (PU), Chandigarh, India
bansal.rakshita@gmail.com, monika25.chd@gmail.com,
{harishk,sakshi}@pu.ac.in

Abstract. Natural Language Processing is a way for computers to explore, analyze, comprehend, and derive significant sense from any language in a smart and useful way. By using NLP, knowledge can be organized and structured to perform tasks such as automatic summarization, translation, named entity recognition, relationship extraction, sentiment analysis, speech recognition, and topic segmentation. Various NLP applications require the identification of events from text documents. The time and event are closely associated with each other. The time dimension is often used to measure the quality and value of events and it has a strong influence in many domains like topic-detection and tracking, query log analysis. In this work, we present an annotation framework to extract temporal information and to specify the temporal relation between extracted events from news corpus by applying the combination of supervised machine learning technique and rule-based method according to the TimeML task std. Artificial neural network (ANN) is trained by the using the TimeBank and AQUAINT TimeML corpus to recognize the events and temporal expressions and for the temporal normalization, part heuristics rules have been used. The efficiency of the proposed work is measured in sense of precision and recall. The system outperformed to the best systems and it is likely that the technique used could be improved further by considering more aspects of the available information when relating the temporal information with events.

Keywords: Temporal information · Information extraction · Event · Artificial neural network

1 Introduction

In the matter of very few years, the web combined itself as a very powerful platform that has changed the manner in which we do business, the way we communicate. It has become the worldwide source of information for many individuals, at home, at school, and at work. It is the most democratic of all the mass media. With its ever-growing importance and utilization, there is increasing demand for the techniques to recognize main events automatically from natural language text. Extraction of data from the text is support for multiple applications involved with data access such as question-answering operations [1], dialogue system, text mining. In the past few years, this

© Springer Nature Switzerland AG 2019
A. Abraham et al. (Eds.): IBICA 2018, AISC 939, pp. 257–268, 2019.
https://doi.org/10.1007/978-3-030-16681-6_26

process has gained some awareness by the TempEval [2]. Machine learning [3], human-engineered rules, used CRF model for English and Spanish language extraction [3, 30] have been used to extract facts from the text. However, despite the achievement of the multilingual TempEval-2 challenge, there is not an efficient system for English text extraction. In this research work, we have determined the relationship between time and event in the newspaper text. Newspaper texts, tales and additional texts represent the time of events, explicitly and inevitably defining the temporal position and events order. Text incorporation needs the capability to recognize the events specified in a text and to position them in time [4]. Normally, a news agency maintains data manually but it is a very difficult task as millions of news is posted daily and hence, the term information extraction came into existence [5]. As the time passed, IE (Information Extraction) popularity increases continuously and become an appropriate tool for a wide array of applications. Initially, the information extraction focused on message judgment in newswires [6]. Despite, due to the incipience of progressively extensive digital data gathering of several natural language text records such as news - messages, items, and web -pages, researchers & practitioners need more high-level techniques, extract higher information with excellent efficiencies and on a real-time basis [7], and perform on larger scales than ever before [8]. The process of extracting the information such as persons and organization from the text is known as event extraction (EE). EE process is mainly used in text mining system, dialogue system and the question answering system. Extracting automatic events from natural texts is an important and challenging task for understanding natural language [9]. Given a group of ontology of event, the aim of event extraction is to recognize references to different event types (event rigger and event nugget) and their parameters from the natural text [10].

1.1   TimeML

Time ML is a framework that is used to encode documents electronically. Time ML forms a standard "markup language" for temporal events in the text document. The main purpose of Time ML is to extract event, time and to find the relationship between them [11]. As per the definition, an event is defined in a general way as a case term for conditions that appear or occur [12].

AN event in the newspaper tells about the occurrence, happening that can be represented by a verb such as killed, acquire, a noun, and an adjective etc. [13]. Time ML is also used to differentiate the event and time in the text. In this research work, Time ML added tags <EVENT> for an event that occurs in the newspaper. The event described the circumstances that hold true [14]. The event created in the proposed research work is listed below.

Once   Colonel   Collins   was   <EVENT   eid="e22" class="I_ACTION">picked</EVENT>   as   a   NASA<EVENT   eid="e52" class="STATE">astronaut</EVENT>,   she   <EVENT   eid="e23" class="OCCURRENCE">followed</EVENT> a normal progression within NASA. Nobody <EVENT eid="e24" class="OCCURRENCE">hurried</EVENT> her up. No one <EVENT eid="e25" class="OCCURRENCE">held</EVENT> her back. Many NASA watchers <EVENT eid="e26" class="REPORTING">say</EVENT> female astronauts have <EVENT eid="e27" class="OCCURRENCE">become</EVENT> part of the agency's routine. But they still have <EVENT eid="e28" class="I_ACTION">catching</EVENT> up to do two hundred and thirty four Americans have <EVENT eid="e30" class="OCCURRENCE">flown</EVENT> in space, only twenty six of them women. Ned Potter, ABC News.

Here the events such as astronaut, held, lack, picked, followed, hurried, say, become, catching, flown are extracted by using (EVENT) tag. The Links used in the proposed work are namedas <SLINK> and <TLINK>. <Tlink> is known as 'temporal link' that introduce to find the relation between time and event. For finding the relation between E-T (event and time) they used some set of rules. Here, they allocate a label whenever a temporal preposition sets a dominion route between an event (E) and a time (T), in which T acts as the temporal modifier of E [15]. <SLINK> is known as 'subordination link' is used to find the relation between two events [16]. E (EVENT)-E (EVENT) sets are certainly marked following two types of rules. The first rule is set as per the dependency path probably existing within the first event (e1) and the second event denoted by e2, and the verb knowledge encoded in e1 [17]. The type of annotation was initially applied to the English language, and after that also used by several researchers to be applied in other languages with a small change in their rules. The annotated corpora of Time ML are also known as 'Timebank' for English 'time bank 1.2' [18] is used. The time in this research is mostly represented in the form of date, hour, month and year [19]. The text document contains a lot of temporal information. The simplest type of 'temporal information' is the time when the document is prepared and when the document is modified. Such type of information is mainly stored in the metadata of the text document [20]. Several domains such as newspaper, Wikipedia, Google contain several temporal expressions in the text document. In this research, we analyzed the English newspaper temporal information and relation between them.

The rest of the paper is organized as follow. Research accomplished in the area of temporal annotation is described in Sect. 2. In Sect. 3 the proposed methodology has been explained in detail. Evaluation result and analysis of the proposed framework are accounted in Sect. 4. Section 5 concludes the work by presenting the outcome on this task based on the experiments and analysis.

2 Related Work

A lot of work has been done in temporal information extraction till date. This section describes the early systems which laid the foundation for recognising temporal expression, events and relation between them.

Strötgen et al. [21] proposed Heidel time system for the temporal expression extraction and normalization. Heideltime system works with some defined rules for temporal extraction. The parameter that is measured for knowing the performance of the system is F-score value and the highest value obtained for F-score is 86%. The pre-processing technique has been applied to remove unwanted words. This work is only applied on pre-processed data for further processing and it is a big disadvantage of this research. UzZaman et al. [22] presented two systems based upon "Markov Logic Networks and Conditional Random Field classifiers" for the extraction of temporal data from the raw document. To overcome the above-defined problem in [11], there is a method used for processing on the pre-processed document as well as raw document. Here, Trips parser has been used to generate logical information from the text without using any grammatical rules. Chang et al. [23] presented SUTIME scheme for recognizing and normalizing the English text expressions. SUMTIME has been used for the annotation of text with temporal knowledge. Text normalization technique is used to overcome the problem of classification using grammatical rule. For knowing the performance of the tool parameters such as precision, recall and F1 have been measured. The precision value and recall values measured for GUTime and SUTime are 0.89, 0.88, 0.79 and 0.96 respectively. The f1 measured for GUTime and SUTime are 0.84 and 0.92 respectively. In this system, the design architecture is only applicable for textual and numeric data and cannot distinguish between the format of duration and date. Mazur et al. [24] proposed a DANTE scheme used for temporal expression extraction DANET has used identification along with normalization of the temporal expressions with the 'TIMEX-2 annotation pattern. DANTE has been formed by using two stages such as guidelines from TIMEX-2 and collecting data from ACE-2005. Basically, this system has been proposed for differentiating the format of duration and date to improve the classification of event and time. JAPE grammar has been utilized implementing the "temporal expression recognizer". The JAPE grammar comprises five sets of rules that consists of 80 numbers of macros and 250 number of the rule set. The value of precision, recall and F-1 to evaluate ACE-2007 data achieved are 54.7, 57.6 and 56.1 respectively. The drawback of this research is that the work is not annotated time and event for raw data. It is only applicable to standard data. Strötgen et al. [25] presented a wide estimation of several existing languages and domains along with the newly formed corpus. The annotation of Timebank is being processed in two different phases: In the 1st phase, 210 documents have been annotated whereas in the 2nd phase 90 documents are annotated. For the first stage data, five numbers of annotators of various profiles have participated whereas for the 2nd phase 45 numbers of undergraduate & graduate students of CSE (Computer science engineering) have participated. Pre-processing has been applied to highlight the negative signals. The average time determined for the annotating 500 number of words is 1 h. The process has been stimulated in TERQAS workshop. <SLINK> a tag has been used to edit and form novel event tags. The problem of annotation has been removed in this research by using <Slink>, which is used to find the relation between two events. Roberts et al. [26] proposed a supervised and unsupervised learning algorithm used for the automatic identification of events, 'temporal expression and has discovered the relation between them in case of clinical text. The best features are selected by using the feature selection algorithm. To know the temporal expression, CRF classifier along with SVM is used. The lexical features, a bag of words, a TIMEX 3 the previously generated features have been used. The experiments have shown the medical events along with their temporal expressions. The authors have used a number of classification algorithms. The accuracy rate of the work obtained is 84%. Chambers et al. [27] presented novel simulations on interlinked events an graph, which includes approximately 10 times more connection per document as compared to the TimeBank. A brief description about the individual learner to a sieve based model has also been discussed. Every sieve inserts labels to the graph of the event at a time and bussing transitive closure previous sieve notifies the later seiver. The event graph shows 14% growth as compared to other approaches. The framework presented in this paper is named as CAEVO (Cascading event ordering architecture). This framework helps to determine the relation between recognition and classification. The framework comprises of a host of classifiers along with various kinds of edges. Velupillai et al. [28] developed a "clearTK" SVM (Support vector machine) that has utilized simple lexical features in context with information from 'rule-based' approach. The challenges faced in 2015 clinical temporal are overcome in this paper with better F-1 value. In this research paper, authors have applied CRF, SVM and rule-based method. The recall for this work is about 80%. Mirza et al. [15] proposed a CATENA system to determine the relation between temporal and causal extraction. The performance of every sieve has been evaluated that demonstrate that the classification algorithms help to obtain better performance in case of TempEVAL 3 and time bank dense data. The CATENA framework consists of different classifiers for both temporal and causal relation. The annotation according to the TimeML standard has been considered. Cheng et al. [29] proposed a method to determine the relation between the closely related tasks. The authors have identified the relation extraction using two approaches (i) LSTM (Long short-term memory) and DP (dependency path). LSTM is a natural selection for processing sequencing dependencies route. For classifying the temporal relation neural network has been used. LSTM have the advantage that it used forward as well as the backward order of information.

3 Proposed Work

This section presents the introduction to proposed work. The overall operation of the proposed temporal annotating framework is as follows. Firstly, the pre-processing of the annotated files in TimeBank and AQUAINT timeML corpus is performed by using the transformation function. It is based on the concept of hashing and it converts the text into the numeric form. After training the feed-forward back propagation neural network with these numeric values, the testing is performed where the occurrence of the event or time is shown by the binary value 1. The extracted event and time are then tagged using <E>-</E> and <T>-</T> tags, respectively. Next, the normalization of the recognised temporal expression is performed using the heuristic rules. Later to analyse the relationship between event-event and event-time, the concept of proximity is used. Finally, the efficiency of the proposed work is measured in sense of precision, recall True positive rate (TPR) and false positive rate (FPR) (Fig. 1).

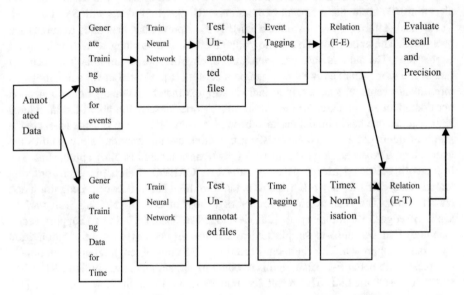

Fig. 1. Flowchart of proposed work

3.1 Simulation Parameters

In the proposed work, a feed-forward back propagation neural network is used where 157 numbers of neurons are provided to the input layer. At hidden layer, 20 neurons are used to adjust the weight to obtain the desired output at the output layer. The Levenberg-Marquard algorithm is used to train the neural the network. The weight and bias values are updated according to Levenberg-Marquardt optimization. It is the most widely used optimization algorithm and is used in many software applications for solving non-linear least-square problems. It is more robust than the Gauss-Newton method because its curve-fitting method is a based upon the combination of gradient descent method with the Gauss-Newton method. Training occurs according to the training parameters in Table 1.

Table 1. Training parameters

Epochs (maximum)	100
Gradient (minimum)	1.00e−07
Mu (maximum)	1.00e+10
Initial mu	0.001
Mu decrease factor	0.1
Mu increase factor	10
Validation check (maximum)	6

The training of the neural network stops when the maximum no. of epochs is reached or when the performance gradient falls below the minimum value of gradient or when mu exceeds the maximum value.

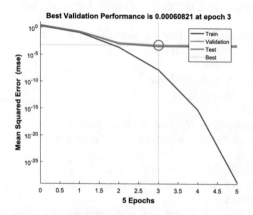

Fig. 2. Performance of ANN

In Fig. 2 the Mean square error (MSE) is shown which occurs during the training process of ANN. If the value of MSE is close to zero at the end of the training phase than it means the desired output and the ANN's output for training set are close to each other. From the figure, it has been observed the ANN is trained at 3rd iteration with 10^{-5} MSE value.

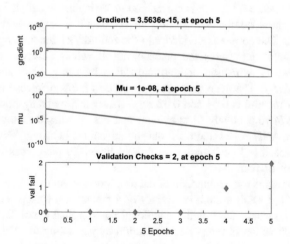

Fig. 3. Training state

In Fig. 3 the training state of the proposed work is displayed. It is shown that the results obtained for 2 validation checks are at 5th iteration where the value of the gradient is about 3.56 and biasing is about 1e−08.

ANN algorithm used for simulation of the work is defined below:

Algorithm 1: Artificial neural network

Input: Training data (T), group (G) and neurons (N)
Output: Classified event and time in English text newspaper
Initialize ANN
Label the training class into training and validation
Find out the parameters of TimeML
Set Training parameters
      Training data=optimized data
      Group= Event and Time
      Epoch=100
      Training algorithm= Levenberg Marquardt
      Performance parameter= Mean square error (MSE)
      Neurons=N
Train (Trained data, Group, neurons)
Classification= Simulate {Event, Time and relation between E-T and E-E; if properties are matched
      Return {Event, Time, E-E and E-T, Precision, Recall, True positive rate, false positive rate}

---

4 Experiment Results

In this section we evaluate the proposed method on publically available news corpus [4] in terms of precision and recall. Precision rate is the relation between truly classified features and all feature set which is considered in testing and recall defined the relation between truly classified features all feature set which is considered in training of proposed work.. The corpus is divided into the ratio of 3:1 for training and testing.

Figures 4 and 5 represents the precision-recall values observed for the task of temporal expression extraction and normalization by using ANN and a rule-based method respectively. The average values observed in terms of precision and recall by using the ANN method is 0.64 and 0.52 *respectively* and by using rule-based method are 0.82 and 0.74 respectively. It is evident that our rule-based method shows better results than our ANN based approach but the advantage of using ANN method over rule-based approach is that it requires minimal linguistic pre-processing and can be applied to out-of-domain data.

Table 2 represents the comparison of the proposed system to results of TIPsem [30] and TRIOS [22]. TRIOS system comprises of a collection of deep semantic parsing, 'Markov logic network' along with conditional random field classifier. It uses TRIPS parser for the extraction of temporal information by generating the deep logical forms for the raw data whereas, TIPSem is based upon data-driven approach and consists of various conditional random field classifier trained using features which are based on semantic information. These systems are dependent on machine learning classifier, which depends on having a training corpus and semantic information. Results are promising of our model which has no input other than word tokens for training in contrast to the systems that requires elaborated linguistic processing.

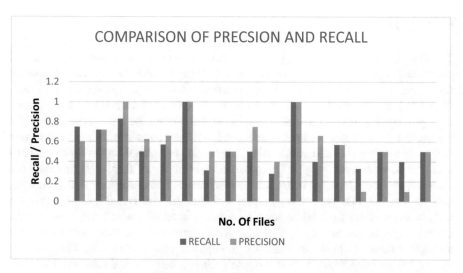

Fig. 4. Comparison of precision and recall (using ANN)

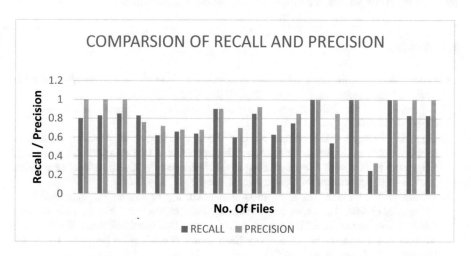

Fig. 5. Comparison of precision and recall (using rule-based approach)

Table 2. Performance comparison of our proposed system and two baseline methods.

System	Precision	Recall	F1 score
TIPSem	0.81	0.86	0.83
TRIOS	0.80	0.74	0.77
Proposed system	0.95	0.94	0.94

5 Conclusion

The objective of the information processing is to build structured information related to events and time to aid contemporary research being carried out in areas of question answering, event extraction, and summarization. In this paper, an annotation framework has been proposed which is based on a neural network named as a feed-forward back-propagation neural network. By using this algorithm the time and events which are present in the news corpus are annotated and the heuristic rules are used to normalize these temporal expressions. Also, the relation between time - event and event – event are identified and tagged. The proposed system shows about 0.95 precision and 0.94 accuracies on the corpus containing news articles. It is also essential to remark that despite the fact that our strategy applies traditional supervised machine learning techniques, it varies from other previous systems in that it doesn't rely upon sophisticated linguistic resources. Specifically, it just uses word tokens and maintains a strategic distance from the utilization of complex syntactic attributes. The proposed system has an impediment in that it doesn't allow integrating the data extracted from different documents. Our future work will be mainly focused on the solution of this inconvenience and utilization of another classification algorithm such as SVM in a hybrid with ANN so that the efficiency can be further increased.

References

1. Pustejovsky, J., Knippen, R., Littman, J., Saurí, R.: Temporal and event information in natural language text. Lang. Resour. Eval. **39**(2–3), 123–164 (2005)
2. Boguraev, B., Ando, R.K.: TimeML-compliant text analysis for temporal reasoning. IJCAI **5**, 997–1003 (2005)
3. Mani, I., Verhagen, M., Wellner, B., Lee, C.M., Pustejovsky, J.: Machine learning of temporal relations. In: Proceedings of the 21st International Conference on Computational Linguistics and the 44th Annual Meeting of the Association for Computational Linguistics, pp. 753–760. Association for Computational Linguistics, July 2006
4. UzZaman, N., Llorens, H., Derczynski, L., Allen, J., Verhagen, M., Pustejovsky, J.: Semeval-2013 task 1: tempeval-3: evaluating time expressions, events, and temporal relations. In: Second Joint Conference on Lexical and Computational Semantics (* SEM), Volume 2: Proceedings of the Seventh International Workshop on Semantic Evaluation (SemEval 2013), vol. 2, pp. 1–9 (2013)
5. Hogenboom, F., Frasincar, F., Kaymak, U., De Jong, F., Caron, E.: A survey of event extraction methods from text for decision support systems. Decis. Support Syst. **85**, 12–22 (2016)
6. Lim, C.G., Choi, H.J.: Efficient temporal information extraction from korean documents. In: 2017 18th IEEE International Conference on Mobile Data Management (MDM), pp. 366–370. IEEE, May 2017
7. Zenasni, S., Kergosien, E., Roche, M., Teisseire, M.: Spatial information extraction from short messages. Expert Syst. Appl. **95**, 351–367 (2018)
8. Fragkou, P.: Combining information extraction and text segmentation methods in Greek texts. Artif. Intell. Res. **7**(1), 23 (2018)

9. Joan, S.F., Valli, S.: A survey on text information extraction from born-digital and scene text images. In: Proceedings of the National Academy of Sciences, India Section A: Physical Sciences, pp. 1–25

10. Wang, S., Yuan, Y., Pei, T., Chen, Y.: A framework for event information extraction from chinese news online. In: Spatial Data Handling in Big Data Era, pp. 53–73. Springer, Singapore (2017)

11. Pustejovsky, J., Ingria, B., Sauri, R., Castano, J., Littman, J., Gaizauskas, R., Setzer, A., Katz, G., Mani, I.: The specification language TimeML. The Language of Time: A Reader, pp. 545–557 (2005)

12. Pustejovsky, J.: ISO-TimeML and the annotation of temporal information. In: Handbook of Linguistic Annotation, pp. 941–968. Springer, Dordrecht (2017)

13. Zhong, X., Cambria, E.: Time expression recognition using a constituent-based tagging scheme. In: Proceedings of the 2018 World Wide Web Conference on World Wide Web, pp. 983–992. International World Wide Web Conferences Steering Committee, April 2018

14. Wei, Y., Singh, L., Buttler, D., Gallagher, B.: Using semantic graphs to detect overlapping target events and story lines from newspaper articles. Int. J. Data Sci. Anal. 5(1), 41–60 (2018)

15. Mirza, P., Tonelli, S.: Catena: causal and temporal relation extraction from natural language texts. In: Proceedings of COLING 2016, The 26th International Conference on Computational Linguistics: Technical Papers, pp. 64–75 (2016)

16. Derczynski, L.R.: Events and times. In: Automatically Ordering Events and Times in Text, pp. 9–24. Springer, Cham (2017)

17. Zhao, S., Liu, T., Zhao, S., Chen, Y., Nie, J.Y.: Event causality extraction based on connectives analysis. Neurocomputing 173, 1943–1950 (2016)

18. Boguraev, B., Pustejovsky, J., Ando, R., Verhagen, M.: TimeBank evolution as a community resource for TimeML parsing. Lang. Resour. Eval. 41(1), 91–115 (2007)

19. Mukkamala, A., Beck, R.: The Development of a Temporal Information Dictionary for Social Media Analytics (2017)

20. Dligach, D., Miller, T., Lin, C., Bethard, S., Savova, G.: Neural temporal relation extraction. In: Proceedings of the 15th Conference of the European Chapter of the Association for Computational Linguistics: Volume 2, Short Papers, vol. 2, pp. 746–751 (2017)

21. Strötgen, J., Gertz, M.: Heideltime: high quality rule-based extraction and normalization of temporal expressions. In: Proceedings of the 5th International Workshop on Semantic Evaluation, pp. 321–324. Association for Computational Linguistics, July 2010

22. UzZaman, N., Allen, J.F.: TRIPS and TRIOS system for TempEval-2: extracting temporal information from text. In: Proceedings of the 5th International Workshop on Semantic Evaluation, pp. 276–283. Association for Computational Linguistics, July 2010

23. Chang, A.X., Manning, C.D.: Sutime: a library for recognizing and normalizing time expressions. In: Lrec, vol. 2012, pp. 3735–3740, May 2012

24. Mazur, P., Dale, R.: The DANTE temporal expression tagger. In: Language and Technology Conference, pp. 245–257. Springer, Heidelberg, October 2007

25. Strötgen, J., Gertz, M.: Multilingual and cross-domain temporal tagging. Lang. Resour. Eval. 47(2), 269–298 (2013)

26. Roberts, K., Rink, B., Harabagiu, S.M.: A flexible framework for recognizing events, temporal expressions, and temporal relations in clinical text. J. Am. Med. Inform. Assoc. 20(5), 867–875 (2013)

27. Chambers, N., Cassidy, T., McDowell, B., Bethard, S.: Dense event ordering with a multi-pass architecture. Trans. Assoc. Comput. Linguist. 2, 273–284 (2014)

28. Velupillai, S., Mowery, D.L., Abdelrahman, S., Christensen, L., Chapman, W.: Blulab: temporal information extraction for the 2015 clinical tempeval challenge. In: Proceedings of the 9th International Workshop on Semantic Evaluation (SemEval 2015), pp. 815–819 (2015)
29. Cheng, F., Miyao, Y.: Classifying temporal relations by bidirectional LSTM over dependency paths. In: Proceedings of the 55th Annual Meeting of the Association for Computational Linguistics (Volume 2: Short Papers), vol. 2, pp. 1–6 (2017)
30. Llorens, H., Saquete, E., Navarro, B.: TIPSem (English and Spanish): evaluating CRFs and semantic roles in TempEval-2. In: Proceedings of the 5th International Workshop on Semantic Evaluation, pp. 284–291. Association for Computational Linguistics, July 2010

PPARM: Privacy Preserving Association Rule Mining Technique for Vertical Partitioning Database

Virendra Dani[(⊠)], Shubham Kothari, and Himanshu Panadiwal

Computer Science and Engineering Department, SVIIT (SVVV), Indore, India
virendradani.cs@gmail.com, Shubhever.cs@gmail.com,
hpanadiwal@gmail.com

Abstract. With the development and penetration of data mining within different fields and disciplines, security and privacy concerns have emerged. The aim of privacy-preserving data mining is to find the right balance between maximizing analysis results and keeping the inferences that disclose private information about organizations or individuals at a minimum. In this paper, we proposed Privacy Preserving Association Rule mining i.e. "PPARM" technique for multiparty computation of privacy preserving data model for aggregation, cryptographic security and association rule mining concept. In this process the data is secured using the cryptographic techniques and for providing the more secure mining technique the server generated random keys are used. Using the proposed technique the data is mined in similar manner as the association rule mining do, but for securing the data sensitivity the cryptographic technique is used at the client end. After mining of data the association rules are recoverable at client end also by the similar keys as produced by the server.

Keywords: Data mining · AES cryptography · Association rule mining ·
Apriori · PPDM · PPARM

1 Introduction

New advances in data mining and knowledge discovery have generated contentious impact in both scientific and technological arenas. On the one hand, data mining is capable of analyzing vast amount of information within a minimum amount of time. On the other hand, the excessive processing power of intelligent algorithms puts the sensitive and confidential information that resides in large and distributed data stores at risk. Providing solutions to database security problems combines several techniques and mechanisms [1, 2].

1.1 Privacy Preserving

Privacy is a matter of individual perception, an infallible and universal solution to this dichotomy is infeasible. The common term of privacy in the general, limits the information that is leaked by the distributed computation to be the information that can be learned from the designated output of the computation. The current state-of-the-art

© Springer Nature Switzerland AG 2019
A. Abraham et al. (Eds.): IBICA 2018, AISC 939, pp. 269–278, 2019.
https://doi.org/10.1007/978-3-030-16681-6_27

paradigm for privacy-preserving data analysis is differential privacy, which allows untrusted parties to access private data through aggregate queries [3].

This Privacy preserving [4] has originated as an important concern with reference to the success of the data mining. Privacy preserving data mining (PPDM) deals with protecting the privacy of individual data or sensitive knowledge without sacrificing the utility of the data. People have become well aware of the privacy intrusions on their personal data and are very reluctant to share their sensitive information. This may lead to the inadvertent results of the data mining. Within the constraints of privacy, several methods have been proposed but still this branch of research is in its early life.

1.2 Association Rule Mining

Association rule mining has been an active research area in data mining, for which many algorithms have been developed. In data mining, association rule learning is a popular and well-accepted method for discovering interesting relations between variables in large databases. Association rules are employed today in many areas including web usage mining, intrusion detection and bioinformatics [4].

This In general, the association rule is an expression of the form $X \rightarrow Y$ where X is antecedent and Y is consequent. An antecedent is an item found in the data. A consequent is an item that is found in combination with antecedent. The main aim is extracting important correlation among data items in the database. Basically it extracts the pattern from the data based on the two measures such as minimum confidence and minimum support. Support it indicates of how frequently the items appear in the database. Confidence indicates the number of times the if/then statement have been found to be true. Support it is the probability of item or item sets given transactional database [5].

1.3 Categories of Privacy Breach

A privacy breach occurs when private and confidential information about the user is disclosed to an adversary. So, preserving privacy of individuals while publishing user's collected data is an important research area. The privacy breaches in social networks can be categorized into three types [6]:

- Identity Disclosure - Identity disclosure occurs when an individual behind a record is exposed.
- Sensetive Link Disclosure - Sensitive link disclosure occurs when the associations between two individuals are revealed.
- Sensetive Attribute Disclosure – Sensitive attribute disclosure takes place when an attacker obtains the information of a sensitive and confidential user attribute.

All these mentioned privacy breaches pose severe threats like annoyance, blackmailing and theft because users expect privacy of their data from the service provider end. Thus data should be anonymised before releasing or publishing to third parties.

2 Literature Review

Here, we are describing some of the prior work which has their own importance in data privacy.

Li et al. [7] have designed an proficient homomorphic encryption scheme and a secure comparison method. Authors proposed a cloud-aided common itemset mining solution, which is used to make an association rule mining mechanism. This solution is intended for outsourced databases that permit various data owners to proficiently distribute their data securely with no compromising on data privacy.

Kantarcioglu et al. [8] addressed secure mining of association rules over horizontally partitioned data. The methods integrate cryptographic techniques to reduce the information shared, while accumulate modest overhead to the mining task.

Hussein et al. [9] proposed a modification to privacy preserving association rule mining algorithm on dispersed homogenous database. This algorithm is faster, privacy preserving and provides precise results. The flexibility for extension to any number of sites can be accomplished without any modify in the implementation. Also any enhance in number of these sites does not add more time overhead, because all client sites execute the mining process in the same time so the overhead is in communication time merely.

3 Proposed Work

This section includes the solution process which satisfies the identified problem or the established goal. Therefore first the straightforward solution is provided and then the entire modeling of the system is defined.

3.1 Methodology

This part provides the detail methodology and basic functional aspects of the proposed PPARM approach and in the next section summarized steps of in algorithmic form.

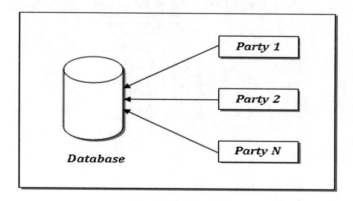

Fig. 1. Database organization

The proposed system can be understood by the Fig. 1. In this diagram the more than one party are wants to store their vertically partitioned data in a centralized data base. Additionally the centralized database is accessed by the application for further use.

In client server model, it is more important to know how the data is transferred to the server end and which format data is delivering to the server. Let consider by an example of list of some transaction on which basis we have to understand the scenario. If we have some itemsets

$$I = \{a, b, c, d\}$$

On the basis of this itemsets, we make list of random trisection for multiple clients. Here we considering an example of two client's transaction list i.e. *client 1* and *client 2*.

Table 1. Clients transaction list

Client 1	Client 2
a, b	d
b, c	a
c	a, d
a	b

Table 1 shows the client transaction on the basis of Item list for two party's databases. Each client's transaction may be same of different fir every item set. After make the transaction of the item sets, this is needs to encode on client side for data privacy preservation. Here Tables 2 and 3 show the encode table transaction list which map the on the basis of transaction list for both individual client. This data will be encoded in binary format at the client end. After encoding, the data client sends it to server end and server mapping this binary data in other format of item list.

Table 2. Encoding of client 1 data

a	b	c	d
1	1	0	0
0	1	1	0
0	0	1	0
1	0	0	0

Table 3. Encoding of client 2 data

Client 2 data encode			
a	b	c	d
0	0	0	1
1	0	0	0
1	0	0	1
0	1	0	0

Here, Tables 2 and 3 list the encoding of the clients transactions. The item will be mapped in other encrypted form e.g. α, β, γ, δ for itemsets a, b, c, d. Hence, new mapped data will be shown in Table 4. This mapped data list with the help of generated above encoded data. Here we list the combine list the client 1 and client 2 data.

Table 4. Mapping table

Client 1				Client 2				Mapping
α	β	γ	δ	α	β	γ	δ	
1	1	0	0	0	0	0	1	α, β, δ
0	1	1	0	1	0	0	0	β, γ, α
0	0	1	0	1	0	0	1	γ, α, δ
1	0	0	0	0	1	0	0	α, β

Finally, this database transfer to the server side in each transmission and on server side this data are combined. For privacy preserving data mining, the combined database resides on server side which is shown in mapping Table 4. Whenever, client receive the generated encrypted rules it will decrypt this by using reverse mapping table and generated client key. The user decrypts only that part of the rule which he want to access information. The proposed system ensures him strong privacy of the user confidential information using mining association rule.

This data is transmitted in binary format to server side. Hence, the transmission of data is happened through memory stream. Memory stream is a region between server and client where server and client read and write data respectively. Firstly clients write the data into memory stream in byte format and server read this byte data and converts it into its original form. The internal process of client and server based of socket programming. Therefore, step by step process is run and generates intermediate output. The server will not do anything until the client requests it to server, as soon as the server starts the process of Request Accept. Hence process is run continuously and privacy preserving data mining approach is successfully applied using proposed methodology.

The key issue in this given model all the parties have some private, confidential information which is not disclose-able to others. But how the decisions are made is required to distribute. Thus between the server and different parties the following

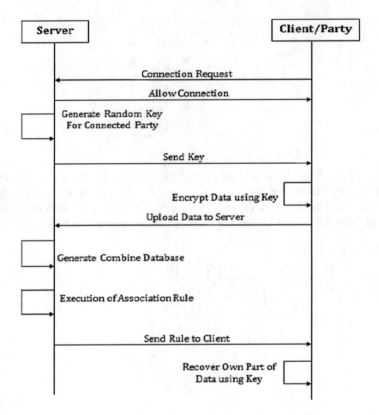

Fig. 2. System processing architecture

operations are need to utilize for effective data processing and knowledge discovery as given in Fig. 2. In this figure, we show that step-by-step process of entire work methodology. The process is happened between client and server. Client and server are the two basic entities by request transmit for different purpose. Initially, clients who want to privacy preserving data mining technique need to prepare connection between client and server. Therefore, firstly client or party sends the request to the server for establishing connection. After that server accept the request and response of enable connection in terms of acknowledgement.

After establishing successful connection between both, server generates the random key for linked parties of client. And this key sends to the client for further use. After receiving this random key of client side and use this for transforming data in to ciphertext. Thereafter, clients upload this data to the server for utilization by server and combine database. Therefore, server calls the apriori algorithm for processing the data and extracts the association rule for the form data. Finally the recovered rules from the server are provided to the end client. The end client applies the key again to the received data for recovering the part of information which is owned by the self. The process shows the data privacy preservation for multiple clients. For data encryption and decryption we have used AES algorithm. Apriori algorithm have been implemented for both scenario i.e. proposed and base method and results generated on the basis of time complexity, memory utilization and number of rule generated by algorithms.

3.2   Proposed Algorithm

This section provides the algorithm in Table 5 steps for the secure and privacy preserving data mining technique. Let the N number of parties $C = C_1, C_2, ..., C_n$ are want to associate their data to find the common decision using different provided attributes. The data from different clients can be defined by the following manner.

$$D = dC^1, dC^2, dC^3, ..., dC^n \qquad (1)$$

In the above *Eq.* (1) show the aggregated data from the different participating clients and dC^n is the data provided by the nth user of the system. Using the provided symbols the following function is used to prepare the privacy preserving model.

Table 5. Proposed PPARM technique algorithms

Input: Number of Clients (C), Clients Data (dC^n)
Output: Rules (R)
Process:
1: Client Send Connection Request to Server
2: connection = client.connect (server)
3: Count number of active connection for single process
4: client send data to server for each active client
5: Server read data of individual client
6: newAttributeList = mapping (merge, AES)
7: newTransection = generateTransection (merge, newAttributeList)
8: Rules = Apriori.getrules(newTransection, min support, min confidence)
9: Return R

4   Result Analysis

4.1   Memory Usage

The amount of main memory required to perform data analysis using the algorithm is termed here as memory usages or space complexity. The estimated comparative memory consumption of the implemented algorithm is reported using Fig. 3.

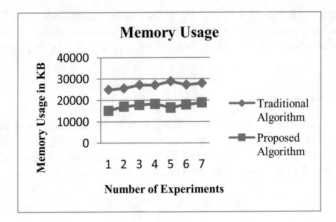

Fig. 3. Memory usage

The memory consumption or space complexity of the implemented algorithms namely proposed and traditional Apriori algorithm is reported using the Fig. 3. In this diagram the X axis contains different experiments performed with system and Y axis shows amount of main memory consumed in terms of KB (kilobytes). Additionally to represent the performance of algorithms red line shows the performance of proposed algorithm and traditional algorithm is denoted using blue line. In most of the experiments the performance of algorithms are fluctuating but it remains adaptable for data analysis in both the cases. In experimentations size of data is increases and their memory consumption is evaluated. During the experimentations that are observed if the number of candidate set generation is large then the memory requirement is higher and otherwise it remains fixed not much fluctuating.

4.2   Time Consumption

The amount of time consumed for developing the Apriori based association rules using the input datasets is termed here as the time consumption of algorithm or time complexity. The time consumption of the implemented algorithms is reported using the Fig. 4. According to the given performance the implemented algorithms proposed Apriori algorithm consumes less amount of time as compared to the traditional Apriori. But the time consumption is increases as the amount of data for association rule development is increases.

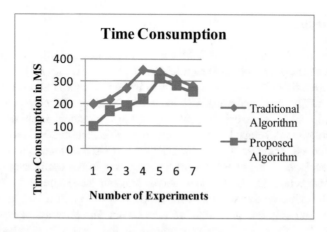

Fig. 4. Time consumption

4.3   Transaction vs. Rules

In order to represent the effectiveness of the proposed technique the comparison of both the implemented algorithms is performed which number of input transactions and developed association rules in both situations. The performed experimentation and their results are provided using Fig. 5.

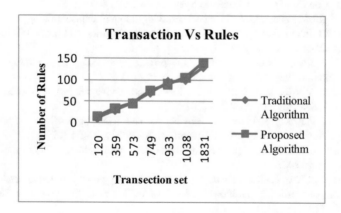

Fig. 5. Transactions vs. rule

To shows the performance red line shows the performance of proposed technique and blue line shows the performance of traditional Apriori algorithm. In most of the time similar numbers of rules are generated with the less amount of time as compared to the traditional algorithm. The obtained observational results are also demonstrated the proposed technique provides high quality rules as compared to traditional algorithm.

5 Conclusion

Tremendous growth in the IT field and the problems addressed during the storage of the huge data is the major problem. Data mining aims to discover secret information from large database although secret data is kept safely when data is allowed to access by single person. The proposed work is intended to develop a privacy preserving technique for extracting the association rules form the transaction database. Therefore a secure technique is developed using the cryptographic data manipulation and reorganization of the association rules. To simulate the proposed technique there are a client system is developed. This client system first connects to the centralized server. After successful connection on client request server transmit the secure key randomly generated. For the entire session we follow the process for two parties. This key is works as a session key and only one time used with the server. The obtained key by the server, client manipulate their own data and upload to the server. The uploaded data is combined with the other party's data and encoded in binary form and finally the Apriori algorithm is implemented to extract the rules. These rules are also deliverable to the end client and the client can recover their own part of data using the obtained key.

References

1. Agrawal, D., Aggarwal, C.C.: On the design and quantification of privacy preserving data mining algorithms. In: Proceedings of the Twentieth ACM SIGMOD-SIGACT-SIGART Symposium on Principles of Database Systems. ACM (2001)
2. Oliveira, S.R.M., Zaiane, O.R.: Privacy preserving frequent itemset mining. In: Proceedings of the IEEE International Conference on Privacy, Security and Data Mining, vol. 14. Australian Computer Society (2002)
3. Tiwari, D., Tiwari, R.G.: A survey on privacy preserving data mining techniques. IOSR J. Comput. Eng. **17**(5), 60–64 (2015). Version 3
4. Michalewicz, V.R.R., Honwadkar, K.N.: Privacy-preserving mining of association rules in cloud. Int. J. Sci. Res. (IJSR), **3**(11) (2014)
5. Fayyad, U., Piatetsky-Shapiro, G., Smyth, P.: The KDD process for extracting useful knowledge from volumes of data. Commun. ACM **39**(11), 27–34 (1996)
6. HajYasien, A.: Preserving privacy in association rule mining. Ph.D. thesis, Griffith University, June 2007
7. Li, L., Lu, R., Choo, K.-K.R., Datta, A., Shao, J.: Privacy-preserving-outsourced association rule mining on vertically partitioned databases. IEEE Trans. Inf. Forensics Secur. **11**(8), 1847–1861 (2016)
8. Kantarcioglu, M., Clifton, C.: Privacy-preserving distributed mining of association rules on horizontally partitioned data. IEEE Trans. Knowl. Data Eng. **16**(9), 1026–1037 (2004)
9. Hussein, M., El-Sisi, A., Ismail, N.: Privacy preserving association rules mining on distributed homogenous databases. Int. J. Data Min. Model. Manag. **3**(2), 172–188 (2011)

Generation of Hindi Word Embeddings and Their Utilization in Ranking Documents Using Negative Sampling Architecture, t-SNE Visualization and TF-IDF Based Weighted Average of Vectors

Arya Prabhudesai[✉]

College of Engineering, Pune, Pune, India
apprabhudesai98@gmail.com

Abstract. Hindi is the official language of India and has over 500 million speakers worldwide. Being a dominant language with a widespread impact, implies the need for development of technologies that cater to its native speakers. In this paper, a text mining based information retrieval model has been developed to generate Hindi word embeddings and their application ranking documents in order of relevance to an input query. Word embeddings are multi-dimensional vectors that can be created by utilizing the linguistic context of words in a large corpus. To generate the embeddings, a corpus was created from the Hindi Wikipedia dump, on which the skip-gram approach was applied using a neural network based negative sampling-architecture. The weighted average of each word embedding along with its tf-idf score generated the embeddings for each individual document. The cosine-similarity was then calculated between each document vector and the query vector. Using these similarity scores, the documents were ranked in descending order of relevance to the query. Highly relevant rankings were obtained in response to a query input. The results of the model were visualized using the t-SNE visualization method. The accuracy of this method proves that in the process of conversion of words to numeric vectors, the semantic context of the words was preserved.

Keywords: Text mining · Natural language processing · Information retrieval

1 Introduction

Most of the latest technologies developed in today's age cater to the English-speaking public, but according to an estimation by Google India, by 2021 content consumption in Hindi will be bigger than consumption in English. Around 90% of new net users today are non-English [2]. According to a report by Google and KPMG India published in April 2017, there were 234 million Indian-language Internet users in 2016 while only 175 million English users, and the gap between the two groups is expected to widen going forward. The report states that nine out of ten new Internet users between 2016 and 2021 will use local languages [3].

© Springer Nature Switzerland AG 2019
A. Abraham et al. (Eds.): IBICA 2018, AISC 939, pp. 279–288, 2019.
https://doi.org/10.1007/978-3-030-16681-6_28

Hence, in such times it becomes necessary to develop technologies that cater to non-English users of net content. In this paper, a model has been developed that will help such users in easier lookups of their text documents in Hindi. The model accepts an input query document, after which it computes the similarity between each document in the database and the query. On the basis of this result, the documents are then recommended in order of relevance to the input query.

In order to build this model, first a list of all the stop words in Hindi language is created. In computing, stop words are words which are filtered out before or after processing of natural language data (text) [4]. After the removal of stop words, word embeddings for each word are generated. The method used in generating these vectors is a Word2vec model which uses a neural network based architecture. Word2vec can utilize either of two model architectures to produce a distributed representation of words: continuous bag-of-words (CBOW) or continuous skip-gram. In the continuous bag-of-words architecture, the model predicts the current word from a window of surrounding context words. The order of context words does not influence prediction (bag-of-words assumption). In the continuous skip-gram architecture, the model uses the current word to predict the surrounding window of context words. The skip-gram architecture weighs nearby context words more heavily than more distant context words [5, 6]. The contextual based meaning of a word is to be preserved in the Word2vec algorithm. Hence the model developed in this paper makes use of a skip-gram based Word2vec approach to generate the word embeddings.

This involves an iterative learning process. The end product of this learning will be an embedding layer in a network – this embedding layer is a kind of lookup table – the rows are vector representations of each word in our vocabulary. Learning this embedding layer/lookup table can be performed using a simple neural network and an output softmax layer. But the problem with using a full softmax output layer is that it is very computationally expensive:

$$P(y = j \mid \text{mid } x) = \frac{e^{x^T w_j}}{\sum_{k=1}^{k} e^{x^T w_k}} \tag{1}$$

From (1) the probability of the output being class j is calculated by multiplying the output of the hidden layer and the weights connecting to the class j output on the numerator and dividing it by the same product but overall the remaining weights. When the output is a 10,000-word one-hot vector, there are millions of weights that need to be updated in any gradient based training of the output layer [7]. A one-hot encoding is a representation of categorical variables as binary vectors where all the values of the vector are 0, except for one which has a value of 1. This gets seriously time-consuming and inefficient. Hence the model trained in this paper makes use of another approach called the negative sampling architecture which was developed by Mikolov et al. [5]. It works by reinforcing the strength of weights which link a target word to its context words, but rather than reducing the value of all those weights which aren't in the context, it simply samples a small number of them – these are called the "negative samples" [7]. The results of the Word2vec algorithm are then visualized using t-distributed stochastic neighborhood embedding method (t-SNE). These word embeddings are used to calculate document embeddings for each document in the database. There are a few approaches that can be used to do the same. The first one is to use solely the tf-idf scores of each word, the second is to take the simple average of the word embeddings and the third is to combine the first two and calculate the weighted average of the vectors. In this paper, the third approach has been used, since in tf-idf weighted vectors, both the semantic context of a word as well as the relative frequency/importance of the word are considered. Now, the model takes a query document as an input whose document embedding is calculated using the same procedure described above. Once the query vector is calculated, it is compared (cosine-similarity) with each document vector to generate a similarity score.

2 Related Work

Word2vec is a very important step in development of any natural language processing application. All computing machines execute their functions on numerical entities, and hence do not understand the difference between a linguistic based letter, word or a paragraph input. This is where the Word2vec algorithm comes into play. It converts a linguistic term into a numerical format that can be processed by the computer and on which mathematical operations can be carried out. For a long time, a latent semantic analysis was carried out in order to achieve this result, but with his work, Mikolov developed the new and efficient Word2vec algorithm [5]. T-distributed Stochastic Neighbor Embedding (t-SNE) is a machine learning algorithm for visualization developed by Laurens van der Maaten and Geoffrey Hinton [12]. In probability and statistics, Student's t-distribution (or simply the t-distribution) is a member of a family of continuous probability distributions that arises when estimating the mean of a normally distributed population in situations where the sample size is small and population standard deviation is unknown. The t-distribution is symmetric and bell-shaped, like the normal distribution, but has heavier tails, meaning that it is more prone to producing values that fall far from its mean [17]. The t-SNE is an application of t-distribution and is a nonlinear dimensionality reduction technique well-suited for embedding high-dimensional data for visualization in a low-dimensional space of two or three dimensions. Once the word embeddings have been calculated, they can be extended to calculate embeddings for entire documents. In his work, "Using TF-IDF to Determine Word Relevance in Document Queries", Juan Ramos has demonstrated how the tf-idf values can be used for this purpose [8]. In the model developed in this work, a combination of tf-idf values and pre-trained word-embeddings to get the document embeddings has been used. Microsoft also developed a "Improving Document Ranking with Dual Word Embeddings" [9] model that aims to tweak the accuracy of current document ranking models. The model in this paper has been developed to cater to the Hindi content users keeping in mind all the prerequisites of processing and preprocessing a Hindi based model. These include removal of stop words present in the Hindi data and a pre-trained lemmatized reference tool which converts each word in the corpus to its lemma in order to avoid repetitions [10].

3 Proposed Model

3.1 Word2Vec Using Negative Sampling Architecture

Algorithm 1, is responsible for pre-processing the input wiki dump file. It first extracts all the files in the wikidump using a wiki-extractor [14]. This will result in a single document containing all the text data, "data_file". A list of the Hindi "stop_words" from the corpus is created, that are to be excluded from processing of the text. The "data_file" is read and stored in a single string called "vocabulary". The string "vocabulary" is tokenized into individual tokens to obtain a list of the individual words in the string "all_words". All the "stop_words" from the list of all_words are removed to get a new processed list called "words". The frequencies of all the words in the list are calculated to get a new list of the most commonly occurring 10,000 words and are stored in variable "dictionary" (Fig. 1).

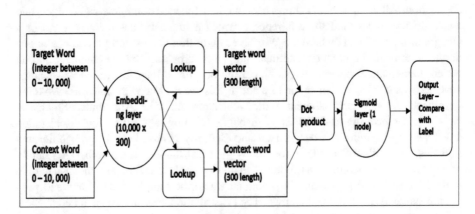

Fig. 1. Word2Vec architecture

In Algorithm 2, the steps to generate word embeddings for each word in variable "dictionary" are iterated. First the variables "window-size" = 4, vector dimension "vector_dim" = 300 and the number of epochs, "epochs" = 200k are defined for the model. The skip-gram function is then called, which generates a similarity value of either 1 or 0 between each pair of words in the vocabulary which are stored in variables "couples" and "labels". "couples" is a list where each its each element is a pair of words from "all_words", while its corresponding similarity value is stored in the array "labels". The embedding layer of the model is initialized with dimensions 10k × 300 to hold random values, "word_embedding". Now, the model accepts two inputs: - the target word and the context word. It then looks-up the embedding of both the words from the embedding table, "word_embedding" that is already defined. The cosine similarity between the two vectors (embeddings) will then be calculated after reshaping them into the predefined dimensions. The model is then compiled with a binary cross-entropy loss between "similarity" and "label". Cross entropy is one of the core principles in information theory that is often used in machine learning to calculate the loss

function [13]. On each iteration of training the model, the values in the embedding
layer will be updated to minimize the loss generated.

Algorithm 1 Preprocessing of wikidump_file

```
1. procedure PROCESS(wikidump_file)
2.     data_file <- wiki_extract(wikidump_file)
3.     set list(stop_words) for Hindi
4.     string(vocabulary) <- read(data_file
5.     list(all_words) <- tokenize(vocabulary)
6.     list(words) <- all_words − stop_words
7.     vocab_size <- 10,000
8.     dictionary <- most_common(vocab_size)(words)
9.     return dictionary, all_words
```

Algorithm 2 Word2Vec Model

```
1. procedure WORD2VEC (all_words, dictionary)
2.     window_size <- 4, vector_dim <- 300, epochs <-
   200000
3.     couples,labels <- skipgram(all_words,vocab_size,
   window_size)
4.     word_target, word_context <- zip(*couples)

5.     procedure MODEL (in_target, in_context, label)
6.         word_embedding <- Embeddings(vocab_size, vec-
   tor_dim, input_length = 1)
7.         target <- word_embedding(in_target)
8.         context <- word_embedding(in_context)
9.         similarity <- cosine_sim(target, context)
10.        compile MODEL with loss <- binary crossentropy
   (label, similarity)
11.        return loss
12.     count <- 0
13.     while count ≠ epochs do
14.         idx <- random(0, length(labels))
15.         arr_1[0,] <- word_target[idx]
16.         arr_2[0,] <- word_context[idx]
17.         arr_3[0,] <- labels[idx]
18.         loss = MODEL(input_target = arr_1, input_con-
   text = arr_2, label = arr_3)
19.         count <- count + 1
20.     return word_embedding
```

3.2 Document Embeddings Using Weighted TF-IDF Average of Word Vectors

In Algorithm 3, the input text_document is preprocessed. First, the text document is preprocessed in order to get a list of all the unique and non stop-words in the text document. The process used for this purpose is similar to the one followed in Algorithm 1 up to step 6. Then a list "doc_words" is generated which contains only those words from the list "words" which are present in the "dictionary". Now, a set called the "set_vocab" is created from the "doc_words" which does not contain more than one copy of the of each word in doc_words. Hence the list "doc_words" is converted to a set. Then, the Inverse Document Frequency (IDF) and Term Frequency (TF) is calculated for each item in the set, set_vocab according to (2) and (3) [11] and stored in arrays TF and IDF.

In Algorithm 4, the document embedding for the input text_document is calculated. This is done using (4) where, the weighted average of each word_embedding is taken. Hence the output generated at the end of Algorithm 4 is the document embedding for the input text document. This algorithm can then be repeated for each document in the corpora. Algorithm 3 and Algorithm 4 are repeated for each document in the corpora to get doc_embeddings for all the documents in the corpora. All the embeddings are then stored in an array doc_vectors.

Algorithm 3 Preprocessing of the text_document

```
 1. procedure PROCESS_FILE(text_document)
 2.      total_docs<- total_document
 3.      data_file <- wiki_extract(wikidum_file)
 4.      set list(stop_words) for Hindi
 5.      string(vocabulary) <- read(data_file)
 6.      list(all_words) <- tokenize(vocabulary)

 7.      list(words) <- all_words — stop_words
 8.      set i <- 0
 9.      set doc_words <- []
10.     while i ≠ length(words) do
11.         if (words[i] belongs to dictionary) then
12.             doc_words <- doc_words + words[i]
13.         i <- i + 1
14.     set_vocab <- set(doc_words)
15.     i <- 0, IDF <- []
16.     while i ≠ length(set_vocab) do
17.         word_present <- frequency(set_vocab[i])

18.         frequency <- doc_words.count(set_words(i))
```

$$IDF(i) = \log\frac{total\_docs}{word\_present} \tag{2}$$

$$TF(i) = \frac{frequency}{length(doc\_words)} \tag{3}$$

```
19. return TF, IDF, doc_words, set_vocab
```

Algorithm 4 Generate the embeddings for the text_document

```
1. procedure DOC2VEC(word_embedding, TF, IDF)
2.     set doc_embedding <- []
3.     for i = 0 to i = length(set_vocab):
```

$$\text{doc\_vector} = \frac{\Sigma(\text{word embedding(i)} * \text{TF(i)} * \text{IDF(i)})}{\text{length(doc words)}} \qquad (4)$$

```
4.     return doc_vector
```

3.3 Cosine-Similarity Between Query Vectors and Document Vectors

Algorithm 5 takes in as an input, "query_vector" and "doc_vectors" both of which are generated using Algorithm 3 and Algorithm 4 described in Section B. "doc_vectors" is an array which contains the vectors for each document in the corpora. Using cosine-similarity formula given in (5), we calculate the similarity score between the query vector and document vector for each document in the database and store the values in an array. The original cosine similarity formula between two vectors.

Algorithm 5 Cosine Similarity

```
1. procedure SIMILARITY(query_vector, doc_vectors)
2.     i <- 0, similarity <- []
3.     while i ≠ total_docs do
```

$$\text{similarity}[i] = \frac{doc\_vectors[i].query\_vector}{||doc\_vectors[i]||\,||query\_vector||} \qquad (6)$$

```
4. i <- i + 1
5. top_n <- (sort)similarity(: -n)
6. return dictionary, all_words
```

$$\cos \Theta = \frac{a \cdot b}{||a||\,||b||} \qquad (5)$$

Using the similarity scores from all the comparisons, the top 'n' documents with the highest scores are ranked as the most relevant documents in as the output using descending order.

4 Results and Future Work

The results of the Word2vec model are be viewed using the t-SNE visualization algorithm. The t-SNE algorithm comprises two main stages. First, t-SNE constructs a probability distribution over pairs of high-dimensional objects in such a way that similar objects have a high probability of being picked, whilst dissimilar points have an extremely small probability of being picked. Second, t-SNE defines a similar probability distribution over the points in the low-dimensional map, and it minimizes the Kullback–Leibler divergence between the two distributions with respect to the locations of the points in the map [12]. While Fig. 2 shows a plot of 2000 words and how close they lie in terms of contextual similarity with each other.

Using the word embeddings, the vectors for each document were calculated using the method described in Subsection B of Sect. 3. Then, an input query was processed and ranked documents in response to the query were obtained as results. In Fig. 3, the query results from a sample input query are tested with three different formulae to measure similarity: the cosine similarity, Euclidean similarity and ts-ss (triangular similarity sector similarity) [15, 16]. While the cosine similarity preserves the relevant results of the search in its successive suggestions, the Euclidean and ts-ss values do not synchronize with the expected values. As can be observed, the values outputted by the other two measures are not in sync with the expected results. Hence, cosine-similarity has been used to develop the model in this paper.

This model can be further developed to cluster/group the documents into similar groups using various algorithms. The accuracy of the model can also be analyzed by using different similarity measures such as the Manhattan and Jaccard similarity formulae. By exploring these alternatives, we might be able to achieve quicker and more accurate results. Other than these mentioned techniques, the accuracy of the model can also be tested by trying various alternatives for the skip-window size and epochs defined in Algorithm 2 in Subsection B of Sect. 3. By increasing the number of epochs, the efficiency of the model can be tweaked, although the model will require more training time.

Similar models can also be developed to design technologies that test the relevance of a heading/title given to a document by comparing the vector embeddings of the document content with those of the provided titles.

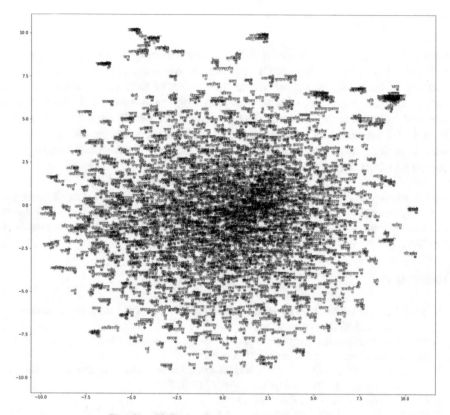

Fig. 2. t-SNE visualization of Word2Vec results

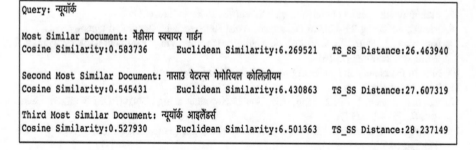

Fig. 3. Query results

5   Conclusion

Hence from Fig. 3 it can be deduced that cosine similarity is the most apt distance calculation technique in this case. The query results prove that the semantic context and meaning of the words was preserved in the process of their conversion to numeric vectors and that the cosine similarity technique for calculation of similarity scores provided the most accurate results. The usage of tf-idf based weighted average also helps in the generation of relevance rankings since the weights are assigned to each word in such a manner that their relative importance in a document is inversely proportional to their frequency in the document.

Acknowledgements. I would like to thank Mr. Vaibhav Khatavkar, ME CSE-IT, Assistant Professor, Department of Computer Engineering and Information Technology at College of Engineering, Pune for the providing the invaluable guidance and advice during the process of conducting this research.

References

1. https://scroll.in/article/884754/surging-hindi-shrinking-south-indian-languages-nine-charts-that-explain-the-2011-language-census
2. https://timesondia.indiatimes.com/people/around-90-of-new-net-users-non-english-google-indias-rajan-anandan/articleshow/58375379.cms
3. https://www.forbes.com/sites/baxiabhishek/2018/03/29/more-indians-access-the-internet-in-their-native-language-than-in-english/#1cec6e474a03LNCS
4. https://en.wikipedia.org/wiki/Stop_words
5. Mikolov, T., et al.: Efficient Estimation of Word Representations in Vector Space. arXiv: 1301.3781
6. Mikolov, T., Sutskever, I., Chen, K., Corrado, G.S., Dean, J.: Distributed representations of words and phrases and their compositionality. In: Advances in Neural Information Processing Systems (2013). arXiv:1310.4546
7. http://adventuresinmachinelearning.com/word2vec-keras-tutorial/
8. Ramos, J.: Using TF-IDF to Determine Word Relevance in Document Queries
9. Nalisnick, E., Mitra, B., Craswell, N., Caruana, R.: Improving Document Ranking with Dual Word Embeddings (2016)
10. http://universaldependencies.org/treebanks/hi_hdtb/index.html
11. http://www.tfidf.com/
12. van der Maaten, L.J.P., Hinton, G.E.: Visualizing data using t-SNE (PDF). J. Mach. Learn. Res. **9**, 2579–2605 (2008)
13. Murphy, K.: Machine Learning: A Probabilistic Perspective. MIT (2012). ISBN 978-0262018029
14. https://github.com/attardi/wikiextractor/wiki
15. http://dataaspirant.com/2015/04/11/five-most-popular-similarity-measures-implementation-in-python/
16. A Hybrid Geometric Approach for Measuring Similarity Level Among Documents and Document Clustering. https://github.com/taki0112/Vector_Similarity/blob/master/TS-SS_paper.pdf
17. Hurst, S.: The Characteristic Function of the Student-t Distribution, Financial Mathematics Research Report No. FMRR006-95, Statistics Research Report No. SRR044-95 Archived February 18, 2010, at the Wayback Machine

Design and Performance of Millimeter Wave Antenna for the 5th Generation of Wireless Communication Systems Application

Giriraj Kumar Prajapati$^{(\boxtimes)}$ and Santhosh Kumar Allemki

Sree Chaitanya Institute of Technological Sciences, Karimnagar, T.S., India
prajapatigiri38@gmail.com, santhosh.allenki@gmail.com

Abstract. The worldwide bandwidth deficiency confronting wireless carriers has propelled the investigation of the underutilized millimeter wave (mm-wave) recurrence spectrum for future broadband cell correspondence systems. Along these lines, the present innovation like 3G, 4G can't bolster this subsequently there is a prerequisite of creating cutting edge portable system which is to be called 5G organize. Millimeter wave (mm-Wave) frequencies in the vicinity of 6 and 100 GHz give requests of size bigger spectrum than current cell designations and permit use of substantial quantities of antennas for abusing beam forming and spatial multiplexing gains. The antenna outline for 5G application is testing assignment. This paper comprises of the antenna outline for the 5G application which will utilize microstrip fix antenna with openings stacked on the emanating patch to enhance the execution of antenna regarding gain, radiation example and bandwidth at 5–6 GHz spectrum. Loading the spaces in the transmitting patch can cause winding of the energized fix surface current ways and bring about bringing down of the antenna's essential resounding recurrence, which compares to the decreased antenna measure for such an antenna. The planned antenna has reverberating recurrence of 25 GHz which is appropriate for 5G antenna. The blueprint is contrived on Fr4-Eproxy material used as a dielectric material. In this paper FR4 material is used for the 1.5 mm thick substrate. The characteristics of the designed structure are investigated by using MoM based electromagnetic solver, IE3D.

Keywords: Slotted patch antenna · Return loss · Bandwidth · VSWR resonating frequency · 5G

1 Introduction

Progressing and future societal advancement will prompt changes in the way correspondence systems are utilized. On-request data and stimulation will progressively be conveyed over portable and wireless correspondence systems. These improvements will prompt a major ascent of portable and wireless movement volume, anticipated to build a thousand-crease throughout the following years [1]. It is additionally anticipated that the present overwhelming situations of human-driven correspondence will be supplemented by an enormous increment in the quantities of conveying machines; there are figures of an aggregate of 50 billion associated gadgets by 2020 [3]. The

© Springer Nature Switzerland AG 2019
A. Abraham et al. (Eds.): IBICA 2018, AISC 939, pp. 289–299, 2019.
https://doi.org/10.1007/978-3-030-16681-6_29

conjunction of human-driven and machine-type applications will prompt a huge decent variety of correspondence attributes. However, mm-Wave signals experience the ill effects of increment in isotropic free space misfortune, higher infiltration misfortune, and propagation constriction because of air assimilation of oxygen atoms and water vapor, resulting in blackouts and irregular channel quality. Therefore, higher antenna gain is required at both transceiver sides, where directional transmissions have impact on radio resource usage, multiple access, and interference qualities, and subsequently affect radio access systems (RANs) and radio asset management (RRM) plan. The fifth era of innovation began from 1G, 2G (GSM), 3G (WCDMA), 4G (LTE) and now going to thrive 5G everywhere throughout the world [2]. Each age are expert with a few developments and several inconsistencies. The fifth era (5G) innovation is stretched out to inside and out the fourth era. Such as limited bandwidth and speed. To have 5G, the antenna must have in any event gain of 28 GHz [3] and band width more than 1 GHz. Microstrip fix antenna are broadly utilized as a part of satellite correspondence, military reason, GPS, etc. due to its compact shape and light weight it is easy to implement [4].

Fig. 1. Some characteristics of recent wireless systems generations.

The vision of 5G can be just accomplish by consolidating different radio access innovation, LTE, HSPA with 5G and not by building up another innovation to supplant them. Microstrip antenna comprise of fix which is thin metallic strip or sheet set over the ground plane isolated by a substrate of dielectric material. The execution of the Microstrip antennas relies upon the tallness of the substrate and dielectric consistent of substrate. The execution of Microstrip antennas are useful for thick substrate with bring down dielectric steady of substrate material. The real constraint of Microstrip antenna is impedance bandwidth is bring down for thin substrate. However, for the handheld gadgets and wireless correspondence, the antenna size ought to be little and for that the stature of substrate ought to be as little as could reasonably be expected. This paper fundamentally contains the segments in which plan conditions of customary rectangular transmitting patch, method to enhance execution of antenna with stacking space on emanating patch, hypothetical figuring of Return misfortune and VSWR and proposed antenna for 5G application are described. MSAs are utilized as a part of a wide scope of uses from correspondence systems to biomedical systems, basically because of a few alluring properties, for example, light weight, low profile, low creation cost, likeness, reproducibility, dependability, and straightforwardness in manufacture and reconciliation with strong state gadgets. The work to be displayed in this paper is likewise a reduced Microstrip antenna by cutting two L openings on the correct side of the fix [5–8]. Our point is to decrease the span of the antenna and additionally increment the working bandwidth. The proposed antenna (substrate with $\varepsilon r = 4.4$) presents a size lessening of 71.14% when contrasted with a customary square Microstrip fix with a greatest bandwidth of 1 GHz. The reproduction has been completed by IE3D programming which utilizes the MOM strategy [9]. Because of the Small size, minimal effort and low weight this antenna is a decent contender for the utilization of EMPS and WiMax innovation.

2 Key Issues for Deployment of 5G Network

2.1 Data Coverage

The indoor coverage is one of the most important hurdles in 5G mobile communication and research has already proved that more than 70% of data traffic and 50% of voice movement starts from indoor zone however the flag scope in indoor region is so poor when contrasted with out-entryway territory [3]. To give a nature of administration, better scope, availability in very thickly populated territory is likewise one of the significant difficulties in versatile correspondence framework. A great deal of arrangement was proposed by different specialists to conquer the scope issues like by separating an incorporate with a little cell, by utilizing repeaters, cell part approach and so on yet the arrangement is muddled with issue like high interest in additional framework, standard support and so on along these lines for all intents and purposes it isn't reasonable.

2.2 Ultra Dense Network

To over-come the issue in high thickly populated region ultra-thick system is proposed. Ultra-thick system keeps up a steady network, information speed in very populated territory.

2.3 Interference Issue in 5G Mobile Communication

The 5G network will be the integration of different technologies and it also use high order modulation technique. Multipath interference, Co-channel interference, PAPR, fading etc. are the key issues need to be solved for the successful implementation of 5G technology. A proper network interference management has to be develop in order to address these issues.

3   Mathematical Formulation of Patch Antenna

In today's wireless communication, the most important requirement is of antenna with low profile. Most probably for the handheld devices, the challenging task is to design antenna which provide improved performance day by day with miniaturized size. The most probably preferred and extensively used antennas are Microstrip antennas because of easy to integrate with circuits the geometry of the square patch is shown in Fig. 1 which is a 20 mm × 20 mm. The antenna is fabricated on a substrate of FR4 epoxy with dielectric constant (εr) = 4.4 and substrate height (h) = 1.6 mm. Co-axial probe feed of radius 0.5 mm. Microstrip patch antennas have many methods of simulation. The rectangular patch is far the most widely used configuration. It is very easy to analysis using the cavity and transmission line mode the designing parameters of Microstrip patch antennas for rectangular patch are length of patch (*L*) and width of patch (*W*). These two parameters are depending on the height of substrate, dielectric constant of material and resonant frequency (Resonant frequency should be same as operating frequency). In the Microstrip antennas, the patch is main radiating element. For rectangular patch, the width of the patch (*W*) is depends on the resonant frequency (*fr*) and dielectric constant (*εr*) of the material. The slot on fix can be broke down by utilizing duality connection between the dipole and the slot [2]. The slot stacked on the fix influence to the execution parameter of the antenna. The slot stacked rectangular Microstrip fix antenna can be considered as parallel mix of capacitance C1, inductance L1 and obstruction R1 of fix and capacitive reactance of slot [2] (Fig. 2 and Table 1).

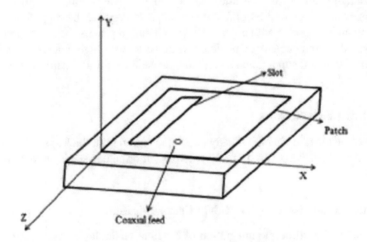

Fig. 2. Slot loaded rectangular Microstrip patch antenna

3.1 Theoretical Calculations of Slot

The slotted Microstrip antenna is investigated in above segment and in this segment the estimation of Return misfortune and VSWR are figured hypothetically utilizing conditions portrayed in (III) for various slot widths and slot lengths. As said the principle target of this part is to locate the ecological effects on the different fix antennas, for example, square and pentagonal fix with different dielectric covers which is straightly spellbound having the resounding recurrence of 2.45 GHz (with 10% bandwidth), utilizing FR4 (epoxy), as the dielectric substrate material of 0.8 mm thickness. The relative permittivity of the dielectric material is 4.4. The primary explanation for choosing this recurrence run is that the antenna is utilized as a part of WLAN (wireless-LAN) systems. As there are numerous ecological elements which influence the typical working of the Microstrip fix antennas. Henceforth it is imperative to think about the execution variety of the Microstrip fix antenna because of different climatic conditions, for example, snow, dust-particles and Plexiglas, NelTec, Glass PTFE and Rogers. As we realize that in blustery conditions the water layer is shaped because of attachment and surface strain and its genuine momentary thickness relies on number of variables, for example, correct introduction of fix surface (if fix surface is marginally disposed, the gravitational power will have its impact in like manner), rate of precipitation, wind condition, moistness and so forth.

4 Designing of Dielectrics Loaded Patch Antennas

To think about the impact of dielectric stacking of various dielectric steady on the execution conduct of square fix antenna, the ideal plan parameters are chosen to accomplish the minimized measurements and in addition the most ideal qualities, for example, high radiation productivity, high gain, directivity and bandwidth. The proposed antenna structure is sustained with 50 Ω coaxial link for impedance coordinating and IE3D has been utilized for the investigation.

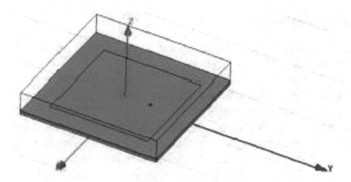

Fig. 3. Structure of square patch antenna

4.1   Design of Square Patch Antenna

The patch antenna that introduces here has made of the conduction material copper (Fig. 3).

Table 1. Rectangular patch antenna

S. no	Parameter	Dimensions (mm)
1	W	3
2	L	1
3	L_{eff}	2
4		3.3
5	(FR4-Eproxy)	4.4

The geometry of square patch antenna having a dielectric cover is shown in Fig. 4.

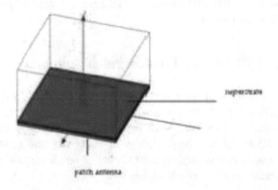

Fig. 4. Structure of antenna with dielectric cover

4.2   Design Specifications

The proposed antenna was planned utilizing following details: Relative permittivity of the substrate $\varepsilon r = 4.4$, Design frequency $f0 = 25$ GHz, Loss digression of substrate $\tan\delta = 0.001$, Height of the substrate $h = 1.575$ mm, Length of fix antenna $L = 1$ mm, Height of the dielectric $h = 1.575$ mm, Relative permittivity of the dielectrics $\varepsilon r = 2.2$, 2.5 and 4.4. Dielectric cover materials NelTec, Glass PTFE, Rogers in actuality, the Microstrip antenna appended to an electronic gadget will be secured by a dielectric cover (dielectric) that goes about as a shield against perilous natural impacts. These protecting materials regularly plastics (lossy dielectric) will diminish the general exhibitions of the antenna working attributes, for example, full frequency, impedance bandwidth and emanating productivity in view of the limited component strategy. The got comes about uncover that dielectric stacking don't change just the reverberation frequency yet additionally influences alternate parameters; gain, directivity and bandwidth. Specifically, the reverberation frequency brings down and move in thunderous frequency increments with the dielectric steady of spreads. Likewise, it has additionally been watched that arrival misfortune and VSWR increments, anyway bandwidth and directivity diminishes with the dielectric steady of dielectrics. Now the requirement for the 5G application about gain is that gain should be more than 5 dB. The gain can be improved by the array of patches but it does not affect to the bandwidth. Thus to increase the bandwidth, one slot is loaded on the radiating patch of the antenna as shown in figure. The length of the slot is 1 mm and width of slot is 2.1 mm. By simulating this slotted antenna resulting in return loss of −11.83 dB at resonant frequency 25 GHz. The bandwidth is 25 GHz and gain is 2.38 dB. Thus bandwidth is increases by slot on patch but it does not affect to the gain of patch (Fig. 5).

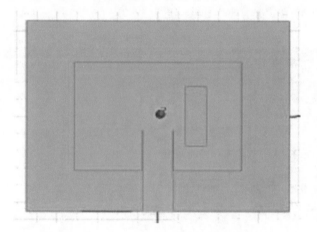

Fig. 5. Perspective view of the rectangular patch Rectangular Patch antenna

By loading the slot, resonant frequency can be decreases and bandwidth of antenna increases. Resonant frequency varies slightly for different slot width as compared patch without slot. The bandwidth is also increases with slot width for given slot length. Gain and bandwidth both can be increased by array of patch with slots on each patch.

5 Results and Discussion

A. Return Loss
The return loss versus frequency of the proposed multiband 4G/5G antenna is depicted in figure. The designed antenna is investigated using Finite Integration Technique. The return loss of the proposed multiband 4G/5G antenna is −42 dB, −28.35 dB, and −35.87 dB at 3.6 GHz, 14.33 GHz and 28.86 GHz, respectively. The antenna resonates with a −10 dB bandwidth of 0.57 GHz, 1.68 GHz, and 3.94 GHz at 3.6 GHz, 14.33 GHz and 28.86 GHz respectively. It can be evident from the return loss that the antenna resonates at three frequency bands i.e. 3.6 GHz, 14.3 GHz, and 28.86 GHz,

respectively. The 3.6 GHz frequency band is used for 4G, while, the rest of the two frequency bands are used for 5G cellular communication systems (Figs. 6 and 7).

B. Voltage Standing Wave Ratio (VSWR)

Figure 3 shows the VSWR of the proposed multiband 4G/5G antenna. The designed antenna has VSWR less than 1.1 for all the three resonant frequencies which displays good matching of the antenna being proposed. The simulated VSWR is 1.01, 1.07, and 1.04 at 3.6 GHz, 14.33 GHz, and 25 GHz, respectively. In Fig. 8 there is comparison between the S-Parameter in rectangular patch antenna and modified antenna. The pink colour graph shows the curve of rectangular patch antenna and the green colour graph shows the curve of modified antenna. The Maximum Resonant frequency, VSWR and Gain for without modified is 25.35 GHz, 27.46 GHz and −2.0 dB and for modified antenna 25.16 GHz Maximum Resonant frequency, 25.46 GHz for VSWR and −1.20 dB for Gain (Fig. 9 and Table 2).

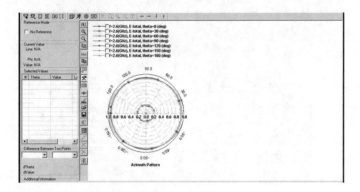

Fig. 6. Azimuth pattern in dielectric 4.4

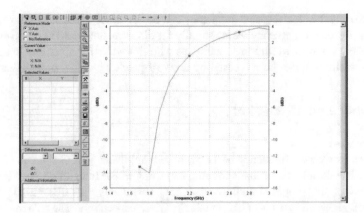

Fig. 7. Graph response between frequency versus dBi in dielectric 4.4

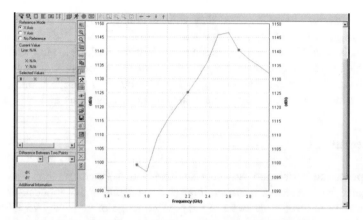

Fig. 8. Graph response between frequency versus gain in dielectric 4.4

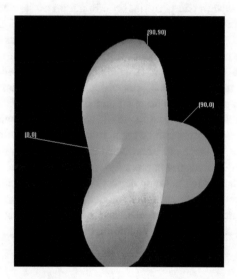

Fig. 9. Far field pattern in dielectric 4.4

Table 2. Comparison table

Parameter	Rectangular patch antenna	Modified rectangular patch antenna
Resonent frequency	25.35 GHz	25.16 GHz
VSWR	25.46 GHz	25.16 GHz
Gain	−2 dB	−1.30 dB

A possible way out is to design networks with change in mind, so that they will be more robust to disruption caused by growing demands and changing user patterns and yet-unimagined applications, and that the risks associated with investment in these kinds of networks will be lower, as they will be more durable and scalable. Networks that are designed with change in mind will also make effective use of resources (e.g., spectrum, bandwidth, power, processing capabilities, backhaul, etc.) and ensure a sustainable future. There is therefore ground to believe that telecommunication systems are evolving from being simple monolithic structures to complex ones, and that complex systems science might prove beneficial in their analysis and design.

6 Conclusion

In this paper the correlation between the Rectangular Patch Antenna and Modified Rectangular Patch Antenna has been improved the situation 5G application. The outcomes have been discovered to be in the coveted range of frequency 25 GHz. In this course of the undertaking, we reasoned that the Modified Rectangular Patch antenna gives the preferable outcome over the rectangular patch antenna.

In this work, the design and performance evaluation of the multiband metamaterial based antenna was proposed and analyzed. A 1.524 mm thicker low loss dielectric material was used as a substrate in the design of the antenna as well as in the metamaterial surface. The overall performance of the proposed antenna was enhanced using metamaterial surface as a reflector. A significant improvement in the gain of the antenna using metamaterial surface was observed. The antenna with metamaterial ground plane radiates with more gain and efficiency in all the three bands. The size of the antenna is such that it can be used in mobile and handheld devices. The antenna can be fabricated and measurements will be taken to validate the simulated results. The year 2020 will be the one of the most important year for telecommunication world. This technology expects to deliver a great quality of service almost in every field of communication. A 5G mm-Wave cellular network has been characterized to have large amount of available bandwidth at higher frequency bands, dense deployed small cells that closely interwork with macro cells, and large antenna arrays with directional antennas at both transceiver sides to enable high beam forming gains.

References

1. Li, Y., Pateromichelakis, E., Vučić, N., Luo, J., Xu, W.: Towards the 5th generation of wireless communication systems. In: Nicola Marchetti CTVR/The Telecommunications Research Centre Trinity College Dublin, Ireland (2016)
2. Cisco: Cisco Visual Networking Index: Global Mobile Data Traffic Forecast, 2015–2020 (whitepaper) (2016). https://goo.gl/yst7Dw. Accessed 23 Nov 2016
3. Fallgrenet, M., et al.: Scenarios, Requirements and KPIs for 5G Mobile and Wireless System (METIS Deliverable D1.1), April 2013. https://goo.gl/Jk7b5U. Accessed 23 Nov 2016
4. Terceroet, M., et al.: 5G systems: the mmMAGIC project perspective on use cases and challenges between 6–100 GHz. In: 2016 IEEE Wireless Communications and Networking Conference, Doha, 2016, pp. 1–6 (2016)
5. Andrews, J.G., et al.: What will 5G be? IEEE J. Sel. Areas Commun. **32**(6), 1065–1082 (2014)

6. FP7 European Project 318555 5G NOW (5th Generation Non-Orthogonal Waveforms for asynchronous signalling) (2012). http://www.5gnow.eu/
7. 5G-Infrastructure Public-Private Partnership (2013). http://5g-ppp.eu/
8. Wang, C.X., et al.: Cellular architecture and key technologies for 5G wireless communication networks. IEEE Commun. Mag. **52**(2), 97–105 (2014)
9. Panda, R.A., Panda, E.P., Sahana, R., Singh, N.K.: Modified rectangular microstrip patch antenna for 5G application at 28 GHz. IOSR-JECE J. **12**, 6–10 (2017)

Path Construction for Data Mule in Target Based Mobile Wireless Sensor Networks

Udit Kamboj, Vishal Prasad Sharma, and Pratyay Kuila$^{(\boxtimes)}$

Department of Computer Science and Engineering,
National Institute of Technology, Sikkim, Ravangla 737139, India
uditkamboj56@gmail.com, vishpsharma161992@gmail.com,
pratyay_kuila@yahoo.com

Abstract. Evolution in computer networks, Wireless Sensor Network (WSN) has played a significant role in sensing, processing and transmission of data from remote location across the network. There are many target points or point of interests (POIs) in a network that need to be monitored periodically or all the time. These POIs are covered by many static sensor nodes and collectively these sensor nodes form clusters. Each of the cluster contains a cluster head (CH) or relay node which collects the data sensed by the static sensors. One of the cluster has higher priority among the other clusters. So, a path is to be constructed for a data mule to collect data from these relay nodes. This path need to be optimized such that the higher priority relay node will be traversed times equal to its weight in one complete cycle. In this paper, Weighted Target Traversing with Stable Visiting Interval (WTT-SVI) algorithm has been proposed that gives the optimized path for a weighted target.

Keywords: Points of interests (POIs) · Mobile mule ·
Wireless sensor network (WSN) · Weight · Cluster head (CH) ·
Relay node

1 Introduction

Wireless sensor network (WSN) is composed of hundred or thousand number of spatially distributed sensor nodes dedicated to sensing and recording the physical parameters of the environment and finally transmit the data to a central location or base station [1–3]. The functionality of a WSN is shown in Fig. 1. WSNs have wide range of potential applications in environment monitoring, disaster warning systems, health care, defense, surveillance systems and target tracking [4–8]. In such applications, the sensor nodes are deployed in the target area either in ad-hoc or pre-planned manner. Ad-hoc deployment of the sensor nodes is useful for the harsh environments which are hard to access such as in deep forest, under water and so on. However, this approach requires a large number of sensor nodes

© Springer Nature Switzerland AG 2019
A. Abraham et al. (Eds.): IBICA 2018, AISC 939, pp. 300–309, 2019.
https://doi.org/10.1007/978-3-030-16681-6_30

to be deployed randomly to ensure full coverage of the target region. The pre-planned or manual deployment is used for easy accessible region.

In many applications, it is required to monitor or collection of data periodically instead of all along. For such applications, the traditional node deployment strategies of placing hundreds of nodes to ensure coverage and connectivity in larger areas turn to be impractical. Since, it incurs an additional cost for larger number of node deployment and their maintenance, with frequent channel contentions, message collisions, and losses. By introducing some degree of mobility, a WSN can overcome the above mentioned issues. In such applications, sensor nodes transmits the sensed data to its cluster head (CH) or relay nodes [9,10], then a mobile data collecting node or mule collects the sensed data from the relay nodes periodically or on demand [11]. Note that the mules are battery operated and they need to be recharged periodically. The energy consumption of the mules are mainly depend on the movement or length of the moving path. Therefore, reducing the total movement path and also periodically collection of the data from the relay nodes is an challenging issue.

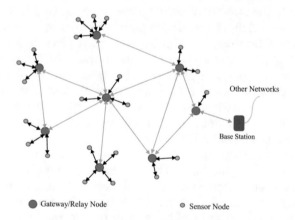

Fig. 1. A wireless sensor network

In this paper, we have proposed an algorithm named as Weighted Target Traversing with Stable Visiting Interval (WTT-SVI), which efficiently constructs the sub paths of the mobile sensor node or mule satisfying the periodical data collection demand of the relay nodes. Moreover, we have also consider the real life scenarios where their may be some relay node which has to be visited more frequently than the other relay nodes. The proposed algorithm have taken care the issue. We perform extensive simulation on the proposed methods to evaluate its performance and the relevant data is shown in a tabular form.

The rest of the paper is organized as follows. Section 2 shows a study of current state of the art and related works. The proposed work is demonstrated with algorithm and illustration in Sect. 4. The Sect. 5 shows simulation result of the algorithm. Finally, we conclude the paper with future directions in Sect. 6.

2   Related Work

Wang et al. [12] proposed a scheme in which coverage hole in neighborhood of deployed sensors is find out by using voronoi diagram. After that, mobile nodes are informed by static sensor nodes to fill the coverage hole. After receiving request, the former makes a decision to whether move to highest coverage hole or not. If the largest coverage hole stated in the request is larger than the coverage hole generated due to its motion then, it moves otherwise not.

Recently, Gupta et al. [13] have proposed a relay node placement algorithm based on GA. The main objective of this algorithm is to place minimum number relay nodes to the given potential positions so that all the sensor nodes (targets) can be k-connected with relay nodes. Here, authors have considered only the connectivity between sensor nodes and relay nodes but connectivity between the placed relay nodes is not considered. Rebai et al. [14] presented a novel genetic algorithm to find the various locations to place minimum number of sensor nodes in such a manner that the set of sensor nodes provide full coverage to area and individual sensor nodes are connected with the network. Fitness value of a chromosome is the number of sensor nodes required to provide the required coverage and connectivity. Here, the proposed crossover operation may generate invalid offspring. Therefore, an offspring correction phase is introduced. Then Gupta et al. [15] have presented a GA based approach for node deployment that provides k-coverage of a given set of targets and m-connectivity of the sensor nodes. Authors have also presented an efficient scheme for chromosome representation and derived an efficient fitness function with three objectives, namely use of minimum number of sensor nodes, coverage and connectivity.

In the scheme proposed by Batalin et al. [16], while patrolling, a DM dispose some sensor nodes in target area. Later disposed sensor nodes find out weakly covered area in their neighborhood. The information about it is conveyed to DM. Based on information gathered from disposed sensor nodes and its past information, DM makes a decision for new patrolling route. Based on local information, more sensors are disposed to provide coverage to new area.

Howard et al. [17] proposed a technique based on potential field. A field is generated which causes nodes to repel each other. When they are at some distance from one another, the repulsive force weakened causing them to be in equilibrium (stabilized). In the scheme proposed by Wang et al. [18], the deployment type of nodes is sparse which creates coverage gap in the region. Using Voronoi diagram, these coverage gaps are identified. So, in order to fill these coverage holes, the sensor nodes are moved from dense region to sparse region.

Cheng et al. [11] proposed centralized and distributed data collection techniques, known as CSweep and DSweep. Since CSweep divides the targets into various clusters and a DM is allocated to each cluster that is responsible for monitoring of that cluster. In DSweep, Data Mule interchanges information locally with other DMs so as to decide the next visiting target. But, neither CSweep nor DSweep incorporates the path construction. Moreover, these does not consider that each target might require different monitoring. Furthermore, these does not acknowledge the recharging problem and instability of visiting intervals (VIs) of each target.

Wu et al. [19] proposed a Convex Hull-Based (CH-Based) scheme in which a delegation node in each subnetwork is selected and then a convex hull of these delegation nodes is constructed. DM traverse the constructed convex hull and periodically visits nodes. But, the CH-Based scheme does not acknowledge the recharging problem and requirement of stable visiting interval. Moreover, it does not consider the different monitoring requirement.

In [20], the DM is equipped with two antennas and uses SDMA for data transmission. It stops at certain positions called polling points to collect the data from nearby sensors. If two sensors are in tune when related to the same polling point, they are treated as compatible pair that transmits data concurrently on arrival of DM. As each DM uses SDMA, the hardware is costly.

Konstantopoulos et al. [21] presumed that the traversing path is a predefined trajectory and DM follows it with a periodic schedule to retrieve data from the rendezvous nodes (RNs). Various clusters are formed by dividing the sensors and each cluster comprises of a cluster head (CH). These CHs perform data filtering by filtering out the redundant data and forwards this data to the corresponding RN. Yet, Konstantopoulos et al. [21] does not take into account the energy utilization of CH and RN. Moreover, CH and RN are likely to be drained out of energy earlier than the other nodes which results in network lifetime reduction.

3   System Model and Terminologies

3.1   Network Model

Here, we assume a network model, where all sensor nodes are placed randomly or manually along with a few relay nodes and all the nodes are stationary after deployment. Clusters are formed using the algorithm as proposed in [22]. The sensor nodes collect the local data and send it to their corresponding relay nodes. On receiving the data, the relay nodes aggregate them to reduce the redundant data within their cluster. There is a mobile data collector, which is also known as data mule. The data mule visits all the relay nodes periodically to collect data, which are subsequently returned to the sink node or base station which is located near to the target area as shown in Fig. 2. The assumed network environment is applicable to a wide range of applications of WSNs, for example, battlefields, a robot or helicopter to periodically monitor a set of predefined targets, such as forts, powder rooms, or marshal rooms.

3.2   Terminologies

The following terminologies are used in the proposed algorithm.

- $\{C_1, C_2, C_3, \ldots, C_N\}$ be the set of N number of relay nodes.

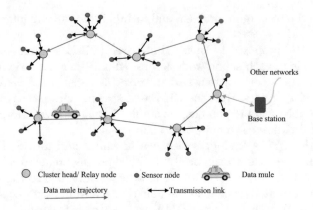

Fig. 2. Cluster based WSNs with data mule

- $\{w_1, w_2, w_3, \ldots, w_N\}$ be the priority weight of the relay nodes. The priority value indicates the minimum number of times a relay nodes to be visited in a cycle. Here, we assume that only w_1 has the higher priority and all other nodes have the same priority, i.e., $w_i = 1, \forall i, 2 \leq i \leq N$ (without loss of generality it may be considered that all priorities are one except w_1).
- P denotes a complete cycle or path obtained via Traveling Salesmen Problem (TSP) solution. For example, given four nodes $\{C_1, C_2, C_3, C_4\}$, the complete path represents a tour $C_1 \rightarrow C_2 \rightarrow C_3 \rightarrow C_4 \rightarrow C_1$.
- SP_i denotes the ith sub path. For example, $C_1 \rightarrow C_2 \rightarrow C_1$, $C_2 \rightarrow C_3 \rightarrow C_4 \rightarrow C_2$, etc. are the two sub paths which can be constructed from four nodes $\{C_1, C_2, C_3, C_4\}$.

3.3   Problem Formulation

Given a set of relay nodes $\{C_1, C_2, C_3, \ldots, C_N\}$ and their corresponding priority values $\{w_1, w_2, w_3, \ldots, w_N\}$, we have to construct efficient patrolling routes enabling the mule to visit all the relay nodes and collect information from those relay nodes within the certain time period.

4   Proposed Work

Now, we present our proposed work as follow. First, from the given set of relay nodes, a complete cycle or path is created using the TSP problem as shown in Fig. 3. Note that, in this path, all the relay nodes are traversed by a mobile mule exactly once in a complete cycle. As it is already mentioned that some relay node may be a critical or privileged node, i.e., it need to be visited more frequently in order to satisfy the monitoring requirements of the corresponding targets. The critical relay node (say C_1) has to be traversed as per the number of times equal to its weight. Therefore, we have to find out the sub cycles of the complete path so that the visiting interval (i.e., the time interval in which a node should be traversed at least once) will be stable for C_1. In order to implement it the proposed algorithm uses a threshold value n_c which is calculated by the following Eq. (1).

$$n_c = \lceil \frac{N+1}{w_1 - i + 1} \rceil \tag{1}$$

where N is the number of relay nodes, w_1 is the priority value of the privileged node (here, C_1) and i denotes i^{th} sub cycle. The pseudo code of the proposed algorithm is shown in Algorithm 1.

Algorithm 1. Weighted Target Traversing with Stable Visiting Interval (WTT-SVI)

Input: (1) Set of nodes, $\{C_1, C_2, C_3, \ldots, C_N\}$
       (2) the priority, w_1 of the privileged node C_1.
Output: Sub cycles $SC_i, \forall i, 1 \le i \le w_1$.

```
 1: GeneralPath()                          /* To obtain a TSP path */
 2: t = 2                                   /* t is a temporary variable*/
 3: for i = 1 to w₁
 4:     SCᵢ = ∅
 5:     if i == 2 then
 6:         N = N − 1
 7:     end if
 8:     n_c = ⌈ (N+1)/(w₁−i+1) ⌉            /* Calculation of threshold value*/
 9:     SCᵢ = SCᵢ ∪ C₁
10:     for j = t to t + n_c − 2
11:         SCᵢ = SCᵢ ∪ Cⱼ
12:     end for
13:     SCᵢ = SCᵢ ∪ C₁
14:     t = j
15:     N = N − n_c + 1
16: end for
```

An Illustration

For better understanding of the algorithm, an example is shown with illustration. Let us consider a path with $N = 10$ and $w_1 = 6$ as shown in Fig. 3. Now, as per the algorithm step wise creation of sub cycles is presented as follow (Fig. 4):
    Initially, for $i = 1$ (as per the for loop of line 3):

$$n_c = \lceil \frac{10+1}{6-1+1} \rceil = 2 \text{ (refer to line 8)}$$
$$SC_1 = \{C_1\} \text{ (refer to line 9)}$$

Now, after the for loop of line 10, $SC_1 = \{C_1, C_2\}$ and finally after line 13, $SC_1 = \{C_1, C_2, C_1\}$. Now, the same process can be followed for $i = 2$ and

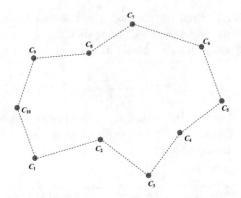

Fig. 3. Generalized path using TSP for $N = 10$ and $w_1 = 6$.

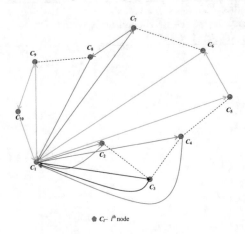

$\bullet\ C_i$- i^{th} node

Fig. 4. Result of WTT-SVI algorithm for Fig. 3.

$SC_2 = \{C_1, C_3, C_1\}$. Subsequently, for $i = 3$ to 6, the corresponding sub cycles are $SC_3 = \{C_1, C_4, C_1\}$, $SC_4 = \{C_1, C_5, C_6, C_1\}$, $SC_5 = \{C_1, C_7, C_8, C_1\}$ and $SC_6 = \{C_1, C_9, C_{10}, C_1\}$.

Therefore, the overall path $P = \{SC_1 \cup SC_2 \cup SC_3 \cup SC_4 \cup SC_5 \cup SC_6\}$ $\equiv \{C_1 \rightarrow C_2 \rightarrow C_1 \rightarrow C_3 \rightarrow C_1 \rightarrow C_4 \rightarrow C_1 \rightarrow C_5 \rightarrow C_6 \rightarrow C_1 \rightarrow C_7 \rightarrow C_8 \rightarrow C_1 \rightarrow C_9 \rightarrow C_{10} \rightarrow C_1\}$. The overall step wise threshold value and sub cycles are also shown in Table 1.

Note that the fixed threshold value of n_c results into formation of the sub cycles $SC_1, SC_2, SC_3, \ldots, SC_{w_1-1}$ of equal length and the last sub cycle SC_{w_1} with long length (as it will contain all the remaining nodes). Therefore, it is necessary to change the threshold dynamically after each iteration. We derive the value of n_c using ceiling function (refer to Eq. (1)) consists of N and i. It can be observed from Algorithm 1, line 8 that the value of n_c changes dynamically along with N and i after each iteration ensuring all the sub cycles $SC_1, SC_2, SC_3, \ldots, SC_{w_1}$ of almost equal length.

Table 1. Value of each variable in each iteration

Iteration no (i)	Threshold value n_c	Sub cycle (SC_i)
1	2	$\{C_1, C_2, C_1\}$
2	2	$\{C_1, C_3, C_1\}$
3	2	$\{C_1, C_4, C_1\}$
4	3	$\{C_1, C_5, C_6, C_1\}$
5	3	$\{C_1, C_7, C_8, C_1\}$
6	3	$\{C_1, C_9, C_{10}, C_1\}$

5   Experimental Results

Experiments are performed on the proposed algorithm using C programming language on a system with an Intel $i7$, 6th generation processor, 2.4 GHz CPU, 8 GB RAM and Microsoft Windows 10 as a platform. The experiments are run on different values of N and w_1. We get the sub cycles and the number of nodes traversed in each sub cycle as a result. The algorithm can give the result for millions of nodes but due to shortage of space we are demonstrating up to 10 nodes. Experimental results are being shown in Table 2.

Table 2. Experimental results

Sl. no.	Input		Output	
	Number of relay nodes (N)	Priority value (w_1)	Sub cycle	Number of nodes traversed
1.	4	2	$C_1 C_2 C_3 C_1$	3
			$C_1 C_4 C_1$	2
2.	6	4	$C_1 C_2 C_1$	2
			$C_1 C_3 C_1$	2
			$C_1 C_4 C_1$	2
			$C_1 C_5 C_6 C_1$	3
3.	10	6	$C_1 C_2 C_1$	2
			$C_1 C_3 C_1$	2
			$C_1 C_4 C_1$	2
			$C_1 C_5 C_6 C_1$	3
			$C_1 C_7 C_8 C_1$	3
			$C_1 C_9 C_{10} C_1$	3

6 Conclusion

The proposed algorithm gives the optimized path for a weighted target. It generates the sub cycles and the number of nodes traversed in each sub cycle. In this paper only a single higher priority node is being considered for path construction. Our future attempt will be to design some algorithm by considering multiple privileged nodes also. Moreover, a mule can also be made rechargeable so that it can recharge itself before running out of energy from any recharging station nearest to it.

References

1. Akyildiz, I.F., Su, W., Sankarasubramaniam, Y., Cayirci, E.: Wireless sensor networks: a survey. Comput. Netw. **38**(4), 393–422 (2002)
2. Kuila, P., Jana, P.K.: Evolutionary computing approaches for clustering and routing in wireless sensor networks. In: Handbook of Research on Natural Computing for Optimization Problems, pp. 246–266. IGI Global (2016)
3. Kuila, P., Jana, P.K.: Heap and parameter-based load balanced clustering algorithms for wireless sensor networks. Int. J. Commun. Netw. Distrib. Syst. **14**(4), 413–432 (2015)
4. Lersteau, C., Rossi, A., Sevaux, M.: Minimum energy target tracking with coverage guarantee in wireless sensor networks. Eur. J. Oper. Res. **265**(3), 882–894 (2018)
5. Singh, D., Kuila, P., Jana, P.K.: A distributed energy efficient and energy balanced routing algorithm for wireless sensor networks. In: 3rd International Conference on Advances in Computing, Communications and Informatics (ICACCI 2014), pp. 1657–1663. IEEE (2014)
6. Kuila, P., Jana, P.K.: Approximation schemes for load balanced clustering in wireless sensor networks. J. Supercomput. **68**, 87–105 (2014)
7. Gupta, S.K., Kuila, P., Jana, P.K.: Energy efficient multipath routing for wireless sensor networks: a genetic algorithm approach. In: 2016 International Conference on Advances in Computing, Communications and Informatics (ICACCI), pp. 1735–1740. IEEE (2016)
8. Kuila, P., Jana, P.K.: Improved load balanced clustering algorithm for wireless sensor networks. In: International Conference Advanced Computing, Networking and Security (ADCONS 2011). LNCS, vol. 7135, pp. 399–404. Springer (2011)
9. Kuila, P., Gupta, S.K., Jana, P.K.: A novel evolutionary approach for load balanced clustering problem for wireless sensor networks. Swarm Evol. Comput. **12**, 48–56 (2013)
10. Kuila, P., Jana, P.K.: Energy efficient load-balanced clustering algorithm for wireless sensor networks. Procedia Technol. **6**, 771–777 (2012)
11. Cheng, W., Li, M., Liu, K., Liu, Y., Li, X., Liao, X.: Sweep coverage with mobile sensors. In: IEEE International Symposium on Parallel and Distributed Processing, IPDPS 2008, pp. 1–9. IEEE (2008)
12. Wang, G., Cao, G., LaPorta, T.: A bidding protocol for deploying mobile sensors. In: Proceedings of the 11th IEEE International Conference on Network Protocols, pp. 315–324. IEEE (2003)
13. Gupta, S.K., Kuila, P., Jana, P.K.: Genetic algorithm for k-connected relay node placement in wireless sensor networks. In: Proceedings of the Second International Conference on Computer and Communication Technologies, pp. 721–729. Springer (2015)

14. Rebai, M., Le berre, M., Snoussi, H., Hnaien, F., Khoukhi, L.: Sensor deployment optimization methods to achieve both coverage and connectivity in wireless sensor networks. Comput. Oper. Res. **59**, 11–21 (2015)
15. Gupta, S.K., Kuila, P., Jana, P.K.: Genetic algorithm approach for k-coverage and m-connected node placement in target based wireless sensor networks. Comput. Electr. Eng. **56**, 544–556 (2016)
16. Batalin, M.A., Sukhatme, G.S.: Coverage, exploration and deployment by a mobile robot and communication network. Telecommun. Syst. **26**(2), 181–196 (2004)
17. Howard, A., Matarić, M.J., Sukhatme, G.S.: Mobile sensor network deployment using potential fields: a distributed, scalable solution to the area coverage problem. In: Distributed Autonomous Robotic Systems 5, pp. 299–308. Springer (2002)
18. Wang, G., Cao, G., La Porta, T.F.: Movement-assisted sensor deployment. IEEE Trans. Mob. Comput. **5**(6), 640–652 (2006)
19. Wu, F.-J., Tseng, Y.-C.: Energy-conserving data gathering by mobile mules in a spatially separated wireless sensor network. Wirel. Commun. Mob. Comput. **13**(15), 1369–1385 (2013)
20. Zhao, M., Ma, M., Yang, Y.: Efficient data gathering with mobile collectors and space-division multiple access technique in wireless sensor networks. IEEE Trans. Comput. **60**(3), 400–417 (2011)
21. Konstantopoulos, C., Pantziou, G., Gavalas, D., Mpitziopoulos, A., Mamalis, B.: A rendezvous-based approach enabling energy-efficient sensory data collection with mobile sinks. IEEE Trans. Parallel Distrib. Syst. **23**(5), 809–817 (2012)
22. Gupta, S.K., Kuila, P., Jana, P.K.: E3BFT: energy efficient and energy balanced fault tolerance clustering in wireless sensor networks. In: 2014 International Conference on Contemporary Computing and Informatics (IC3I), pp. 714–719. IEEE (2014)

Learning to Solve Single Variable Linear Equations by Universal Search with Probabilistic Program Graphs

Swarna Kamal Paul[(✉)], Prince Gupta, and Parama Bhaumik

Jadavpur University, Salt Lake, Kolkata 700098, India
swarna.kpaul@gmail.com

Abstract. The motivation for general AI is to overcome the problem of specificity in traditional AI approaches. Universal search guarantees to solve this problem with asymptotic optimality. However, the constant factor associated with the search time exponentially depends on solution size and can be immensely large. Evidently this reduces practical interest. Transfer learning can help reducing this constant factor for a series of related problems. We propose a dataflow graph-based programming model which evidently helps improving transfer learning between related tasks. We built a universal search based agent using our programming model which learned to solve single variable linear equations with minimal prior knowledge about list operations and arithmetic expression evaluation. The agent returned the solution as a program graph and was able to find general solution for a set of equations. Experimental results reveal the efficiency of the agent. The experiments demonstrate how universal search can be deployed in solving some practical problems like algebra equations.

Keywords: Universal search · Dataflow graph · Artificial general intelligence

1   Introduction

Traditional AI methods focus on solving one or few domain specific problems. The challenge for general AI is to solve problem in wide variety of environments without loosing optimality compared to current best methods. Universal search [1] in true sense is asymptotically optimal for all machine inversion and time limited optimization problems. This means time and space requirement for solving these problems is a constant factor away from the current best. However, though this constant factor is independent of problem size, yet it can be immensely large and depends on the solution size. This constant factor is the one which is holding back Universal search from being a game changer in real world AI applications.

Universal search searches for solution in a program space on some machine model. The complexity of the search process is invariant on the machine model or the language chosen. Thus, the choice of a grammar for the programs only effects the search by a constant factor. However, for practical application of universal search we need to focus on dampening the constant factor. Transfer learning is a way to dampen the combinatorial

© Springer Nature Switzerland AG 2019
A. Abraham et al. (Eds.): IBICA 2018, AISC 939, pp. 310–320, 2019.
https://doi.org/10.1007/978-3-030-16681-6_31

explosion in the program space which enables knowledge gained in solving one task to be transferred to another related task. Given the focus is to dampen the program search space incrementally, the choice of a programming model is of high importance.

In this article we propose a programming model, based on dataflow graphs to be used for universal search. Dataflow graphs [2] are good at implicitly capturing data dependencies and independencies among different functions in a program. The relative independencies among different functions allows the program graph to be divided into logical independent subgraphs which solves separate specific subtasks. These subgraphs can be directly reused in solving subsequent related problems, thus making the transfer learning process efficient. We developed an agent using our proposed programming model and tested it in a problem environment of solving single variable linear equations. This is to demonstrate how universal search can be efficiently used to solve practical problems. The experimental results show the efficiency of transfer learning with our programming model and the agent was able to quickly find general solutions for solving simple equations. This type of programming models may also be adopted in existing universal search based AGI agent designs [3, 8, 9].

2 Agent Model

Figure 1 shows the schematic diagram of a solver agent. At the core it contains a searcher program which runs universal search to find solution in the program space. The problem definition, current state and goal state are encoded in the environment. The environment accepts perception requests and returns output based on the current state. It also accepts action request and modifies its current state as per some axiom. The transducer components within the agent are responsible for bridging the communication between the searcher program and the environment. For a given problem environment the agent runs its searcher program to find solution in the program space. Once it finds a solution the searcher program stops and the solution program is given as output.

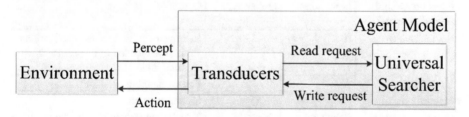

Fig. 1. Schematic diagram of an agent model.

3 Programming Model

A dataflow graph based programming model is proposed which would be used to generate solution program using universal search. Programs are represented as directed acyclic graphs where nodes represent functions to be applied on data received as input arguments. Edges represent dataflow from one node to another. Each node is of specific function type and number of input ports of a node depends on the number of input arguments of the function it contains. Each input port of a node corresponds with each input argument of the function and is type-casted. Only one output port is associated with every node. Edges can run between type compatible node ports only, such that an edge e can connect the output port of a node n with input port i of node n' if datatype of output port of node n is equal to the datatype of input port i. Only one edge can connect to an input port of a node. Multiple outgoing edges can connect to an output port of a node. Each node has data storage capability to store the output data returned by node function. However, output data is immutable during a program evaluation. On evaluation of a node function, it requests for data from its parent nodes (if any). If the parent nodes are already executed, then they just copy the output data to the concerned outgoing edge. If output data is not available at the parent nodes, then they are executed and the same process follows. A valid executable program graph should end with a single terminal node and start with a single initial node. The program graph excluding the terminal node is called as the parent program graph of the terminal node. Several primitive node functions are created which are to be used by the universal search to construct program graphs.

3.1   Node Functions

The node functions are the set of primitive functions which can be composed to create programs to solve simple algebraic equations algorithmically. These functions can be easily extended to be applicable for other problem domains. Table 1 lists the primitive functions to be used for solution program generation. The symbols A, B denotes data of any types from number, character, Boolean, function and list. List(A) denotes a list type with elements of type A and List() denotes empty list. func: denotes a function type. 'a' denotes a specific value. A/B denotes either type A or type B.

Table 1. List of primitive functions

Function symbol (Function name)	Input types	Output type	Operation
K_a: (constant)	A	a	Returns a constant value. The constant value is hardcoded in a specific constant node
i: (identity)	A	A	Passes the input to output
c: (create list)	A	List()	Creates an empty list
I: (invert operator)	character	character	Takes the mathematical operators as input and inverts it. Allowed input characters are: '+', '−', '*', '/'. '+' is inverted to '−'and '*' is inverted to '/' and vice versa
A: (append)	A/List(A), List(A)	List(A)	Appends first argument to the list object received as the second argument. If first argument is a list type, then it joins the two lists by appending elements of first list after the second

(continued)

Table 1. (*continued*)

Function symbol (Function name)	Input types	Output type	Operation
h: (head)	List(A)	A	Returns the first element of the input list
t: (tail)	List(A)	List(A)	Returns the remaining list after removing the first element from the input list
r: (reverse)	List(A)	List(A)	Reverse the order of elements in the input list
s: (split list)	List(A), A	List(List(A))	Splits a list based on second argument. The splitting element is excluded from the split lists
j: (join List)	List(List(A)), A	List(A)	Join operation is the opposite of split. It joins multiple lists within a list with the element at the second argument added in between
F: (Fmap)	List(A), func:	List(B)	Applies a function received as second argument to all elements in the list received as first argument
a: (apply)	func:, A	func:/B	Applies the input function on the element at the second argument. Returns either a partially evaluated function (if number of input arguments required by the input function is more than one) or the complete evaluated result
R: (repeat)	A	B	Repeat execution of the program by copying and re-executing the parent program graph after replacing the initial node with identity node and connecting its input port with the parent node of the repeat node. All intermediate repeat nodes are deleted from the copied parent program graph
E: (evaluate expression)	List(character)	List(character)	Simplifies a mathematical expression within a list. If first element of the simplified expression is not a mathematical symbol, then '+' symbol is added as the first element. Example: For Input expression $x + 5 + 2$ it will return $+ x + 7$
G: (Goalchecker)	List(A)	Boolean	Takes a list input and sends it to the environment to check if the goal is reached. If goal is reached it returns True else, False
IR: (Initial Input reader)	None	List(character)	Reads initial data from the environment

The last two functions namely, Goalchecker and Initial Input reader in Table 1 are part of transducers. Input reader sends a percept request to the environment to read the initial state. Goalchecker sends an action request to update the current state followed by percept request to check the goal status. Every valid program should start with an IR: node. G: node is added after every program terminal node satisfying type compatibility. No other occurrences of transducer nodes will be there within the program.

3.2   Composite Node Functions

Primitive functions can be composed in arbitrary ways, satisfying type compatibility to create arbitrary composite functions. We will create such composite functions and store it as available library functions to be used by the universal search. These functions are specifically created to aid the searcher program in finding solutions for algebraic equations. This can be considered as pushing some domain knowledge into the searcher process in the sense mentioned by Schaul and Schmidhuber [7].

Cut-Invert (CI:). The function CI: cuts the last two elements of a list and inverts the mathematical operator present at the second last position. Thereafter it returns the two elements in a list in the same order. It takes an input of type list and returns a list. Figure 2 illustrates the program graph of the CI: function. In all diagrammatic representations the input port order of a node is in clockwise direction, starting from the output port.

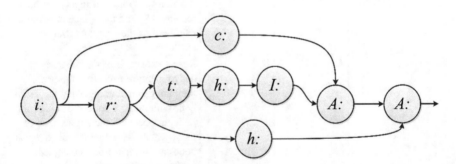

Fig. 2. Illustration of the program graph of composite function CI:

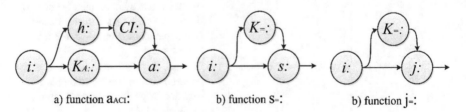

a) function a_{ACI}:   b) function $S_=$:   b) function $j_=$:

Fig. 3. Program graphs of composite functions. (a) program graph of function a_{ACI}:. (b) program graph of function $s_=$:. (b) program graph of function $j_=$:.

Partial Append of Cut-Invert (a_{ACI}:). The function a_{ACI}: takes a list of lists as input. It applies CI: function on the first element of the input list and thereafter partially applies append function on the result. The partially evaluated function is returned as output. Figure 3a illustrates the program graph of this function.

Split on Equality Symbol ($s_=$:). The function $s_=$: takes a list of character as input. It applies the s: function on the input list and splits the list based on "=" symbol as an element. Figure 3b illustrates the program graph of this function.

Join on Equality Symbol ($j_=$:). The function $j_=$: takes a list of lists as input. It applies the j: function on the input list and joins the lists within list with "=" symbol as an element added in between. Figure 3c illustrates the program graph of this function.

3.3 Node Configurations

In order to plug in the programming model with universal search the nodes need to be configured to store some data apart from the function output data. We would follow a probabilistic strategy for universal search where each program is allocated a specific runtime proportional to the program probability during testing phase. Program probability is calculated my multiplying all conditional probabilities between program nodes and the prior probability of the initial node. The prior probability of the initial node is set to 1 as all programs start with the initial node. Conditional probability is associated with each edge in a program graph. Given a program graph F ending with terminal node t, the conditional probability $P(n_i|t)$ denotes the probability of connecting the output port of t with input port i of node type n, where node type n denotes any of the available in-built functions. The default value of $P(n_i|t)$ is calculated by the formula $1/\sum_{n \in \text{all node types}} \sum_{i \in \text{all input port of } n} i_n$, such that $i_n = 1$ if n_i is type compatible with t else, 0. While extending a program F and adding new nodes after terminal node t the default probability is assigned to each edge. This is stored in the parent node as the child initial probability. Each node also stores the factored program probability of the program for which the concerned node is the terminal node. The factored program probability is the list of probabilities associated with all unique edges of a program graph. The total program probability can be calculated by multiplying all probability values within the factored probability list.

In universal search each program needs to be aborted when some execution time limit is crossed. To implement this, each node stores a list of execution times of all the nodes of the program graph for which the concerned node is the terminal node. Total execution time is calculated by adding up all these values in the list. Each primitive function which does not take other functions as input, performs atomic operations and consumes 1 unit of time. Functions which takes other functions as input, consumes time equal to the number of atomic operations done by its input functions. Composite functions consume time equal to sum of atomic operations done by all its composing functions. During test phase each node is given access to a global time limit. Before executing a node, it checks if the current runtime exceeds the global time limit. If yes, then execution stops else it continues.

4 Implementing Universal Search

The search process proceeds in phases. Programs are generated and tested in the order of levins complexity [10]. In phase 1 a search graph is created with an initial node IR: and all possible function nodes from the list of available functions are added after the initial node. After adding each node, it is executed. Each node is allocated a runtime *Phase* $\times P/C$, where P is total program probability of the program graph of which the executed node is the terminal node. If *Phase* $\times P$ exceeds 1 then the program is tested until it halts, or time runs out. Unexecuted programs are tried in next phase. Only successfully executed programs are extended further. Extension occurs by adding all possible combination of function nodes available, satisfying type compatibility, as child node. A function with multiple input argument may connect one of its input ports after such an executed node and other input ports may connect to any combination of available executed nodes in the search graph, satisfying type compatibility. Programs which generate an error are not extended further. After successfully executing a program the output is passed to the transducer node G: (goalchecker). If the goalchecker node returns True, the search process stops and the program graph is returned. If no programs can be extended further in a phase, the phase is doubled and the same process repeats.

Universal search needs two predefined parameters to execute the search process, namely *Phase* and *C*. The value of the *Phase* should be chosen as the maximum possible integer such that doubling that *Phase* value would cause the universal search process to fail due to hardware constraint. The value of C should be chosen as a fraction between 0 and 1. A low value of C would allocate more time to each program which are allowed to execute is a specific *Phase*. In general, this would allow a possible solution program to get fully executed in the *Phase* in which it is first allowed to execute, thus avoiding an unnecessary search in the next *Phase*. However, making the C too low would cause infinitely or long running irrelevant programs to take up too much time, thus delaying the overall search process. Thus, a tradeoff needs to be made while choosing the value of C.

5 Applying Incremental Learning

A simple incremental learning strategy is used to update the conditional probability distribution of the edges in the search graph. The strategy is similar to adaptive Levin search [6]. However, the programming model in our case is different as we are using a dataflow graph based programming model instead of sequential symbol based programming. After reaching the goal state, the probabilities of all edges of the solution program is updated as $P(n_i|t) = P(n_i|t) + \alpha(1 - P(n_i|t))$, $P(n_i|t)$ is an edge probability and α is a constant factor known as learning rate. The increment in an edge probability is balanced by decrementing a total equal amount of probability mass from all outgoing distinct type of edges of node t. A distinct edge is identified by the parent node function and the port number of the child node function. Thus, edges connecting from t to port i of two instances of same node function n will be assigned same probability though they might be part of separate programs. After updating the probability distribution, the search graph is reset to remove stored output data, program runtime and flags to identify node state (executed, failed or unexecuted) from each node. Thereafter the same search graph is used by the agent to search for solutions of subsequent problems.

The incremental learning process needs one predefined parameter to operate, namely the learning rate α. α should be chosen as a fraction between 0 and 1. A high value of α would cause a quick convergence to a learned solution thus causing overfit. Whereas a low value of α would make the learning process too slow to reap the benefits. Thus, a tradeoff needs to be made while choosing α. We found choosing α as 0.5 gave us near optimum results.

6 Solving Single Variable Linear Equations

We experimented with our agent by giving it simple algebra problems to solve. The agent was given three different set of single variable linear equations in a sequence and it returned three different program graphs for each set, solving every equation in that set. We encoded the problem specification and the goal state in the environment. The agent can communicate with the environment with the transducer nodes, namely IR: and G:. The IR: node returns the initial equation in a list where each symbol in the equation is an element of character type in the list. The list elements of the equation x + 15 = 20 will be 'x', '+', '15', '=', '20'. Due to hardware constraints and to keep the search space manageable we choose the following subset of node functions as the list of available functions which would be used to construct programs by the search process. IR:, $K_{E:}$, $s_{=:}$, $a_{ACI:}$, F:, $j_{=:}$, R:, r:. The constant node $K_{E:}$ occurs only once in the search graph after the IR: node. Putting $K_{E:}$ node in other places would create redundant programs as it always returns constant value irrespective of the input.

The first set contain the equations x + 5 = 15, x + 80 = 100, x + 6 = 3. The agent is required to find a solution program which solves all the above equations. Solution program should return a list containing the elements in the form +x = +a or +x = −a. After solving the equations in the set, incremental learning is applied and probability distribution is updated. Learning rate is set to 0.5 and C is set to 0.001. Figure 4 shows the solution program graph found by the agent for solving the above set of equations.

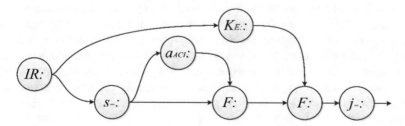

Fig. 4. Solution program graph found by the agent for solving equation set 1.

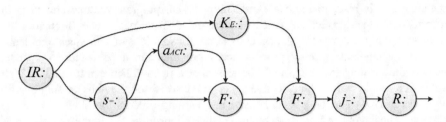

Fig. 5. Solution program graph found by the agent for solving equation set 2.

The second set contain the equations x * 2 + 5 = 15, x * 8 + 20 = 100, x * 3 + 6 = 3. The agent started with the search graph generated by the earlier search process. Figure 5 illustrates the solution program found by the agent which solves all the equations in the set. Incremental learning is applied after solving the set.

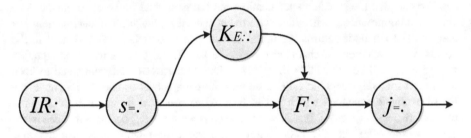

Fig. 6. Solution program graph found by the agent for solving equation set 3.

The third set contain the equations x = 5 + 15, x = 20 + 100, x = −2 + 3. Figure 6 illustrates the solution program found by the agent which solves all the equations in the set.

Table 2 compares the efficiency of the search process with and without incremental learning. Total runtime denotes the number of all atomic operations performed in the search graph during the search process. The program probability of solutions for later tasks increased due to incremental learning and thereby reduced the search time. The solutions for subsequent problems reused a section of previous solution graphs due to presence of common subproblems between the tasks. For example, the third equation set is just a subproblem of first and second equation set. The program graph in Fig. 4 is able to implicitly segregate these dependent or independent subproblems due to program representation as graph. The independencies among different functional nodes allows to divide the solution graph into several independent subgraphs, each solving a specific subtask. These subgraphs can be directly re-used for related tasks as it happened for solution of equation set 3 in Fig. 6. This makes the transfer learning process efficient in program representation as data flow graphs.

Table 2. Comparison of search process with and without incremental learning

Equation set	With incremental learning		Without incremental learning	
	Program probability	Total runtime	Program probability	Total runtime
1	3.93675988 * 10^−6	77809	3.93675988 * 10^−6	77809
2	0.000780794	1261	6.561266481 * 10^−7	534080
3	0.01589209	48	0.000283446	1356

The main performance issue faced while experimenting with our agent is due to exponential growth of the program search space, initially when the agent has no learnt memory of the environment. Many of the programs in the search space are over represented programs. Some equivalent program pruning method in the sense mentioned by Paul et al. [5] can be applied to handle the problem of over representation in program space [4]. However, the pruning method proposed by Paul et al. will not work in agents with memory and incremental learning, like in our case. Thus, we would need a different pruning method based on program equivalence. That would dampen the exponential constant factor even further.

7 Conclusion

Making universal search practically useful in designing a general intelligent agent is a tough challenge. We presented a dataflow graph based programming model to be used by universal search to solve practical problems. Our designed agent efficiently learnt to solve single variable linear equations with minimal initial bias. We demonstrated how program representation as dataflow graph helps improving transfer learning. However, the problem of program overrepresentation and chaotic execution still exists in our designed agent. The programming model can be modified to follow functional programming paradigm which would ease reasoning about programs to handle the problem of over representation and chaotic execution. The programming model can be enhanced to make it more general to be applied in multiple different environments. The agent can be tested with more complex equations or other mathematical problems and if required enhancements can be made.

References

1. Li, M., Vitányi, P.: An Introduction to Kolmogorov Complexity and Its Applications. Springer, New York (2013)
2. Sousa, T.B.: Dataflow programming concept, languages and applications. In: Doctoral Symposium on Informatics Engineering, vol. 130 (2012)
3. Özkural, E.: Gigamachine: incremental machine learning on desktop computers. arXiv preprint arXiv:1709.03413 (2017)
4. Looks, M., Goertzel, B.: Program representation for general intelligence. In: Proceedings of AGI, vol. 9, May 2009

5. Paul, S.K., Bhaumik, P.: A fast universal search by equivalent program pruning. In: International Conference on Advances in Computing, Communications and Informatics, ICACCI, pp. 454–460 (2016)
6. Schmidhuber, J., Zhao, J., Wiering, M.: Shifting inductive bias with success-story algorithm, adaptive Levin search, and incremental self improvement. Mach. Learn. **28**(1), 105–130 (1997)
7. Schaul, T., Schmidhuber, J.: Towards practical universal search. In: Proceedings of the Third Conference on Artificial General Intelligence, Lugano, June 2010
8. Hutter, M.: A gentle introduction to the universal algorithmic agent AIXI. In: Artificial General Intelligence (2003)
9. Schmidhuber, J.: Gödel machines: fully self-referential optimal universal self-improvers. In: Artificial General Intelligence, pp. 199–226. Springer, Heidelberg (2007)
10. Jankowski, N.: Applications of Levin's universal optimal search algorithm. Syst. Model. Control **95**, 34–40 (1995)

PKI Model Optimisation in VANET with Clustering and Polling

S. Arul Thileeban[✉], C. Sathiya Narayan, J. Bhuvana,
and V. Balasubramanian

Department of Computer Science and Engineering,
Sri Sivasubramaniya Nadar College of Engineering, Chennai, India
arulthileeban023@gmail.com, sathiyagunners@gmail.com,
{bhuvanaj,balasubramanianv}@ssn.edu.in

Abstract. The growth of Ad-Hoc systems due to the need of decentralised networks has seen the advent of MANET and VANET that makes the communication quicker. VANET is a real time application arena which requires high level and lightweight security protocols. In particular, detection and removal of malicious nodes are essential. Since third parties are involved for the revocation of digital certificates for each nodes, it doesn't truly act decentralised. This increases the time delay which is the main advantage in an Ad-Hoc system. We propose a clustering model with a polling based Digital Certificate revocation system. This will let the existing model function swiftly and remove the latency using third party revocations.

Keywords: PKI · VANET · Clustering · Digital certificate · Ad-hoc security · Decentralization

1 Introduction

With new technologies evolving every day, security and privacy of a person comes into picture drastically. Communication between two parties should be made heavily secure since otherwise it would allow for tampering or eavesdropping of data. Communication between parties is not only privy to cellular, server interaction but also in the field of Vehicular Node Communication. In Vehicular Adhoc network (VANET), tampering of messages can prove fatal causing road accidents, traffic congestion etc. VANET is the facet which reflects the vehicles of modern world. With each growing day, Vehicle to Vehicle (V2V) communication is established in more cars for the assistance it heavily provides in realtime. The applications, characteristics and challenges in VANET are described briefly in [1]. The main motivation behind these works and research about VANET and its security issues is the growing use of vehicular technologies around the globe. Driving systems are becoming automated with time and it is extremely important provide a model that wouldn't be affected because of the false messages sent by malicious intruders into the vehicular network. Also, multiple vehicular issues like traffic congestion, accident alerts, etc.

VANET can be described similar to that of a Mobile Ad Hoc Network (MANET), but in case of VANET, vehicles act as nodes to create an Ad Hoc network. The vehicles

© Springer Nature Switzerland AG 2019
A. Abraham et al. (Eds.): IBICA 2018, AISC 939, pp. 321–329, 2019.
https://doi.org/10.1007/978-3-030-16681-6_32

participating in VANET act as wireless nodes. This enables vehicles around the specified proximity to create a vehicular network and communicate among them on the road. VANET has become a propitious and optimistic field of research in wireless communications paving the way to safety of vehicles. VANET enables communication between the Vehicles also known as the V2V Communication and between the Vehicles and Road-side Units or Infrastructures also called as V2I communications [2]. Communication in VANET also includes Inter-roadside unit communication (R2R) and Roadside unit to vehicle communication (R2V).

The characteristics of VANET such as mobility, topology, and network size lay both challenges and opportunities in achieving the goals of security. VANETs require a set of security and privacy policies that needs to be fulfilled in order to ensure a successful use and public acceptance. Related to information security, there are three main needs. The primary need is Confidentiality which ensures that messages will only be accessed by the intended parties is required in some private services, for example location based services. Apart from confidentiality, Data integrity ensures that they have not been altered since their inception. Even beyond, the third need is related to guaranteeing data trust, i.e. data is fresh, updated and reliable (in case of dynamic clustering). The last and the most important basic need is availability, which implies that every node must be able to timely process and send the required information. All these factors help in building a safe and steady network. The new Dedicated Short Range Communication (DSRC) offers the potential to effectively support vehicle-to vehicle and safety communications, which has become known as Vehicle Safety Communication (VSC) technologies. Dedicated Short Range Communication (DSRC) is used as communication medium and it operates on 5.9 GHz frequency band. DSRC is based on IEEE 802.11a standard and IEEE 1609 working group is being standardized as IEEE802.11p for special vehicular communication [14].

Vehicles which act as the nodes can detect each other with the help of various sensors and GPS (Global Positioning System) inbuilt in vehicles that have the ability to detect vehicles around them. Even when vehicles leave the network or become unreachable, other vehicles can join the network resulting in a mobile network with nodes as vehicles connecting to one another and thus this depends on the network size. As a result, this forms a cluster of vehicles which can be classified as two types. The first type is called the Static Clustering. In this type, stable cluster is formed. These clusters also contain Road Side Units (RSU). In this case cluster works within the range of RSU. There is no need of reconfiguration of cluster in the case of static clustering. These clusters are not scalable. Cluster formation and maintenance is easy for static clustering. Routing protocols are easily designed. But scalability and other factors decrease the performance of this network [3]. The other one is known as Dynamic Clustering. In this type, cluster formation done dynamically in real time. Due to the dynamic nature of the network cluster reconfiguration is required. Clusters heads change at high frequency because of high mobility and random speed of vehicles. Cluster reconfiguration and range of cluster head depends on the density of the area. These clusters are easily scalable as discussed in [4].

The process of communicating among the vehicles in VANET is at greater risk because the messages are broadcasted and a wireless medium is used for data propagation. As the vehicles invoke a dynamic nature, vehicular network is likely to face many entries and thus result in congestion. In order to avoid these kind of problems many solutions have been proposed of which clustering is one of the solution technique. Clustering brings lucidity to the network and increases the connectivity among the vehicles. A mobile node in MANET can move in haphazard and random directions but in the case of VANET it can only move along the street. There are many types of clustering algorithms and protocols for establishing efficient communication between the vehicles. Thus for secure communication among these vehicles, clustering algorithms are preferred. Some of them discussed in [6] are Mobility based clustering schemes, direction based clustering, lane based clustering schemes.

A RSU is a predominant assist mechanism in VANET Architecture help in efficient communication between the vehicles in the network. Road Side units are typically placed at equal distances along the Roads. The distance varies based on the density of Nodes on an average day. RSUs primarily assist in the revocation of a digital certificate of a node if they are found to send maliciously intended messages to other vehicles. The need for decentralization is growing drastically and usage of RSUs restrict that need, also prohibiting VANET from functioning purely ad-hoc. Here, we propose a solution by eliminating the RSUs in the network and by usage of a proposed algorithm which utilizes clustering followed by polling, the need for RSUs are eliminated to make the network decentralized.

In this paper, we simulate the algorithm proposed using NS2 (Network Simulator) with an intermediate trace file generated and obtain various parameters like packet transmission rates between nodes, etc.

2 Related Works

In this section, we give a brief discussion of work related to security in Vehicular Ad hoc Networks. The detailed literature survey is done, to get preliminary knowledge on clustering, attacks possible on the cluster and methods to overcome them. Also, we have analysed existent algorithms to detect malicious nodes which invoke the usage of RSU to revocate the digital certificate. There are numerous works carried in the field of VANET using clustering. Many models, schemes and techniques are proposed for VANET in clustering. We have also covered some well known techniques for message authentication and certificate revocation which are commonly used for preventing the attacks possible on a network.

Clustering is the process where the vehicles otherwise called as the nodes group together to form an agglomeration of vehicles. Cluster size varies from one cluster to another and is mostly dependent on the transmission range of the wireless communication device that a node uses [11]. One of the most principal requirements when it comes to clustering is the importance of a cluster head. Cluster head in VANET communication plays a very important role in passing information and also carry information about other vehicles and their positions. In [10], Lo et al. discussed a multi-head clustering algorithm where the proposed algorithm intends to create stable clusters by prolonging cluster lifetime and shortening the average distance between cluster heads and the other cluster members. Node mobility will frequently destroy existing cluster structures. So, reclustering overhead becomes an important cost metric.

Mohammad et al. [5] proposed a novel technique where cluster head is selected depending on the lane having the highest traffic flow. Clustering algorithms are designed to make cluster process efficient and secure.

In [13], Kaur et al. discussed two types of clustering algorithms which are followed in order to develop a cluster. The first one is called the Cluster Formation algorithm. A cluster is a small group of vehicles containing a cluster head, an entry node and more than one member. To enable communication, selecting the cluster head and choosing the gateway node, formation algorithm is preferred. The second one is the Cluster Maintenances algorithms. These algorithms are used to recover the links and cluster from any type of failure such as network breakdown or GPS malfunctioning. A member node is dead when the cluster head does not receive messages send by that node. A node rejoins the cluster when it stops receiving the messages send by the cluster head.

Ensuring a secured network is one of biggest challenges in VANET. Zhou et al. [7] discuss that if a new vehicle node wants to join the network, we need to validate the new vehicle node to improve the security of the VANET. They proposed a security authentication method based on trust evaluation. So basically, when a vehicle wants to access the internet through the roadside base station, they evaluate the access node by employing the direct trust evaluation. They look for the history of records of the vehicle requesting to join by looking at their event records and other key factors. Thus based on all these direct security degree of the new vehicle is determined. When a group of vehicles form a wireless network to communicate information with each other, they adopt the indirect trust evaluation mechanism to evaluate the new vehicle node.

Security attacks in this vehicular network by pranksters are becoming growing cause of concern nowadays. Some of the most common attacks such as DoS (Denial of Service), DDoS (Distributed Denial of Service), Sybil attack, blackhole attack, wormhole attack, Alteration attack have been discussed in detailed in [12, 15]. Upma et al. [8] discussed a method which deploys an IDS (Intrusion Detection System) based genetic algorithm which determines a prankster node which tries to disrupt the system by feeding improper inputs to the network. This improves the performance of the network to a very large extent. Factors like throughput, energy, delay graphs were plotted and it showed results which proved that the performance of the network is improved to a very large extent.

Thus the best way to overcome this is to revocate the vehicles with the help of third party authentication. In [9], Zhang et al. discuss a certificate revocation status validation scheme. It uses the concept of clustering in the domain of data mining. In VANET, certificate validation is more time sensitive as every vehicle is prone to receive large number of messages in a minimum amount of time. The discussed scheme employs the K-means clustering technique to boost up the efficiency of certificate validation. There by, this enhances the security of vehicular ad hoc network. Credibility and the issued date are added to certificates to improve the security of the system. Another common way for authenticating the messages sent by the vehicles is through Public key Infrastructure (PKI). This is possible by using asymmetric cryptography where each vehicle has a public-private key pair. Any vehicle can generate a digital signature for its outgoing messages using its unique private key. On the other hand the receiver of the message verifies the digital signature using the sender's public key. This digital signature verification ensures message authentication. For entity authentication of a vehicle, its public key must be authentic to all other vehicles in the network. Thus PKI is required for securing VANETs. Thus to get the ownership of keys they need to be assessed by the Regional Transportation Authority (RTA).

Upon analyzing the above techniques, we observe that there is a huge latency due to need to use a RSU and the delayed revocation of certificates upon detecting a malicious node. This process is highly dangerous since the impact of malicious messages can be life threatening. Holding these thoughts in mind, we propose a new way to detect the malicious attacker and remove him without the usage of RSUs.

2.1 Existing Model

PKI (Public Key Infrastructure) is the existing security system being used in V2V Communications. Road Side Units are setup from place to place for communication with vehicles and between themselves. On detection of a malicious intent vehicle using various detection algorithms, a certificate revocation process is initiated by a Road Side Unit and it revocates at a third Party Site and then transfers the revocation message to all Road Side Units. This is seemingly a very lengthy and time consuming process. V2V Spectrum is also very sparse. The spectrum is already shared by other industries, hence allowing only usage of less bandwidth, thus making lightweight components more worthy for V2V model. But PKI is not lightweight. It is quite heavy since the data transfer is of voluminous size (Fig. 1).

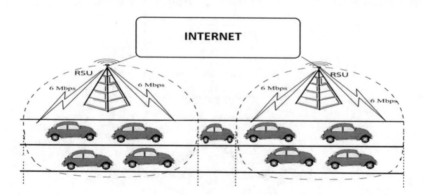

Fig. 1. VANET architecture.

3 Proposed Methodology

In our proposed system, a model based on clustering and user polling based revocation is deployed. There are a few assumptions under which this model is most optimal. The assumptions are the number of vehicles operating under VANET should be huge enough to accommodate clustering, RSUs aren't deployed at a required number and there is a need for suitable other arrangements to support revocation of Certificates at a faster rate to prevent further damage based on fake information provided by the malicious node.

The basic algorithm is as follows:

(1) Cluster the cars based on distance from one of the clustering mechanisms used. Allocate a cluster head on random or center basis.
(2) When a false message is received, a car would know about it based on user visual or detection using deviation off path – closest node to the malicious node detects through visual contact that the message is false and wouldn't take appropriate action according to the message. This deviation in action is tagged and used as an initializer automatically.
(3) Then the node initiates a polling process based on user action within the cluster.
(4) Cluster head receives the initialization notification and collects data (vote) from each node based on their individual action and based on the ratio of votes, the certificate of the node is revoked.
(5) The message of revocation is passed onto other clusters. This message growth will be exponential and hence will be spread completely under the assumption that the number of cars operating under VANET will be huge in all areas.

This model would help distort the existing PKI solution where a Road Side Unit (RSU) is required for revocation along with detection of a malicious node and hence speeding up the process and reducing effort from third parties. This can be considered as a purely decentralised model since each cluster can act independently and stores a set of data which is then transferred to other blocks aka clusters (Fig. 2).

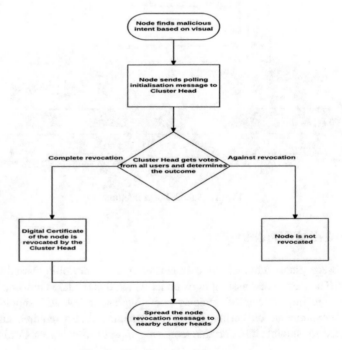

Fig. 2. Flowchart for node revocation.

4 Implementation Details

The simulation for the proposed model is done using NS2 and Python. The clustering mechanism was built over NS2 using DSRed Queue. Double Slope Random early detection (DSRed) is the new queue management technique which improves the performance of Random Early Detection (RED). RED suffers from various problems like instability in the network, throughput, large delay which are considerably reduced in the case of DSRed queuing mechanism. The data collected based on the results were stored in a trace file. The trace file contains data like the event, time, from node, to node, packet size, flags, source address, destination address. It was used to run the Python Simulation for user based action which is nothing but the voting mechanism by vehicles to identified upon sending a false message. This action was completely randomized and the results were successful for the above proposed model. The data collected contains critical information of the configuration used for the cluster. The components like packet size, rate and other parameters were varied to find the peak value of successful packet transmission. The rate is fixed at approximately 30000 bytes/s with peak performance and the packet size is then varied and the data is tabularized in Table 1 and the graph for the corresponding data is plotted in Fig. 3.

Table 1. Analysis of successful packet transmission and packet size.

Packet size	Bytes transmitted	Lost bytes	Success ratio
10	9998719	658734	93.4119
100	999870	65558	93.443
500	199970	12995	93.5016
1000	99983	6423	93.576
5000	19993	1310	93.447
10000	9994	866	91.334
20000	4995	512	89.749

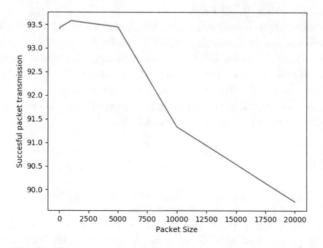

Fig. 3. Graph for packet transmission rate against packet size.

328 S. Arul Thileeban et al.

The above proposed simulation has been tested with variety number of clusters and the results tend to the values plotted.

$$Success\ Ratio = (Bytes\ Transmitted - Lost\ Bytes)/Bytes\ Transmitted \qquad (1)$$

It can be inferred that packet transmission rate is efficient enough to handle the polling model proposed under clustering. The results obtained above under corresponding scenario indicate the parameters involved in a cluster formation along with user voting system. Here, a general clustering algorithm is used. But in real-time, various parameters like number of vehicles, etc. could decide the clustering mechanism and if the clustering mechanism varies based on parameters, it would make the whole system more efficient.

5 Conclusion

VANET is a real time model which requires security measures to be lightweight and without latency. Hence, by counteracting the high latency produced while deploying third party digital certificates, the proposed model produces a major advantage over existing models. In addition to this, it allows users to take action when deemed necessary if the car escapes other detection algorithms involved. Digital Certificates might be advantageous in places where there is proper connectivity established, but not in scenarios like roads, where connectivity needs to be established through Road Side Units. The results also show us that the cluster model with the voting algorithm could act alone. This also has the added advantage of acting decentralized thus reducing the need for Road Site Units which will reduce the cost and energy needed to implement VANET on a global level. Decentralised networks, especially block-chains are on a growing trend due to speed, no central point of failure. All these factors affect our need for smarter and quicker security models in developing technologies. VANET is a rapidly growing network which demands such kind of solutions and the proposed model is a good solution to the need.

Future works would include extending the model with data analysis over vehicle patterns in real time to automatically deploy the user voting pattern hence making it automated and faster. It can be achieved using streaming classifiers. Considering both the above mentioned issues, PKI model which is already used in V2V is not suitable as the technology grows. Advent of autonomous cars already makes most of V2V communications for human usage redundant. Usage of our proposed model can be extended to autonomous cars with minor variations.

References

1. Chaubey, N.K.: Security analysis of vehicular ad hoc networks (VANETs): a comprehensive study. Int. J. Secur. Appl. **10**(5), 261–274 (2016)
2. Yang, K., Ou, S., Chen, H.-H., He, J.: A multihop peer-communication protocol with fairness guarantee for IEEE 802.16-based vehicular networks. IEEE Trans. Veh. Technol. **56** (6), 3358–3370 (2007)

3. Singh, A., Kaur, M.: A novel clustering scheme in vehicular ad hoc network. Res. Int. J. Appl. Inf. Syst. (IJAIS) **10** (2015)
4. Kakkasageri, M.S., Manvi, S.S.: Connectivity and mobility aware dynamic clustering in VANETs. Int. J. Future Comput. Commun. **3**, 5 (2014)
5. Mohammad, S.A., Michele, C.W.: Using traffic flow for cluster formation in vehicular ad-hoc networks. In: IEEE 35th Conference on Local Computer Networks, LCN, pp. 631–636 (2010)
6. Priyanka, T., Sharma, T.P.: A survey on clustering techniques used in VANET. In: 11th IRF International Conference (2014)
7. Zhou, A., Li, J., Sun, Q., Fan, C., Lei, T., Yang, F.: A security authentication method based on trust evaluation in VANETs. Springer (2015)
8. Zhang, Q., Almulla, M., Ren, Y., Boukerche, A.: An efficient certificate revocation validation scheme with k-means clustering for vehicular ad hoc networks. In: IEEE Symposium on Computers and Communications, ISCC, pp. 862–867 (2012)
9. Upma, G., Tanisha, S.: Defense against prankster attack in VANET using genetic algorithm. Indian J. Sci. Technol. **9**(35) (2016) https://doi.org/10.17485/ijst/2016/v9i35/98094
10. Lo, S.-C., Lin, Y.-J., Gao, J.-S.: A multi-head clustering algorithm in vehicular ad hoc networks. Int. J. Comput. Theory Eng. **5**(2), 242 (2013)
11. Vodopivec, S., Bester, J., Kos, A.: A survey on clustering algorithms for vehicular ad-hoc networks. In: 2012 35th International Conference on IEEE Telecommunications and Signal Processing, TSP (2012)
12. Al-Kahtani, M.S.: Survey on security attacks in vehicular ad hoc networks (VANETs). In: 6th International Conference on Signal Processing and Communication Systems, ICSPCS (2012)
13. Kaur, P., Bhagat, N.: A review on clustering in VANET. Int. J. Innov. Res. Comput. Commun. Eng. (IJIRCCE) **4**(5) (2016)
14. Jadhao, A.P., Chaudhari, D.N.: Study of various reliable routing protocols and related attacks in vehicular adhoc network. Int. J. Adv. Found. Res. Sci. Eng. (IJAFRSE) **1**(3) (2014)

Optimizing Probability of Intercept Using XCS Algorithm

Ravindra V. Joshi[✉] and N. Chandrashekhar

NI-University, Kanyakumari District, TN, India
rvjoshi18@hotmail.com, drncshekhar60@yahoo.com

Abstract. In battle scenarios, the amount of RF signals/energy intercepted by a sensor like Radar Warning Receiver is characterized by a metric called Probability of Intercept. Due to increase in density and complexity of radars, achieving optimal POI within given cost-constraints is challenging. The basic problem arises since radars scan in spatial (and hence time domain), whereas RWR scan in frequency domain. Synchronizing RWR's reception frequency with same of Radar and, at the time when radar is illuminating Aircraft will require more than analytical techniques. Situation becomes more complicated with multiple radars and radars with Low Probability Intercept (LPI) signatures. In this paper, we explore how to exploit eXtended Classifier System (or XCS) to tackle this problem.

Keywords: EW · POI · XCS

1 Introduction

1.1 Probability of Intercept

Since ancient times, battle awareness especially of the adversarial side has played a critical role in success of the combat missions. In modern battles, radar and electronic warfare have assumed roles of paramount significance. It is key differentiator of F35 against F16 aircraft [2]. One important indicator of how much of enemy is observed is given by POI. POI is defined as the probability that the EW system will detect the presence of a particular threat signal between the time it first reaches the system's location and the time at which it is too late for the system [3]. Thus, to perfectly, know enemy radiations, RWR has to receive all the energy incident on it. This gives 100% POI. However, in practice most of narrow band receiver based systems will perform around 90%.

The basic bottleneck in maximizing POI is orthogonal scanning of radar and RWR domains. Radars essentially search across the space by different scan patterns, generally configured by Mission Requirements [4, 5]. So, it will be illuminating the aircraft (on which RWR is mounted) for brief period of time, called Dwell Time [3]. RWR, due to its hardware limitations, can sample only certain bandwidth of frequency in fixed time (1 MHz/1 μs). The challenge is tune RWR's reception-frequency to illuminating radar, exactly when it is dwelling on RWR. For this it has to predict and estimate radars scan

© Springer Nature Switzerland AG 2019
A. Abraham et al. (Eds.): IBICA 2018, AISC 939, pp. 330–338, 2019.
https://doi.org/10.1007/978-3-030-16681-6_33

pattern, rate and sometimes signatures also. We use Extended Classification System Real (XCSR) [5] for scheduling RWR in bands so that interception and hence POI is optimal.

2 Background

Mono Pulse Radars are very common Radars used in modern battles. They are called "Mono Pulse" since angle and other parameters of the target is determined by echo of single transmitted pulse. Though transmitted pulse is one, echoes will be many depending on number and locations of targets. Even atmospheric factors like snow, fog create clutter and contribute some echoes. To cut unnecessary clutter, and more importantly for determining angle each target, Radar has to use directional antennae. It means at a given instance of time, it transmits pulses and receives echoes only in certain direction and is insensitive to echoes from all other directions.

The direction (we will consider only azimuth dimension here, though in practice elevation also plays equally important role) in which Radar's sensitivity will be maximum is called "Main Beam Azimuth" (and/or elevation).

Its Sensitivity will fall as the target moves away from main beam azimuth. Very soon, it reaches "near-zero" and becomes invisible to Radar. Note that this is angular difference. Range (linear distance may be same).

Diagram below shows a spatial representation of Radar's Gain w.r.t. to its Main Beam Azimuth (Fig. 1).

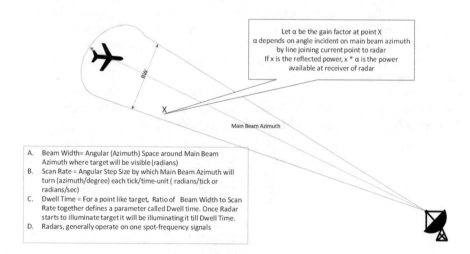

Fig. 1. Basics of radar pulse echo measurement

Let Radar complete its scan in SCAN ticks. Also let it visit Target only once during SCAN TIME. Then SCAN = DWELL + NON_DWELL. NON_DWELL is nothing but SCAN - DWELL. For most of our experiments DWELL, NON_DWELL is sufficient (Fig. 2).

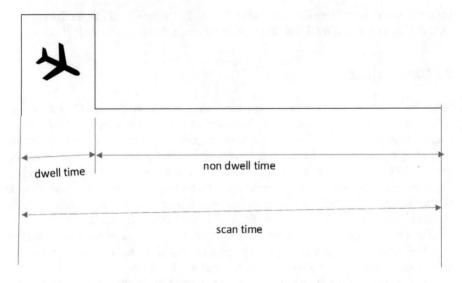

Fig. 2. Dwell time is small slice of scan time

Radar Warning Receiver (RWR in short) is a Passive Warning System that warns about Radars present in environment to Pilot. It is mounted on Attack Aircraft. It is passive because, unlike Radars, it will not transmit pulses. Instead, it receives pulses transmitted by Radars and then uses that information to infer the presence of radars. Its antenna can be arranged so that, it can receive from all 360°, there is no need to scan. However, its receivers (due to cost constraints) will be of Narrowband type. It means it can receive signals of certain bandwidth. RWR Band specifies the frequency range over which it can receive simultaneously.

Band allocation is arbitrary and depends on RWR designer. RWR will have table called Search Regime Table which specifies the sequence in which the Radars have to be searched. A typical search table may look like this (Table 1).

Table 1. RWR search configuration table

Band	Duration
P	20 ticks
Q	10 ticks
R	10 ticks
S	10 ticks

It implies that for 20 ticks reception (search) will be made in Band P, then 10 ticks in Band Q, R, S each.

THE BASIC CONSTRAINT: Let Radar be operating in Band X (any of P to S). Then, to intercept Radar of Band X, RWR should be receiving in Band X when Radar is Illuminating Aircraft (and hence RWR). Thus illumination (spatial overlap) and frequency (spectral overlap) both should happen for RWR to detect Radar (Fig. 3).

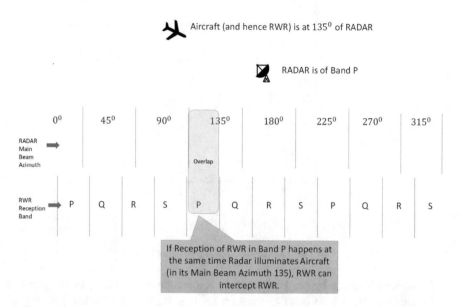

Fig. 3. Radar Azimuth and RWR frequency overlap is necessary for Radar detection by RWR

See figure above RADAR is scanning from 0 to 360° in steps of 45°. Let the Radar be of Band P. Let us say, Aircraft is at 135° azimuth w.r.t. Radar. Similarly RWR will be receiving in P, Q, R, S bands cyclically. When RADAR main beam illuminates aircraft at 135°, RWR is receiving in B and P and hence it can detect Radar. If this overlap was not there, it would have missed the RADAR.

No of instances of [RWR seeing Radar] will be always less that number of instances [Radar sees RWR] as Radar Illumination of RWR is pre-requisite for detection of RADAR by former. But it is desirable that whenever Radar illuminates Aircraft, RWR intercepts Radar. This can be expressed quantitatively with ratio [RWR sees Radar] to [Radar sees RWR]. This Ratio is known Probability Of Intercept POI.

It can be easily seen that $0 \leq POI \leq 1$ (from discussion above) [or 0% to 100%].

To be more precise, 100% POI will happen only when RWR reception will completely be receiving in "Radar Band" as long is Radar is illuminating the Aircraft. Check the figure below. As long as Radar is illuminating RWR, RWR should Receive in the band of RADAR (Fig. 4).

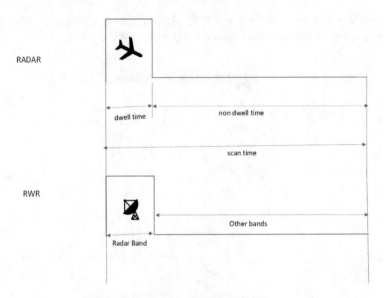

Fig. 4. Hit scenario for RWR (Best Case)

Now, the challenge before RWR is this. Radars are not working in co-operative mode with it. Then which band to open up for next reception? Problem get tougher when multiple Radars are there. As, the reception of Bands should be sequenced in such a way that on average maximum interception is achieved (for now, in case of multiple radars, average of poi with reach radar can be chosen) (Fig. 5).

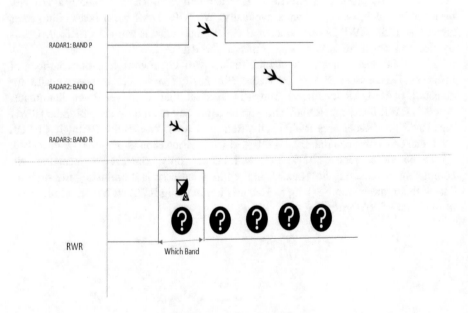

Fig. 5. Miss case for RWR

To this, probabilistic behavior should be added to account for radar dynamics. Since, this problem cannot be solved by direct analytical methods, we have attempted a machine learning approach. We have selected Extended Classifier System or XCS as a Machine Learning algorithm. XCS has been very popular and has been widely used in past 10 years. It is a type of Ensemble learning machine combining both Reinforcement and Genetic Algorithm. This is ideal for this problem as it associates payoff directly with domain rather than with rulesets.

3 POI Optimization Using XCS

3.1   Description of the Model

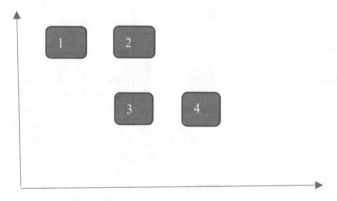

Fig. 6. Time vs frequency mapping of emitters

It is useful to model the radar (scenario) in a landscape of Time vs Frequency. See the figure above. The radars 1 and 2 are near-by in frequency and 3 and 4 are another set. Since 2 and 3 occur at the same time, either we can receive 1, 2 or 3, 4 but not both. This problem is similar to covering all red square with minimum rectangles, subject to the constraint that height of the rectangle does not exceed a limit (Fig. 6).

Overall Approach
Given a set of Radars in battlefields (with their position coordinates) and Timed coordinates of Flight Trajectory for the mission. Now, RWR will generate some random candidate Time-Frequency Windows of various height, width (respecting the constraints). Alternatively, it can do sequential scan to find initial set of emitters. Once the initial radars are found, rectangles are tightened around them as necessary. From then onwards, Reinforcement and Genetic Algorithms will take-over to keep right TF Windows in the stage.

Radars are generated using BattleSim model as outlined in [5].

Mapping XCS to POI

XCS Model describes clear set of parameters for mapping domain concepts. The key concepts are Classifier, Population, Matching Set, Action Set and Prediction Arrays [5]. These are mapped as follows (Table 2).

Table 2. Mapping of RWR-POI problem with XCS algorithm parameters

Sl. No.	XCS parameter	POI parameter	Specification
1.	Classifier	Time-Frequency Window	**Condition:** Radar is in Time Frequency Window **Action:** Tune Frequency to Height of TF window **Prediction:** POI of its window
2.	Input from environment (sigma)	Illuminations from Radars	
Major sets			
1.	Population [P]	Set of TF Windows	Generated randomly or through sequential scan
2.	Matching Set [M]	TF-Windows with Radars in them	These are candidate windows for reception in next window
3.	Action-Set	TF-Windows that will be satisfied frequency reception parameter	Overlapping and Collinear TF Windows will be members of Action-Set
4.	Acton-Set-Prev	Action set of previous cycle	Since this is a multi-step process, action set of previous iteration will be stored here

Other parameters are initially as defined by [8], and tuned experimentally. One walkthrough the execution is shown below.

Post Reception Optimizations

XCS algorithm is applied in two stages to RWR Reception Band. In first half, which happens prior to the reception, classifiers in population are compared to input radar parameters (sigma). Based on the Matched Classifiers (Match Set), a sequence of Actions (band, durations) are determined using which reception is made.

After action is executed, that is reception is done, the emitters are processed to improve the fitness of classifiers of customers. Fitness of a classifier can be defined as percentage of time in which the reception was made by one or more of the conditions of the classifier (i.e. Time Frequency Windows). Genetic Algorithm is used to improve overall fitness of classifiers.

Candidates for offspring are selected by using Roulette Wheel Selection based on Fitness criteria defined above. Using 2-point crossover criteria, conditions of the classifiers are exchanged. This completes crossover. Replacing a randomly chosen condition with don't care and vice-versa realized a form of mutation.

4 Results

Basic Results are observed, where correct Time Frequency windows were generated. Simulation was run on a configuration where 8 emitters were located randomly with respect to the trajectory of the aircraft. Figure Below shows "Time Frequency Window as observed initially (Fig. 7).

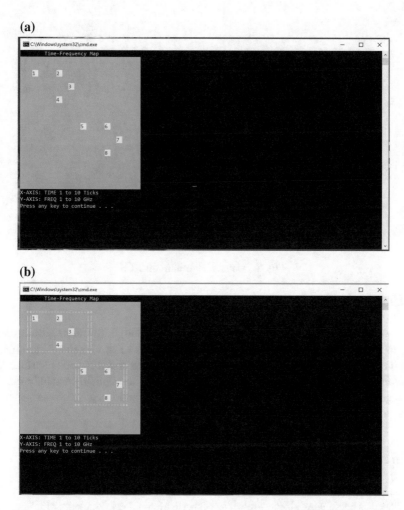

Fig. 7. (a) Raw distribution of inputs. (b) Same inputs surrounded by rectangles

When we run XCS algorithm, classifier with appropriate condition is selected.

This shows how genetic algorithm is used to optimize two classifiers into one. Each of classifier CL1 and CL2 searches in their dedicated band. A simple 2-point crossover resulted in a schedule in which both A and B can be searched in same duration. It is not guaranteed that such cross-over will be of high fitness. This will work only if BAND-A and BAND-B are distributed throughout reception duration. If not, it will get poor fitness score and will get purged out in due course. Only high-fitness classifiers (off-springs) will survive, thus increasing overall fitness of population (in line with genetic algorithm philosophy) (Fig. 8).

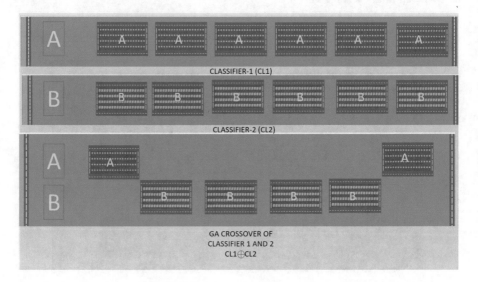

Fig. 8. Genetic algorithm on XCS

5   Conclusion

This investigation has clearly shown suitability of XCS algorithm for Optimization of POI. In addition to get benefit from basic reinforcement and Genetic Algorithm, the overall set of parameters also readily adapted. How-ever the scenarios considered are basic. Further scenarios like LPI signatures, windows to explicitly exclude emitters, coverage of high-priority emitters etc., make important and interesting candidates. We intend to take up these issues in future.

References

1. Tzu, S.: Art of War, 1st edn. Jaico, Mumbai (2010)
2. Wikipedia. https://en.wikipedia.org/wiki/Lockheed_Martin_F-35_Lightning_II
3. Adamy, D.: EW 101: A First Course in Electronic Warfare (Artech House Radar Library)
4. Joshi, R.V., Chandrashekhar, N.: Discrete time vs agent based techniques for finding optimal radar scan rate-a comparative analysis. In: International Conference on Soft Computing Systems (2018). First and Corresponding Author
5. Butz, M.V., Wilson, S.W.: An algorithmic description of XCS
6. Holland, J.H.: Introduction to Genetic Algorithm. http://www2.econ.iastate.edu/tesfatsi/holland.gaintro.htm

Reducing the Negative Effect of Malicious Nodes in Opportunistic Networks Using Reputation Based Trust Management

Smritikona Barai[1]([✉]), Nupur Boral[2], and Parama Bhaumik[2]

[1] Department of Computer Science and Engineering,
Heritage Institute of Technology, Kolkata, India
smritikona.barai@heritageit.edu
[2] Department of Information Technology, Jadavpur University, Kolkata, India
nupurboral@gmail.com, parama@jusl.ac.in

Abstract. One of the major security issues common in any trust and reputation management mechanism is Bad mouthing in which malicious participants victimize nodes with high reputations by giving lower or negative feedbacks in order to undermine them. Opportunistic Networks are more vulnerable to this type of attacks as there is no pre defined path between the source and the destination. Hence it is important to ensure the trustworthiness or reliability of the next-hop nodes as the nodes communicates by forwarding the message to the next hop using opportunistic connectivity due to node proximity. Otherwise there is the risk of malicious nodes interrupting secure transmission, thus affecting the performance of the network. To deal with this issue of Bad mouthing due to the presence of malicious nodes in the network, we have proposed and implemented a Reputation based Trust Management Model using Friendship vector (RTMF). We have compared the performance of our algorithm with an existing Trust management system based on Ontology and it has been observed that RTMF has successfully reduced the negative effect of the presence of malicious nodes and improved the performance of the network.

Keywords: Opportunistic networks · Oppnets · Trust · Reputation · Security

1 Introduction

Opportunistic networks are wireless networks of mobile nodes where message delivery is carried out in a hop-by-hop fashion using the devices in proximity. Due to the absence of any concrete path between source and destination, these types of networks suffer from a number of security issues. It is difficult to determine the true nature of the intermediate nodes which may have malicious intentions of breaching privacy, dropping data packets, compromising data confidentiality etc. Therefore establishing good reputation and trust is of utmost importance in OppNets in order to interact securely.

Reputation based trust management systems have been effectively used in MANETS to deal with security concerns [1]. Nodes are judged as trustworthy based on their reputations established through their past behaviors in the network. Hence, the trust value of a node is considered to be an important parameter to decide with which nodes

© Springer Nature Switzerland AG 2019
A. Abraham et al. (Eds.): IBICA 2018, AISC 939, pp. 339–349, 2019.
https://doi.org/10.1007/978-3-030-16681-6_34

it should communicate to improve the message delivery in a network [1]. The trust-worthiness of a node can be quantified by associating each node with a trust value. Every time a node helps in delivering a message to the destination, its trust value is incremented.

Different forms of trust, either based on consciously defined social connection (friend) [3], frequent encounter (familiar) [4] or similar interests [5], can be used in order to best suit the different requirements. In our proposed work, we have drawn reference from the real life scenario of human interactions based on reputation. In our day-to day activities, we tend to trust those people whom we know either directly or indirectly via someone whom we already trust.

One of the major attacks common in any trust based reputation mechanism is Bad mouthing in which the malicious nodes in the network conspire and malign the rep-utation of an otherwise innocent node. They attack the nodes with high reputations by reducing their trust values in order to undermine them. As a result, the number of nodes with poor trust values increases in the network and the number of successful message delivery decreases, thus having a negative effect on the overall performance of the network. The objective of our work is to show that our proposed Trust based reputation mechanism, considering reputation from both direct friends and indirect friends of a node, is capable of reducing the negative effect of the presence of malicious nodes in the network.

2 Related Works

In the field of computer networking, reputation of a node is a quantifier of its behavior in the network it is a part of. As in real world, a node's reputation is not a constant value that it assigns itself but a combination of other nodes' feedback about its performance/behavior in the network. The reputation of a particular node can in turn help other nodes to judge how much 'trustworthy' that node is as a potential next hop to deliver a message.

Several works has been proposed for building reputation system in MANETs to protect routing from attackers and increasing the network performance. In [6, 7], direct measurements and watchdog mechanism were proposed to be used in building repu-tation values. The concept of indirect measurements to build reputation values, in addition to direct measurements obtained using watchdog mechanism, were proposed in [8, 9]. The ACK messages sent by the destination nodes are used to construct the reputation values in several proposed works like [10–14].

However, these protocols are not suitable for dynamic networks like DTNs and Oppnets where an end-to-end path does not exist between the source and destination and all nodes are moving at a certain velocity. Here, an intermediate node needs to store and carry the transfer packets and wait for opportunities to transfer them, thus loosing the proximity with the node it wants to monitor, making the watchdog mechanism impractical. Similarly, the ACK messages may not follow the same intermediate path back to the sender that was used for forwarding the message towards the destination, thus rendering these messages unreliable.

Goncalves et al. [2] proposed a node's behavior evaluation in two parts by considering both its 'direct' and 'indirect' reputation. Direct reputation is based on the previous direct communications with the requester whereas indirect reputation represents the opinion of other nodes about who needs the resource.

A distributed malicious node detection mechanism was proposed by Ayday and Fekri [1] that enabled every node to evaluate other nodes based on their past behavior, without the intervention of any central authority. They considered two major adversaries faced by trust and reputation management mechanisms, namely Bad mouthing and Ballot stuffing. A fading parameter was also used to consider the effect of time on the trust value of a node.

In our work, we have drawn inspiration from these above works based on direct and indirect reputation with a fading factor. Our attempt is to develop a trust based reputation mechanism that reduces the declining effect of bad mouthing by malicious nodes on the overall performance of the network.

3 Working Principles

In the field of Opportunistic networking, it is quite clear that trust and reputation has a strong relation between them since a node's good reputation promotes the trust in that user [15]. This relation is the inspiration of our proposed work.

As already mentioned, our work draws analogy from the real life process of trusting someone based on his/her reputation. If a person X has interacted directly with person Y, X will judge the reputation of Y according to the nature of their interaction (good or bad) and will use this reputation for future interactions with Y. A third person Z who knows X directly but not Y, can ask X about the reputation of Y and decide whether to trust Y for interactions or not.

Using this analogy, in our work we have considered trust to be based on level of friendship ties by creating friendship vectors for each node along with their respective trust values. When a node encounters another, their friend lists are shared. To evaluate the trust value of a particular node P which is not in the direct friend list of node A (say), our proposed algorithm uses both Direct Reputation (DR) value and Indirect Reputation (IR) value. DR means the trust value of the node who is a direct friend of the sender node. IR is the trust value of a node which is a friend of friend (FOF) or friend of friend of friend (FOFF) of the sender. Here the indirect reputation may go beyond two levels.

Understanding the above idea of human reputation further, we can see that the degree of trust decreases as the level of 'indirectness' increases. Obviously person X will trust Y more than Z will trust Y as X is a direct friend of Y but Z knows Y via X. Similarly, the trust value of a node cannot be calculated from direct reputation alone and this value gets decremented as the level of indirect friendship increases.

To reflect this idea in our work, a decay factor is considered with the IR to reflect the impact of 'hop distance', i.e. the more indirectly a node is reachable from the sender node, the lesser impact it have on the final result. If there are multiple possible paths between a pair of sender and receiver nodes, then the average of all the calculated trust values corresponding to each path is considered.

4 Proposed Algorithm

Our Reputation based Trust Management Model using Friendship vector (RTMF) begins with the creation of friendship vectors for each node in the network. To create the friendship vector, each node first sends a dummy packet throughout the network. Interested nodes accept the packets and replies through a Response message with its trust value. The nodes from which Response messages are received are treated as the 'friend' nodes of the sender and added to the sender's friendship vector. With each friend node is associated its corresponding trust value. Now, this trust value is an indicator of how reliable this node is in delivering a message to the next hop. Initially all nodes are assumed to have a trust value of 0. Each time a node helps in delivering a message; its trust value is incremented by 1. Thus, each node in the network creates a friendship vector that consists of the 'friend' nodes and their corresponding trust values.

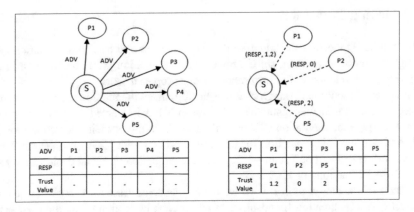

Fig. 1. Friendship vector created by Node S.

As shown in Fig. 1, Node S sends out dummy packets throughout the network but gets a response message back from only P1, P2 and P5. So the initial friend list of S will consist of its friends P1, P2 and P5 with their respective trust values.

	P1	P2	P5
S	1.2	0	2

So, for normal direct communication, the sender checks whether the destination node is present in its friend list with a trust value greater than 0. If yes, the message can be reliably sent to that destination, assuming that the destination node has good reputation. However, situations may arise where a node X wants to send a message to such a node which is not in its direct friend list. In such cases, to verify the trustworthiness of the receiver node, node X has to search for it in the friend lists of its friends, i.e. if the receiver node is a "friend-of-friend" (FOF). We can extend this logic further and make the sender search for the receiver node in the list of "friend-of-friend-of-friend" (FOFF) and so forth, depending on the level of trust the communication may allow.

But as in real world, we tend to believe those people's word of mouth who are themselves trustworthy to us, in these scenarios of indirect communication, the trust values of the intermediate nodes also need to be considered as the receiver is not a direct, but an indirect 'friend' of the sender. Here comes the idea of Direct Reputation (DR) and Indirect Reputation (IR). To evaluate the trust value of a particular node P which is not in the friend list of A (say), our proposed model is using both Direct reputation (DR) value and Indirect Reputation value (IR). DR means the trust value of the node who is a direct friend of the sender node. IR means the trust value of a node which is a friend-of-friend (FOF) or friend-of-friend-of-friend (FFOF) of the sender.

The Indirect reputation of a node P from A's point of view (i.e. when P is not in the friend list of A) is given by the following equation [2]:

$$IR(A, P) = 1/N \sum_{(k=1 \text{ to } L)} R(A, B_k, P)\lambda_k, \ L > 0 \tag{1}$$

where $R(A, B_k, P)$ is the reputation of node P according to the friend list of node B_k, who is the FOF (i.e. L = 1) or FOFF (i.e. L = 2) and so on, of node A. Thus L (Level) denotes 'how much' indirect node P is with respect to the sender node A. The second term λ_k is the decay factor that measures the decreasing relevance of the kth level reputation of the destination node. In other words, the more 'indirect' the destination is with respect to the source, the less is the impact of the intermediate friends' recorded reputation about the destination on the final trust value. If there are N number of multiple paths between a pair of sender and receiver, then the average of the trust values calculated from individual paths is considered.

The decay factor is an adaptation of the Exponential decay formula [16] and is given as:

$$\lambda_L = e^{-0.2L} \tag{2}$$

where λ_L is the decay factor at level L and 0.2 is the assumed decay constant.

For our proposed algorithm, assuming that the trust value follows the same decay curve, we have considered maximum up to 3 levels, i.e. L = 1, 2 or 3. In that case, the decay constant for FOF is 0.81873, for FOFF is 0.67 and for FOFFF is 0.5488 (using Eq. (2)). As already stated, if there are multiple indirect paths, the final trust value of the receiver node is the average of its (DR + IR). A node can be considered to be trustworthy enough to the sending node for message delivery if the final trust value of that node is greater than zero.

Say for example a node N1 wants to send a message to N12. Let us denote the destination node as $N12_D$. Following is the friend list of N1 along with their corresponding trust values.

	N2	N4	N10	N43	N56
N1	1	2	2	1	3

From N1's friend list, it can be observed that $N12_D$ is not a direct friend of N1. Hence, N1 has to search the friend lists of its friends (FOF) for the presence of $N12_D$. Say following are the friend lists of N1's friend nodes.

N2	N7	N14	N33
	2	1	2

N4	N2	$N12_D$	N20	N43
	1	2	2	1

N10	N23	N43	N46	N54	N62
	3	2	2	1	3

N43	N4	$N12_D$	N44
	1	2	2

N56	N3	N4	$N12_D$	N40
	1	2	1	1

It is observed that $N12_D$ is a friend of N4, N43 and N56. So N1 can send message to $N12_D$ via N4 or N43 or N56. Again, node N43, who is a friend of $N12_D$, is present in the friend list of N10. So to calculate the trust value of $N12_D$, we will consider both DR (0 in this case) and IR. Appropriate decay factors will be applied depending on the level L of friendship.

As already discussed, trust-based reputation mechanisms face the adversity of bad-mouthing by malicious nodes. In the process of bad-mouthing, the malicious node's intention is to reduce the reliability or credibility of a harmless node by falsely reducing its reputation or trust value. This intentional maligning of a non-malicious node effects its reputation badly, affecting the overall performance of the network. After creating their respective friendship lists, malicious nodes may replace the original trust values of its friends with some reduced values, even negative values. In our proposal, we have assumed that the malicious nodes set the trust values of their friends with negative values.

Let us consider the same above scenario with bad nodes. Assume that N43 is a bad node. So, it can set the trust value of $N12_D$ in its friend list with a negative value to degrade its reputation. In normal situation, when a node with a negative trust value is encountered, that node will not be further used for message communication, assuming it to be a distrustful node. So, if the malicious nodes forcibly reduce the trust value of a number of nodes, more messages will be dropped, thus affecting the overall delivery ratio of the network. But using RTMF, the paths with bad nodes (falsely indicated) will not be ignored but taken into consideration for the average trust value calculation. Say the bad node N43 has set the trust value of its friends as follows:

	N4	N12$_D$	N44
N43	-1	-1	-2

So, now the possible paths will be same as before, with the bad nodes indicated inside circles:

P1:   N1 → N4 → N12$_D$
P2:   N1 → (N43) → N12$_D$
P3:   N1 → N56 → N12$_D$
P4:   N1 → N10 → (N43) → N12$_D$

Here, only 50% of the paths consider N12$_D$ as trustworthy, the other half considers it to be untrustworthy.

Here, using the same method of calculation as before, the average trust value still comes as a positive value (0.242). As the final value is a positive value, hence irrespective of N43's attempt to make N12$_D$ untrustworthy, N12$_D$ will still be treated as a trustworthy node and packet forwarding will take place.

5 Results

We have used ONE simulator (version 1.5.1) [17, 18] for the implementation of our algorithm. ONE is a simulation environment that is capable of generating node movement using different movement models. Using the same set of parameters as shown in Table 1; we have compared the performances of our algorithm (RTMF) and the Trust management system based on Ontology proposed in [2] (we will indicate it as TMO in short) in the presence of bad nodes. The number of nodes in the simulation network has been taken as 25, 50 and 75. For each case, we have observed the performances of both algorithms in the presence of 2%, 5% and 10% of bad nodes.

Table 1. List of simulation parameters

Parameter	Values
Simulation time	43200 s
Movement model	Random waypoint
Router	Epidemic router
Transmit range	10 m
Transmit speed	250 Kbytes/s
Buffer size	5M bytes

As the performance of the network can be remarkably low if the percentage of bad nodes becomes more than 10%, we have not considered cases with more than 10% bad nodes in our work.

5.1   Effect on the Number of Messages Delivered

The following graph depicts the effect of both algorithms (RTMF and TMO) on the number of messages delivered to destination in the presence of bad nodes. It can be observed that the message delivery decreases quite naturally as the number of bad nodes increase in the network from 2% to 5% to further 10%. But in spite of that, keeping all the system parameters same, the number of delivered message is much more in case of RTMF compared to TMO. Even when the number of nodes in the network is quite high (75 nodes), RTMF handles the negative effect of the increasing bad nodes in the network better than TMO. Thus, it can be observed that RTMF can better handle the presence of bad nodes in the network, as compared to TMO.

The following graph compares the number of delivered messages using RTMF and TMO for 25, 50 and 75 nodes, varying the percentage of bad nodes in the network from 2% to 5% to 10% in each case (Fig. 2).

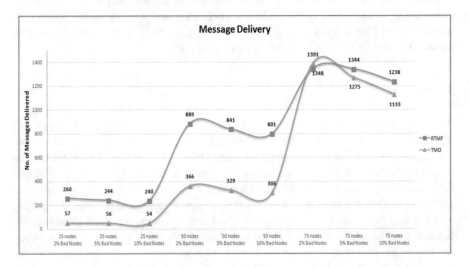

Fig. 2. Comparison of the number of messages delivered using RTMF and TMO

5.2   Effect on the Rate of Decrease in Message Delivery

We have already observed in the above section that as the percentage of bad nodes increases in the network, the number of delivered messages is decreasing, using both the algorithms. This behavior is quite natural as we already know the negative effect of malicious nodes on the performance of a network. But if the rate of decrease in delivered messages is compared, it can be seen that as the number of nodes increases (25 → 50 → 75), the rate of decrease in delivered messages is more using TMO compared to RTMF. This indicates that RTMF is more suited to perform in the presence of malicious nodes.

The following graph compares the rate of decrease in delivered messages (in %) using RTMF and TMO for 25, 50 and 75 nodes, varying the percentage of bad nodes in the network from 2% to 5% to 10% (Fig. 3).

Fig. 3. Comparing the decrease in rate of message delivery using RTMF and TMO

Though in case of small number of nodes (i.e. 25 nodes), this difference is not very significant, but as the number of nodes increases (and we can assume that real life networks will have more than 25 nodes), it can be observed that RTMF is much more efficient in nullifying the bad mouthing by malicious nodes, as the decrease rate of delivered messages is much lesser compared to TMO, as the number of bad nodes increases in the network.

Depending on how much weightage we want to give to direct trust and indirect reputation of a node, our algorithm can be tuned by changing the number of levels (L) of friendship vectors to be considered. Thus the complexity of RTMF will be of the order $O(n^L)$ where n is the maximum number of nodes in the friendlist and L is the level upto which the friend lists will be searched.

6 Conclusions

It is a challenge to ensure a trustworthy path between the source and the destination(s) for secured message transmission in Opportunistic networks due to absence of any predefined route. The scenario becomes more difficult when there is bad-mouthing of innocent node by malicious nodes, thus affecting the message delivery rate.

In this paper, we have proposed a Reputation based Trust Management Model using Friendship vector (RTMF) for message transmission in Oppnets, drawing inspiration from the real life scenario of reputation build-up of a particular entity. In our algorithm, the creation of friendship vectors by all sources in the network ensures that all communications are being carried out among 'known' non-enemy nodes, which are being considered as friend nodes for the time being. Feedbacks from such known nodes ought to be more reliable compared to feedbacks from any random node. Thus the creation of friendship nodes at the very beginning of our algorithm is setting up the first step of avoiding malicious intrusions. Also, as the trust value of each node is then determined by a combination of both its Direct and Indirect reputation instead of its direct reputation only, a message is not being straightaway discarded when a malicious node is encountered, thus improving the network performance.

Several works like [2, 6, 7] etc. has been proposed based on the concept of direct and indirect reputation. In this paper, we have compared our work with [2]. It can be compared with the other similar algorithms in future work. It has been observed that our proposed work is able to handle the negative effect of bad mouthing on the delivery rate better than [2]. Though the presence of bad nodes will naturally decrease the number of delivered messages, but the rate of decrease was much less using the RTMF algorithm. Also as already discussed, our algorithm has the flexibility of increasing or decreasing the levels of reputation to be considered, depending on how much weightage we want to give to indirect reputation and direct trust respectively.

References

1. Ayday, E., Fekri, F.: An iterative algorithm for trust management and adversary detection for delay-tolerant networks. IEEE Trans. Mob. Comput. 11, 1514–1531 (2012)
2. Goncalves, M.R.P, Moreira, E.S., Martimiano, L.A.F.: Trust management in opportunistic networks. In: 9th International Conference on Networks. IEEE, France (2010)
3. Capkun, S., Hubaux, J.-P., Buttyan, L.: Mobility helps peer-to-peer security. IEEE Trans. Mob. Comput. 5, 43–51 (2006)
4. Newman, M.E.J.: The structure and function of complex networks. SIAM Rev. 45(2), 167–256 (2003)
5. Liu, J., Issarny, V.: Enhanced reputation mechanism for mobile ad hoc networks. In: Jensen, C.D., Poslad, S., Dimitrakos, T. (eds.) Proceedings of the 2nd International Conference on Trust Management (iTrust). LNCS, vol. 2995, pp. 48–62. Springer, Heidelberg (2004)
6. Marti, S., Giuli, T., Lai, K., Baker, M.: Mitigating routing misbehavior in mobile ad-hoc networks. In: Proceedings of the 6th ACM MobiCom and Networking, Boston, pp. 255–265 (2000)
7. Paul, K., Westhoff, D.: Context aware detection of selfish nodes in DSR based ad-hoc networks. In: Global Telecommunications Conference GLOBECOM. IEEE, Taiwan (2002)
8. Buchegger, S., Boudec, J.: Performance analysis of CONFIDANT protocol: cooperation of nodes: fairness in dynamic ad-hoc networks. In: Proceedings of 3rd ACM International Symposium on Mobile Ad Hoc Networking & Computing MobiHoc, Switzerland, pp. 226–236 (2002)
9. Buchegger, S., Boudec, J.: A robust reputation system for P2P and mobile ad-hoc networks. In: Proceedings of Second Workshop the Economics of Peer-to-Peer Systems (2004)
10. Ayday, E., Fekri, F.: Using node accountability in credential based routing for mobile ad-hoc networks. In: Proceedings of 5th IEEE International Conference on Mobile Ad-Hoc and Sensor Systems, Atlanta (2008)
11. Ayday, E., Fekri, F.: A protocol for data availability in mobile ad-hoc networks in the presence of insider attacks. Ad Hoc Netw. 8(2), 181–192 (2010)

12. Dewan, P., Dasgupta, P., Bhattacharya, A.: On using reputations in ad-hoc networks to counter malicious nodes. In: Proceedings of 10th International Conference on Parallel and Distributed Systems, ICPADS 2004 (2004)
13. Liu, K., Deng, J., Varshney, P.K., Balakrishnan, K.: An acknowledgment-based approach for the detection of routing misbehavior in MANETs. IEEE Trans. Mob. Comput. 6(5), 536–550 (2007)
14. Yu, W., Liu, K.R.: Game theoretic analysis of cooperation stimulation and security in autonomous mobile ad-hoc networks. IEEE Trans. Mob. Comput. 6(5), 507–521 (2007)
15. Trifunovic, S., Legendre, F., Anastasiades, C.: Social trust in opportunistic networks. In: INFOCOM IEEE Conference on Computer Communications Workshops, USA (2010)
16. Hobbie, R.K., Roth, B.J.: Exponential growth and decay. In: Intermediate Physics for Medicine and Biology, pp. 31–47. Springer, New York (2007)
17. Keranen, A., Ott, J., Karkkaiene, T.: The ONE simulator for DTN protocol evaluation. In: ICST Proceedings of the 2nd International Conference on Simulation Tools, pp. 1–10 (2009)
18. ONE Homepage. https://www.netlab.tkk.fi/tutkimus/dtn/theone/

Features Extraction: A Significant Stage in Melanoma Classification

Savy Gulati[✉] and Rosepreet Kaur Bhogal

Lovely Professional University, Phagwara, India
savygulati99@gmail.com, rosepreetkaurl2@gmail.com

Abstract. Skin cancer is by far the most common cancer. Out of all the skin cancer types, Malignant melanoma is the deadliest one which can lead to mortality. Melanoma can be cured if detected at initial stage, so its early diagnosis is of quite importance. For this purpose Computer Aided systems are preferred which are based on image processing techniques. Among these techniques feature selection stage is of utmost significance as it directly affects the accuracy and robustness of CAD systems. In this work, it has been diagnosed that which kind of features out of color, texture and shape contribute admirable in melanoma classification task. In first phase, pre-processing steps are performed to refine images which involves hair exclusion and other noise removal techniques. Then segmentation is performed using otsu method followed by various morphological operations. After this color, texture and shape features are extracted and given to SVM classifier individually. Experimental results clearly show that color features are prominent in classification having accuracy of 93.5%. And among all color features, distinction ability of features obtained using HSV and CIE L * a * b color space is higher (91%).

Keywords: Melanoma · Non-melanoma · Color features · Texture features · Shape features

1 Introduction

Skin cancer is a state in which unnatural growth of skin cells takes place and it has power to diffuse to other body parts. Skin cancer is majorly caused due to exposure to ultraviolet radiations. These UV rays harm DNA and leads to its variations, resulting in improper growth of skin cells [1]. Skin cancers are mainly categorized into three classes relying upon the fact that abnormal growth of which type of cells has been occurred. First, Basal cell carcinoma (BCC), which occurs because of uncontrolled growth of basal cells found in epidermis (uppermost layer of skin). BCC is least dangerous among all the three types as it rarely spreads to other parts of body [2]. Second, Squamous cell carcinoma (SCC), occurs due to uncontrollable increment of squamous cells found in epidermis. This is severe as compared to the BCC's [3]. Third, Malignant melanoma occurs from uncontrollable emergence of melanocytes. It is deadliest of all the three types as it can spread to other parts of the body easily and ultimately can lead to person's death [4]. Basal cell carcinoma and Squamous cell carcinoma both come under the category of non-melanoma, whereas malignant melanoma come under category of melanoma. Melanoma being the evillest, leads to death of one person every hour and number of new melanoma patients has increased by 53% in last decade that is from 2008–2018 [5]. A study carried out in [6] investigated that even developed areas of globe such as Australia, North America and Europe are greatly under the burden of melanoma. Melanoma has taken such an evil face that it has become mandatory to deal with it effectively and efficiently also it is well established fact that burden and mortality rate of melanoma can be reduced if diagnosed early [4, 7].

For early detection of melanoma both invasive and non-invasive methods are present. Biopsy is an invasive method which involves peeling off portion of skin for diagnosis. Dermoscopy is non-invasive method in which clear and magnified image of lesion area is taken to suspect the lesion. In spite of these techniques, CAD systems are in vogue because these systems do not require human intervention. All they just need is image of skin lesion which is processed, and diagnosis is made by the system itself. These computer aided systems are laid on the concept of image processing. This approach mainly involves five steps: image acquisition, image pre-processing, image segmentation, feature extraction, feature selection and classification. Features extraction plays important role for detection of melanoma and non-melanoma, as it directly affects the recognition rate of CAD systems. Feature extraction is a process of conversion of input data into the set of distinctive features. These distinctive features are given to the classifier to classify melanoma and non-melanoma. Reduction of image into features is performed to increase accuracy and efficiency, decrease feature measurement cost. If extracted carefully and wisely, this reduction in dimensionality can easily serve the expected purpose of the melanoma classification instead of taking full image. Furthermore, diversity can be seen in this step of feature extraction as different authors use different types of features to solve the desired purpose. Most commonly used features are colour, texture and shape features. Colour features are taken because melanoma lesion generally exhibit larger colour variations than the benign lesions, so this factor can act as distinction property in classification task. As textural properties of both melanoma and non-melanoma varies widely so these are also considered for classification purpose. Shape features are too prevalent as shape of melanoma is quite irregular, blurred and ragged than that of the benign lesions thus results fruitful in classification.

In this work, melanoma classification is carried out using three different types of features. Initially by using colour features then by using texture features and lastly by using shape features. Results obtained using these three features are compared and prominent features are decided for classification task. Section 2 is dedicated for literature survey. Section 3 includes methodology which has been followed. Section 4 is devoted for experimental results and Sect. 5 includes conclusion.

2   Literature Survey

In [8], Waheed, Zafar, Waheed and Riaz carried out classification of melanoma from dermoscopic images based on color and texture features. Color features extracted from HSV color space and texture features extracted using GLCM. Task of classification performed using SVM. Munia, Alam, Neubert and Fael-Rezai [9] firstly applied segmentation using K-means clustering and Otsu method followed by post processing using morphological operations and guided filter. Then color, texture, border, asymmetry and non-linear features calculated to perform classification of melanoma and non-melanoma. Color features obtained using RGB, HSV and YCbCr color spaces and texture features extracted using GLCM. SVM, KNN, decision tree and random forest classifiers used for melanoma classification. Kotian and Deepa [10] performed gray scale conversion of images then segment lesion from background with otsu method and detected border with sobel operator. Afterwards asymmetry, border, color, standard deviation and texture features (using GLCM) calculated and provided to three classifiers: Naive Bayes, KNN and MLP to detect disease type and class. In [11], Mustafa and Kimura carried out segmentation using GrabCut algorithm followed by extraction of shape, geometry and border features. Then lastly, classification of malignant and normal mole takes place using RBF-SVM. Mahagaonkar and Soma [12] executed segmentation of skin lesion images using global thresholding then Harlick GLCM and CSLBP features calculated and given to KNN and SVM classifiers. In [13], fusion of textural and structural features used by Adjed, Gardezi, Ababsa, Faye and Dass to perform classification task using SVM. Textural features obtained using LBP whereas structural features include wavelet and curvelet.

Jain, Jagtap and Pise [14], executed rescaling, contrast enhancement to pre-process images and detected melanoma using geometric features and these features are compared with the features obtained using ABCD rule to classify images. Pathan, Lakshmi and Siddalingaswamy [15] developed methodology for hair exclusion and perform segmentation using Fuzzy C-means clustering then extracted shape features, texture features, and color variation. Here shape features include asymmetry and border irregularity, pigment transition include mean and gradient, color variation calculated using CIE-LAB space and texture features obtained using GLCM. Artificial neural network with back propagation training algorithm used for classification purpose. In [16], Do et al. carried out hierarchical segmentation (using otsu and Minimum Spanning Tree method) followed by extraction of lesion's color, border, asymmetry and texture features to classify melanoma from skin lesion images using SVM. Takruri and Abubakar [17] employed resizing operation followed by denoising to remove unwanted objects. k-means clustering for image segmentation and then extracted Wavelet transform, Curvelet transform and texture features using GLCM to form feature vector. This feature vector passed to three parallel SVM for classification of skin lesion images. Output of these three SVM classifiers is pooled by using Bayesian classifier. Mahmoud, Abdel-Nasser and Omer [18] excluded hairs using bottom hat and median filter. Afterwards features using gabor filters, gray level co-occurrence matrix (GLCM), histogram of oriented gradients (HOG), local binary pattern (LBP) and local directional number pattern (LDN) extracted. And classification carried out with help of these features by utilizing multilayer perceptron layer. In [19], Chatterjee, Dey and Munshi carried out pre-processing using median filtering and hair removal using bottom hat. Then segmentation is carried out with use of morphological operations. Further shape, color and texture features extracted. Color features calculated from

RGB, HSV and LAB color space whereas texture features calculated using GLCM. Then these features fed to SVM for melanoma and non-melanoma classification. Kanimozhi and Murthi [20] performed pre-processing using rescaling, gray scale conversion and binarization followed by segmentation by utilizing background sub-traction, edge detection and masking. Then geometric features extracted with use of the segmented images and given to ANN classifier. In [21], Sheha, Mabrouk and Sharawy firstly perform resizing and gray scale conversion of images than extracted texture features using GLCM. Based on fisher scoring effective features selected from the obtained GLCM features and given to traditional and Automatic MLP for cancerous and non-cancerous classification. Chakraborty et al. [22] conducted discrete wavelet transform and then extracted texture features using GLCM. These texture features passed on to NN-NSGA-II to perform melanoma classification.

3 Methodology

The paper aims to inspect results obtained using different types of features: color, texture and shape features. And identify that which features among these three con-tribute more towards accurate classification of melanoma and non-melanoma images.

3.1 Data Acquisition

In this work, PH^2 database [23] has been used which is free of cost and publicly available. This dataset contains total of 200 colored images out of which 160 images are of benign lesion and 40 images are of melanoma. Among benign images, 80 are of common nevi and rest 80 images are of atypical nevi.

3.2 Pre-processing

Pre-processing of images is carried out to remove unwanted substances from images such as hairs, reflection artefacts, air bubbles and uneven distribution. Firstly, to exclude hairs from dermoscopic images DullRazor [24] is used. Then these images with removed hairs are converted to gray scale. Afterwards, contrast enhancement of these gray scaled images is carried out followed by noise removal using median filter. Results of mentioned pre-processing steps are given in Fig. 1.

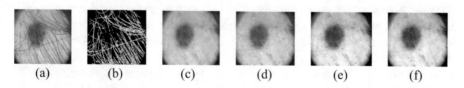

| (a) | (b) | (c) | (d) | (e) | (f) |

Fig. 1. Pre-processing steps. (a) Original skin lesion image, (b) Hair mask, (c) Image with hairs removed, (d) Gray scale image, (e) Contrast enhanced image, (f) Noise free image obtained using median filter

3.3 Segmentation

Segmentation involves distinction of skin lesion from rest of the skin area, so that region of interest can be obtained. Here otsu method for segmentation has been used. It is a global thresholding method in which threshold is chosen in such a way that within-class variance is minimum of background and foreground pixels. For this purpose, best suited value present in between two peaks is considered from a bimodal histogram. After applying segmentation, results are further enhanced with the use of morphological operations. Image filling is applied to remove unwanted background pixels within the region of interest then Image opening is carried out to remove unwanted background pixels present outside the region of interest and to smoothen out the lesion edges. Finally, the largest object in image is detected and cropped. Results of segmentation are given in Fig. 2.

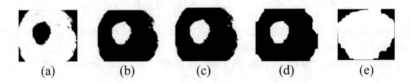

(a)   (b)   (c)   (d)   (e)

Fig. 2. Post segmentation steps. (a) Segmented image using otsu method, (b) Inverted image, (c) Image filling operation applied, (d) Image opening operation applied, (e) Cropped image

3.4 Feature Extraction

Feature extraction stage includes collection of discriminative features from the image so that only effective and useful data is fed to the classifier. Here color, texture and shape features are calculated as follows:

Color Features. Color of skin lesion usually changes abruptly for melanoma, this is the reason that color features are considered of great importance for melanoma classification. For color features we computed color moments for each channel of all seven color spaces. These color spaces are RGB, HSV, NTSc, OPP, YCbCr, CIE L*u*v and CIE L*a*b. Color moments are four popular statistics: moment 1- mean, moment 2 - standard deviation, moment 3 - skewness and moment 4 - variance. From individual color space total of 12 features are extracted. Thus, total of $12 \times 7 = 84$ features are extracted.

Texture Features. These features provide information about spatial arrangement of color and intensities in an image. To calculate these features various texture descriptors are available. Descriptors used here are: GLCM and LBP. Using Gray level co-occurrence method (GLCM) four texture features are computed: contrast, energy, homogeneity and correlation as these four have powerful impact in classification [25]. Furthermore, using local binary pattern (LBP) 59 features are extracted.

Shape Features. As shape of melanoma is quite irregular so shape features also prove fruitful in effective classification. Shape features are calculated with help of segmented images. Extracted shape features of skin lesion are [14]: Area (A), perimeter (P), Major

Axis Length (GD), Minor Axis Length (SD), circularity index (CRC), irregularity index A (IrA), irregularity index B (IrB), irregularity index C (IrC) and irregularity index D (IrD). Area is the total number of pixels in a skin lesion. Perimeter is the number of edge pixels. Major Axis Length is length of line passing through centroid of skin lesion and connecting two farthest points. Minor Axis Length is length of line passing through centroid of skin lesion and connecting two nearest points. Circularity index referred in Eq. (1) gives the measure of uniformity of skin lesion. Equations (2), (3), (4) and (5) gives irregularity index A, B, C and D respectively.

$$CRC = \frac{4A\pi}{p^2} \tag{1}$$

$$IrA = \frac{P}{A} \tag{2}$$

$$IrB = \frac{P}{GD} \tag{3}$$

$$IrC = P\left[\frac{1}{SD} - \frac{1}{GD}\right] \tag{4}$$

$$IrD = GD - SD \tag{5}$$

Apart from the above nine features, two more features are extracted. These two features are Asymmetry and border irregularity. To check symmetricity, lesion is divided horizontally into two halves and then one half of the lesion is folded over the other hypothetically, as given in Fig. 3. Asymmetry index [26] is calculated with the formula given in Eq. (6), here AD denotes difference in pixels and A corresponds to area of skin lesion. Border irregularity [26] can be calculated with the help of compactness index provided in Eq. (7) for this area (A) and perimeter (P) of a lesion are utilized.

$$Asymmetry\ Index = \frac{AD}{A}10 \tag{6}$$

$$Compactness\ Index = \frac{p^2}{4\pi A} \tag{7}$$

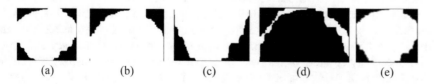

(a)   (b)   (c)   (d)   (e)

Fig. 3. Asymmetry index calculation and border irregularity (a) Otsu segmented image, (b) Upper half of skin lesion, (c) Lower half of skin lesion, (d) Folded image, (e) Border detected of skin lesion

3.5   Classification

At classification stage numeric properties of extracted features are accessed and based on this skin lesions are categorized into melanoma and non-melanoma. Here, Support vector machine is used for classification purpose. This classifier is based on supervised learning that is prior knowledge about data is already available. SVM finds the hyperplane that best classifies melanoma and non-melanoma classes.

4   Experimental Results

4.1   Evaluation Measures

Evaluation is performed using four parameters: Accuracy, Sensitivity, Specificity and Area under ROC curve (AUC). Accuracy is the recognition rate of melanoma and non-melanoma that is the correct identified examples of both cases. Sensitivity is the measure of recognition rate of melanoma. Specificity is the measure of recognition rate of non-melanoma. These three can be calculated using Eqs. (8), (9) and (10) respectively. AUC is the area under the Receiver operating characteristic curve.

- TP (true positive) = positive cases recognised as positive.
- TN (true negative) = negative cases recognised as negative.
- FP (false positive) = negative cases recognised as positive.
- FN (false negative) = positive cases recognised as negative.

$$Accuracy = \frac{TP + TN}{TP + TN + FP + FN} \tag{8}$$

$$Sensitivity = \frac{TP}{TP + FN} \tag{9}$$

$$Specificity = \frac{TN}{TN + FP} \tag{10}$$

4.2   Results

In this work, SVM with 3 cross validation is used to perform classification of melanoma and non-melanoma. First, colour features are inspected. Colour features using all seven colour spaces (RGB, HSV, YIQ, OPP, YCbCr, CIE L * u * v and CIE L * a * b) are extracted and their individual performance is assessed in Table 1. It is can be seen that among all the seven colour spaces, colour features obtained using HSV and CIE L * a * b exhibit better performance than others in terms of accuracy and sensitivity. Specificity is effective with use of colour features obtained using OPP colour space. In terms of AUC, colour features obtained using RGB and HSV colour spaces outperformed others.

Table 1. Comparison of color features using different color spaces

Colour features	Accuracy (%)	Sensitivity (%)	Specificity (%)	AUC
RGB	90.5	70	95.6	**0.96**
HSV	**91**	**75**	95	**0.96**
YIQ	88	65	93.7	0.94
OPP	86	40	**97.5**	0.92
YCbCr	89.5	65	95.6	0.94
CIE L * u * v	89.5	72.5	93.7	0.94
CIE L * a * b	**91**	**75**	95	0.95

Then classification of Malignant melanoma and non-melanoma is carried out using texture features with help of GLCM and LBP texture descriptors. Among these two GLCM outperformed LBP features in terms of three evaluation parameters given in Table 2. It can be clearly seen that LBP features gave worse performance when the turn for melanoma detection came. Then shape features are computed to perform classification task as shown in Table 3.

Table 2. Comparison of texture features obtained using two texture descriptors

Texture features	Accuracy (%)	Sensitivity (%)	Specificity (%)	AUC
GLCM	**86**	**52.5**	94.3	**0.89**
LBP	81	17.5	**96.8**	0.73

Table 3. Shape features obtained using geometric features, asymmetry and border

Shape features	Accuracy (%)	Sensitivity (%)	Specificity (%)	AUC
Geometric features + asymmetry and border	88	47.5	98.1	0.85

In Table 4 comparison of all the three types of features: color, texture and shape has been carried out and confusion matrix of each is given in Table 5. Classification of cancerous and non-cancerous images is accomplished using these three features types individually and it is evident that color features proves more prominent in recognition in terms of all the four considered performance measures that is accuracy, sensitivity, specificity and AUC as seen in Fig. 4(a), (b), (c) and (d) respectively. In case of specificity only, performance of shape features matches with that of color features. If comparison of texture and shape features is made later exhibit better performance in terms of accuracy and specificity.

Table 4. Comparison of color, texture and shape features used for melanoma classification

Features type	Accuracy (%)	Sensitivity (%)	Specificity (%)	AUC
Color features (all 7 color space)	**93.5**	**75**	**98.1**	**0.98**
Texture features (using GLCM)	86	52.5	94.3	0.89
Shape features	88	47.5	**98.1**	0.85

Table 5. Confusion matrix of color, shape and texture features

	TP	TN	FP	FN
With color features	30	157	3	10
With shape features	19	157	3	21
With texture features	21	151	9	19

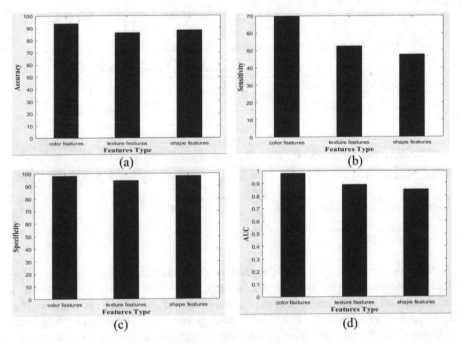

(a) (b) (c) (d)

Fig. 4. Performance measures for different features. (a) Accuracy v/s features type plot, (b) Sensitivity v/s features type plot, (c) Specificity v/s features type plot, (d) AUC v/s features type plot.

5   Conclusion

Feature selection plays significant role in classification of melanoma and non-melanoma. That's why this paper consider different feature types: colour, texture and shape for classification task. Firstly, classification performed by utilising colour features from all seven colour spaces and from these seven spaces it has been seen that colour features obtained using HSV and CIE L * a * b spaces showcase better performance of 91%. Then Texture features calculated using two texture descriptors: GLCM and LBP. Out of these two GLCM outperform the other. Lastly, Shape features also extracted to classify cancerous and non-cancerous images. Results of melanoma classification using all three colour, texture and shape features are compared. And it has been concluded that colour features lead to effective and efficient classification of melanoma and benign lesions with accuracy of 93.8%.

References

1. WebMed. https://www.webmd.com/melanoma-skin-cancer/causes-skin-cancer#1
2. Skin Cancer Foundation. https://www.skincancer.org/skin-cancer-information/basal-cell-carcinoma
3. Skin Cancer Foundation. https://www.skincancer.org/skin-cancer-information/squamous-cell-carcinoma
4. Skin Cancer Foundation. https://www.skincancer.org/skin-cancer-information/melanoma
5. Skin Cancer Foundation. https://www.skincancer.org/skin-cancer-information/skin-cancer-facts
6. Karimkhani, C., Green, A.C., Nijsten, T., Weinstock, M.A., Dellavalle, R.P., Naghavi, M., Fitzmaurice, C.: The global burden of melanoma: results from the Global Burden of Disease Study 2015. Br. J. Dermatol. **177**, 134–140 (2017). https://doi.org/10.1111/bjd.15510
7. Geller, A.C., Swetter, S.M., Weinstock, M.A.: Focus on early detection to reduce melanoma deaths. J. Invest. Dermatol. **135**, 947–949 (2015). https://doi.org/10.1038/jid.2014.534
8. Waheed, Z., Waheed, A., Zafar, M., Riaz, F.: An efficient machine learning approach for the detection of melanoma using dermoscopic images. In: 2017 International Conference on Communication, Computing and Digital Systems (C-CODE), pp. 316–319. IEEE Press, Islamabad (2017). https://doi.org/10.1109/c-code.2017.7918949
9. Munia, T.T.K., Alam, M.N., Neubert, J., Fazel-Rezai, R.: Automatic diagnosis of melanoma using linear and nonlinear features from digital image. In: 39th Annual International Conference of the IEEE Engineering in Medicine and Biology Society (EMBC), pp. 4281–4284. IEEE Press, Seogwipo (2017). https://doi.org/10.1109/embc.2017.8037802
10. Kotian, A.L., Deepa, K.: Detection and classification of skin diseases by image analysis using MATLAB. Int. J. Emerg. Res. Manage. Technol. **6**, 779–784 (2017)
11. Mustafa, S., Kimura, A.: An SVM-based diagnosis of melanoma using only useful image features. In: 2018 International Workshop on Advanced Image Technology (IWAIT), pp. 1–4. IEEE Press, Chiang Mai (2018). https://doi.org/10.1109/iwait.2018.8369646
12. Mahagaonkar, R.S., Soma, S.: A novel texture based skin melanoma detection using color GLCM and CS-LBP feature. Int. J. Comput. Appl. **171**, 1–5 (2017). https://doi.org/10.5120/ijca2017915024

13. Adjed, F., Gardezi, S.J.S., Ababsa, F., Faye, I., Dass, S.C.: Fusion of structural and textural features for melanoma recognition. IET Comput. Vis. **12**, 185–195 (2018). https://doi.org/10.1049/iet-cvi.2017.0193

14. Jain, S., Jagtap, V., Pise, N.: Computer aided melanoma skin cancer detection using image processing. In: International Conference on Computer, Communication and Convergence (ICCC 2015), pp. 735–740 (2015). Procedia Computer Science. https://doi.org/10.1016/j.procs.2015.04.209

15. Pathan, S., Siddalingaswamy, P.C., Lakshmi, L., Prabhu, K.G.: Classification of benign and malignant melanocytic lesions: A CAD tool. In: International Conference on Advances in Computing, Communications and Informatics (ICACCI), pp. 1308–1312. IEEE Press, Udupi (2017). https://doi.org/10.1109/icacci.2017.8126022

16. Do, T.T., Hoang, T., Pomponiu, V., Zhou, Y., Chen, Z., Cheung, N.M., Koh, D., Tan, A., Hoon, T.: Accessible melanoma detection using smartphones and mobile image analysis. IEEE Trans. Multimed. **20**, 2849–2864 (2018). https://doi.org/10.1109/tmm.2018.2814346

17. Takruri, M., Abubakar, A.: Bayesian decision fusion for enhancing melanoma recognition accuracy. In: International Conference on Electrical and Computing Technologies and Applications (ICECTA), pp. 1–4. IEEE Press, Ras Al Khaimah (2017). https://doi.org/10.1109/icecta.2017.8252063

18. Mahmoud, H., Abdel-Nasser, M., Omer, O.A.: Computer aided diagnosis system for skin lesions detection using texture analysis methods. In: 2018 International Conference on Innovative Trends in Computer Engineering (ITCE), pp. 140–144. IEEE Press, Aswan (2018). https://doi.org/10.1109/itce.2018.8327948

19. Chatterjee, S., Dey, D., Munshi, S.: Mathematical morphology aided shape, texture and colour feature extraction from skin lesion for identification of malignant melanoma. In: 2015 International Conference on Condition Assessment Techniques in Electrical Systems (CATCON), pp. 200–203. IEEE Press, Bangalore (2015). https://doi.org/10.1109/catcon.2015.7449534

20. Kanimozhi, T., Murthi, A.: Computer aided melanoma skin cancer detection using artificial neural network classifier. Int. J. Sel. Areas Microelectron. **8**, 35–42 (2016)

21. Sheha, M.A., Mabrouk, M.S., Sharawy, A.: Automatic detection of melanoma skin cancer using texture analysis. Int. J. Comput. Appl. **42**, 22–26 (2012). https://doi.org/10.5120/5817-8129

22. Chakraborty, S., Mali, K., Chatterjee, S., Banerjee, S., Mazumdar, K.G., Debnath, M., Basu, P., Bose, S., Roy, K.: Detection of skin disease using metaheuristic supported artificial neural networks. In: 8th Annual Industrial Automation and Electromechanical Engineering Conference (IEMECON), pp. 224–229. IEEE Press, Bangkok (2017). https://doi.org/10.1109/iemecon.2017.8079594

23. Mendonça, T., Ferreira, P.M., Marques, J.S., Marcal, A.R., Rozeira, J.: PH^2 - a dermoscopic image database for research and benchmarking. In: 35th International Conference of the IEEE Engineering in Medicine and Biology Society, pp. 3–7. IEEE Press, Osaka (2013). https://doi.org/10.1109/embc.2013.6610779

24. Lee, T., Ng, V., Gallagher, R., Coldman, A., McLean, D.: DullRazor: a software approach to hair removal from images. Comput. Biol. Med. **27**, 533–543 (1997)

25. Kolkur, S., Kalbande, D.R.: Survey of texture-based feature extraction for skin disease detection. In: 2016 International Conference on ICT in Business Industry and Government (ICTBIG), pp. 1–6. IEEE Press, Indore (2016). https://doi.org/10.1109/ictbig.2016.7892649

26. Firmansyah, H.R., Kusumaningtyas, E.M., Hardiansyah, F.F.: Detection melanoma cancer using ABCD rule based on mobile device. In: International Electronics Symposium on Knowledge Creation and Intelligent Computing (IES-KCIC), pp. 127–131. IEEE Press, Surabaya (2017). https://doi.org/10.1109/kcic.2017.8228575

Development of an Innovative Mobile Phone-Based Newborn Care Training Application

Sherri Bucher[1], Elizabeth Meyers[2],
Bhavani Singh Agnikula Kshatriya[2], Prem Chand Avanigadda[3],
and Saptarshi Purkayastha[2(✉)]

[1] Department of Pediatrics,
Indiana University School of Medicine,
Indianapolis, USA
[2] School of Informatics and Computing, IUPUI, Indianapoli, USA
saptpurk@iupui.edu
[3] Computer and Information Science, IUPUI, Indianapolis, USA

Abstract. Mobile infrastructure in low - and middle-income countries (LMIC) has shown immense potential to reach the unreachable. Healthcare providers (HCP) are one such group who are at the frontline of the fight against infant mortality in LMICs. Mortality among newborn infants (birth to 28 days) now accounts for around 45% of all under 5-years child mortality. Birth asphyxia is one of the three leading causes of newborn death; neonatal resuscitation training, among health care providers, reduces mortality from birth asphyxia. We have developed a mobile phone-based training app, called mobile Helping Babies Survive (mHBS), to support the training of health care providers on neonatal resuscitation. mHBS is integrated with the District Health Information System (DHIS2) platform, which is used in over 60 countries around the world. The mHBS/DHIS2 training app is a part of an application suite which includes another DHIS2-linked data collection app, mHBS tracker. The mHBS training application has the potential to scale-up integration with other neonatal training apps. Ultimately, the mHBS training suite will provide new insights into healthcare worker education along with the necessary tools for effective care of newborn babies.

Keywords: eLearning · mHealth · mHBS · DHIS2 · Healthcare providers · Newborn care · SDG

1 Background

Life-threatening realities for infants in Low- and Middle-Income Countries (LMICs) include birth asphyxia, infections, and birth trauma, contributing to 2.6 million annual preventable neonatal deaths occurring in the first month of an infant's life [1, 2]. Early recognition of neonatal complications, management of care, and delivery of interventions by skilled healthcare providers (HCPs) are critical components for reduction of global rates of infant mortality [3].

A. Abraham et al. (Eds.): IBICA 2018, AISC 939, pp. 361–374, 2019.
https://doi.org/10.1007/978-3-030-16681-6_36

In the past few years, there has been recognition that mortality which occurs during the neonatal period – birth to 28 days of life among liveborn infants – underlies a large proportion of all infant and under 5-years child mortality. Low/middle-income countries bear a disproportionate burden for newborn death. Accordingly, global health partners, such as the American Academy of Pediatrics have made significant progress in formalizing and disseminating evidence-based guidelines for newborn care in low/middle-income countries. The flagship program of the Helping Babies Survive suite of newborn care interventions, Helping Babies Breathe (HBB), has been implemented in 80 countries and reached more than 300,000 birth attendants [4–6]. However, program implementation barriers exist. These include a lack of quality assurance for initial and refresher trainings [7], limited access to training materials and resources, which are currently scattered across various partner websites, [8], and knowledge gaps regarding engagement metrics on self-motivated learning.

At the same time, mobile penetration rates in LMICs amount to 70% of the world's mobile phone subscriptions [9]. There is increasing accessibility to mobile device ownership and access bolstered by reduced pricing for smartphones and tablets, improved connectivity, and enabling technologies such as mobile banking and social platforms, which encourage previously disconnected populations to acquire mobile phones and tablets and engage with these devices more frequently [10, 11]. In accordance with this changing ecosystem, a plethora of mobile health (mHealth) and electronic learning (eLearning) applications have been developed to address the pernicious issue of maternal and neonatal mortality [12]. Furthermore, data analysis and health management information systems such as the open-source District Health Information System (DHIS2) [13], which is already used in over sixty countries [14], have created a platform to support data collection specifically tailored to build support health programs. Steadily, use of the DHIS2 platform is improving health system capacity for the delivery of evidence-based health interventions, and monitoring and evaluation of healthcare programs, in LMICs. At this opportune time, our team has developed the Mobile Helping Babies Survive (mHBS) app to support healthcare workers as they obtain facilitated training for neonatal resuscitation, and maintain their knowledge, skills, and competencies over time. Through the integration of mHBS training and mHBS tracker app with DHIS2 [15], we link our training and data collection suite with an already scaled system for health information within LMICs.

This strategy supports the potential for scale-up and sustainability of mHBS. In our development efforts, we employ agile development, best practices from eLearning, and strategies from user-centered design. Furthermore, our efforts adhere to the Principles for Digital Development [16]. We have created an open-source, Android-based mobile training application called mHBS [17].

2 Methodology

The mHBS training application followed human-centered design to maximize adoption potential and create a simple experience for users across all technical skill levels. First, the technology team, consisting of 7 members, engaged in an extensive literature review regarding existing mHealth applications to understand the burdens faced by nurses, skilled birth attendants (SBAs), community and facility-based health care workers, and midwives in low- and middle-income countries (LMICs). Then, four subsequent phases of design, programming, beta-evaluation, and re-design occurred. Throughout the development cycle, Iterative, agile development processes were employed, GitHub was used for documentation and collaboration to align with the open-source doctrines. Weekly scrum meetings were held for monitoring progress and addressing program bottlenecks.

2.1 Design Phase

Three members of the technology team participated in the design phase, which lasted for 3 months. This phase built upon two previous iterations of the mHBB application [17] that were developed by two previous teams, each of which used a different technology stack, and did not have integration with DHIS2. Our current application is based on the rationale that low-and-middle-income countries (LMICs) have made substantial investments in deploying the DHIS2 infrastructure where human resources in health system already use DHIS2 for routine data collection and program management. By utilizing user-centered design processes, and linking mHBS with DHIS2, we support each of the nine Principles for Digital Development.

Upon completion of the comprehensive literature review, and during the design phase, meetings were held with domain experts in the fields of Pediatrics and Bio-Health Informatics to further inform the technology team of the health system landscape, potential use cases, and to discuss technology constraints related to app development for low resourced settings [18]. A convergence of ideas led to the identification of core features, including the need to establish a link between the existing mHBS tracker and mHBS training app, utilizing the tracker app as a central point of entry, as a proof of concept for creating a seamless, linked suite of digital tools. Identification of this link as a critical component is due to its potential in extending the tracking of a healthcare provider (HCP) who is already registered in the mHBS tracker/DHIS2. This linkage also allows a mechanism for any number of custom mobile apps in this problem space to be modularly attached to the same portal. A key innovation of this approach is that, traditionally, DHIS2 has been used to monitor health service delivery by a facility or track entities that receive care at the facility. mHBS innovates by using entities at the level of individual healthcare provider.

In addition, it was decided that, to facilitate the user experience, the two apps should appear homogeneous, in terms of the interface, as a user toggles between them. This purposeful design of a seamless transition was expected to decrease the learning curve for a new user, as the style across applications would be more familiar. Therefore, design choices included the use of Google's Material design and a hamburger menu, along with other elements already present in the mHBS tracker. Educational content for the mHBS training app constituted the second key app deliverable, which included standardized videos and PDFs from global health partners, and sequential learning modules (courses). The videos were gathered from partner websites and tested for ideal compression, which led to WebM as the chosen file format.

An added restriction on educational content was that it must be stored and retrieved through a DHIS2 web API to allow full integration with the existing system. The mHBS training app was also designed for extensibility, to accommodate potential future resources, such as images. Other desired features included the need for the off-line mode to support intermittent WiFi, which is prevalent in LMICs [19], along with the ability to capture user-resource interaction metrics (including how often a user viewed content, and for what length of time).

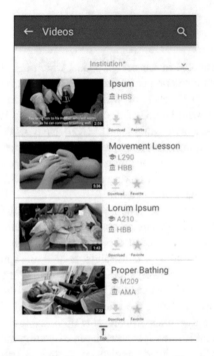

Fig. 1. Initial prototype showing video in a reusable list

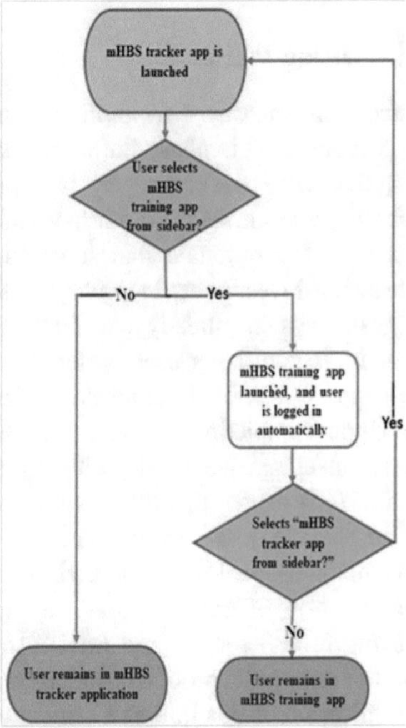

Fig. 2. The process flow for linking reusable list with DHIS2 login

Following feature identification, user experience tools Balsamiq [20] and Marvel [21] were used to making two prototypes which were presented to the project team for re-evaluation. Finally, a high-fidelity prototype (Fig. 1) was created that synthesized features of the original mockups and development using Android Studio followed.

2.2 Programming Phase

2.2.1 Dhis2 SDK

Four members from the technology team participated in the programming phase, which lasted for ten months, and resulted in a deployable Android mobile app. Once the basic skeleton for the training app was created using the DHIS2 Software Development Kit (SDK), tasks were prioritized, (2) functionality improvements, and (3) appearance to support beta-test release. Our open-source project can be found at https://github.com/iupui-soic/mHBS-Tracker.

2.2.2 Plugins

We used *com.github.barteksc:android-pdf-viewer:3.0.0-beta.5* and *fabric.io* plugins to display the PDF documents in the training app, send crashes and events to fabric server.

2.2.3 Offline and Online Modes

DHIS2 SDK doesn't have any existing API to support retrieving resources such as videos and PDFs that are stored in the DHIS2 platform. Health systems use the DHIS2 resource feature to share documents related to health policies, documentation, and other types of related content. But in our project, we utilize the DHIS2 resource feature to store training videos, images and PDF documents that can be used by the Healthcare providers (HCPs) for the self-directed, individually paced learning of knowledge, skills, and competencies in newborn care content. As for the database structure on the web server, a file or URL resource is stored under the single field "document," which supports multiple resource types. Each document contains information such as a tag, title, content type, URL, file permission status, and last update date, all of which is stored in Extensible Markup Language (XML) format. Even though DHIS2's primary usage is like a Data warehouse, it provides a general- purpose RESTful API for communication from external applications. Both the tracker and training app, as well as the DHIS2 Android SDK, uses the DHIS2 REST API for communication. The DHIS2 stored resources are available at the document's endpoint.

A collection of documents (/api/documents) returns an XML or JSON array, and depending on the Accept HTTP Header, a list of tags and titles which links to each document. Since many types of content (PDFs and Videos) can be stored together, the documents array needed to be parsed based on their content types. To handle this, an asynchronous background task was created to access the DHIS2 Web API and implement an Android XML Pull Parser interface to obtain relevant XML tags. The Pull Parser was used to iterate through the documents collection and gather document IDs and tags. An ID was then added to the URL path (api/documents/ID) to retrieve the content type and URL related to every document stored in DHIS2 resources.

2.2.4 Integrating Internal and External Applications/Resources

The mobile Helping Babies Survive (mHBS) training app needed to have integration with internal applications such as mHBS tracker app (data collection and reporting app) and mHBS Aggregate app (data visualization), to switch among the applications seamlessly. The mHBS training app also needed to have integration with external applications supportive of maternal and newborn care program implementation, such as a Safe Delivery app [22], REDCap Mobile [23], Essential Care for Small Babies, and external resources such as DHIS2, to retrieve the training material.

The first integration activities involved linking the mHBS training and tracker apps. While the training app can be launched quickly from the tracker app, the user session must be preserved between the two separate applications. To accommodate this, a *BroadcastReceiver* [24] was implemented.

A broadcast is sent to the tracker app whenever the training app is launched, which allows the receiving of an intent that contains the username, password, and DHIS2 URL of the HCP from the tracker application. DHIS2 login is handled by the

DHIS2 SDK through private, unmodifiable functions. Therefore, a custom login was created to logging in through the app to DHIS2 (Figs. 2, 3 and 4).

Once the content types are retrieved, the documents are parsed into groups. Using the same background task, video thumbnails are gathered as bitmaps using a *MediaMetadataRetriever* [25]. Finally, content is downloaded to folders on the device using a *DownloadManager* [26]. The downloaded content is then displayed as part of an updated dynamic list similar to the prototype mockup shown in Fig. 1. Users can click on the download icon and download the resources for offline viewing or "star" the content so that these can be marked as favorite. Prior to the beta-test release, "work in progress" code was disabled. Notably, this included course functionality and certain PDFs.

For purposes of the proof-of-concept, the Helping Babies Breathe (HBB) Second Edition neonatal resuscitation training guide for Providers Guide was deliberately left as the only visible PDF on the application, as it was of primary interest to gain feedback on font size and navigation within a single PDF, rather than displaying a breadth of options. The HBB Provider Guide was also identified as a critical resource that must always be available to a user.

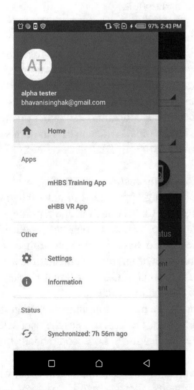

Fig. 3. mHBS tracker app showing a link

Fig. 4. mHBS training app

2.2.5   Privacy and Security

Initially, to address the privacy and security concerns of the training app users, the code was written to allow launching of the training app only from within the mHBS tracker app. Specific user credentials containing username and password were provided, allowing HCPs to launch the mHBS tracker securely.

2.3   Beta Testing Phase

The mHBS training app was released alongside the mHBS tracker for beta testing on Google Play Store as part of a public beta on March 26, 2018. The app was downloaded over 200 times during this phase, which lasted until May, 20, 2018. A user guide for beta testers was distributed, which included screenshots and instructions for app use, as well as a structured survey to capture the System Usability Scale (SUS) [24]. A more elaborate semi-structured feedback questionnaire was subsequently created in PowerPoint for qualitative usability testing of the mHBS Trainer and Tracker apps. This questionnaire was sent to our beta testers, which included global health partners, clinicians, and international neonatal resuscitation training experts. End-user feedback from the beta testers was discussed in our weekly scrum meetings.

The user feedback was generally positive, and included recommendations to increase the number of links between deeply nested pages to allow for more navigational control, along with improvements to app appearance through the inclusion of professional, uniform images, and implementing additional privacy and security features.

3   Results

3.1   Development

3.1.1   Cordova

Based on user recommendations from the beta-testing phase, we removed DHIS2 SDK as underlying architecture for the mHBS training app and implemented a new mobile app development framework. This decision was in response to the fact that DHIS2 SDK has constraints which preclude our ability to efficiently and effectively modify the training app to address the suggestions. We selected Apache Cordova as a new platform for development to replace the DHIS2 SDK. Cordova is an open-source mobile development framework used to develop cross-platform mobile applications using HTML, CSS3, and JavaScript. We also used Framework7, which is a hybrid applications development HTML framework, to develop the mHBS Trainer app, which finally improved user interaction and experience.

3.1.2   Plugins

Along with the implementation of Cordova, we used eight new plugins to address the beta-testing recommendations. *cordova-plugin-whitelist*: For security and configurability of the training app by implementing a whitelist policy for navigating the application WebView. *com-darryn-campbell-cordova-plugin-intent*: For sending and receiving of intents. *cordova-plugin-network-information*: For gaining information on the device's internet connection. *cordova-plugin-dialogs*: For getting native dialog user interface elements in the training app. *cordova-plugin-secure-storage*: For storing confidential credentials such as username, passwords, tokens, certificates, etc. *cordova-fabric-plugin*: For sending fabric events and crash analytics data to the fabric server. *cordova-custom-config*: For maintaining all plugins' version to be limited to initial Cordova application version. *com.lampa.startapp*: For launching integrated applications from the current training app.

3.1.3   Offline and Online Modes

Due to the implementation of *cordova-plugin-network-information*, it is now easy to verify if an HCP's device is offline or online. This further lead to coding the training app to collect and store app usage data as local application data whenever the app is used in offline mode, and sending of those data to DHIS2 server whenever a cellular or WiFi connection is activated.

3.1.4   Integrating Internal and External Applications/Resources

Two plugins were implemented for the integration of internal applications, i.e., mHBS training app and mHBS tracker app. We have used c*om-darryn campbell-cordova-plugin-intent* to receive intent from mHBS tracker which contains the username, password, server name, and pin numbers to access mHBS training. The intent also contains information about healthcare provider organization units and their id's (known as tracked entity instance id) for pushing training app usage data to the DHIS2 server. *com.lampa.startapp* plugin implementation leads to scaling up of training app integration with additional external newborn care applications. This feature also allows us to include links to resources of external partners; for the demonstration version of mHBS Trainer, we have included links to the American Academy of Pediatrics (Helping Babies Survive page) and PATH (Reprocessing Guidelines for Basic Neonatal Resuscitation Equipment in Resource-Limited Settings), as shown in Fig. 5. This feature provides HCPs and other end-users and partners with ease of access to a wide variety of partner resources by which to strengthen knowledge at both the individual (self-directed learning) and institutional (program) levels.

3.1.5   Privacy and Security

Vital features were added to the training app to address privacy and security concerns expressed during the beta-test, and to address evolving standards around privacy and security for mobile health apps. Firstly, a 4-digit pin-based user login was added as shown in Fig. 4. Secondly, a feature which allows the user to login training app only if the phone security lock is enabled. Finally, a feature which invokes user to re-login if the user wants to resume the training app after switching out of it was implemented. To comply with new GDPR privacy guideline, we have included a Privacy policy in the home screen of a training app (Fig. 4). The privacy policy was created based on the GDPR template.

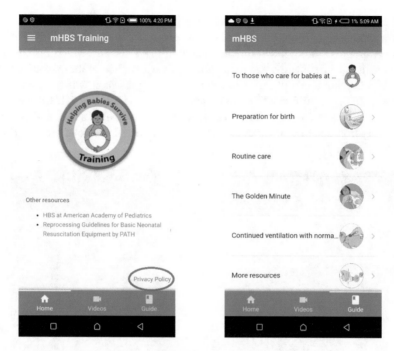

Fig. 5. mHBS training home screen and the HBB guide

3.2 User Interface Improvement

We received important recommendations to improve stability features from beta-testers. Currently, the training app is available only on Android, even though we use Cordova to build the app. The PDF guide in the training app was digitized into HTML. The guide was divided into eight sections based on the neonatal resuscitation training phases, each section containing respective pages (Fig. 5). This modification helps in better navigation control with feasible use of the guide. It also provided us with the ability to capture usability metrics which consequently would be used to evaluate user behavior.

In order to provide HCPs with additional flexibility in viewing, and the ability to resize images in the HTML guide, a zoom function was programmed using the inbuilt class in framework7 called swiper-zoom-container (Fig. 6).

3.3 App Usage Metrics

We utilize Crashlytics as a platform to capture app usage, even though its most intended use is to detect bugs, and app crashes. We created Crashlytics Events to address a variety of questions, including: (1) At what point does an HCP exit a video? (2) How often does an HCP re-watch particular videos? (3) How many times and what pages of a PDF have been viewed? (4) What types of trainer app content have been viewed most frequently and over what period?

These activities are tracked for each user so that the team obtains critical mHBS training app usage information regarding how about, for example, how the healthcare providers progress through self-directed learning modules, and which resources they interact with most frequently, and most thoroughly. Collection of these metrics through the mHBS training app will facilitate future eLearning research which analyzes HCPs behavior patterns for different types of training materials, i.e., video, HTML guide, etc., and how these patterns might influence subsequent knowledge, skills, and competencies, among frontline health workers in LMICs, for maternal and newborn care (Fig. 7).

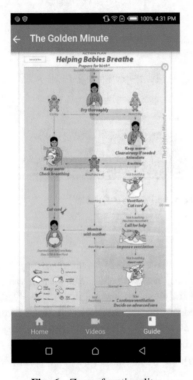

Fig. 6. Zoom functionality

Fig. 7. mHBS training app linking to the mHBS tracker and eHBB app.

4   Discussion

4.1   Insights

Although not formally analyzed, anecdotally, conducting a comprehensive literature review, and consulting with domain experts prior, to programming and design of the app(s), seemed to assist the technology team in more rapid and effective project progression, and appeared to strengthen interdisciplinary team cohesion among faculty and students. Obtaining an extensive understanding of the myriad non-technical challenges faced by healthcare providers in LMICs lent to more informed technical design. For instance, an HCP employed in a low- supply Neonatal Care Intensive Unit operating at full capacity might need to access an educational module when a birth is occurring. Meanwhile, another user who is currently employed at a higher equipped, low burden hospital may wish to merely review PDFs to supplement their formal training while at home. Understanding the uses and nuances of nurse/midwife-mobile app interaction in low- and middle-income countries (LMICs), such as brownouts that affect Wi-Fi, socio-culturally induced stress, overwork and lack of supplies, and availability of mobile charging stations allowed for design considerations that minimized app size, scrolling complexity, and battery usage.

4.2 Considerations

Video and documents downloaded in the application are accessed from outside via gallery and file manager in standard Android devices. This does not allow the tracking of this particular type of content usage as accessed by the user. Discussion led to the decision to disable this feature.

4.3 Limitations

There are a few limitations which should be mentioned. First, the mHBS application suite is currently built on the assumption that healthcare providers are literate, and have reasonable proficiency, in the English language. Future plans include development of trainer and tracker materials in other languages. Modifications of mHBS/DHIS2 for community-based health care workers with lower levels of literacy may require an increase in the graphical interface, and a decrease in the text. Second, although video thumbnails were successfully gathered and displayed on the training app during one iteration of development, the execution time to generate thumbnails proved to be highly problematic for the user experience, and thus non-viable.

Finally, a lower quantity of screen-by-screen beta-tester feedback was acquired than desired. This may have been due to scheduling issues, and the fact that we actively recruited a variety of beta-testers from LMICs – however, due to their extremely busy clinical schedules, and locations across time zones, they were unable to deliver the substantive feedback we requested within a relatively short time period. As a result, most of the feedback received was from high-level global health and international clinical training experts. This may limit the generalizability of the recommendations regarding utility for frontline health workers. We are currently addressing this concern via an extensive, on-going field test of the mHBS/DHIS2 app among African healthcare providers.

5 Conclusion

To the best of our knowledge, this is the first time DHIS2 has been used to support video and PDF content for an Android-based mobile app. Our app has improved potential for global scaling and sustainability through the integration with DHIS2. One exciting innovation of our work is that we are employing DHIS2 to track individual health care providers; in the past, DHIS2 has been primarily used for tracking and measuring program or facility-based outcomes. This paper primarily describes the mHBS/DHIS2 Trainer app. As a linked tool, the full mHBS suite of applications (Trainer, Tracker, and Aggregate) provides users with access to integrated functionalities that support self-directed learning, seamless access to external partner resources, and data collection, reporting, and visualization features. This in turn, allows HCPs in LMICs to access a wide variety of high-quality learning materials and training content, for which access metrics are captured. This may inform future research efforts, and lead to improved, more efficient acquisition and retention of knowledge, skills, and competencies for newborn care, thereby leading to a reduction in neonatal mortality in low/middle-income countries.

Acknowledgment. We would like to acknowledge the contributions of previous developers of the mHBS/DHIS2 application, in particular, Siva Addepally, Olakunle Oladiran, and Taylor Childers. The work reported in this paper was completed as a partial requirement for an undergraduate Capstone Honors Thesis (Informatics) by Ms. Elisabeth Meyers (Drs. Saptarshi Purkayastha and Sherri Bucher, co-Mentors). Ms. Meyers was awarded *"Best Scientific Presentation"* for this work by the IUPUI Center for Research and Learning on April 6, 2018. We also acknowledge the support of collaborators from University of Washington, Oxford University, Moi University (Kenya) and University of Lagos (Nigeria), as well as the Bill and Melinda Gates Foundation Funding for this work was provided by two grants from the IUPUI Center for Research and Learning, Multidisciplinary Undergraduate Research Initiative (MURI), to Dr. Sherri Bucher (Summer, 2017; Academic year 2017–2018).

References

1. Children: reducing mortality, World Health Organization (2018). http://www.who.int/mediacentre/factsheets/fs178/en/. Accessed 20 Apr 2018
2. Pathirana, J., Muñoz, F.M., Abbing-Karahagopian, V., Bhat, N., Harris, T., Kapoor, A., Keene, D.L., Mangili, A., Padula, M.A., Pande, S.L., Pool, V., Pourmalek, F., Varricchio, F., Kochhar, S., Cutland, C.L.: Neonatal death: case definition & guidelines for data collection, analysis, and presentation of immunization safety data. Vaccine **34**(49), 6027–6037 (2016)
3. Ersdal, H., Mduma, E., Svensen, E., Perlman, J.: Early initiation of basic resuscitation interventions including face mask ventilation may reduce birth asphyxia related mortality in low-income countries. Resuscitation **83**(7), 869–873 (2012)
4. Hoban, R., Bucher, S., Newman, I., Chen, M., Tesfaye, N., Spector, J.M.: "Helping babies breathe" training in sub-Saharan Africa: educational impact and Learner impressions. J. Trop. Pediatr. **59**(3), 180–186 (2013)
5. Belllad, R.M., Bang, A., Carlo, W.A., McClure, E.M., Meleth, S., Goco, N., Goudar, S.S., Derman, R.J., Hibberd, P.L., Patel, A., Esamai, F., Bucher, S., Gisore, P., Wright, L.L., HBB Study Group: A pre-post study of a multi-country scale up of resuscitation training of facility birth attendants: does helping babies breathe training save lives? BMC Pregnancy Childbirth **16**(1), 222 (2016). https://doi.org/10.1186/s12884-016-0997-6
6. Abwao, S., Bucher, S., Kaimenyi, P., Wachira, J., Esamai, F., Wamae, A.: Kenya case study. In: Helping Babies Breathe: Lessons learned guiding the way forward. A 5-year report from the HBB Global Development Alliance, 8 June, pp. 99–102 (2015)

7. Mabey, D., Sollis, K., Kelly, H., Benzaken, A., Bitarakwate, E., Changalucha, J., Chen, X., Yin, Y., Garcia, P., Strasser, S., Chintu, N., Pang, T., Terris-Prestholt, F., Sweeney, S., Peeling, R.: Point-of-care tests to strengthen health systems and save newborn lives: the case of syphilis. PLoS Med. **9**(6), e1001233 (2012)

8. Tilahun, D., Hanlon, C., Araya, M., Davey, B., Hoekstra, R., Fekadu, A.: Training needs and perspectives of healthcare providers in relation to integrating child mental health care into primary health care in a rural setting in sub-Saharan Africa: a mixed methods study. Int. J. Ment. Health Syst. **11**(1), 9 (2017)

9. Mobile Applications for the Health Sector, 1st ed., p. 85. World Bank, Washington (2018)

10. Medhi, I., Gautama, S., Toyama, K.: A comparison of mobile money-transfer UIs for non-literate and semi-literate users. In: Proceedings of the 27th International Conference on Human Factors in Computing Systems - CHI 2009, p. 1742 (2009)

11. Mukherjee, A., Purkayastha, S., Sahay, S.: Exploring the potential and challenges of using mobile based technology in strengthening health information systems: experiences from a pilot study. In: AMCIS, p. 263 (2010)

12. Hall, C., Fottrell, E., Wilkinson, S., Byass, P.: Assessing the impact of mHealth interventions in low- and middle-income countries – what has been shown to work? Glob. Health Action **7**(1), 25606 (2014)

13. Collect, Manage, Visualize and Explore your Data, Dhis2.org (2018). https://www.dhis2.org/. Accessed 21 Apr 2018

14. Purkayastha, S., Braa, J.: Big Data Analytics for developing countries – using the cloud for operational BI in health. Electron. J. Inf. Syst. Dev. Ctries. **59**, 7 (2013)

15. Ending newborn deaths, ensuring every baby survives, Save the Children, London, p. 48 (2012)

16. Waugaman, A.: From principle to practice: implementing the principles for digital development. In: Proceedings of the Principles for Digital Development Working Group (2016)

17. Bucher, S.L.: mHBB: using mobile phones to support Helping Babies Breathe in Kenya. In: Levine, R., Corbacio, A., Konopka, S., Saya, U., Gilmartin, C., Paradis, J., Haas, S. (eds.) mHealth Compendium, vol. 5, pp. 46–47. African Strategies for Health, Management Sciences for Health, Arlington (2015)

18. Braa, K., Purkayastha, S.: Sustainable mobile information infrastructures in low resource settings. Stud. Health Technol. Inf. **157**, 127 (2010)

19. Oluoch, T., de Keizer, N.: Evaluation of health IT in low-income countries, vol. 222, p. 324 (2018). IOS Press

20. Balsamiq.Rapid, effective and fun wireframing software. Balsamiq (2018). https://balsamiq.com/. Accessed 21 Apr 2018

21. Free mobile & web prototyping (iOS, Android) for designers – Marvel, Marvel Prototyping (2018). https://marvelapp.com/. Accessed 21 Apr 2018

22. Lund, S., Boas, I.M., Bedesa, T., Fekede, W., Nielsen, H.S., Sørensen, B.L.: Association between the safe delivery app and quality of care and perinatal survival in Ethiopia: a randomized clinical trial. JAMA Pediatr. **170**(8), 765–771 (2016)

23. Harris, P.A., Taylor, R., Thielke, R., Payne, J., Gonzalez, N., Conde, J.G.: Research electronic data capture (REDCap)—a metadata-driven methodology and workflow process for providing translational research informatics support. J. Biomed. Inform. **42**(2), 377–381 (2009)

24. Broadcasts overview. Android Developers: Android Developers (2018). https://developer.android.com/guide/components/broadcasts. Accessed 27 Apr 2018

25. XmlPullParser Android Developers: Android Developers (2018). https://developer.android.com/reference/org/xmlpull/v1/XmlPullParser. Accessed 27 Apr 2018

26. MediaMetadataRetriever. Android Developers: Android Developers (2018). https://developer.android.com/reference/android/media/MediaMetadataRetriever.html.   Accessed 27 Apr 2018
27. DownloadManager. Android Developers: Android Developers (2018). https://developer.android.com/reference/android/app/DownloadManager. Accessed 27 Apr 2018
28. Brooke, J.: SUS-A quick and dirty usability scale. Usabil. Eval. Ind. **189**(194), 4–7 (1996)

An Image Encryption Method Using Henon Map and Josephus Traversal

K. U. Shahna$^{(\boxtimes)}$ and Anuj Mohamed

School of Computer Sciences, Mahatma Gandhi University, Kottayam, India
shahnakakkatt@gmail.com

Abstract. Image security is a vital issue in multimedia communications. Chaos based image encryption has become very popular now a days due to the properties of chaotic systems such as unpredictability, ergodicity and sensitivity to initial conditions. This paper proposes a chaos based image encryption technique for gray scale images. Image pixels are scrambled using Josephus permutation sequence and Henon map is used in diffusion process. Encrypted image is obtained through permutation and diffusion process. Security analysis shows that the proposed encryption scheme can accomplish good encryption results.

Keywords: Henon map · Josephus permutation sequence · Image encryption · Security analysis

1 Introduction

With the rapid development of communication technology a large amount of digital image is transmitted over the network. These images can be accessed and modified by unauthorized users. So it is necessary to protect them from unauthorized access to ensure image security. Conventional encryption techniques are not suitable for digital image encryption because of the large size of image and high correlation among the pixels. Recently, image encryption algorithms using chaos theory has become an active research area as the chaotic system offers an excellent combination of speed and security. They also provide better properties in security and complexity. Chaos based cryptosystem can be implemented either using one-dimensional or higher dimensional chaotic maps. In each of the encryption schemes diverse chaotic maps are used. Recently a lot of image encryption techniques applied one or two dimensional chaotic map [1–11] to ensure confusion and diffusion in image encryption. However, most of them are proved to be insecure. Some other encryption techniques utilize hyper chaotic systems [12–14] for encrypting images. Hyper chaotic systems possess more complex structure and better chaotic performance. Novel image encryption techniques are also implemented on multiple one-dimensional or higher-dimensional chaotic systems based on the fact that applying more than one chaotic map can achieve better key space and higher security.

The concept of Josephus permutation sequence has been used in several image encryption techniques [15–18]. In [15, 16] Josephus traversal is used only for pixel

© Springer Nature Switzerland AG 2019
A. Abraham et al. (Eds.): IBICA 2018, AISC 939, pp. 375–385, 2019.
https://doi.org/10.1007/978-3-030-16681-6_37

permutation process to ensure confusion. Pixel permutation will guarantee the scrambling of pixels but the pixel values are not changed. Such pixel permutation based encryption is vulnerable to known plaintext attack, cipher text only attack and statistical attacks [19]. In order to achieve higher security, we have proposed a new image encryption technique which includes both confusion and diffusion process. In this paper combination of both Josephus permutations sequence and chaotic maps are employed to ensure security.

The remaining sections of the paper are organized as follows. The basic concepts about Henon map and Josephus traversal is given in Sect. 2. The proposed image encryption method is introduced in Sect. 3. Section 4 analyses the security issues of the proposed method. Concluding remarks are given in Sect. 5.

2 Preliminary Concepts

2.1 Henon Map

Henon map is a chaotic map [20] that shows chaotic behavior. It is a two-dimensional discrete dynamic system and is described by

$$x1_{n+1} = 1 - \alpha x1_n^2 + y1_n$$
$$y1_{n+1} = \beta y1_n$$

(1)

Here, the map turns into chaotic nature when the parameters α and β takes the values 1.4 and 0.3 respectively. Figure 1 shows an example of Henon map attained from discrete number of iterations starting from initial values of x_1 and y_1 as (0.4838, 0.0038).

2.2 Josephus Permutation Problem

Josephus problem is a hypothetical problem in mathematics and computer science. It is associated with a counting-out game that start by having x people standing in a circle, sequentially labeled from 1 to x. The counting can start from a predestined individual. Once the pth person is reached, take him out of the circle and the remaining members form a new circle. Repeat the process until only one person remains and he/she wins the game. Recording the sequence in which the persons are taken out forms a Josephus permutation.

The different parameters used in this problem are: total number of persons initially present in the circle (x), counting starting position (pos) and the counting period (p). Based on these Josephus permutation sequence can be denoted as JPS(x, pos, p).

For example, JPS(12, 1, 5) = [5, 10, 3, 9, 4, 12, 8, 7, 11, 2, 6, 1]
JPS(12, 6, 9) = [2, 11, 9, 8, 10, 1, 5, 3, 4, 12, 6, 7]
JPS(12, 4, 9) = [12, 9, 7, 6, 8, 11, 3, 1, 2, 10, 4, 5]

The Josephus problem always create a random number sequence of length x and for different pairs of (pos, p) it produces different permutation sequences. But this sequence's first number will disclose the counting period p if starting position (*pos*) is 1. Also in certain cases the randomness feature of this permutation sequence tends to degrade while using fixed parameters x and p. For example, in JPS(12, 6, 9) and JPS(12, 4, 7), the difference between two JPS values (in circular manner) shows their starting positions difference. To overcome these flaws in Josephus permutation the proposed method performs one more scrambling process using the concept of matrix diagonal.

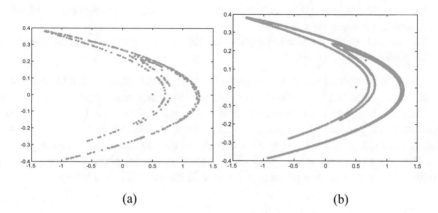

(a)   (b)

Fig. 1. Henon map with (a) 1000 iterations and (b) 10000 iterations with x1 = 0.4838 and y1 = 0.0038

3   The Proposed Method

The technique adopted here is a symmetric cryptosystem, which involves pixel permutation, key generation and diffusion of scrambled image. In the confusion phase of our cryptosystem, Josephus traversal is used for pixel scrambling. In order to enhance the permutation process we have introduced a second round of scrambling using the concept of matrix diagonal. The confusion state changes the pixel positions only. A novel method is proposed for diffusion and for generating the secrete key. In diffusion phase the value of each pixel is altered through a sequence of operations using the secrete key and the random numbers generated from Henon map. In this technique the confusion and diffusion process ensures better security against various attacks.

3.1   Pixel Permutation

Image pixel scrambling is performed as follows:

1. Read the image I of size M × N and convert it into a one dimensional vector of the same size. Using Josephus permutation sequence scramble the image I to obtain I'. The three parameter values of JPS(x, pos, p) for this scrambling process can be taken as follows:

(i)   The total number (x) can be taken as the size of the image M × N.
(ii) Starting position (*pos*) is randomly selected as 3.
(iii) Counting period *p* can be calculated as follows:

$$P = \text{floor}(\text{SUM}/\text{DIG\_SUM}) \tag{2}$$

Here SUM and DIG_SUM represents sum of all the pixels and sum of diagonal pixels in the original image. Figure 2 shows an example for pixel permutation using Josephus permutation sequence.

2. The second round of scrambling proposed for enhancing the pixel permutation can be performed as follows.

Take the scrambled image I′ and convert in to the matrix form of size M × N. Now scan I′ diagonally from lower right to upper left. Similarly scan upper half portion of the diagonal and then lower half portion of the diagonal. Now simply swap the upper and lower half portions of this image based on the new diagonal elements. Let S_IMG denote the final scrambled image. Figure 3 shows this process using an example.

The above two scrambling steps will reduce the adjacent pixels correlation in the original image and gives a better performance in the permutation process.

3.2   Image Encryption Algorithm

Step 1: Perform pixel scrambling of original image I of size M × N using the above pixel permutation steps and S_IMG be the final scrambled image. Figure 4(a) shows original image of Lena and Fig. 4(b) the scrambled image after pixel permutation.

Step 2: Generate the secrete key for encryption using the following steps.

a. The initial values of Henon map x1 and y1 can be calculated as

$$x1 = SUM/(DIG\_SUM * M * N) \tag{3}$$

$$Y1 = DIG\_SUM/(M * N) \tag{4}$$

b. The parameter λ can be set using the equation

$$\lambda = floor(SUM/DIG\_SUM) * M * N \tag{5}$$

Step 3: Perform the diffusion process as follows:
Iterate the Henon map ((M * N)/2) + 1000 times and simply throw away the first 1000 values for getting better randomness. Generate a random matrix RM using x and y values obtained from Henon map by the following equations.

$$V1 = abs(\lfloor x_k * x_k * x_k * \lambda \rfloor) mod\ M * N \tag{6}$$

$$V2 = abs(\lfloor y_k * y_k * y_k * \lambda \rfloor) mod\ M * N \tag{7}$$

Here k can take values 1, 2, 3......$M * N/2$. The parameter λ can be obtained from Eq. (5)

The final encrypted image can be obtained as follows:

$$E\_IMG = NEW\_IMG \oplus RM \tag{8}$$

Here NEW_IMG can be obtained using the equation

$$NEW\_IMG(i,j) = S\_IMG(i,j) + SUM + DIG\_SUM \tag{9}$$

Also,

$$SUM = \sum_{i=1}^{M} \sum_{j=1}^{N} I(i,j) \tag{10}$$

And

$$DIG\_SUM = \sum_{i=1}^{M} \sum_{\substack{j=1 \\ i==j}}^{N} I(i,j) \tag{11}$$

From Eq. (9) we can see that the value of any pixel in the encrypted image not only depends up on the corresponding original image pixel value but also on the entire pixel values of the original image. Figure 4(c) shows the final encrypted image of Lena using the proposed technique.

1	2	3	4	5
6	7	8	9	10
11	12	13	14	15
16	17	18	19	20
21	22	23	24	25

5	10	15	20	25
6	12	18	24	7
14	22	4	16	1
11	23	13	3	21
19	2	9	17	8

(a)   (b)

Fig. 2. Example of first level pixel scrambling using Josephus traversing. (a) Original matrix (b) Scrambled matrix using Josephus traversing with JPS(25, 1, 5)

5	10	15	20	25
6	12	18	24	7
14	22	4	16	1
11	23	13	3	21
19	2	9	17	8

8	3	4	12	5
21	16	18	10	1
24	15	7	20	25
17	13	22	6	9
23	14	2	11	19

8	21	24	15	17
3	16	13	22	23
4	12	7	14	2
5	18	10	6	11
1	20	25	9	19

(a)   (b)   (c)

Fig. 3. Example of second level pixel scrambling (a) Original matrix (b) Diagonal scan from lower right to upper left (c) Swap upper and lower half portions of the diagonal.

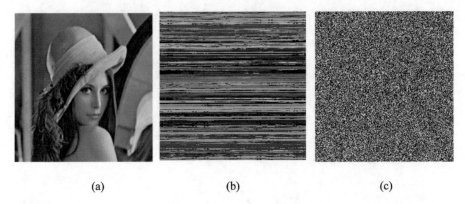

(a)   (b)   (c)

Fig. 4. Lena image. (a) Original image. (b) Scrambled image. (c) Encrypted image

4   Results and Discussion

Here, experiments are conducted on gray scale images of various sizes. The platform used is Matlab8.1 (R2013a) for performing experiments.

4.1   Statistical Analysis

The likeness and diversity between the plain and encrypted images are performed using histogram analysis and correlation analysis.

Correlation Analysis of Adjacent Pixels
Correlation analysis is used to evaluate the quality of the proposed encryption method. For performing this analysis adjacent pixel pairs are randomly selected from original and encrypted images. Then correlation between pixels (x_n, y_n) is calculated using the formulas give below:

$$r(x_n, y_n) = \frac{\text{cov}(x_n, y_n)}{\sqrt{D(x_n)}\sqrt{D(y_n)}} \tag{12}$$

Here,

$$\text{cov}(x_n, y_n) = \frac{1}{N}\sum\nolimits_{i=1}^{N}\left((x_{n_i} - E(x_n))(y_{n_i} - E(y_n))\right) \tag{13}$$

$$E(x_n) = \frac{1}{N}\sum\nolimits_{i=1}^{N} x_{n_i} \tag{14}$$

$$E(y_n) = \frac{1}{N}\sum\nolimits_{i=1}^{N} (x_{n_i} - E(x_n))^2 \tag{15}$$

In Fig. 5 correlation coefficient of two adjacent pixels in original and encrypted Lena image is shown. Table 1 shows the different correlation coefficient values of various original and encrypted images. The experimental results shown in Table 1 indicates that the two adjacent pixels in the plain image are highly correlated (very close to 1), while that of an encrypted image is very low (close to 0), means they resist statistical attack.

Histogram Analysis of Image
An image histogram shows how pixels in an image are distributed by graphing the number of pixels at each color intensity level. From Fig. 6 it is seen that histogram of a plain image has an irregular distribution while that of an encrypted image has a uniform distribution. This shows that the cipher image and original image has no statistical relationship with each other and the cipher image defend against all statistical assaults.

Differential Attack Analysis
Differential analysis evaluates how a slight change in the plain image reflects in the encrypted image. NPCR (Number of Pixels Change Rate) and UACI (Unified Average Changing Intensity) are the two criteria to resist against differential attack. It can be calculated as follows:

$$NPCR = \frac{\sum_{i_1,j_1} f(i_1,j_1)}{M \times N} \times 100\% \tag{16}$$

Where, $f(i_1,j_1) = \begin{cases} 0, E_1(i_1,j_1) = E_2(i_1,j_1) \\ 1, E_1(i_1,j_1) \neq E_2(i_1,j_1) \end{cases}$

$$UACI = \frac{1}{M \times N} \left[\sum_{i_1,j_1} \frac{E_1(i_1,j_1) - E_2(i_1,j_1)}{255} \right] \times 100\% \tag{17}$$

Here, M and N represent the width and height of the cipher image. Also, E_1 and E_2 are two encrypted images with a single bit variation in original images. Table 2 shows the NPCR and UACI values obtained with the proposed method. In general, the upper bound of NPCR is 100% and UACI is about 33.46%. The results show that the NPCR value is greater than 99.6% and the UACI is nearly 33.46%, which means the proposed algorithm is very sensitive to minute changes in the plain image. If there is only one bit difference between two plain images, the decrypted images will be completely different.

Information Entropy Analysis
Analysis of Information entropy can be determined as follows

$$H(y) = \sum_{j=0}^{2^n-1} p(y_j) log_2 \frac{1}{p(y_j)} \tag{18}$$

Here, y be the information source, $p(y_j)$ indicates the y_j's probability. 2^n is the total number of states. The information entropy of various images is given in Table 3. From these results, it can be seen that the information entropy value of all encrypted images are much closer to the theoretical maximum value 8. This shows that the cipher images have better random distributions and proposed method is more secure against the entropy attack.

Table 1. Correlation coefficient evaluation

Correlation	Lena plain image	Lena cipher image	Baboon plain image	Baboon cipher image	Pepper plain image	Pepper cipher image
Horizontal	0.9684	0.0006	0.9499	0.0039	0.9469	0.0123
Vertical	0.9860	−0.0023	0.9395	−0.0021	0.9546	0.0074
Diagonal	0.9516	0.0025	0.9061	−0.0004	0.9138	0.0047

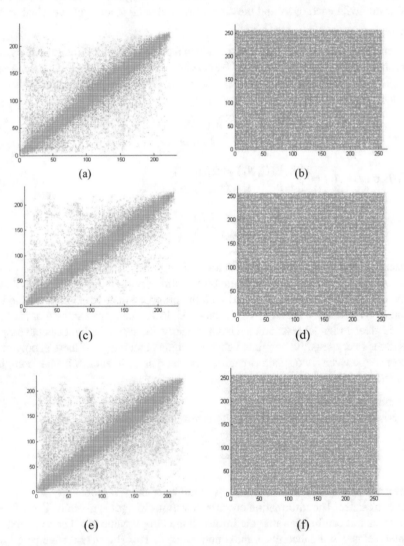

(a)

(b)

(c)

(d)

(e)

(f)

Fig. 5. Correlation distribution of original and cipher image. (a) Horizontal correlation - original image. (b) Horizontal correlation - cipher image. (c) Vertical correlation-original image. (d) Vertical correlation - cipher image. (e) Diagonal correlation - original image. (f) Diagonal correlation - cipher image.

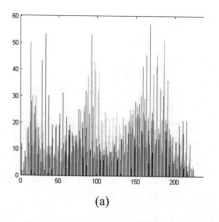
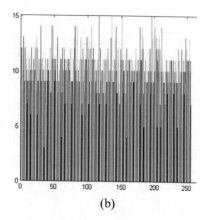

(a)   (b)

Fig. 6. Histogram representation (a) Lena original image. (b) Lena encrypted image

Table 2. NPCR and UACI values of cipher images in proposed technique

Image name	NPCR	UACI
Lena	99.6615	33.5321
Baboon	99.6731	33.4614
Pepper	99.6422	33.5221

Table 3. Information entropy of original and cipher images

Images	Original image	Encrypted image
Lena	7.3879	7.9975
Baboon	7.1287	7.9972
Pepper	7.5931	7.9974

Performance Comparisons of Proposed Technique with Other Schemes

Table 4 shows comparison of the values of NPCR, UACI, entropy and correlation coefficients in proposed method with some other schemes. These comparisons are made on Lena image of size 256 × 256. From the analysis table, one can see that our scheme achieves better NPCR, UACI, entropy and low correlation values. Comparisons shows the proposed method has better performance and can more effectively resist differential attacks.

Table 4. Performance comparison of proposed method with other schemes

Encryption schemes	Correlation coefficients			NPCR	UACI	Entropy
	Horizontal	Vertical	Diagonal			
Proposed	0.0006	−0.0023	0.0025	99.6615	33.5321	7.9975
[3]	0.0321067	0.0271879	0.0383929	99.617	33.4933	7.9982
[4]	0.02507	−0.02071	−0.00796	99.561	33.546	–
[7]	0.0028	0.0032	0.0100	99.71	37.63	7.9969
[8]	0.003503	0.000213	0.000728	99.7421	33.5278	7.9976
[18]	−0.0029	−0.0017	0.0004	99.5986	33.4561	7.9971

Analysis of Speed

Experiments are carried out on desktop PC with Intel(R) Core(TM) i3-6100U 2.30 GHz CPU, 4.00 GB RAM and 64-bit Microsoft Windows 7 operating system. The computational platform is Matlab 8.1 (R2013a). The average encryption time taken by the algorithm for processing various images is given in Table 5. A good encryption technique is expected to have faster running speed.

Table 5. Encryption time consumption (sec)

Name of image	Encryption speed (sec)
Lena	0.423016
Baboon	0.441534
Pepper	0.432375

5 Conclusion

A novel encryption technique using Josephus permutation sequence and Henon map to encrypt gray scale images is proposed in this paper. Initially, pixel permutation is performed using Josephus permutation sequence. In order to reduce the adjacent pixels correlation in the scrambled image second round of scrambling is performed based on the diagonal elements. Henon map is employed for generating random values, which is used in diffusion process. To guarantee the security of this new method various statistical analysis are carried out. The results show that the proposed method ensures the highest security level in terms of NPCR, UACI and entropy of the cipher images. So the proposed technique is appropriate for real-time applications.

6 Future Work

This paper proposes an encryption technique for gray scale images. In future we want to implement a cryptosystem for color image encryption using the concept of hyper chaos.

Acknowledgments. The authors acknowledge the support extended by DST-PURSE (Phase II), Government of India.

References

1. Xu, L., Li, Z., Li, J., Hua, W.: A novel bit-level image encryption algorithm based on chaotic maps. Opt. Lasers Eng. **78**, 17–25 (2016)
2. Zhang, W., Wong, K., Yu, H., Zhu, Z.: An image encryption scheme using reverse 2-dimensional chaotic map and dependent diffusion. Commun. Nonlinear Sci. Numer. Simul. **18**(8), 2066–2080 (2013)
3. Wang, X., Jin, C.: Image encryption using game of life permutation and PWLCM chaotic system. Opt. Commun. **285**(4), 412–417 (2012)
4. Ye, G., Huang, X.: An image encryption algorithm based on autoblocking and electrocardiography. IEEE MultiMed. **23**(2), 64–71 (2016)
5. Wu, X., Zhu, B., Hu, Y., Ran, Y.: A novel color image encryption scheme using rectangular transform-enhanced chaotic tent maps. IEEE Access **5**, 6429–6436 (2017)
6. Hua, Z., Zhou, Y.: Image encryption using 2D Logistic-adjusted-Sine map. Inf. Sci. **339**, 237–253 (2016)
7. Ye, R., Ma, Y.: A secure and robust image encryption scheme based on mixture of multiple generalized Bernoulli shift maps and Arnold maps. Int. J. Comput. Netw. Inf. Secur. **5**(7), 21 (2013)
8. Hu, Y., Zhu, C., Wang, Z.: An improved piecewise linear chaotic map based image encryption algorithm. Sci. World J. (2014)
9. Nasir, Q., Abdlrudha, H.H.: High security nested PWLCM chaotic map bit-level permutation based image encryption. Int. J. Commun. Netw. Syst. Sci. **5**(09), 548 (2012)
10. Abdlrudha, H.H., Nasir, Q.: Low complexity high security image encryption based on nested PWLCM chaotic map. In: 2011 International Conference for Internet Technology and Secured Transactions (ICITST), pp. 220–225. IEEE (2011)
11. Fu, C., Zhang, Z., Chen, Y., Wang, X.: An improved chaos-based image encryption scheme. In: International Conference on Computational Science, pp. 575–582. Springer, Heidelberg (2007)
12. Niyat, A.Y., Moattar, M.H., Torshiz, M.N.: Color image encryption based on hybrid hyper-chaotic system and cellular automata. Opt. Lasers Eng. **90**, 225–237 (2017)
13. Norouzi, B., Mirzakuchaki, S.: A fast color image encryption algorithm based on hyper-chaotic systems. Nonlinear Dyn. **78**(2), 995–1015 (2014)
14. Zhu, C.: A novel image encryption scheme based on improved hyperchaotic sequences. Opt. Commun. **285**(1), 29–37 (2012)
15. DeSheng, X., Yueshan, X.: Digital image scrambling based on Josephus traversing. Comput. Eng. Appl. 10 (2005)
16. Ye, G., Huang, X., Zhu, C.: Image encryption algorithm of double scrambling based on ASCII code of matrix element. In: 2007 International Conference on Computational Intelligence and Security, pp. 843–847. IEEE (2007)
17. Yang, G., Jin, H., Bai, N.: Image encryption using the chaotic Josephus matrix. Math. Probl. Eng. (2014)
18. Wang, X., Zhu, X., Zhang, Y.: An image encryption algorithm based on Josephus traversing and mixed chaotic map. IEEE Access **6**, 23733–23746 (2018)
19. Furht, B., Kirovski, D. (eds.): Multimedia Security Handbook. CRC Press, Boca Raton (2004)
20. Milnor, J.: On the concept of attractor. In: The Theory of Chaotic Attractors, pp. 243–264. Springer, New York (1985)

Revealing Abnormality Based on Hybrid Clustering and Classification Approach (RA-HC-CA)

C. P. Prathibhamol[✉] and Asha Ashok

Department of Computer Science and Engineering, Amrita Vishwa Vidyapeetham
Amritapuri, Kollam, India
{prathibhamolcp,ashaashok}@am.amrita.edu

Abstract. Abnormality Detection is the process of locating abnormal instances within the data. In this work, we have applied Abnormality Detection to the domain of detection associated with Credit Card Fraud. This problem is actually attributed to demonstrating those credit card transactions which have occurred in the earlier times, with the presence of awareness related to those instances, which are actually fraud ones. Applying this model, we can use it to predict if a new transaction is a fraud based or not. In this proposed work, we have utilized a combination framework of data mining clustering algorithms so as to solve the problem of credit card fraud detection to a particular extent. The proposed work Revealing Abnormality Based on Hybrid Clustering and Classification Approach (RA-HC-CA) consists of two stages namely a clustering phase followed by a detection phase. In the clustering phase, we have employed a combined clustering approach initiated by k-means clustering algorithm followed by hierarchical clustering algorithm. Prior to Hierarchical/Agglomerative clustering, the whole data set is clustered into meaningful 'k' knots by k-means clustering procedure. The output of 'k' groups is then inputted to Agglomerative clustering algorithm to merge the already obtained 'k' clusters from the previous phase, into more meaningful clusters. This is continued until 70–75% of data falls on one large group, which is the Normal group. The remaining data instances may converge in various other abnormal groups. The strong assumption made here is that such clusters with less instances, than a particular threshold are considered to be groups pertaining to fraud ones. Then, so as to check for the presence of an instance as fraud one, we initially identify the proximate gathering to which it fit into. Then, within that identified cluster, LDA (Linear Discriminant Analysis) is carried out. It has been observed that the proposed approach (RA-HC-CA) achieved 80.5% accuracy in comparison with various other existing methods.

Keywords: Agglomerative · k-means · Credit-card-fraud detection · LDA · Data mining

© Springer Nature Switzerland AG 2019
A. Abraham et al. (Eds.): IBICA 2018, AISC 939, pp. 386–394, 2019.
https://doi.org/10.1007/978-3-030-16681-6_38

1   Introduction

In recent times, there is a tremendous shift in the shopping behaviour of the public's belonging to various "societies". Prior to the change, people used to carry money along with them in order to purchase anything from which ever shop they required. But in today's age, we find majority people carrying a lot of credit/debit cards so that it can be utilized for shopping appropriately. One main advantage associated with this trend is that people can fearlessly travel without cash. The whole world is moving to a digital or a cashless economy. But along with this improvement, comes a major obstacle. There can be a possibility of theft of these digital cards and possibly a series of fraudulent actions associated with these cards. Inevitably, detection of fraudulent actions associated with these credit/debit cards is an important task interrelated with high relevance. The credit or debit cards may be used at various places such as hospitals, banks, shopping malls etc. In the banking sector, any bank manager can associate the decision so as to give or not a loan to any customer who wishes to avail it. Based on the past or previous records, a bank manager can predict so as to decide whether to give or discard the loan facility whenever a new fresh applicant approaches. This proposed work focuses on detection of fake activity associated with credit cards.

There are mainly two different categories of scam linked with credit cards. The former kind is Application fraud which occurs when a deceptive person tries to issue duplicate or replacement cards. So, the further illegal transactions are then carried out with these duplicate cards. The latter one belongs to deceitful activities like theft via email or password etc.

In this proposed work (RA-HC-CA), we aim to perform credit card fraudulent detection based on the German dataset. The whole work discussed in this paper can be divided into two parts. The first part deals with the training step where the entire data is inputted to k-means algorithm, where the total dataset is concentrated into 'k' distinct clusters. Now to check for real closeness between clusters, Agglomerative clustering [1] is then used. As the output for any clustering algorithm is a number of clusters, similar is the output in this step too. But how may ever clusters are obtained, a fixed threshold is used to determine which all are the sparse clusters. It is assumed to be sparse because they are having lower number of instances than the certain brink. So, with this, training phase has finished until majority of the data, already partitioned as k-clusters, gets converged into one group and then the training part ends with calculating the Fisher coefficients of LDA (Linear Discriminant Analysis) for the only abnormal or small clusters obtained. In the testing phase, which ever test instance's label has to be predicted, it is evaluated to be lying near to which cluster. After the nearest cluster is located, if that cluster happens to be a normal one, then the instance is considered as a 'Valid' one. If the test instance is lying close to an abnormal cluster, then in that cluster, LDA is performed so as to evaluate the class label of the instance under consideration.

The paper is organized as, different sections with the literature survey or related works presented in Sect. 2, followed by Sect. 3 for proposed approach and

C. P. Prathibhamol and A. Ashok

it's analysis on Sect. 4. And, finally, we move on to Sect. 5 for the conclusion and the future enhancements.

2   Related Works

There is a lot of research work carried out in the domain of credit card fraud detection. In [2], the authors have focused on the usage of innovative data mining techniques and also neural network mixture context to obtain a better detection algorithm in this domain.

In [3], the researchers have developed a data mining system, CARDWATCH, for the sole determination of credit card fraud detection. This system is built on a learning module by using neural network. A comfortable graphical user interface is also supported and many test cases with results are also depicted. These test results indicates very good fraud detection rates.

In [4], the scholars have modelled the order of credit-card transactions as a Hidden Markow Model (HMM). Comprehensive investigational results are presented to prove the efficacy of their approach used in this work.

In [5], the researchers have done an extensive study on the modern practices such as Artificial intelligence, Fuzzy logic, Machine learning, data mining, Genetic Programming etc. for detection of frauds associated with credit cards. In their effort, the authors also estimate each technique based on firm design standards.

Authors in [6] had demonstrated the working of LDA (Linear Discriminant Analysis) by giving insight in to its mathematical formulation. This is an important algorithm and it plays a dominant role in our proposed system too.

Works in [7] and [9] deals with a kind of anomaly detection in the domain of the bench mark, KDD dataset and cardiac arrhythmia respectively. Both these works have validated the use of unsupervised clustering process to form groups of entities with related or almost similar behaviour.

In [10] and [11], the authors have devised a new variant of k-medoid clustering algorithm in the domain of heterogeneous datasets and a mining algorithm based on Apriori rule for association rule mining regarding high dimensional data.

The result confirmation of our proposed work RA-HC-CA is compared with numerous results and approaches discussed in [8].

3   Proposed System

In this planned work, we have taken German Credit Data set from UCI Machine Learning Repository. The entire dataset contain 20 vital attributes and class label whether an applicant belongs to good or bad credit risk. The total dataset comprises information about 1000 applicants. In general, the German Credit Data comprises or contains data on 20 variables and the classification whether an applicant is considered a Good or a Bad credit risk for 1000 loan applicants. Of this 20 important features, only 13 of them are categorical in nature. The remaining 7 attributes are purely numerical in nature. The complete dataset can be divided into attribute/feature space and class label space. Initially, the dataset is first clustered using conventional k-means algorithm to get "partial" clusters. This clustering process is conducted within the feature space. Any conventional k-means algorithm tries to minimize or reduce the sum of squared error criterion. Then the clustering phase is not yet over as the partial or intermediate clusters obtained in the initial phase is then passed to Hierarchical clustering algorithm. The main advantage of clubbing this two clustering algorithms is that Hierarchical algorithm does not have to start building clusters from individual data instances. Instead, the output of k-means can be the starting phase for it. It is a well-known fact both k-means and agglomerative clustering algorithm have their own merits and demerits.

By doing k-means clustering alone we can get two clusters, normal and abnormal with k = 2. But this is not always possible, subsequently clustering is an unsubstantiated method. After merging of two clustering algorithms we initiate a stopping criteria such that hierarchical algorithm stops only after 70–75% of data gets aggregated in one normal group. The remaining data may fall in more than one abnormal groups. By trial and error only it is found out around 75% normal data fell in the correct group. Towards the end of the training phase itself, already the Fisher coefficients as pertained to LDA are computed and kept for use in the Testing stage (Fig. 1).

Once the training period is over, then starts the testing phase. In the testing step, any test data instance is compared with the already obtained optimal clusters. Here the test instance, to which ever cluster is found to be near, that cluster is checked whether it is a normal or abnormal cluster. If the cluster is having much number of instances, then the considered test instance is of normal behaviour. If it is lying near to an abnormal cluster, where the number of instances is less than a specified brim value, then it is assumed to be of abnormal nature. That is if the test data instance is found to be lying near an abnormal cluster, then LDA is applied inside that particular obtained cluster to confirm its class label once again. "Figure 2" exhibits the testing phase with the input test data under consideration.

Both in the training and testing phase depicted below, German dataset is utilized. It is a two class dataset with class label Good and Bad risk class. We have only considered 70% for training and 30% for testing purposes respectively.

3.1   Linear Discriminant Analysis (LDA)

LDA is an algorithm which hunts for linear combinations between variables that can be utilized best so as to separate any two classes. To describe the concept of seperability, there are score functions well-defined by Fisher [6]. Assume for two distinct or separate classes, they are having mean and standard deviation for the respective classes as follows: $\mu 0$ and $\mu 1$ respectively. The linear combination of features can be computed as

$$\vec{w} \cdot \vec{x} \tag{1}$$

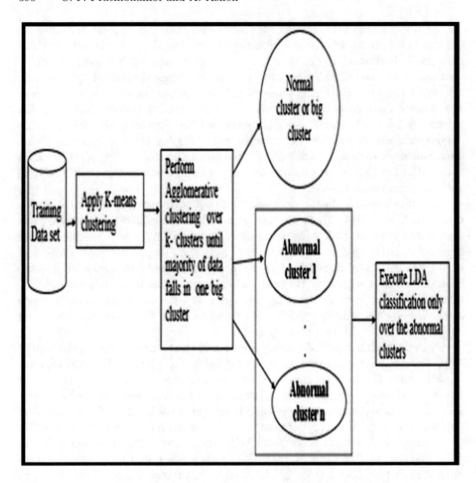

Fig. 1. Training phase (RA-HC-CA)

And moreover variances will be

$$\vec{w} \sum_i w, \ i = 0, 1, ...$$ (2)

As, given by the definition of Fisher, the separation between two classes will be the ratio of variance between the classes to that of variance within each individual class as

$$S = \frac{\sigma^2_{between}}{\sigma^2_{within}}$$ (3)

For solving the accuracy of any two class problem in terms of YES and NO, for both the authentic and anticipated class, it is done by means of the Confusion matrix [7]. Assume the normal class is NO and the presence of fraud is marked by YES. So, eventually the sum of TP+TN is the number of fraud instances detected

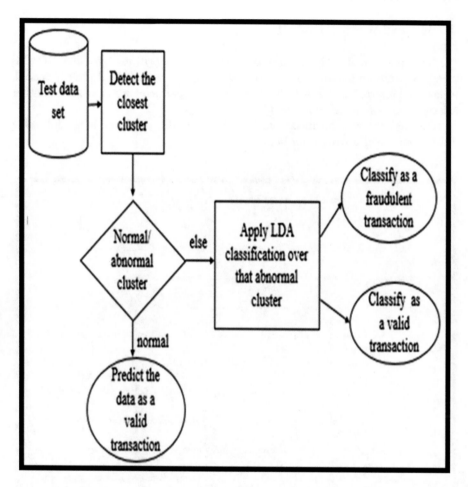

Fig. 2. Testing phase (RA-HC-CA)

correctly whereas FP+FN indicates instances that were predicted incorrectly. Tuples that has class label: YES and the model projected those instances as YES is denoted as TP. Similarly, tuples that pertains to class label: NO class and the model evaluated them as NO class is symbolized as TN. Also, the sum of FP+FN denotes the total amount of erroneously classified instances. This is because, tuples that has actually class label: YES but for which the model estimated as NO label is pertaining to FP. Whereas, if the actual class is NO but the mode assessed to YES class is represented as FN. So, Accuracy can be measured or computed as

$$Accuracy = \frac{TP + TN}{TP + TN + FP + FN} \tag{4}$$

4   Experimental Results

We have performed various analyses of different algorithms with RA-HC-CA, using the German data set. Thus, from the large set of tests that are conducted, it has been inferred that RA-HC-CA overtakes in performance than several other existing state-of-the-art discussed in [8].

This fact can be confirmed from the following table displayed in the following Table 2 and in the give graph below.

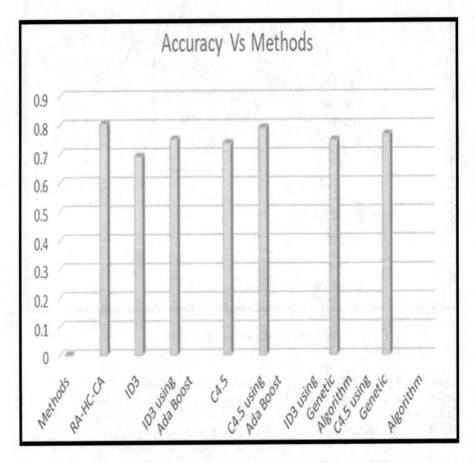

Fig. 3. Result analysis (RA-HC-CA)

From Table 1, it is evident clearly that RA-HC-CA works better in German Dataset, for the combination of 70% training and 30% testing set, when compared with various other existing methods performed in the same combination (Fig. 3).

Table 1. Accuracy of existing approaches and RA-HC-CAC

Variant approaches	Accuracy
RA-HC-CA	80.5%
ID3	69%
ID3 using Ada Boost	75%
C4.5	73.67%
C4.5 by AdaBoost	79%
ID3 by genetic algorithm	74.67%
C4.5 by genetic algorithm	76.67%

5 Conclusion

There is a lack of efficient mechanism to track fraud transaction. The main objective proposed in this paper was to find an effective solution for credit card fraud detection. The main emphasis used in this approach was to combine an unsupervised machine learning technique so that comparable and related entities were combined together. For this purpose, initially the data set taken into consideration was clustered using k-means clustering algorithm. From this stage, the output of k-clusters obtained was then passed to Agglomerative clustering algorithm so as to identify correct grouping between the k clusters. Further, for any test data instance, if it was found to be close to any optimal anomalous cluster, then only for verification LDA (Linear Discriminant Analysis) was executed within that nearest optimal cluster found. It was observed that RA-HC-CA was able to achieve 80.5% accuracy with German dataset taken in the combination of 70% training and 30% testing instances respectively.

References

1. Guha, S., Rastogi, R., Shim, K.: CURE: an efficient clustering algorithm for large databases. In: SIGMOD 1998 Proceedings of the 1998 ACM SIGMOD International Conference on Management of Data, vol. 26, pp. 35–58 (2001)
2. Brause, R., Langsdorf, T., Hepp, M.: Neural data mining for credit card fraud detection. In: Proceedings of the 11th IEEE International Conference on Tools with Artificial Intelligence (2002)
3. Aleskerov, E., Freisleben, B., Rao, B.: CARDWATCH: a neural network based data base mining system for credit card fraud detection. In: Proceedings of the IEEE/IAFE 1997 Computational Intelligence for Financial Engineering (2002)
4. Srivastava, A., Kundu, A., Sural, S., Majumdar, A.: Credit card fraud detection using hidden Markow model. IEEE Trans. Dependable Secure Comput. **5**, 37–48 (2008)
5. Raj, S.B.E., Portia, A.A.: Analysis on credit card fraud detection methods. In: International Conference on Computer, Communication and Electrical Technology (2011)

6. Welling, M.: Fisher linear discriminant analysis. Department of Computer Science, University of Toronto (2005)
7. Ashok, A., Smitha, S., Krishna, M.H.K.: Attribute reduction based anomaly detection scheme by clustering dependent oversampling PCA. In: International Conference on Advances in Computing, Communications and Informatics (2016)
8. Rao, V.M., Singh, Y.P.: Decision tree induction for financial fraud detection using ensemble learning technique. In: Proceeding of the International Conference on Artificial Intelligence in Computer Science and ICT (2013)
9. PrathibhaMol, C.P., Suresh, A., Suresh, G.: Prediction of cardiac arrhythmia type using clustering and regression approach (P-CA-CRA). In: International Conference on Advances in Computing, Communications and Informatics (ICACCI) (2017)
10. Harikumar, S., Surya, P.V.: K-medoid clustering for heterogeneous datasets. Proc. Comput. Sci. **70**, 226–237 (2015)
11. Harikumar, S., Dilipkumar, D.U.: Apriori algorithm for association rule mining in high dimensional data. In: International Conference on Data Science and Engineering (ICDSE), pp. 1–6. IEEE (2016)

Medical Recovery System Based on Inertial Sensors

Silviu Butnariu$^{(\boxtimes)}$ ⬥, Csaba Antonya ⬥, and Petronela Ursu

Transilvania University of Brasov, 500036 Brasov, Romania
butnariu@unitbv.ro

Abstract. This paper presents an equipment that can be used in the medical recovery activity, including for people with special needs. The equipment is composed of two human body motion tracking systems. The first system is based on video tracking technology and uses a Kinect sensor. The data captured by this video sensor will be used to animate an avatar from a dedicated software application, after which the patient can analyze the accuracy of real-time movements on a display. The second tracking system uses inertial sensors mounted on the patient's lower limbs. These sensors return Euler angles that are processed, recorded and used for later evaluate. For this analysis, a human body model, using non-deformable segments and joints, able to reproduce human body movements, was developed. This skeleton model can be superimposed on the CATIA software mannequin and can be animated using a macro based on the Euler angles read by the sensors so that it can analyze the patient's movements using data records gathered during recovery session.

Keywords: Biomechanics · Human kinematics · Health monitoring devices

1 Introduction

One of the goals of medical recovery activity is the development of systems and applications based on human body tracking technologies that can be used either in the medical cabinet or at the patient's home. This concept involves following a treatment plan for certain conditions, with/without the direct supervision of the specialist. Exercises can be done at the patient's home or in any other location, with the possibility of strict self-control (due to modern technologies implemented), recording the data of all movements so that they can be further analyzed by the doctor (video or recorded data).

Physiotherapy is represented by: (i) physical exercise, (ii) posture, (iii) occupational therapy, and (iv) massage, ways to help the individual maintain an optimal state of health or treat the various illnesses that occur during life.

Posture is a means by which the body or some body segments are positioned by charging or being voluntarily maintained periods of time with therapeutic or prophylactic ends [1].

Medical recovery is defined as the problem-solving and education process, during which a person with disabilities is helped to reach the best possible level of physical, functional, social and emotional life, with the least restriction [2]. Medical recovery and

© Springer Nature Switzerland AG 2019
A. Abraham et al. (Eds.): IBICA 2018, AISC 939, pp. 395–405, 2019.
https://doi.org/10.1007/978-3-030-16681-6_39

kinetotherapeutic procedures take place in specialized offices with specialized equipment, under the supervision of a specialist, following a physician's treatment procedure. Activity involves the allocation of a long time, moving to the medical office where time-limited exercises involve cost of health care. For a series of exercises, the physiotherapist can use a goniometer or other static devices/equipment that measures angles between human body segments, values that can be compared to other records (especially if there is an affected part and a healthy part); at the same time, there is no solution that involves analyzing the trajectory of the movement. The evaluation process depends on the target joint and pathology (for example, in the absence of neuromotor control, the movement can be fragmented, with a different trajectory and a different rhythm that does not indicate the quality in neuromotor control).

For some illnesses, medical recovery involves physical exercise, according to the specialist's treatment plan, for a longer period of time, which exceeds the time spent in specialized medical practice sessions. In some of these cabinets there are a number of modern equipment: the ERIGO system - a lower limb recovery robot [3]; ARMEO SPRING robot - for upper limbs [4]; ACTIV-K – system for the recovery of knee pathology [5], but which are mechanical, complex, cumbersome devices that can emotionally affect patients with special needs, some of which require fixation, immobilization of limbs [2]. Therefore, it is intended that part of the exercises should be performed in a familiar environment with which the patient is accustomed.

There are some treatments done at the patient's home, usually motricity exercises, based on a recovery plan, executed following tutorials, which may contain a control element and verify the accuracy of the movements. In order to track the movements of the human body, having the possibility to record the data on the trajectories of the body segments, various types of sensors were analyzed. Thus, over the last decade, various researchers have had approaches to research on systems used to track the movement of the human body in various applications: sports, medicine [6–10], the position in day-to-day activities [11].

Several types of sensors used in body tracking have been identified: mechanical, optical, inertial, electromagnetic, ultrasonic [12]. Among these, inertial sensors have a lot of advantages: small size, low power consumption, high precision, possibility to be part of portable equipment [13–15]. This type of sensors can be used in commercial equipment used in various activities: sports (fitness, cycling, running) [16]; medical, recovery [17]; games (Optic - Kinect®, LeapMotion®, inertia - Wii®) [18].

A system based on inertial sensors used for medical recovery exercises was also identified in the Re.Flex system [19]. It consists of a pair of sensors that are applied to the limbs of the human body either on the hand (forearm + arm) or on the leg (thigh and calf), having the ability to measure the angle between the two sensors.

The software runs on a smartphone and contains various preset exercises that can be done and tracked by the patient. However, one can notice the simplicity of this system, by which one parameter can be recorded - the angle between the two sensors. Therefore, in the field of kinetotherapy there are a number of exercises that can complete the treatment plan and can be performed outside the medical cabinet but which must be executed correctly, so controlled. The solution consists of human body tracking systems with real-time recording and control.

2 Equipment

The proposed equipment can be used in recovery/rehabilitation activities, with primary target group patients whose treatments include lower limb/locomotor exercise kinetotherapy exercises.

The equipment contains two equally functioning systems (Fig. 1): the first motion tracking system is based on a Kinect sensor, which will generate a movement of the avatar character from the image on the display, and the second one is able to record the angular values of the patient's movements, viewing them on the smartphone and sending the data to the server for storage/distribution to healthcare professionals in order to verify the accuracy of the exercises.

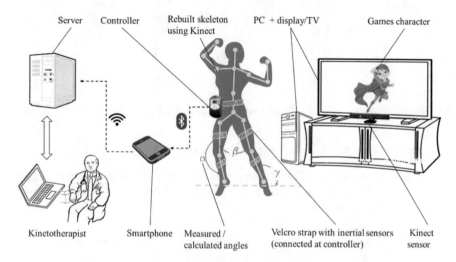

Fig. 1. Developed equipment

2.1 Operating Principle

In this paper, we will focus only on the second system, based on inertial sensors. The human body is considered a skeleton built by unidimensional segments connected to each other by various degrees of coupling (Fig. 2a).

It is considered a system based on 5 sensors: 2 on the thighs, 2 on the calf and one on the basin area. These sensors are fixed to the surface of the body, following its geometry. Prior to use, a configuration of these sensors is required to enable positioning over the virtual skeleton towards them. Each sensor provides information on the inclination of a segment of the locomotor system (thigh - femur or calf), measured in three axes (Fig. 2b).

Euler angles are measured by the sensors around the following axes (Fig. 2): X - denotes pitch (nodding), Y denotes roll and Z denotes yaw (shaking) [20].

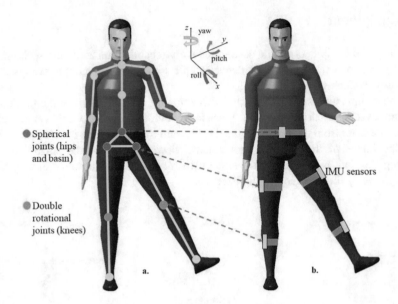

Fig. 2. Skeleton with joints (a); inertial sensors and its correspondence with human joints (b)

2.2   System Architecture, Developed Equipment and Mode of Operation

The system can work in two different ways: the first involves connecting the equipment to a PC - used for setting and calibration (Fig. 3), and the second (Fig. 4) - used for medical recovery exercises.

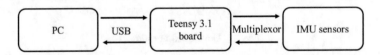

Fig. 3. Data flow of the communication for calibration

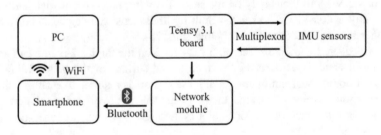

Fig. 4. Data flow of the communication for exercises

Using the equipment shown in Fig. 5, the tests are carried out. The motion tracking system based on the Kinect sensor only serves to provide a dynamic image of an avatar on the display. Inertial sensors from the second motion tracking system detect the inclinations of the human body segments and the data returned by them is in the form of Euler angles.

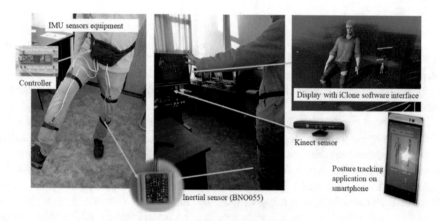

Fig. 5. The developed prototype

The second motion tracking system includes five inertial sensors (IMU: accelerometer, magnetometer and gyroscope) attached to the body through Velcro tapes, according to a controlled scheme and connected to a controller that can communicate wirelessly with a software application on smartphone or tablet. Near the sensors can be mounted buzzers, which can generate small vibrations for warning. The system thus constructed is fully mechanically determined; the inclinations sensed by the sensors, together with the patient's anthropometric data, are enough to define the posture of the locomotor system. Using a WiFi network, the smartphone can download data to a server with a chosen frequency.

The therapeutic function of the device consists in continuous monitoring of patient exercises and alerting the user when needed by a warning signal (e.g. acoustic or luminous signals emitted by the smartphone). On the other hand, an important factor is saving the data on a server, so that the doctor/kinetotherapist can access and study throughout the entire treatment implementation period.

The portable computer (smartphone/tablet) receives information in angles of magnitude in the three directions from the sensors mounted on the human body and reconstructs a virtual parametric model of the patient's posture at a particular moment. This posture is compared with the reference post extracted from the knowledge base represented by certain geometric features, i.e. exercises imposed by the physician/physical therapist. If the differences between the two positions exceed a predetermined tolerance, the system warns the user to correct their exercises.

To facilitate measurements via sensors, they will be carefully attached to the patient's body and calibrated for each use. The kinematics of the virtual model of the locomotor system is reproduced based on the data received from the sensors by creating virtual elements representing the segments of the locomotor system connected by kinematic joints. The patient's posture will not be identified point-by-point, but will be fully determined using a mathematical model based on the entire set of points acquired by the sensors. This mathematical algorithm will be included in the system calibration module.

2.3 Mathematical Model

A kinematic model and a mathematical algorithm that aims to calculate the coordinates of points that represent the articulations of the human locomotor system, having initially measured angles with the help of IMU sensors placed on the legs is presented below.

This model is a development of an older model that did not consider spherical joints at the basin-spine joint and hip joint and the knee was just a simple rotation joint [21]. In order to determine the transformation matrices in the locomotor system's joints, we need to identify the type of these joints (basin to spine, hips and knees). In the references literature several types of kinematic models of the human body were described. For example, according to [22], the hip joint is considered as a spherical coupler and the knee joint as a simple rotation coupler. However, taking into account how the IMU sensors collect the data, we can choose the following joints: connection between basin and spine – spherical joint, hips - spherical joint and knees – double rotational joints (Fig. 6).

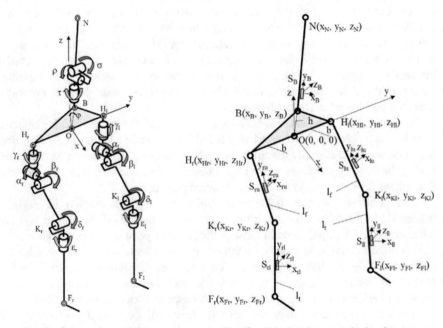

Fig. 6. Kinematic model   **Fig. 7.** Calculated geometrical points

Some notations are made: H – hip, N – neck, B – basin; K – knee, F – foot, S – sensor; at the same time, the following indices will be used: r – right leg, l – left leg.

The scheme with all the joints and nodes coordinates is shown in Fig. 7. The calculus algorithm of the points coordinates is presented only for the right side of the kinematic scheme. On the left side, the calculation is symmetrical.

The human body model is dimensioned based of the anthropometric features of the subject, after the measurements of the following dimensions: body length (l_0), femur length (l_f), tibia length (l_t), the basin's half-width (b), the height of the basin (h), the basin angle from the vertical (φ).

The calculation of the coordinates of K_r point in global coordinates system use the following transformation matrices (see Fig. 7): translation in the Y direction of the point H_r against the center of the global coordinate system O - $Trans\ (y, -b)$; rotation around the X, Y and Z axis in the point H_r - $Rot\ (x, \alpha_r)$, $Rot\ (y, \beta_r)$, $Rot\ (z, \gamma_r)$; translation in the X direction of the point H_r in the point K_r - $Trans\ (x, l_f)$ (Eq. 1).

$$K_r = Trans(y, -b) + Rot(x, \alpha_r) \cdot Rot(y, \beta_r) \cdot Rot(z, \gamma_r) \cdot Trans(x, l_f) \qquad (1)$$

The coordinates of the point F_r will be calculate starting from the coordinates of the point K_r, where there are two rotation couples around the Y and Z-axis, which take into account the distance on the X axis between F_r and K_r (Eq. 2).

$$F_r = K + Rot(y, \delta_r) \cdot Rot(z, \varepsilon_r) \cdot Trans(x, l_t) \qquad (2)$$

The coordinates of the point N will be calculated using the following matrices of rotations and translations: $Rot\ (x, \rho)$ - rotation around the X axis in the point B; $Rot\ (y, \sigma)$ - rotation around the Y axis in the point B; $Rot\ (z, \tau)$ - rotation around the Y axis in the point B: $Trans\ (z, l_0)$ - translation in the Z direction of the point B in the point N (Eq. 3).

$$N = Rot(y, \varphi) \cdot Trans(z, h) + Rot(x, \rho) \cdot Rot(y, \sigma) \cdot Rot(z, \tau) \cdot Trans(z, l_0) \qquad (3)$$

The calculated coordinates of all the points corresponding to the posture of the patient's legs, depending on the position of the basin are presented below (Table 1). These coordinates were calculated by solving Eqs. (1), (2) and (3) in matrices form.

Later, we used these coordinates in various software applications for viewing and analyzing movements (either on the computer or on the smartphone). Note that all these coordinate values contain known data (skeletal dimensions measured before the tests) and measured data (angles measured with inertial sensors).

Table 1. Coordinates of the points

	x	y	z
H_r	0	$-b$	0
H_l	0	b	0
B	$h\sin\varphi$	0	$h\cos\varphi$
N	$h\sin\varphi + l_0\sin\sigma$	$-l_0\sin\rho\cos\sigma$	$h\cos\varphi + l_0\cos\rho\cos\sigma$
K_r	$l_f\cos\beta_r\cos\gamma_r$	$-b + l_f(\sin\alpha_r\sin\beta_r\cos\gamma_r + \cos\alpha_r \sin\gamma_r)$	$l_f(-\cos\alpha_r\sin\beta_r\cos\gamma_r + \sin\alpha_r \sin\gamma_r)$
K_l	$l_f\cos\beta_l\cos\gamma_l$	$b + l_f(\sin\alpha_l\sin\beta_l\cos\gamma_l + \cos\alpha_l \sin\gamma_l)$	$l_f(-\cos\alpha_l\sin\beta_l\cos\gamma_l + \sin\alpha_l \sin\gamma_l)$
F_r	$l_f\cos\beta_r\cos\gamma_r + l_t \cos\delta_r\cos\varepsilon_r$	$-b + l_f(\sin\alpha_r\sin\beta_r\cos\gamma_r + \cos\alpha_r \sin\gamma_r) + l_t\sin\varepsilon_r$	$l_f(-\cos\alpha_r\sin\beta_r\cos\gamma_r + \sin\alpha_r \sin\gamma_r) + l_t\sin\delta_r\cos\varepsilon_r$
F_l	$l_f\cos\beta_l\cos\gamma_l + l_t \cos\delta_l\cos\varepsilon_l$	$b + l_f(\sin\alpha_l\sin\beta_l\cos\gamma_l + \cos\alpha_l \sin\gamma_l) + l_t\sin\varepsilon_l$	$l_f(-\cos\alpha_l\sin\beta_l\cos\gamma_l + \sin\alpha_l \sin\gamma_l) + l_t\sin\delta_l\cos\varepsilon_l$

3 Use of Equipment and Validation

The steps of using the equipment are performed in the following order:

1. Measurement of anthropometric data of the patient (total height, height to waist, femur length, tibia length) by a specialist (physician or physical therapist) using specialized measuring device.
2. Dimensioning of the generic CAD model based on patient measured data.
3. Transferring the treatment data to the smartphone and calibrate the sensors according to the anthropometric characteristics of the patient.
4. Patient training and mounting the equipment on the patient.
5. Starting applications on PC and smartphone, selecting some pre-installed exercises, running them under the automatic control of the sensor system. The data recorded during the exercises are sent for saving on the server.
6. Analyzing the records on the server by the specialist doctor or physiotherapist.

The inertial sensor system will record Euler angles as shown below.

```
|| Euler angles:      X5: 108.00 Y5: -7.23 Z5: -98.07     CALIBRATION: Sys5=3 Gyro5=3 Accel5=3 Mag5=2  ^R
                                                                                                        ^R
```

These data are retrieved and processed to obtain the alpha, beta ... values for the hip, knee and basin joints in all three directions. The equipment requires calibration at startup.

After the system collected the angular joints values, is used a CATIA mannequin assembled according to the above requirements as a skeleton composed of non-deformable bodies and joints between them (Fig. 8). Using the CAD programming language in which the angular values of the IMU sensors were the input data, respecting the reading frequency, we created a macro. Thus, a movie is recorded that reproduces the movements of the human subject.

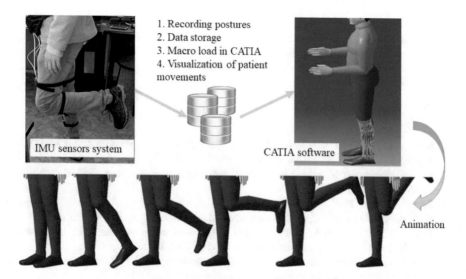

Fig. 8. Mannequin and succession of images obtained by running the macro in CATIA

A sequence of images obtained in CATIA, following the running of a program that contains the recordings made with the equipment based on inertial sensors is presented in Fig. 8. We can see a succession of a knee flexion movement. Motion recording of the CATIA macro application is a movie type. Its accuracy depends on the frequency with which the sensors work.

4   Conclusions

Two important achievements have been completed: (1) theoretically - the kinematic model and the algorithm for determining the posture of the human body and (II) experimental - a prototype of equipment consisting of two systems of human body motion tracking.

The kinematic model refers to the locomotor system, where the human skeleton was considered to be made up of rigid segments connected by spherical or rotating couplings. The calculation algorithm is based on matrices transformations and calculates the coordinates of the knee and ankle points, based on input data (patient anthropometric dimensions and measured angles).

The measured data by sensors are recorded in a *txt* file. In *CATIA* software a mannequin was configured based on the anthropometric measurements of the patient. A macro application has been created to view the movements of this mannequin, having input data the records with measured angle values.

We can say that this method of storing information about a patient's exercise is very cheap, being a simple txt file that can be analyzed as often as needed.

Acknowledgments. The publishing of this paper was supported by the project no. 1804/2018, entitled "SIM-TACK/Real-time motion tracking system for physiotherapy exercises for people with special educational needs" financed by Transilvania University of Brasov, programme "Grants for interdisciplinary teams", competition 2018.

References

1. Means of Modern Physical Therapy. https://kinetogym.wordpress.com/2011/02/18/mijloacele-kinetoterapiei-moderne/. Accessed 01 Jan 2018
2. Medical Recovery (RO - Recuperarea medicala). http://www.recuperaremedicala.com/Page-10.html. Accessed 01 Jan 2018
3. System ERIGO PRO – robot for lower limb recovery (RO - robot pentru recuperarea membrelor inferioare). https://www.hocoma.com/solutions/erigo/. Accessed 01 Jan 2018
4. Robot ARMEO SPRING – for hands (RO - Robot pentru membrele superioare). https://www.hocoma.com/solutions/armeo-spring/. Accessed 01 Jan 2018
5. ACTIV-K – system for recovery of knee pathology. (http://www.axone-med.com/mobilisation-/326-artromot-activ-k-.html). Accessed 01 Jan 2018
6. Ciuti, G., Ricotti, L., Menciassi, A., Dario, P.: MEMS sensor technologies for human centred applications in healthcare, physical activities, safety and environmental sensing: a review on research activities in Italy. Sens. (Basel) **15**, 6441–6468 (2015)
7. Kim, J.-N., Ryu, M.-H., Choi, H.-R., Yang, Y.-S., Kim, T.-K.: Development and functional evaluation of an upper extremity rehabilitation system based on inertial sensors and virtual reality. Intl. J. Distrib. Sens. Netw. **2013**, 1–7 (2013)
8. Leardini, A., Lullini, G., Giannini, S., Berti, L., Ortolani, M., Caravaggi, P.: Validation of the angular measurements of a new inertial-measurement-unit based rehabilitation system: comparison with state-of-the-art gait analysis. J. NeuroEng. Rehabil. **11**, 2–7 (2014)
9. Li, H.T., Huang, J.J., Pan, C.W., Chi, H.I., Pan, M.C.: Inertial sensing based assessment methods to quantify the effectiveness of post-stroke rehabilitation. Sens. (Basel) **15**, 196–209 (2015). https://doi.org/10.3390/s150716196
10. Mohamed, A.A., Baba, J., et al.: Comparison of strain-gage and fiber-optic goniometry for measuring knee kinematics during activities of daily living and exercise. J. Biomech. Eng. **134**, 084502 (2012). https://doi.org/10.1115/1.4007094
11. Moncada-Torres, A., Leuenberger, K., Gonzenbach, R., Luft, A., Gassert, R.: Activity classification based on inertial and barometric pressure sensors at different anatomical locations. Physiol. Meas. **35**, 1245–1263 (2014). https://doi.org/10.1088/0967-3334/35/7/1245
12. Qi, Y., Soh, C.B., Gunawan, E., Low, K.S., Thomas, R.: Lower extremity joint angle tracking with wireless ultrasonic sensors during a squat exercise. Sens. (Basel) **15**, 9610–9627 (2015). https://doi.org/10.3390/s150509610
13. Attal, F., Mohammed, S., Dedabrishvili, M., Chamroukhi, F., Oukhellou, L., Amirat, Y.: Physical human activity recognition using wearable sensors. Sens. (Basel) **15**, 31314–31338 (2015). https://doi.org/10.3390/s151229858
14. Papi, E., Osei-Kuffour, D., Chen, Y.M., McGregor, A.H.: Use of wearable technology for performance assessment: a validation study. Med. Eng. Phys. **37**, 698–704 (2015). https://doi.org/10.1016/j.medengphy.2015.03.017
15. Patel, S., Park, H., Bonato, P., Chan, L., Rodgers, M.: A review of wearable sensors and systems with application in rehabilitation. J. NeuroEng. Rehabil. **9**, 2–17 (2012). https://doi.org/10.1186/1743-0003-9-21

16. Activity Tracker Comparison Chart. http://www.bestfitnesstrackerreviews.com/comparison-chart.html. Accessed 01 Jan 2018
17. Wearable Motivates Physical Activity, http://healthtechinsider.com/tag/exercise/. Accessed 01 Jan 2018
18. Perception Neuron Sensor. http://www.fitness-gaming.com/news/markets/home-fitness/perception-neuron-brings-motion-capture-technology-to-average-consumer.html. Accessed 01 Jan 2018
19. Sistem RE.FLEX bazat pe senzori inerțiali. http://reflex.help/. Accessed 03 Feb 2018
20. Zhao, Y., et al.: An orientation sensor-based head tracking system for driver behaviour monitoring. Sensors 17, 2692 (2017). https://doi.org/10.3390/s17112692
21. Butnariu, S., Mogan, G., Antonya, C., Using inertial sensors in driver posture tracking systems. In: Proceedings of the 4th International Congress of Automotive and Transport Engineering AMMA, October 2018. https://doi.org/10.1007/978-3-319-94409-8_2
22. Otani, T., Hashimoto, K., Miyamae, S., Ueta, H., Natsuhara, A., Sakaguchi, M., Kawaka-mi, Y., Lim, H.-O., Takanishi, A.: Upper-body control and mechanism of humanoids to compensate for angular momentum in the yaw direction based on human running. Appl. Sci. 8, 44 (2018). https://doi.org/10.3390/app8010044

A Comparative Study of Performance and Security Issues of Public Key Cryptography and Symmetric Key Cryptography in Reversible Data Hiding

S. Anagha$^{(\boxtimes)}$, Neenu Sebastian, and K. Rosebell Paul

SCMS School of Engineering and Technology, Ernakulam, India
anaghaajith1995@gmail.com,
neenusebastian@scmsgroup.org, rosebellpaulk@gmail.com

Abstract. Security of data is the main aspect to be considered in the digital network. Data transmission can be made secure by performing reversible data hiding in images. Here the data can be hidden and transmitted inside a host image. Security to the image can be provided by various algorithms like symmetric key algorithm and public key algorithms. This paper provides a comparative study of AES and RSA algorithms for image encryption and reversible data hiding. Data embedding in both cases is done by histogram shifting method. The RSA algorithm can be used for encrypting the image to provide higher security but consumes more time whereas the security of image in AES algorithm is comparatively small but consumes only small amount of time for both encryption and decryption.

Keywords: Encryption · Data hiding · Data embedding

1 Introduction

Communication through digital media plays a vital role in today's network. While digital data transmission security of the data should be strictly maintained. The security can be achieved by data hiding and cryptographic techniques. Data hiding is the process in which the existence of some information is hidden in an another media. The additional secret information can be hidden in a cover media and can be securely transmitted over the network. Cryptography is am method in which the information is converted to another format which is not easily recognisable. The various types of data hiding techniques include digital watermarking, reversible data hiding and steganography. Digital watermarking is a method for verifying authenticity of a user. Additional data can be embedded into the cover media without changing the cover media. Reversible data hiding is a method of embedding secret information into media by modifying its contents. It is a method in which both the media and also the secret data can be retrieved perfectly.

© Springer Nature Switzerland AG 2019
A. Abraham et al. (Eds.): IBICA 2018, AISC 939, pp. 406–412, 2019.
https://doi.org/10.1007/978-3-030-16681-6_40

Reversible data hiding can be performed in images according to two approaches:

(a) Vacating room before encryption (VRBE)
(b) Vacating room after encryption (VRAE).

In the VRBE method the data hiding scheme takes place in the plain text domain. That is before performing image encryption some preprocessing is done on the images and extra space is created. Data hiding takes place to these vacant spaces.

The VRAE method first encrypts the image using some algorithms and create sparse space in encrypted images. So here data hiding takes place in the encrypted domain. The image encryption can be performed using several cryptographic algorithms.

There are symmetric and asymmetric cryptographic algorithms. AES algorithm is one of the example of symmetric key cryptographic algorithm. A shared secret key i used in the system for both encryption and decryption. AES consists of different rounds of operation. The size of the shared key in AES depends on the number of rounds of operation. AES is a block cipher based algorithm where data are processed in blocks of data. Typical key sizes of AES ranges from 128, 192 and 256 bits. RSA is a asymmetric key cryptographic algorithm. It used two keys for encryption and decryption: that is private key and public key [1]. The private key of a user is strictly confidential and it cannot be accessible by any other persons. RSA works based on the prime number. The chosen prime numbers should also be made secret. Compared to AES the RSA algorithm is relatively slow but provided high level of security.

[1] proposes a method in which reversible data hiding is done on encrypted images. Here public key cryptographic algorithm like RSA is used to encrypt the image. Data is hidden into the encrypted image after expanding and shifting the histogram. Using the private key the receiver can decrypt the image and extract the secret data.

[2] A combined lossless and reversible data hiding technique was proposed using public key cryptography. The image is encrypted by the content owner and additional data is hidden using the data hiding key. Wet paper coding is used to replace the cipher text values by new ones. The embedded data can then be extracted from the encrypted domain and the image can be recovered perfectly.

[3] proposed a reversible data hiding technique which is based on two dimensional histogram modification and also difference pair mapping (DPM). we calculate the differences of each pixel pairs. A two dimensional difference is then generated. Secret bits are then embedded using DPM pair.

[4] presented a scheme in the original image is encrypted using stream cipher algorithm. Additional data is embedded by modifying a proportion of encrypted data. The receiver can then decrypt the image to obtain original image and data extraction can be done according to the encryption key.

[5] proposed a method with difference expansion for embedding data. Here the original image is first grouped into pairs of pixels. Quality degradation is very low for this method even after embedding the data. Payload capacity limit, visual quality and complexity can be used to measure the systems performance. The differences between two neighboring pixels are utilized in this system for embedding data.

[6] proposed a technique based on histogram. From the image histogram the peak points and the zero points are found. The range between these points are shifted by one

position to make space for additional data. The receiver can scan the image for data extraction and image recovery.

[7] proposed a lossless generalized LSB technique for data embedding. By using this method the original image can be recovered exactly the same upon extracting the secret data. The data-embedding method used here is the generalization of least significant bit (LSB) modification.

[8] proposed another system in which each pixel in the original image is encrypted by using the XOR operation. The encrypted image is then converted into chunks and each chunk embeds one bit of secret data into the LSB of the image. Side match method is used for finding the smoothness of the image.

[9] proposed a method in which the sender encrypt the uncompressed image with the help of an encryption key. The least significant bit of the image is compressed to makes space for hiding data. The receiver can extract the data using that key but cannot obtain original image.

[8] proposed a different system for reversible data hiding. In this method a pre-processing is done to create room for data embedding before the data encryption. The receiver with the data hiding key can directly extract the data and having encryption key can decrypt the image to obtained original image.

Encryption Using RSA and AES

For efficient reversible data hiding we perform image encryption and data hiding through histogram shifting operation. Image encryption is done using AES algorithm first and the a comparative study is done after encrypting the original image with RSA.

2   Proposed Scheme

The proposed method is a reversible data hiding technique which performs data hiding in encrypted images. For a comparative study image is encrypted using symmetric cryptographic algorithm like AES and asymmetric cryptographic algorithm like RSA. After performing image encryption the image is sent to the data hider who embeds additional secret information using histogram shifting operation. The receiver the on the reception of marked encrypted image can decrypt the image and extract the additional data.

2.1   Image Encryption Using AES Algorithm

The original plain text image is converted into an encrypted image using AES algorithm. The size of the key used in AES algorithm is 128 bits in which there are 10 rounds of operation. The key used for encryption and decryption is shared secret key and is available only to sender and receiver. AES is a block cipher algorithm in which data processing is done as certain blocks [11]. AES algorithms operates on two dimensional array of bytes called state. The state array consists of 4 rows of bytes. Each of the state contains NB bytes where Nb is the block length divided by 32. During the state of encryption process the input data is copied into the state array. After performing encryption operation the result is then transformed into output array. Each round of AES operation consists of four operations.

- Subbytes transformation
- Shift row
- Mix columns
- Add round key.

2.2   Data Embedding

For each encrypted image the histogram is computed. The additional data is embedded into the histogram of the host image. To embed data into the host image, the histogram of the original image was expanded to have space for data embedding. If each pixel value of the original image is doubled, the range of the histogram will be expanded from [0, 255] to [0, 511]. For every histogram the maximum and minimum frequency values are then stored into different arrays. For each row and column the values of histogram are shifted to one position right so that there is enough space for the additional secret data. Then the additional data is embedded into these shifted positions.

2.3   Data Extraction and Image Recovery Using AES Algorithm

Decryption can be performed into the original image to obtain the original image. For decryption inverse operations are performed. The decryption operation also consists of four different operations

- Add round key
- Inverse shiftrows
- Inverse mix column
- Inverse subbyte transformation.

2.4   Image Encryption Using RSA Algorithm

RSA is mainly used for secure data transmission. The image is encrypted using public key and decrypted using the private key. The RSA algorithm mainly includes three stages:

- Key generation
- Encryption
- Decryption.

2.5   Data Embedding

After encrypting the image additional data is inserted to the additional space by histogram shifting operation.

2.6   Data Extraction and Image Recovery

The encrypted image can be decrypted to obtain the original image. The image can be decrypted using the following equation

$$m \equiv c^d (\text{mod } n)$$

3   Experimental Results

Images of different sized is used for data hiding. The data which is to be hidden are binary sequence generated by pseudo random number generator. The amount of time required for encryption and decryption (Figs. 1, 2, 3, 4 and 5).

Fig. 1. Original image [10]

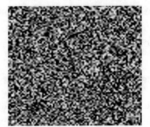

Fig. 2. Embedded image

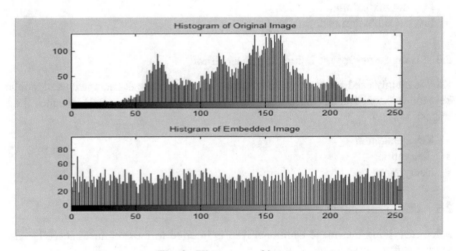

Fig. 3. Histograms of images

Fig. 4. Decrypted image using RSA   **Fig. 5.** Decrypted image using AES

Comparison of AES and RSA Algorithm

From the comparative study we can see that the time taken by RSA algorithm is greater when compared to that of AES. Even though the RSA algorithm takes more time security of images encrypted using the algorithm is higher (Table 1).

Table 1. Comparative study of AES and RSA algorithm

Image size	Time taken in seconds			
	RSA encryption	RSA decryption	AES encryption	AES decryption
28 × 16	15.645578	19.534178	5.148213	2.8319605
50 × 28	28.394322	32.544784	10.155426	7.391726
100 × 100	146.97067	166.428108	53.818500	50.262932
512 × 512	3699.19099	3705.182341	1327.71771	1231.56112

4 Conclusion

Image encryption can be performed using either RSA or AES algorithms. The amount of time required to encrypt and decrypt images while using RSA is higher. AES consumes much lower time. But the security of the image is properly maintained only when using RSA. Decryption of image using AES algorithm produces more distortion to the image.

Acknowledgment. We would like to thank our mini project coordinator without whom this project would not have completed. We would also like to thank SSET for providing the opportunity to make this project successful.

References

1. Li, M., Li, Y.: Histogram shifting in encrypted images with public key cryptosystem for reversible data hiding. Signal Process. **130**, 190–196 (2017)
2. Zhang, X., Wang, J., Cheng, H.: Lossless and reversible data hiding in encrypted images with public key cryptography. IEEE Trans. Circuits Syst. Video Technol. (2015). http://dx.doi.org/10.1109/TCSVT.2015.2433194

3. Lee, X., Zhang, W., Gui, X., Lang, B.: A novel reversible data hiding scheme based on two-dimensional difference-histogram modification. IEEE Trans. Inf. Forensics Secur. **8**(7), 1091–1100 (2013)
4. Zhang, X.: Reversible data hiding in encrypted images. IEEE Signal Process. Lett. **18**(4), 255–258 (2011)
5. Tian, J.: Reversible data embedding using a difference expansion. Trans. Circuits Syst. Video Technol. **13**(8), 890 (2003)
6. Ni, Z., Shi, Y., Ansari, N., Su, W.: Reversible data hiding. IEEE Trans. Circuits Syst. Video Technol. **16**(3), 354–362 (2006)
7. Celik, M.U.: Lossless generalized LSB data embedding. IEEE Trans. Image Process. **14**(2), 253–266 (2005)
8. Hong, W., Chen, T., Wu, H.: An improved reversible data hiding in encrypted images using side match. IEEE Signal Process. Lett. **19**(4), 199–202 (2012)
9. Zhang, X.: Separable reversible data hiding in encrypted image. IEEE Trans. Inf. Forensics Secur. **7**(2), 826–832 (2012)
10. Standard picture collection: gray scale images and color images. http://www.media.cs.tsinghua.edu.cn/_ahz/digitalimageprocess/benchmark.htm. Accessed 1 Jan 2016
11. Advanced Encryption standard (AES). https://nvlpubs.nist.gov/nistpubs/fips/nist.fips.197.pdf

Haptic Device with Decoupled Motion for Rehabilitation and Training of the Upper Limb

Csaba Antonya$^{(\boxtimes)}$ ⓘ, Silviu Butnariu ⓘ, and Claudiu Pozna ⓘ

Transilvania University of Brasov, 500036 Brasov, Romania
antonya@unitbv.ro

Abstract. In this paper, a haptic device is proposed for rehabilitation and training for the upper limb. The proposed design for the haptic device has the main feature of decoupling two type of displacement present in the forward-backward motion: the pulling/pushing of the handle and the downwards displacement. The latter displacement is precisely controlling the strength of the stroke. The proposed haptic device is intended for motor-skill rehabilitation, learning new skills or training alternative. The computation of the actuating torques of the haptic device can be achieved with the hybrid dynamic model, based on the Newton - Euler formalism with Lagrange multipliers. In this modeling approach the position of one of the generalized coordinates (the handle's vertical position, imposed by the user) is known and the corresponding haptic feedback (the pushing/pulling force) is computed. The fast computation of the actuation is increasing the quality of the haptic feedback, allowing dynamic consistency over time.

Keywords: Rehabilitation · Training · Haptic device · Decoupling

1 Introduction

Applications of robotics and virtual reality as therapy aid for patients with post-stroke recovery, spinal cord or brain injury, cerebral palsy, musculo-skeletal deficits play an increasing role in physical rehabilitation [1]. Over the years, several haptic devices were developed for the upper and lower extremity rehabilitation, for improving the shoulder, elbow, wrist, knee, ankle, foot, such as the two degree of freedom MIT Manus robot [2], the ARM-Guide [3], the Mirror Image Movement Enabler [4] etc.

Majority of stroke survivors continue to experience functional limitations in their limbs for several months. Clinical trials demonstrated that therapy and assessment haptic devices for the upper-limb movements following brain injury improves the recovery process [2] and effectively promotes fine motor rehabilitation [5]. Also, compared to motor skill training with just 3D visual feedback, the haptically trained users for kinesthetic tasks presented significantly higher performance [6].

The haptic devices are important training alternative for users wishing to increase their skills, performing repetitive practice of a motor skill [7]. The quality of the

© Springer Nature Switzerland AG 2019
A. Abraham et al. (Eds.): IBICA 2018, AISC 939, pp. 413–422, 2019.
https://doi.org/10.1007/978-3-030-16681-6_41

feedback is an important aspect [8], the main contribution to the overall system quality is given by the mechanical and control elements.

The proposed haptic device is intended for training and rehabilitation exercises, it has bi-directional decupled motion for precisely controlling the pulling/pushing haptic feedback and it is maximizing the range of motions for the wrist and elbow throughout the exercise. Adding other virtual reality features, like visual and acoustic feedback or posture tracking, the rehabilitation and training can be made more interesting, because the process can be seen as a form of interactive video game-play. The desired virtual reality environment is intended for motor-skill rehabilitation, learning new skills or training alternative for novices.

2 Haptic Interfaces for Rehabilitation and Training

A haptic device is a human-computer interface using a force-reflecting instrument for tactile sensation of interaction with virtual objects. Several applications have been developed in the contexts of medical and dental training, education, flight simulators, rehabilitation, etc. Medical training simulators are becoming an important training alternative [8], especially those with haptic guidance, where the users are physically guided through the ideal prerecorded path by the haptic interface [9], adding a supplementary perception channel for trainees.

New technologies such as virtual reality and haptics are becoming useful in post-stroke rehabilitation performed on daily bases or other brain injury motor skills recovery. Interactive games with haptic feedback enhance motivation levels and are useful in application in the field of neuro-rehabilitation [10, 11].

The performance of a haptic device is characterized by the degree of freedom, resolution, workspace, friction, inertia, and force. The degree of freedom (DOF) is depending on the number of directions they supposed to move, most common are the 3 (three translation) and 6 DOF haptic devices. High translational and rotational resolution, measured as the amount of feedback per distance, provides good performance, while low inertia and high structural stiffness enables quick movements. The force-feedback generation can be achieved with impedance and admittance control, depending on how the device is reacting to force or displacement [12]. The decoupled motion control is dividing the control into several independent axis motions [13], especially for cooperation tasks. The elements of the haptic interface are a computer controlled virtual reality engine, data acquisition boards, motor drivers, and a linkage with the handle which constitutes the manipulated part. The angles of the drivers are measured and the kinematic model is computing the avatar's position on the computer screen, while the controlled motor torque is responsible for generating the proper interaction force with the virtual objects.

3 Functional Requirement and Design of the Haptic Device

The functional and design requirement for the proposed haptic device are the following:

- Two degrees of freedom (DOF) for the mechanical system controlled with separate electric motors to decouple the force feedback in two directions.
- The trajectory of the handle has to be close to linear: mainly linear trajectory for the handle, but also rotation for the handle in the vertical plane (less than 10°) for ergonomic purpose.
- The dimension of the device should be as small as possible to decrease the inertia of the elements.
- Friction in joint should be minimized, by introducing only limited number of mechanical joints and mainly rotational joints.
- The force feedback should be reflecting and reacting to the force input in two directions: the pushing/pulling forward and backward, and vertical pushing. These two forces have to be decoupled – the bigger the downward displacement, the pulling/pushing motion will be harder to achieve.
- The position of the handle has to maximize the range of motions for the wrist and elbow throughout the exercise.
- The passive phase of the training or rehabilitation process can be achieved with a virtual fixture (known in the literature also as active constraint or motion constraint) [14].

Usually, the haptic devices are linkage or string based. The string-based devices have larger workspaces, low inertias, but are not suitable for big force input. The haptic device can be designed as a serial robot with 2 or more DOF, but this has the disadvantage of not being decoupled in the mentioned two directions. For the haptic feedback in two directions, a mechanism with closed kinematic chain is required. A mechanism with 2 DOF will need five mechanical joints with one degree of freedom (revolute joints or translation joints). The possible solutions are presented in Fig. 1 (is possible also to interchange the revolute and translational joint on the same link). From these solutions, according to the requirements and for a quick response, the best design is the one with just revolute joints, Fig. 1(a).

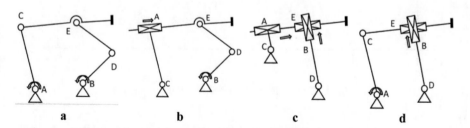

Fig. 1. Design solutions with mechanisms with 2 DOF and five joints

The control of the actuating torques of the haptic device can be achieved with the hybrid dynamic model [15] (imposing the force in one DOF and the displacement in the other) or with classic control theory.

4 The Hybrid Dynamics of the Decoupled Model

The haptic feedback generation requires a high rate of update [16]. The hybrid dynamic [15] is using the Newton-Euler formalism with Lagrange multipliers model and can be applied in case the torque acting on one element of the mechanical system of the haptic device is unknown, but the variation of the generalized coordinate (the displacement) of one element is imposed by the user.

The proposed haptic device is fulfilling the requirement for the development of the hybrid dynamic model: it has two degrees of freedom, the variation of one applied force is known (this is the required force feedback), and the time history of the displacement of one element is imposed (this is the displacement in the vertical plane).

The multi-body model of the haptic device (Fig. 2) has 4 mobile elements, connected with five revolute joints (A, B, C, D, E). The position in the global reference frame xOy of each element is defined by the coordinates of the origins of the local reference frames $x_i O_i y_i$ and the angle of local axes Ox_i and global axis Ox (i = 1...4), so the dynamic model of the haptic device can be described with 12 generalized coordinates (q):

$$[q] = [q_1, \ldots q_{12}] = [x_{O1}, y_{O1}, \varphi_1, \ldots, y_{O4}, \varphi_4]. \tag{1}$$

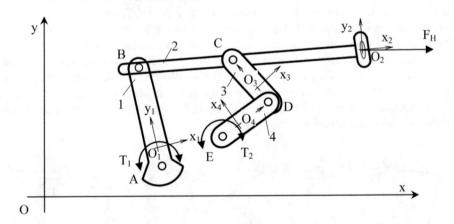

Fig. 2. The multi-body model of the haptic device

The system of equation (with 10 equations, two for each of the five revolute joints) of the geometrical constraints of the relative motion of the connected elements introduced by the revolute joints has the general form (2):

$$F_i(q_1, q_2, \ldots, q_{12}, t) = 0, \quad with\, i = 1\ldots 10. \tag{2}$$

To simulate the motion of this mechanical system and to ensure the desired feedback and trajectories, the Newton-Euler equation with Lagrange multipliers of the dynamic model can be used:

$$[M][\ddot{q}] + [J]^T[\lambda] = [F^A] \tag{3}$$

where $[M] = \begin{bmatrix} m_1 & 0 & \cdots & 0 \\ 0 & m_1 & \cdots & 0 \\ \vdots & \vdots & \ddots & \vdots \\ 0 & 0 & \cdots & I_4 \end{bmatrix}$, $[\ddot{q}] = \begin{bmatrix} \ddot{q}_1 \\ \ddot{q}_2 \\ \vdots \\ \ddot{q}_{12} \end{bmatrix}$, $[J] = \begin{bmatrix} \frac{\partial F_1}{\partial q_1} & \cdots & \frac{\partial F_1}{\partial q_{12}} \\ \vdots & \ddots & \vdots \\ \frac{\partial F_{10}}{\partial q_1} & \cdots & \frac{\partial F_{10}}{\partial q_{12}} \end{bmatrix}$, $\tag{4}$

$$[\lambda] = \begin{bmatrix} \lambda_1 \\ \vdots \\ \lambda_{10} \end{bmatrix}, [F^A] = \begin{bmatrix} F^A_1 \\ \vdots \\ F^A_{12} \end{bmatrix} \tag{5}$$

In these Eqs. (3, 4, 5) m_i is the mass, I_i is the moment of inertia, λ_i is the Lagrange multiplier, F^A_i is the generalized applied force corresponding to the generalized coordinate q_i.

This dynamic Eq. (3) has 12 generalized coordinates and the 12 generalized applied forces containing the gravitational forces, friction forces, the force applied by the user on the handle (F_H) and the torques from the electric motors (T_1 and T_2) responsible for generating the requested feedback.

According to the hybrid dynamic model, the dynamic equations (3) can be rearranged and partitioned into the following system of Eq. (6) where the first i equations incorporate the known applied forces, the next j equation the known generalized coordinates variations, and the remaining k equations the geometrical constraints:

$$\left([M_{ii} \quad M_{ij} \quad M_{ik}] - [J^T_{qi}][J^T_{qk}]^{-1}[M_{ki} \quad M_{kj} \quad M_{kk}] \right) \begin{bmatrix} \ddot{q}_i \\ \ddot{q}_j \\ \ddot{q}_k \end{bmatrix}$$
$$= [F^A_i] + [J^T_{qi}][J^T_{qk}]^{-1}[F^A_k] \tag{6}$$

The hybrid dynamic model of the haptic device will have:

- $i = 4$ (body 2, x direction), is representing the equation with the known applied force, which is the force exerted by the user on the handle (F_H) and is the 4th equation in (3), with $F^A_i = F_H$,
- $j = 5$ (body 2, y direction), is representing the equation with the known generalized coordinate variation (y_{O2}) - the downward displacement of the handle,
- $k = 1, 2, 3, 6, 7, \ldots 12$, the remaining equations from (3).

The hybrid dynamic model (6), combined with (2) - the geometrical constraint and the time history of the vertical displacement of the handle, forms 12 equations with 12 unknowns. Solving these equations will lead to the acceleration of the elements and the unknown applied forces, including T_1 and T_2 - the torques of the actuating motors, and will separate the pulling/pushing movement and the vertical displacement of the handle.

5   Simulation and Evaluation of the Haptic Device

The simulation and evaluation of the haptic device design (Fig. 2) were accomplished using the MSC ADAMS™ (http://www.mscsoftware.com/product/adams) multi-body dynamics software. In the simulation (Fig. 3) the considered lengths of the links measured between two revolute join axes were: $l_1 = 512$ mm, $l_2 = 691$ mm, $l_3 = 349$ mm, $l_4 = 442$ mm.

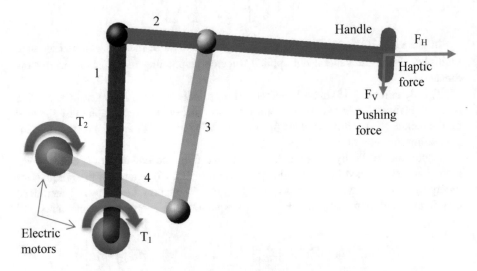

Fig. 3. The ADAMS simulation model of the haptic device

Two simulations were performed for the evaluation of the design of the haptic devise:

- Kinematic simulation, for verifying the correct position of the handle, the required stroke and path.
- Inverse dynamic simulation, for computing the actuating torques in the two electric motors.

In the kinematic simulation, the displacement of the handle was imposed in two directions. In the horizontal plane the displacement is increasing linearly with time, and the vertical displacement is almost constant 80% of the stroke, followed by an increase of approximately 80 mm for ergonomic purpose.

The haptic device is designed to have decoupled motion - the vertical position of the handle is also influencing the pulling/pushing haptic force. From the same start and end point, the vertical displacement of the handle (Fig. 4) will increase the haptic force linearly (for example, 10 mm variation will increase 10% the force).

The variation of the handle's angle (in the vertical plane) is ±5° during the linear movement and at the end of the path – as it is required for ergonomic purpose.

In the inverse dynamic simulation, the movement of the handle was imposed as in the kinematic case, but this time the haptic force was added on the handle. The imposed force variations are presented in Fig. 5 in three scenarios, when the user is increasing the haptic feedback while pushing downwards the handle. The rotation angle variation of the two electric motors was computed from the kinematic simulation for the required path of the handle, and these values were imposed in the inverse dynamic simulation. In this simulation, the torques required for generating the imposed motion could be computed.

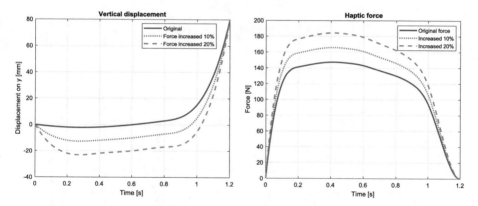

Fig. 4. Vertical displacement of the handle   **Fig. 5.** The haptic force variation during various vertical displacements

The actuating motor torques (T_1 and T_2) are presented in Figs. 6 and 7. The three cases presented are: the original force (with the magnitude shown in Fig. 5), 10% and 20% increase of the original force. It can be seen that proportional to the reduction of the haptic force, the torques of the motors are reduced too.

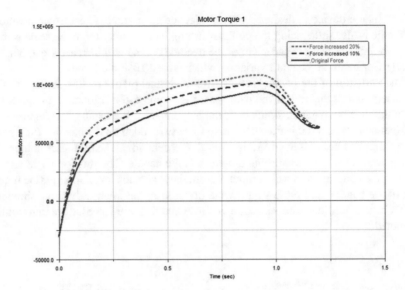

Fig. 6. Actuating motor torque (T$_1$)

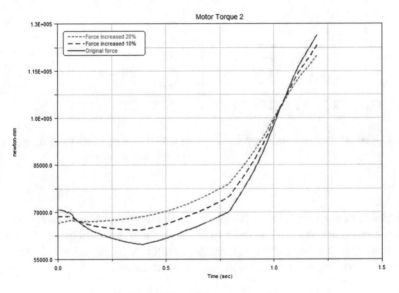

Fig. 7. Actuating motor torque (T$_2$)

6   Conclusions

The proposed structure of the haptic device, as a mechanism with two degrees of freedom, is fulfilling the requirements for a haptic device with decoupled motion for rehabilitation and training of the upper limb. From several solutions, the mechanism with only revolute joint was chosen. The force feedback is reflecting and reacting to the force input in two directions. The force for pulling/pushing of the handle is controlled by the vertical displacement imposed by the user. The hybrid dynamic model is allowing quick computation of the feedback force in the haptic system. This model is using the Newton-Euler formalism with Lagrange multipliers and can be used for computing the actuating torque when the variation of the generalized coordinate (the displacement) of one element is imposed by the user. Because of the simplified dynamic model, the frequency of aliased harmonics can be increased, together with the bandwidth of the force, allowing dynamic consistency over time. The force feedback is generated with two electric motors, where the torques are proportional to the user's force. The proposed haptic device is intended for motor-skill rehabilitation, learning new skills or training. Future work will be to find the ideal position of the revolute joints in order to improve the quality of the force-feedback.

Acknowledgements. The publishing of this paper was supported by the project no. 1804/2018, entitled "SIM-TACK/Real-time motion tracking system for physiotherapy exercises for people with special educational needs" financed by Transilvania University of Brasov, programme "Grants for interdisciplinary teams", competition 2018.

References

1. Burdea, G.: The role of haptics in physical rehabilitation. Haptic Rendering, pp. 517–529 (2008)
2. Krebs, H.I., Ferraro, M., Buerger, S.P., Newbery, M.J., Makiyama, A., Sandmann, M., Hogan, N.: Rehabilitation robotics: pilot trial of a spatial extension for MIT-Manus. J. NeuroEng. Rehabil. **1**(1), 5 (2004)
3. Kahn, L.E., Zygman, M.L., Rymer, W.Z., Reinkensmeyer, D.J.: Robot-assisted reaching exercise promotes arm movement recovery in chronic hemiparetic stroke: a randomized controlled pilot study. J. Neuroeng. Rehabil. **3**(1), 12 (2006)
4. Lum, P.S., Burgar, C.G., Shor, P.C., Majmundar, M., Van der Loos, M.: Robot-assisted movement training compared with conventional therapy techniques for the rehabilitation of upper-limb motor function after stroke. Arch. Phys. Med. Rehabil. **83**(7), 952–959 (2002)
5. Yeh, S.C., Lee, S.H., Chan, R.C., Wu, Y., Zheng, L.R., Flynn, S.: The efficacy of a haptic-enhanced virtual reality system for precision grasp acquisition in stroke rehabilitation. J. Healthc. Eng. **2017**, 9 (2017)
6. Singapogu, R.B., Sander, S.T., Burg, T.C., Cobb, W.S.: Comparative study of haptic training versus visual training for kinesthetic navigation tasks. Stud. Health Technol. Inform. **132**, 469–471 (2008)
7. Shea, C.H., Lai, Q., Black, C., Park, J.H.: Spacing practice sessions across days benefits the learning of motor skills. Hum. Mov. Sci. **19**(5), 737–760 (2000)

8. Escobar-Castillejos, D., Noguez, J., Neri, L., Magana, A., Benes, B.: A review of simulators with haptic devices for medical training. J. Med. Syst. **40**(4), 104 (2016)
9. Feygin, D., Keehner, M., Tendick, R.: Haptic guidance: experimental evaluation of a haptic training method for a perceptual motor skill. In: IEEE Proceedings of the 10th Symposium on Haptic Interfaces for Virtual Environment and Teleoperator Systems, HAPTICS 2002, pp. 40–47 (2002)
10. Ozgur, A.G., Wessel, M.J., Johal, W., Sharma, K., Özgür, A., Vuadens, P., Dillenbourg, P.: Iterative design of an upper limb rehabilitation game with tangible robots. In: ACM/IEEE International Conference on Human-Robot Interaction (HRI), pp. 241–250 (2018)
11. O'Malley, M.K., Gupta, A., Gen, M., Li, Y.: Shared control in haptic systems for performance enhancement and training. J. Dyn. Syst. Meas. Control **128**(1), 75–85 (2006)
12. Lam, P., Hebert, D., Boger, J., Lacheray, H., Gardner, D., Apkarian, J., Mihailidis, A.: A haptic-robotic platform for upper-limb reaching stroke therapy: preliminary design and evaluation results. J. NeuroEng. Rehabil. **5**(1), 15 (2008)
13. Liu, G., Lu, K., Zhang, Y.: Haptic-based training for tank gunnery using decoupled motion control. IEEE Comput. Graph. Appl. **33**(2), 73–79 (2013)
14. Bowyer, S.A., Davies, B.L., y Baena, F.R.: Active constraints/virtual fixtures: a survey. IEEE Trans. Robot. **30**(1), 138–157 (2014)
15. Antonya, C.: Hybrid dynamic model for haptic systems with planar mechanisms. In: 6th IEEE Conference on Robotics, Automation and Mechatronics (RAM), pp. 174–178 (2013)
16. Carignan, C.R., Cleary, K.R.: Closed-loop force control for haptic simulation of virtual environments. Haptics-e **1**(2), 1–14 (2000)

LSTM Approach to Cancel Noise from Mouse Input for Patients with Motor Disabilities

Soham Harnale[✉], Ashwin Vaidya, Aditya Bhide, Aniket Sanap,
and Mangesh V. Bedekar

Department of Computer Engineering,
MAEERs Maharashtra Institute of Technology, Pune, India
soham.harnale@gmail.com, ashwinnitinvaidya@gmail.com,
adityabhide4@gmail.com, aniket.s2509@gmail.com,
mangesh.bedekar@mitpune.edu.in

Abstract. In this paper, we attempt to suppress noise from mouse input due to involuntary tremors from patients with motor disabilities. In order to achieve this task, we used techniques like breakpoint detection and mean filtering to preprocess our data which was then fed into a deep recurrent neural network. Additionally, we tested various architectures and explored existing hardware and software based assistive technologies.

Keywords: LSTM · Motor disabilites · Noise reduction ·
Assistive technologies

1 Introduction

Noise cancelling has applications in various domains. Its major application is in the audio industry. The algorithm relies on adaptive filtering and Fourier transforms. Additionally, neural network models can now cancel noise from audio. However, these depend on preprocessing of the signals using techniques such as short-time Fourier transforms [1].

While various neural network architectures have shown success in noise elimination in audio, there is a distinct lack of available models for removing jitters from computer mouse movements. Such a system would have applications in assisting people with neurodegenerative diseases.

Those with certain neurodegenerative diseases face involuntary tremors in their hands. The tremors make doing everyday tasks a challenge. This becomes an obstacle especially when using a computer. The need to depend on a pointing device such as a mouse or a stylus makes it difficult for people with involuntary tremors to effectively use their computers. For such people, reducing the noise from the pointing device will help them use a computer effectively.

In this paper, we analyze various Neural Network based models and present our findings as well as our hypothesis.

© Springer Nature Switzerland AG 2019
A. Abraham et al. (Eds.): IBICA 2018, AISC 939, pp. 423–430, 2019.
https://doi.org/10.1007/978-3-030-16681-6_42

2   Related Work

Most systems, even in this day and age are built for able-bodied people. For people with motor disorders, the existing solutions can be broadly classified into two parts-hardware and software. Hardware solutions, although effective, may be bulky and lack portability which may lead to additional problems for a person with motor disabilities. Commonly, these hardware devices also tend to be expensive which may be a problem for people who cannot afford such solutions, for example: one of the recent famous mouse adapters AMAneo Assistive Mouse Adapter, costs around $190 [5]. Software solutions, on the other hand, may be free or come at a lower cost as compared to hardware solutions and can be downloaded easily through the internet or from another machine. Some existing software and hardware solutions include:

2.1   Adaptive Path Smoothing Technique via B-Spline [2]

This procedure consists of primarily three different stages:

1. Preliminary stage: Hashem et al. starts off with RBPD (Real-time BreakPoint Detection). Breakpoints provide several characteristics about the user input and are key in smoothing out the user input. Breakpoints can be derived from the x and y coordinates of the user input. The more the number of breakpoints, the larger is the magnitude of the tremor. A breakpoint has to satisfy two conditions:
   (a) There must be a change in the direction of the cursor in the last 0.5 s.
   (b) There must be a large enough change in position in the last 0.5 s.
       The second condition can be satisfied by making use of the following equation:
       $|bx - ax| > m$ or
       $|by - ay| > m$
       (ax, ay) and (bx, by) are two coordinates which are to be compared to determine whether a breakpoint exists. The value of m is variable depending on the degree of the tremor but is set to 4 by default.
2. Automatic adaption stage: Where this method excels in comparison to several other methods is during this stage. While most of the other methods like steadymouse [3] require you to set the degree of tremors yourself, this method automatically adapts to the user input in the first few seconds (less than 5 s). The total number of breakpoints are considered in the first few seconds to decide the value of m. This action is repeated throughout the user interaction and hence the value of m may change over time depending on the number of breakpoints detected.
3. Smoothing stage: This stage consists of two steps:
   (a) Mean filtering: In this step, the aim is to reduce the larger tremors and bring them down to workable levels. As the patients tried to keep the mouse steady, a zig-zag pattern was introduced. This pattern can be reduced significantly using mean filtering. This method uses the previously detected breakpoints- say (ax, ay) and (bx, by) and calculates their mean.

$Bx = (ax + bx)/2$
$By = (ay + by)/2$
(Bx, By) is now taken as the new point and the two previous breakpoints can be ignored.

(b) B-spline: As mean filtering will provide a set of discrete non-continuous points, B-splines are applied to create a smooth curve.

This entire process together is called adaptive path smoothing via B-spline-APSS. After some preliminary testing by a few hospital patients, it was observed that APSS yielded slightly better results in direct comparison with existing solutions like steady mouse. These results, paired with the ease of use of APSS make it a fitting piece of software for patients of Parkinson's disease to adopt, making it easier for them to use a computer.

2.2 A Mouse Adapter for People with Hand Tremor [4]

In hardware-based filtration, a filter can be added in the mouse or in the adapter between mouse and computer to suppress tremors and enable easy use of mouse. The specialised mouse idea is not considered because it is inconvenient and uneconomical to create a special mouse based on each person's individual preference. The adapter also has options to enable and disable the filter. The adapter also has the filter logic so the filtering load does not add to the computer processing. In the filter, the amplitude versus frequency response was calculated using the given equation.

$$A(f) = 1/\sqrt{(1 + (f/f_1)^2)}$$

The adjustment parameter at which filter response is 0.707 of zero frequency response is denoted by f_c. This filter continuously keeps decreasing the extra frequency. The adapter also takes care of the accidental clicks to justify the difference between single and double click, which are quite likely for people with motor disabilities. A small-time delay is introduced during the computation in the adapter. A usability test was conducted on ten people aged thirty-six to seventy-nine. The adapter was tested for eight different mice manufactured by different companies and on different versions of windows and the outcome was satisfactory.

3 Gathering Data

In order to gather the training data for the neural network, we created a simple point tracking game in Unity Engine. The test subjects were made to follow a moving square. The square moved randomly on the screen. This was done in order to increase unpredictability. A log file was generated of the actual coordinates and the recorded mouse coordinates. Before saving, the coordinates were normalized to ensure that the predictions are independent of the resolution of the test and production systems.

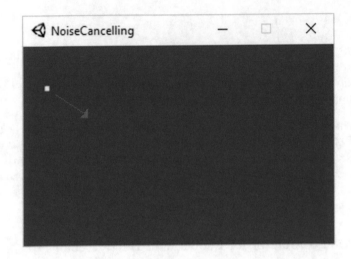

Fig. 1. Game to gather data

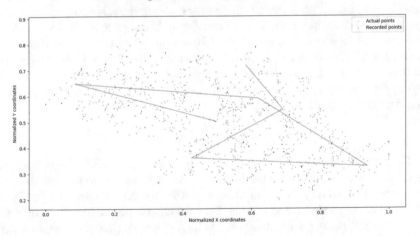

Fig. 2. Sample dataset with actual mouse coordinates vs recorded mouse coordinates

The arrow in the Fig. 1 represents the direction of movement of the square. In Fig. 2 the red line represents the actual movement of the object. The blue points represent the recorded data from the mouse.

3.1   Structure of Log File

ActualX, ActualY refers to the x and y coordinates of the square on the screen. These represent the ideal movement. RecordedX, RecordedY refers to the data captured from the mouse movements (Fig. 3).

ActualX	ActualY	RecordedX	RecordedY
0.5005337	0.5038428	0.49375	0.5
0.5032454	0.5233665	0.60625	0.255
0.5033728	0.5242839	0.60625	0.255
0.5037308	0.5268621	0.60625	0.255
0.5040857	0.5294171	0.60625	0.255
0.5045484	0.5327484	0.60625	0.255
0.5051134	0.5368164	0.615625	0.245
0.5054444	0.5391997	0.61875	0.245
0.5059372	0.5427474	0.628125	0.23

Fig. 3. Log file

4   Architecture

We hypothesise that the neural network can handle the task of noise reduction without the need for employing techniques such as adaptive filtering and Fourier transforms. We tested our hypothesis on various architectures. Furthermore, the reduction of noise in Recurrent Neural Network and Feedforward Network architectures were compared (Fig. 4).

4.1   Single RNN Architecture

The single model predicts both the x and the y coordinates. Our architecture consists of three LSTM layers with 128 hidden units. The recorded x and y coordinates are preprocessed and then fed into the network as the inputs. The last layer is made of a TimeDistributed dense layer with two outputs to predict x and y coordinates. Between each layer, a dropout of 0.2 was applied to prevent overfitting. We tested on multiple architectures. One of the RNN architecture consisted of 3 hidden layers having 128 LSTM cells in each layer. The other consisted of 1024 LSTM units having 4 hidden layers. The architecture for the input and the output followed the single RNN architecture. There were also multiple time steps considered from 4 time steps to 100 time steps.

4.2   Independent Architecture

In this architecture, we trained two separate RNN models. One model predicts the x coordinates whereas the other predicts the y coordinates. Both the recorded x and y coordinates are provided as input to the two models. The hypothesis is that the Neural Network architecture will learn the relationship of only one variable. Thus eliminating the need to perform any transformations.

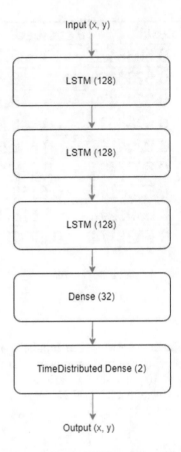

Fig. 4. Single RNN architecture

4.2.1   Recurrent Neural Network Architecture
This model followed the same architecture as the single RNN except for having separate models to predict each coordinate.

4.2.2   Feedforward Architecture
Multiple architectures were used to test the Feedforward. We tested 3 layer network with 128 hidden units and a 4 layer network with 1024 hidden units. Input and the output architecture was same as the others.

5   Data Preprocessing

The data was gathered in the required format making it easy to use and reducing the required preprocessing. However, to increase our accuracy, we took some steps to reduce noise before feeding the data to our model. These steps consisted of:

1. Breakpoint detection: The raw data was accepted as input by this function and it returned all the breakpoints in the data. The breakpoints were detected by the sudden change in the direction of movement of the input device (mouse) as well as the magnitude of the change in movement.
2. Mean filtering: After the breakpoint detection stage, the midpoints between two breakpoints were identified. These midpoints help greatly in reducing the larger tremors. The midpoints were then connected via straight lines which resulted in a path which while reducing the noise from the bigger tremors was jagged and not usable (Fig. 5).

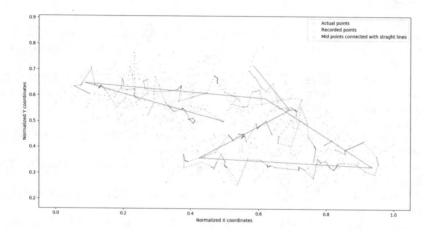

Fig. 5. Output after mean filtering

The output after mean filtering was then fed into our model.

6   Results

We observed that the single RNN model outperformed both the independent RNN and Feedforward models. This shows that predicting both the x and y coordinates from the same model is advantageous. Further, the results indicate that the maximum noise is reduced when the Recurrent Neural Network architecture is employed. This indicates that the sequence of previous mouse coordinates are an essential feature when reducing the noise. We additionally found that there was no significant difference in the accuracy of prediction between the independent RNN and Feedforward models. This re-emphasizes the need to predict both the coordinates from a single model.

7   Conclusion and Future Scope

Our current model does not make use of noise reduction techniques such as Fourier transforms or adaptive filters. Instead, we implemented an End-to-End

Neural Network to minimize the noisy data from the input. Our results indicate that further preprocessing is required before feeding it into the Neural Network. Additionally, Convolutional Neural Networks were not explored during our research. We plan to expand our architecture to include such models. Further, we believe that our model would benefit by preprocessing the data using existing signal processing algorithms and techniques. Our eventual aim is to develop an application which will include live noise cancellation which will enable patients of motor disabilities to successfully interact with computers.

References

1. Simpson, A.J.R., Roma, G., Plumbley, M.D.: Deep karaoke: extracting vocals from musical mixtures using a convolutional deep neural network. arXiv:1504.04658 [cs.SD]
2. Bani Hashem, S.Y., Mat Zin, N.A., Mohd Yatim, N.F., Mohamed Ibrahim, N.: Improving mouse controlling and movement for people with Parkinson's disease and involuntary tremor using adaptive path smoothing technique via B-Spline. Assistive Technol. **26** (2014). https://doi.org/10.1080/10400435.2013.845271
3. Gottemoller, B.: The SteadyMouse project-mouse accessibility software for people with essential tremor (2011). http://www.steadymouse.com/
4. Levine, J.L., Schappert, M.A.: A mouse adapter for people with hand tremor. IBM Syst. J. **44**(3), 621–628 (2005)
5. AMAneo Assistive Mouse Adapter. http://www.inclusive.co.uk/amaneo-assistive-mouse-adapter
6. Stella, M., Begusic, D., Russo, M.: Adaptive Noise Cancellation Based on Neural Network. An adaptive noise canceller adaptively filters a noise reference input to maximally match and subtract out noise or interference from the primary (signal plus noise) input. https://ieeexplore.ieee.org/document/4129919

Advancements in Medical Practice Using Mixed Reality Technology

Adam Nowak[(✉)], Mikołaj Woźniak, Michał Pieprzowski,
and Andrzej Romanowski

Institute of Applied Computer Science, Lodz University of Technology,
Stefanowskiego St. 18/22, 90-924 Lodz, Poland
{203151,210893}@edu.p.lodz.pl, mikolaj@pawelwozniak.eu,
androm@iis.p.lodz.pl

Abstract. Mixed reality technology is one of the most emerging fields
of interactive visual interfaces in recent years. It's evolution from virtual reality systems brings outstanding performance in highly immersive experiences. Furthermore, the MR addresses the most pressing issue
among VR - the space and context awareness of the user. The core of
MR, being a holographic projection displayed onto real-world objects,
highly increases the number of possible applications. In this paper we
discuss the openings for using the leading commercial device for MR -
Microsoft HoloLens - in numerous contexts among medical training and
practice. We also propose a novel way to facilitate the home-based therapy of the lazy eye syndrome. The proposed solution combines the principles of traditional medical methods with increased ease of use and user-
friendliness of exercising. Finally, we discuss the developments required
for the method to become valid for medical practice.

1 Introduction

The recently introduced mixed reality (MR) headsets, such as the Microsoft
HoloLens, became a significant turn from traditional 2D screen-based displaying
towards 3D holographic interfaces. On-head displays proved themselves to be far
more usable and efficient for deep-immersion applications [5]. Mixed reality approach combines virtual and real objects coexisting in a common 3-dimensional
space seamlessly, enabling development of spatial relationships and abstract concepts [7]. In this paper we discuss the directions and openings provided by MR
technology, with focus set on applying those into medical practice. Among the
medical field, MR devices are likely to find broad audience among education.
The holographic displays could be used as a didactic tool, for virtual surgery
and internal organ simulation. Further, the surgical simulation could play piv-
otal role, offering medical adepts an opportunity to practice skills outside the
operating theatre, in a safe and controlled environment [23]. Considering using
HoloLens in real operating room, there is no risk compromising environment
sterility, which is essential on-site [17], since the device is operated through

© Springer Nature Switzerland AG 2019
A. Abraham et al. (Eds.): IBICA 2018, AISC 939, pp. 431–439, 2019.
https://doi.org/10.1007/978-3-030-16681-6_43

hand gestures and voice commands, without employing touch. Training sessions with mixed reality headsets may improve and develop spatial understanding of the 3-Dimensional relationships of musculature, ligamentous supports and neurovasculature, which cannot be efficiently learned from 2D figures [28].

2   Mixed Reality in Common Medical Practice

Extended reality has been investigated by scientists among various branches of medicine. The holographic displays were used as a tool to support diagnostics and an aid in for training purposes on medical universities. One of the most promising didactic applications are simulated surgeries [4,24].

Examination of the liver cancer has also been performed with the use of mixed reality glasses. Shi et al. [21] developed a three-dimensional model of liver cancer, which was created through using HoloLens at the surgical theatre. Magnetic resonance measurements were employed to create a virtual surgery model. Furthermore, the 3D models generated were considered useful both prior and during the surgery.

Anatomic pathology is another field in which the use of mixed reality technology is a promising advancement. The autopsy procedures might be aided through introducing HoloLens, as the enhanced display is desired for enabling more detailed analysis. Hanna et al. [8] proposed a system which enables the pathologist to get access to additional data in real-time, providing annotations and diagrams for extensive analysis. Furthermore, 3D scans of gross specimen can be operated. Providing optimal display usability, the device contributed to more successful reporting of pathological findings.

Monitoring of vital functions can also be supported through employing MR. McDuff presents a prototype of a measurement and visualisation system for analysis of blood flow and vital signs [14]. Real-time sensing is performed in a contactless way, being a blend of image analysis techniques with ballistocardiography and photoplethysmography.

Kuhlemann et al. [12] developed a solution for aiding X-ray endovascular interventions through on-line holographic visualisation. As the catheter tracking is required, while the impact of radiation is considered harmful, there is strong demand for new methods to overview vascular system. Patient's surface and vascular system is mapped and saved through magnetic tracking. The tool proposed has been tested using artificial organism and brought promising results.

Another concept of holographic data visualisation was developed by the Artanim Foundation [3]. Debarba et al. presents system for rehabilitation and medical data analysis based on optical motion capture and virtual human skeleton. Augmented reality technology enables displaying artificial bones in real-time, obtaining the result similar to "X-ray vision". Presented system brings possibility to record the data and save for future examination and analysis. Therefore, the whole rehabilitation process might be supervised and altered accordingly to patient's progress. Assessment of numerous records may help in prevention of injuries and hazardous movements performed. The system may gather broad audience among sport medicine.

3   Proposed Openings

3.1   Towards Patient-Friendly Amblyopia Treatment

The lazy eye syndrome, professionally referred as amblyopia is a vision impairment of a single eye, which is caused by reduced activity of the eyeball. The syndrome is most likely to occur during first years of life. One of the most common reasons of amblyopia is strabismus. A brain rejects the vision from the "lazy eye" despite it is anatomically healthy. Thus the crippled eye executes less and less movements and the disorder gets intensified [10]. However, it is a difficult task to properly diagnose the syndrome. If detected early enough, the chance to inhibit the symptoms and gain visual outcomes from treatment is relatively high. However, disregarding the syndromes can lead to full attenuation of the eye [25]. Among main afflictions, the incorrect depth perception is one of the most cumbersome - preventing the patients from numerous daily activities.

The traditional treatment employs covering the healthy eye with small patch, forcing the brain to use the "lazy one". Further, extraocular muscles shall be exercised. Such approach is not very convenient for the patient and may lead to discouragement and considered tiring [9,22]. Another approach involves atropine eye drops instead of an eye patch [2]. One drop is placed in child's healthy eye each day. Atropine blurs vision in the good eye, which forces patient to use the amblyopic eye more. The novel approaches are being developed thanks to rapid progress of technology. Electronic glasses, filled with LCD lenses can be configured to block out vision in either eye [1]. More advanced solution involves Oculus Rift, an on-head virtual reality device [18,29]. Displayed scenes are split into two images - one for the strong eye and one for the weak eye. Despite numerous drawbacks of the discussed methods, they pose new openings for amblyopia therapy including full control of the distinct displays for each eye (scene, objects, contrast, blur, etc.), see Fig. 1.

Fig. 1. VR Vivid's app: operator control panel view (left), exemplary treatment app screenshot taken during a therapy session (right)

This paper presents another endeavour towards creating an entertaining manner of lazy-eye exercising, that can be easily performed in home environment.

3.2   The HoloLens Implementation

Several solutions for amblyopia therapy are commercially available. From the traditional eye-patch, through LCD glasses, up to cutting edge VR and AR headsets [15]. Each approach leads in several aspects, while falling behind with some others. However, the Microsoft HoloLens device is considered promising as a technological support for lazy-eye exercise. Being the first standalone mixed reality device, it allows the user to move freely in almost any environment. User remains fully aware of the surroundings, thus the risk of collision with real objects is reduced to minimum. It is advantageous over VR headsets, as they might be born with improper calibration and lose of the focus when applied to children [16]. Moreover, the displayed content is adjusted to the environment conditions - the opacity and intensity of the holograms may be changed, but also the position of artifacts might be affected by real objects and obstacles. Yet, the HoloLens provides the most crucial feature required for amblyopia treatment - independent display for each eye. Therefore, it stimulates the lazy eye with appropriate balance, while the healthy eye is still used, so to develop the stereo-vision abilities of the patient (Fig. 2).

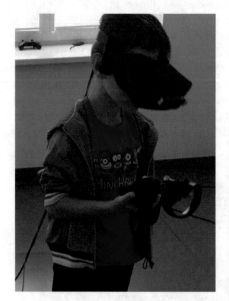 

Fig. 2. Patients wearing VR solution (left) vs MR solution (right)

Another essential aspect, in terms of medical use, is the ability to monitor the therapy, in terms of both the live supervision during treatment sessions, as well as post-performance analysis. Microsoft HoloLens enables streaming the holographic content directly to any web browser through Windows Device Portal web page. If glasses and other computer are connected to the same WiFi network, the live preview is available. Therefore, the patient's progress can be monitored and the tool configured accordingly to the medical requirements of particular user.

The mixed reality devices are capable of scanning and understanding user's environment - detection and recognition of walls, tables, floors and other obstacles and constraints. Therefore, active interaction of virtual and real objects can be performed. Further, the 3D space awareness is maintained - objects put beneath or behind others will be hidden. This features support unconventional approaches, which might be beneficial not only in alleviating the amblyopia itself, but also making the experience of exercising more fun, which is crucial considering the procedure is to be performed mostly by children.

3.3   The Exercise Application

A simple game has been implemented, in which the user tries to build shown construction with use of given pieces of blocks. In the top left part of user's display, the construction scheme can be found. The blueprint is rather opaque not to disturb and annoy user during work, whilst readable and visible. The composition follows user's head movements - it is always visible to the user, no matter in which direction one is gazing. The game implements five different blocks' models (height, width, depth). Each kind of block has individual color to simplify the game play. All available blocks that should be used to build given shape are displayed below. Every construction requires usage of all available blocks. When user selects one block, it starts to follow the user's gaze. Clicking again on the object makes the gravity act for this model and the block falls down, if not set properly on the ground level. The game is divided into levels, containing various constructions. Prior to the gameplay, the device is scanning the surroundings to find planes and surfaces on which user might build. Just after all pieces are placed in correct position with respect to each other, the current level is finished and user proceeds to the next task. Since the accuracy of users' actions is somewhat limited, the system interprets the positions of blocks with sufficient tolerance. The user interface is minimized and kept as simple as possible to concentrate user's attention on the task (Fig. 3).

The artifacts displayed have been divided accordingly to the importance and awareness requirements to the user. Most of the building blocks and scheme are displayed by single eye projection only - the one set to operate for the amblyopic eye, whereas the rest, such as few blocks and navigation panel, are displayed in the healthy-eye part. Therefore, the user's brain is forced to operate the lazy eye to participate in the game. The characteristics of the game imposes high activity of the eye treated. Simultaneously, the healthy eye is not excluded from the perception, unlike in the traditional treatment approach. Therefore, the game acts as an exercise for restoring the proper stereo vision. Moreover, the exercise is presented as a form of entertainment, which makes the treatment process more pleasant, encouraging to greater compliance to the prescribed training routine [13]. Regarding that the lazy-eye syndrome therapy concerns mostly children, it is even more desired to provide an enjoyable manner of exercising. Due to

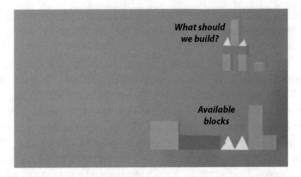

Fig. 3. Main part of the interface, the scheme of the objective and available blocks

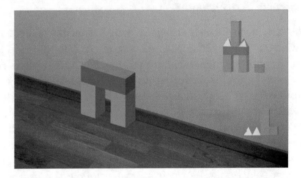

Fig. 4. Few blocks being already placed on the ground level.

long-lasting treatment needed to honestly evaluate our application, we cannot present any structured nor longitudinal study results so far neither in terms of app assessment nor in terms of therapeutic matters (Fig. 4).

4   Discussion

Mixed reality technology becomes increasingly popular among various branches of entertainment, art, science and medicine. Applications among medical field are especially promising, since the interface offers novel ways of stimulating the perception, which were not used in treatment procedures so far. Doctors as well as patients may benefit from wide usage of the mixed reality. Especially, planning and performing surgeries seem to be an invaluable help for professionals. Further, 3D models can facilitate analyzing human's anatomy and make it more accessible, having no constraint of producing costly physical models.

The proposed solution presents novel approach to amblyopia treatment with use of MR. Employing independent display for each eye creates a novel manner of treating lazy eye without harming healthy eye nor stereoscopy, unlike in traditional treatments. Entertaining activity increases the attractiveness of the system, thus regular sessions and examinations are more enjoyable. Furthermore patient is aware of the surroundings and does not require constant supervision during participation.

Convenient way of sharing the projected display and streaming it onto server provides an outstanding aid, whenever the clinician supervision is desired. This feature provides new openings for additional guidance for more complex workout programs, as well as extensive analysis and precise assessment of patient's performance during the exercise. Combined with the possibility of recording the projected sight, it enables further opportunities of exercise task design and accurate examination of conducted treatment.

5   Future Work

Since today, there are no verified clinical results of applications efficiency available. Therefore, the main objective is to perform user studies in order to assess the system proposed. Children participation is required for those studies, both to assess treatment effects and the offered user experience. To provide reliable results, a long-term study with plentiful sample should be performed. Cooperation with doctors could significantly increase efficiency of the system - only regular and supervised medical studies may have quantifiable and credible results in terms of treatment capability.

Further, the game itself may be improved in terms of gameplay as well as in terms of variety. Usage of more engaging and entertaining elements should contribute towards avoiding monotony and boredom during long-period treatments. Keeping in mind that the app is dedicated mainly for children this approach deserves broad consideration. Yet another possibility to extend usability and applicability of such system is to add gaze tracking system to the HoloLens device. Such feature could help doctors to gather more data and understand the behavior of the patient, thus the clinician may plan the therapy better and more accurate.

Following the described methodology of home-based treatment can be beneficial also for other types of impairments, such as coordination issues. Involving movement recognition peripherals, such as EMG sensing bands can broaden the field of possible therapy assistance [27]. we are planning to use eye-tracking systems for further adjustment and tracking of therapy progress, since it could be fairly interesting to see how the eyes perform [26]. Another interesting direction to utilize AR technology and Hololens would be the monitoring and control of industrial reactors to visualize crucial parameters [6,11,19,20].

6   Conclusions

The mixed reality technology is considered promising as an enhancement tool for numerous areas of life, from industry, through business and education, up to medical applications. The proposed amblyopia treatment and exercising method the MR directly, at its current state of technological development. Further medical openings focus on training medical staff and supporting doctors during examinations or surgeries. Numerous advantages of using mixed reality technology, such as lack of sterility's deprecation, visualization stimulating imagination and spatial awareness offer prospects for even greater developments among health care and medical training. Proposed HoloLens-driven software provides novel tool for lazy eye syndrome treatment process. The methodology offers a patient-friendly manner of exercising and supports stereoscopic vision, therefore may be considered as an alternative to the traditional approach. An advanced clinical study is necessary to prove the assumed efficiency, yet the technology seems promising for addressing the issue.

References

1. BenEzra, O., Herzog, R., Cohen, E., Karshai, I., BenEzra, D.: Liquid crystal glasses: feasibility and safety of a new modality for treating amblyopia. Archi. Ophthalmol. **125**(4), 580–581 (2007)
2. de Zárate, B.R., Tejedor, J.: Current concepts in the management of amblyopia. Clin. Ophthalmol. **1**(4), 403–414 (2007). 19668517[pmid]
3. Debarba, H.G., de Oliveira, M.E., Ladermann, A., Chague, S., Charbonnier, C.: Augmented reality visualisation of joint movements for physical examination and rehabilitation. In: Proceeding of 2018 IEEE Conference on Virtual Reality and 3D User Interfaces. IEEE (2018)
4. Gasques Rodrigues, D., Jain, A., Rick, S.R., Shangley, L., Suresh, P., Weibel, N.: Exploring mixed reality in specialized surgical environments. In: Proceedings of the 2017 CHI Conference Extended Abstracts on Human Factors in Computing Systems, CHI EA 2017, New York, NY, USA, pp. 2591–2598. ACM (2017)
5. Grigore, P.C., Burdea, C.: Virtual Reality Technology. Wiley, London (2003)
6. Grudzien, K.: Visualization system for large-scale silo flow monitoring based on ECT technique. IEEE Sens. J. **17**(24), 8242–8250 (2017)
7. Guo, W.: Improving engineering education using augmented reality environment. In: Zaphiris, P., Ioannou, A. (eds.) Learning and Collaboration Technologies. Design, Development and Technological Innovation, pp. 233–242. Springer, Cham (2018)
8. Hanna, M., Ahmed, I., Nine, J., Prajapati, S., Pantanowitz, L.: Augmented reality technology using microsoft hololens in anatomic pathology. Arch. Pathol. Lab. Med. **142**, 01 (2018)
9. Hess, R.F., Mansouri, B., Thompson, B.: A new binocular approach to the treatment of amblyopia in adults well beyond the critical period of visual development. Restorative Neurol. Neurosci. **28**, 1–10 (2010)
10. Holmes, J.M., Clarke, M.P.: Amblyopia. Lancet **367**(9519), 1343–1351 (2006)
11. Grudzien, K., Romanowski, A., Sankowski, D., Williams, R.A.: Gravitational granular flow dynamics study based on tomographic data processing. Part. Sci. Technol. **26**(1), 67–82 (2008)
12. Kuhlemann, I., Kleemann, M., Jauer, P., Schweikard, A., Ernst, F.: Towards X-ray free endovascular interventions - using hololens for on-line holographic visualisation. Healthc. Technol. Lett. **4**, 184–187 (2017)
13. Li, R.W., Ngo, C., Nguyen, J., Levi, D.M.: Video-game play induces plasticity in the visual system of adults with amblyopia. PLOS Biol. **9**(8), 1–11 (2011)

14. McDuff, D.J., Hurter, C., González-Franco, M.: Pulse and vital sign measurement in mixed reality using a hololens. In: VRST (2017)
15. Nowak, A., Wozniak, M., Pieprzowski, M., Romanowski, A.: Towards amblyopia therapy using mixed reality technology. In: 2018 Federated Conference on Computer Science and Information Systems (FedCSIS), pp. 279–282, September 2018
16. Pan, Z., Cheok, A.D., Yang, H., Zhu, J., Shi, J.: Virtual reality and mixed reality for virtual learning environments. Comput. Graphics **30**(1), 20–28 (2006)
17. Pratt, P., Ives, M., Lawton, G., Simmons, J., Radev, N., Spyropoulou, L., Amiras, D.: Through the hololens™ looking glass: augmented reality for extremity reconstruction surgery using 3D vascular models with perforating vessels. Eur. Radiol. Exp. **2**(1), 2 (2018)
18. Qiu, F., Wang, L., Liu, Y., Yu, L.: Interactive binocular amblyopia treatment system with full-field vision based on virtual realty. In: 2007 1st International Conference on Bioinformatics and Biomedical Engineering, pp. 1257–1260, July 2007
19. Romanowski, A.: Big data-driven contextual processing methods for electrical capacitance tomography. IEEE Trans. Ind. Inform. **15**(3), 1609–1618 (2019)
20. Romanowski, A., Grudzien, K., Chaniecki, Z., Wozniak, P.: Contextual processing of ECT measurement information towards detection of process emergency states. In: 2013 13th International Conference on Hybrid Intelligent Systems (HIS), pp. 291–297 (2013)
21. Shi, L., Luo, T., Zhang, L., Kang, Z., Chen, J., Wu, F., Luo, J.: Preliminary use of hololens glasses in surgery of liver cancer. Zhong nan da xue xue bao. Yi xue ban = J. Central South Univ. Med. Sci. **43**, 500–504 (2018)
22. The Pediatric Eye Disease Investigator Group: A randomized trial of atropine vs patching for treatment of moderate amblyopia in children. Arch. Ophthalmol. **120**(3), 268–278 (2002)
23. Turini, G., Condino, S., Parchi, P.D., Viglialoro, R.M., Piolanti, N., Gesi, M., Ferrari, M., Ferrari, V.: A microsoft hololens mixed reality surgical simulator for patient-specific hip arthroplasty training. In: De Paolis, L.T., Bourdot, P. (eds.) Augmented Reality. Virtual Reality, and Computer Graphics, pp. 201–210. Springer, Cham (2018)
24. Vaughan, N., Dubey, V.N., Wainwright, T.W., Middleton, R.G.: A review of virtual reality based training simulators for orthopaedic surgery. Med. Eng. Phys. **38**(2), 59–71 (2016)
25. Waddingham, P., Cobb, S., Eastgate, R., Gregson, R.: Virtual reality for interactive binocular treatment of amblyopia. In: Proceedings of 6th International Conference on Disability, Virtual Reality & Associated Technologies, Esbjerg, Denmark (2006)
26. Wojciechowski, A., Fornalczyk, K.: Exponentially smoothed interactive gaze tracking method. In: International Conference on Computer Vision and Graphics, pp. 645–652. Springer, Cham (2014)
27. Woźniak, M., Pomykalski, P., Sielski, D., Grudzień, K., Paluch, N., Chaniecki, Z.: Exploring EMG gesture recognition-interactive armband for audio playback control. In: 2018 Federated Conference on Computer Science and Information Systems (FedCSIS), pp. 919–923, September 2018
28. Wu, H.-K., Lee, S.W.-Y., Chang, H.-Y., Liang, J.-C.: Current status, opportunities and challenges of augmented reality in education. Comput. Educ. **62**, 41–49 (2013)
29. Žiak, P., Holm, A., Halička, J., Mojžiš, P., Piñero, D.P.: Amblyopia treatment of adults with dichoptic training using the virtual reality oculus rift head mounted display: preliminary results. BMC Ophthalmol. **17**(1), 105 (2017)

Contouring the Behavioral Patterns of the Users of Social Network(ing) Sites and the Need for Data Privacy Law in India: An Application of SEM-PLS Technique

Sandeep Mittal[1]([✉])[iD] and Priyanka Sharma[2][iD]

[1] NICFS (MHA), New Delhi, India
sandeep.mittal@gov.in
[2] I.T. & Telecommunication, Raksha Shakti University, Ahmedabad, India

Abstract. In an era of online social networking, the data privacy is becoming an important topic for researchers to explore. To regulate collection, use and processing of personal data, one needs to understand the behavioral attitude of users of internet so that their need for a law can be assessed. Several studies in this regard have been conducted in other parts of the world where data privacy law is in place, but in India the data privacy was not even a right till recently, when Supreme Court of India declared data privacy as part of the fundamental right of privacy in accordance with constitution of India. The present study is a maiden attempt in India wherein structural equation modelling using partial least square method (SEM-PLS) has been used to analyze the relationship between the attitude of the Indian users of social networking sites (SNSs) and their perception regarding need for data privacy law. The study has validated the model for the correlation between attitude of the users of SNSs and their perception regarding the need for a data privacy law in India.

Keywords: Data privacy · Privacy attitudes · SEM-PLS

1 Introduction

The concern about the privacy of the members of the civil society has been pondering over the minds of the citizens, thinkers, intellectuals, governments and lawmakers alike during the historical past and present, and perhaps would continue to be an important social and individual concern in future in any civil society. The general privacy beliefs are results of complex interaction of social norms and moral value beliefs [1] often mediated in space and time by a number of social variables at individual and collective levels. In real-life social interactions, the individuals have a control over the personal information shared amongst each other. The personal information thus shared in physical world has a limited and slow flow to others and generally dissipates with time with no trace after a relatively reasonable timespan. Its impact on a person's reputation is also relatively limited to a relatively close social-circle.

The rise of the Internet, Web 2.0 and easy availability of smart devices has resulted in an era of privacy development where the use of social network(ing) sites (SNSs) like

A. Abraham et al. (Eds.): IBICA 2018, AISC 939, pp. 440–451, 2019.
https://doi.org/10.1007/978-3-030-16681-6_44

Facebook, LinkedIn, Twitter etc. for exchanging information in virtual space has become the norm [2]. The user generated content, mostly beyond the knowledge and comprehension of SNSs' users, and algorithms of the web aggregating services further worsens the privacy scenario today with SNSs sensing the every breath and the every step one takes in real life.

2   The Literature Review

In course of social interactions in the physical world, while an individual uses his physical senses to perceive and manage threats to his privacy, he has no such social and cultural cues to evaluate the target of self-disclosure in a visually anonymous online space of SNSs. Therefore, while the cognitive management of protection of privacy in offline world is performed unconsciously and effortlessly, deliberate actions are required for effective self-protection are required on SNSs [3]. These deliberative actions can be understood in terms of the "Theory of Planned Behavior" (TPB) which stipulates that "an individual's intention is a key factor in predicting his or her behavior and the intentions are shaped based on attitude, subjective norms and perceived behavioral control" [4–6].

As a general rule, "the more favorable the attitude and subjective norm, and greater the perceived control, the stronger is the person's intention to perform a behavior" [7, 8]. The TPB has been proposed to be a useful and relevant theoretical frame work for online privacy protection on SNSs [4]. However, it is not out of the context to mention that several situational, contextual and demographic factors influence he privacy related attitude, belief and behavior of an individual.

Several theoretical and empirical studies across disciplines have been conducted to understand the attitudes on privacy and data privacy protection laws in jurisdictions worldwide. A few findings relevant to the present work are enumerated below:

1. Information disclosures by users of SNSs are associated with their levels of concern for privacy [9].
2. Users of SNSs are aware of privacy settings and change default settings as per their needs [9].
3. The privacy policies of SNSs help in protecting the privacy of users of SNSs' [10].
4. Disclosure of personal information on SNSs is a bargaining process wherein the perceived benefits and gratifications of networking outweigh the privacy [10].
5. Demographic factors influence the privacy behaviour of users of SNSs [11, 12].

In India, in a maiden study, the scholars tested several hypotheses to understand the thought process on the data privacy law in India, and significant associations have been detected [13]. The high level of concern for privacy was found to be statistically significantly associated with the,

1. Perception that privacy is embedded in the honor and dignity of a person.
2. Perception that privacy should be a constitutional fundamental right.
3. Perception that data privacy should be part of the constitutional fundamental right of privacy.

4. Perception that the right to be forgotten should be a fundamental right.
5. Perception that the right to data privacy is difficult to enforce in a social networking environment.
6. Perception that the stratified right to data privacy is difficult to implement.
7. The "right to data privacy as part of the fundamental right to privacy" seems to be not only statistically significant but also the most important among all the constructs of the thought process on the data privacy debate in India. The right to be forgotten, though statistically highly significant, ranks last among all the constructs of the thought process on the data privacy law in India.

3   The Research Methodology

The population for the present study is the users of the SNSs in India grouped into five strata, namely, Law Enforcement Officers, Judicial and Legal Professionals, Academicians, Information Assurance and Privacy Experts and the Internet Users (other than listed in strata above) in India adopting disproportionate, stratified, purposive, convenience mixed sampling technique, and a statistically adequate sample size of 385 having 95% Confidence Level, 5% Margin of Error (Confidence Interval), 0.5 Standard Deviation and 1.96 Z-score was calculated. A questionnaire was designed for this study by incorporating modified questions based on the Eurobarometer and modified in Indian context and limited to the objectives of the present study [14]. The variables included in the tool can be categorized as nominal and ordinal variables. A pilot study was conducted and reliability of instrument was checked by running reliability analysis which returned a Cronbach Alpha value of 0.700, and modified to adjust the scale and a Cronbach Alpha value of 0.795 obtained which is well within the acceptable norms (>0.700) [15]. All the 401 respondents gave their informed explicit consent signifying their willing participation in this study.

The objective of the present study is to assess theoretical framework for attitudes of users of internet in India and validate the same. For this purpose, the Structural Equation Modelling (SEM)-Partial Least Squares Analysis (PLS) model was used which facilitates the estimation of causal relationships, defined according to a theoretical model, linking two or more latent complex concepts (i.e. the composite indicators), each measured through a number of observable indicators. The basic idea is that complexity inside a system can be studied taking into account a whole of causal relationships among latent concepts, called Latent Variables (LV), each measured by several observed indicators usually defined as Manifest Variables (MV). More specifically, a path model is a relational model with direct and indirect effects among observed variables. SEM has the ability to assess latent variables at the observation level (outer or measurement model) and to test relationships between latent variables on the theoretical level (inner or structural model) [16]. The fully functional trial version of SmartPLS3 was used for the modelling in the present study.

As the study relied upon disproportionate, stratified, purposive, Convenience Sampling, the study may have limitation of non-generalization to wider population, and not taking into account the children presumptively below 18 years of age using the SNSs with fake accounts.

4   Results and Discussions

The cognitive management of the protection of privacy in the offline world is performed unconsciously and effortlessly, deliberate actions are required for effective self-protection on SNSs [4]. These deliberative actions can be understood in terms of the "Theory of Planned Behavior" (TPB) which stipulates that "an individual's intention is a key factor in predicting his or her behavior, and the intentions are shaped on the basis of attitude, subjective norms, and perceived behavioral control [5–7]. The Theory of Planned Behavior was used as the conceptual model (superimposed in red circles) to formulate a 'priori PLS-path model' in present study. The manifest variables (MV) and the latent variables (LV) are shown in yellow solid rectangles and blue solid circles respectively. The part of model shown in blue rectangles is the measurement model while the part shown in green triangle is the structural model (Fig. 1).

The PLS-algorithm was run to make various factors interact and generate the measurement and structural model data. The algorithm run was followed by Boot-strapping which was followed by Blindfolding with standard data settings. The resultant measurement and structural model is shown in Fig. 1 [13].

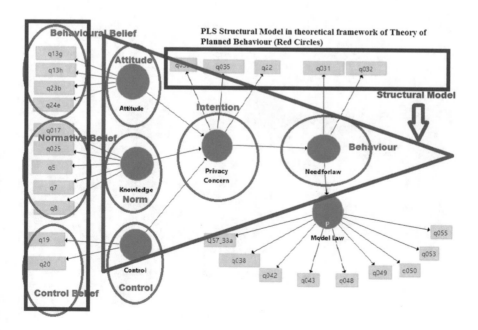

Fig. 1. Priori PLS-PATH model based on Theory of Planned Behavior

The measurement model evaluates the relations between manifest variables (observed items) and latent variables (factors). The measurement model was tested through assessment of validity and reliability of the construct measures in the model. This ensured that only reliable and valid constructs' measures were used for assessing the nature of relationships in the overall model [17]. Structural model specifies relations between latent constructs. Estimating and analyzing the path coefficients between the constructs test the structural model. Path coefficients are indicators of the model's predictive ability [14]. PLS algorithm was applied and the resultant relationships, coefficients and values of loadings are shown in Fig. 2.

4.1 Assessment of PLS Model

4.1.1 Assessment of Measurement Model

Internal Consistency Reliability: As PLS-SEM prioritizes the indicators according to their individual reliability, Cronbach's Alpha generally tends to underestimate the internal consistency reliability due to its presumption that all indicators have equal outer loadings in the construct, the 'composite reliability' is used as a measure of internal consistency reliability in present study [17, 18]. All the latent variables in present study have composite reliability of above 0.60 (Table 1) indicating higher levels of reliability in exploratory research [17, 18].

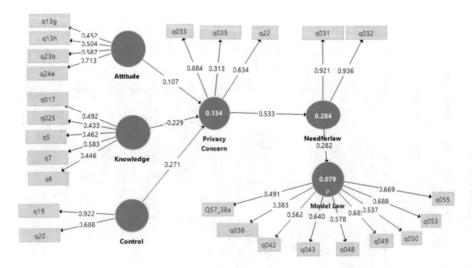

Fig. 2. PLS measurement and structural path model

Table 1. Construct reliability and validity

Latent variable	Composite reliability	Average Variance Extracted (AVE)
Attitude	0.654	0.328
Control	0.793	0.662
Knowledge	0.605	0.237
Model law	0.823	0.347
Need for law	0.926	0.862
Privacy concern	0.661	0.427

Convergent Validity: All the outer loadings, though weak, are significant at 5% significance level (p < 0.05) as is common in Social Science Studies [15]. As earlier noted, the latent variables are not true measurements rather they are treated as observations, therefore, the Average Variance Extracted (AVE) values indicates the percentage of variance explained by construct which are significant for present exploratory study.

Discriminant Validity: Discriminant Validity is the extent to which a construct is truly distinct and unique, capturing phenomena not represented by other constructs in the model. Traditionally, three criteria viz., cross-loading, Fornell-Larcker Criterion and Heterotrait-Monotrait Ratio (HTMT) of correlations are used to establish discriminant validity. Cross-loading fail to indicate a lack of discriminant validity when two constructs are perfectly correlated thus rendering it ineffective in empirical studies [19], therefore, Fornell-Larcker Criterion and Heterotrait-Monotrait Ratio (HTMT) of correlations are used to establish discriminant validity in the present study.

The discriminant validity is well established based on Fornell-Larcker Criterion as the square root of each construct's AVE (on the diagonal of table shown in shaded box of Table 2) is larger than its correlation with the other constructs.

The Heterotrait-Monotrait Ratio (HTMT) of correlations is the ratio of 'between-trait correlation' to the 'within- trait correlations' [19]. An HTMT value above 0.90 indicates lack of discriminant validity. The Heterotrait-Monotrait Ratio (HTMT) values in present study (Table 3) indicate that discriminant validity is well established as all HTMT values are below 0.90 [19].

4.1.2 Structural Model Evaluation

Collinearity Statistics (VIF): First, the structural model has to be checked for collinearity issues by examining the VIF values of all sets of predictor constructs in the structural model. The Table 4 shows the VIF values of all combinations of endogenous constructs (represented by the columns) and corresponding exogenous constructs (represented by rows) [17]. All VIF values are clearly below the threshold of 5. Therefore, collinearity among the predictor constructs is not a critical issue in the structural model and further analysis can be proceeded with.

Let us examine the R^2 values of endogenous latent variables. The R^2 values in are rather weak in the present model (Table 5).

Table 2. Fornell-Larcker Criterion for discriminant validity

	Atti-tude	Control	Knowl-edge	Model law	Need for law	Privacy concern
			Fornell-Larcker Criterion			
Attitude	**0.573**					
Control	0.093	**0.813**				
Knowledge	-0.170	0.142	**0.486**			
Model law	0.196	0.016	-0.167	**0.589**		
Need for law	0.152	0.146	-0.121	0.282	**0.928**	
Privacy concern	0.171	0.249	-0.209	0.440	0.533	**0.654**

Table 3. Hetrotrait-Monotrait Ratio (HTMT)

	Attitude	Control	Knowledge	Model law	Need for law	Privacy concern
Attitude						
Control	0.332					
Knowledge	0.639	0.709				
Model law	0.419	0.140	0.512			
Need for law	0.276	0.217	0.197	0.353		
Privacy concern	0.539	0.581	0.628	0.864	0.855	

Table 4. Collinearity statistic-Inner VIF values in structural model

	Attitude	Control	Knowledge	Model law	Need for law	Privacy concern
Attitude						1.045
Control						1.035
Knowledge						1.057
Model law						
Need for law				1.000		
Privacy concern					**1.000**	

The effect sizes F^2 values for all structural model relationships are tabulated in Table 6. Let us see the effect sizes F^2 for all structural model relationships. While privacy concern has a large effect size on need for law (0.397); the attitude (0.013); control (0.082) and knowledge (0.057) have small effect size on privacy concern. The need for law also has a small effect size on Model Law (0.086).

Table 5. R square values of structural model

	R square adjusted
Model law	0.077
Need for law	0.282
Privacy concern	0.127

Table 6. F square values of structural model

	Attitude	Control	Knowledge	Model law	Need for law	Privacy concern
Attitude						0.013
Control						0.082
Knowledge						0.057
Model law						
Need for law				0.086		
Privacy concern					0.397	

Table 7. Mean, STDEV, T-values, and P-values of path coefficients of structural model

	Original sample (O)	Sample mean (M)	Standard deviation (STDEV)	T Statistics (O/STDEV)	P values
Attitude -> Privacy concern	0.107	0.127	0.045	2.381	0.018*
Control -> Privacy concern	0.271	0.266	0.051	5.308	0.000**
Knowledge -> Privacy concern	−0.229	−0.239	0.051	4.529	0.000**
Need for law -> Model law	0.282	0.281	0.045	6.313	0.000**
Privacy concern -> Need for law	0.533	0.537	0.044	12.143	0.000**

**Significant at 1% level (p < 0.01), *Significant at 5% level (p < 0.05)

Path coefficients: The results for path coefficients are tabulated in Table 7 and shown in Fig. 3. The significance of these path coefficients can be deducted from the p values. All path coefficients are significant at 1% significance level (p < 0.01) except for attitude and privacy concern which is significant at 5% significance level (p < 0.05).

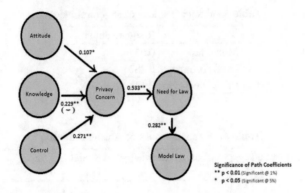

Fig. 3. Significance of path coefficients in structural model

Total Effect: The P-values gives the significance of effect size. All constructs have strong effect significant at 1% (p < 0.01) except for attitude vis a vis model law, which is insignificant and while both attitude vis a vis need for law and attitude vis a vis privacy concern being significant at 5% level of significance (p < 0.05). Looking at the outer weights of the constructs indicators of the structural model, it is (total effects boot strap) seen that all the relationships of the indicators are significant at 5% significance level except q17 (p = 0.161), q25 (p = 0.053), q13 g (p = 0.175) which are not significant.

Predictive Relevance of Model: To ascertain the predictive relevance of structural model blindfolding was run to find Construct Cross-validated Redundancy the results being tabulated in Table 8. The Q2 values of all endogenous latent variables are above zero indicating support for model's predictive relevance regarding the endogenous latent variables.

Table 8. Total construct cross-validated redundancy

	SSO	SSE	Q^2 (=1 − SSE/SSO)
Attitude	1,604.000	1,604.000	
Control	802.000	802.000	
Knowledge	2,005.000	2,005.000	
Model law	3,609.000	3,519.172	0.025
Need for law	802.000	615.517	0.233
Privacy concern	1,203.000	1,155.000	0.040

Based on the results of PLS-SEM studies, the following inferences are drawn (Fig. 3),

1. Existence of a statistically significant positive relationship (significant at 5% level) between 'attitude' of the user of SNSs and his 'privacy concern', i.e., behavioral beliefs of the user improves his concern for privacy.

2. Existence of a statistically significant positive relationship (significant at 1% level) between 'control' of the user of SNSs and his 'privacy concern', i.e., better control of the user improves his concern for privacy.
3. Existence of a statistically significant negative relationship (significant at 1% level) between the 'privacy knowledge' of the user of SNSs and his 'privacy concern', i.e., improved knowledge of the user diminishes his concern for privacy.
4. Existence of a statistically significant positive relationship (significant at 1% level) between the 'privacy concern' of the user of SNSs and his 'perceived need for a data privacy law', i.e., more the privacy concern, greater is the perceived need for a data privacy law.
5. Existence of a statistically significant positive relationship (significant at 1% level) between the 'perceived need for a data privacy law' of the user of SNSs and a 'model law', i.e., more the privacy concern is, the greater is the advocacy for a model data privacy law.

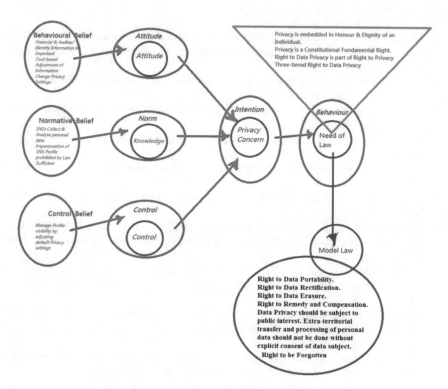

Fig. 4. Summary of SEM-PLS studies

5   Conclusion

To summarize, the PLS-SEM for correlation between behavioral attitude of Indian users of internet, their perceived need for a data privacy law and perception about model data privacy law was evaluated for both measurement and structural models and the proposed model in the present study is well supported in accordance with the established principles of PLS-SEM [17]. This model corroborates well with the earlier findings of present authors [13]. The behavioral, normative and control beliefs of Indian users of social network(ing) sites have significant relationship with their attitude, norms and controls which, in turn, affects their privacy concern to behave in a manner feeling a need for data privacy model law. The summary of the model is depicted in Fig. 4.

References

1. Smith, H.J., Dinev, T., Xu, H.: Information privacy research: an interdisciplinary review. MIS Q. **35**, 989–1016 (2011)
2. Westin, A.F.: Social and political dimensions of privacy. J. Soc. Issues **59**, 431–453 (2003)
3. Rosen, J.: The web means the end of forgetting. In: The New York Times. The Sunday Edition, New York (2010)
4. Yao, M.Z.: Self-protection of online privacy: a behavioral approach. In: Privacy Online, pp. 111–125. Springer (2011)
5. Ajzen, I.: From intentions to actions: a theory of planned behavior. In: Action Control, pp. 11–39. Springer (1985)
6. Ajzen, I.: The theory of planned behavior. Organ. Behav. Hum. Decis. Process. **50**, 179–211 (1991)
7. Ajzen, I., Fishbein, M.: The influence of attitudes on behavior. Handb. Attitudes **173**, 31 (2005)
8. Mittal, S.: Understanding the human dimension of cyber security. Indian J. Criminol. Crim. **34**, 141–152 (2015)
9. Gross, R., Acquisti, A.: Information revelation and privacy in online social networks. In: Proceedings of the 2005 ACM Workshop on Privacy in the Electronic Society, pp. 71–80 (2005)
10. Debatin, B., Lovejoy, J.P., Horn, A.-K., Hughes, B.N.: Facebook and online privacy: attitudes, behaviors, and unintended consequences. J. Comput.-Med. Commun. **15**, 83–108 (2009)
11. Stutzman, F., Kramer-Duffield, J.: Friends only: examining a privacy-enhancing behavior in Facebook. In: Proceedings of the SIGCHI Conference on Human Factors in Computing Systems, pp. 1553–1562 (2010)
12. Tufekci, Z.: Grooming, gossip, Facebook and MySpace: what can we learn about these sites from those who won't assimilate? Inform. Commun. Soc. **11**, 544–564 (2008)
13. Mittal, S., Sharma, P.: A study of the privacy attitudes of the users of the social network(ing) sites and their expectations from the law in India. In: Abraham, A., et al. (eds.) Intelligent Systems Design and Applications, New Delhi, 2018, ISDA 2017, AISC, vol. 736, pp. 1038–1051 (2018). https://doi.org/10.1007/978-3-319-76348-4_100
14. Special Eurobarometer, "359. 2011," Attitudes on Data Protection and Electronic Identity in the European Union, p. 42 (2011)

15. Serbetar, I., Sedlar, I.: Assessing reliability of a multi-dimensional scale by coefficient alpha. Revija za Elementarno Izobrazevanje **9**, 189–195 (2016)
16. Bollen, K.A.: Structural equation models. In: Encyclopedia of Biostatistics, vol. 7 (2005)
17. Hair Jr., J.F., Hult, G.T.M., Ringle, C., Sarstedt, M.: A Primer on Partial Least Squares Structural Equation Modeling (PLS-SEM). Sage Publications, Thousand Oaks (2016)
18. Hulland, J.: Use of partial least squares (PLS) in strategic management research: a review of four recent studies. Strateg. Manag. J. **20**, 195–204 (1999)
19. Henseler, J., Ringle, C.M., Sarstedt, M.: A new criterion for assessing discriminant validity in variance-based structural equation modeling. J. Acad. Mark. Sci. **43**, 115–135 (2015)

Classification of Melanoma Using Different Segmentation Techniques

Savy Gulati$^{(\boxtimes)}$ and Rosepreet Kaur Bhogal

Lovely Professional University, Phagwara, India
savygulati99@gmail.com, rosepreetkaurl2@gmail.com

Abstract. Malignant melanoma is the most dangerous type of skin cancer which has capability to spread to other parts of body and proves deadly for a person. Mortality rates of melanoma are quite high. Also, number of melanoma patients is inclining at a faster pace. In such scenarios, Computer aided systems can act as savior because early detection of melanoma increases chances of survival. These CAD systems are based on image processing techniques. Among these techniques, Segmentation is the most challenging task because melanoma moles appear in random patterns having ragged and irregular shapes. In this work, classification of melanoma and non-melanoma is performed using four different segmentation techniques namely otsu, k-means clustering, maximum entropy and active contour individually. Comparison of results obtained using all these techniques has been carried out and it is seen that among these techniques active contour-based method proves eminent in efficient classification of melanoma and non-melanoma by showing accuracy of 94.5%.

Keywords: Melanoma · Non-melanoma · Segmentation techniques · Classification

1 Introduction

Melanoma is a type of skin cancer which occur in melanocytes. Melanocytes are the cells which produce melanin pigment, this brown coloured pigment is responsible for giving skin its colour [1]. Exposure to Ultraviolet rays is the major reason for occurrence of Malignant melanoma. When sunlight strikes melanocytes, they release excessive melanin which leads to formation of dark, tan moles [2]. These moles set out of control and results in cancerous mass which has the capability to spread to other parts of body [3]. Burden of melanoma is extremely high that death of one person takes place every hour [4]. Although, relieving fact is that melanoma can be treated if detected early [5]. For detection purpose, Computer aided systems can be used which can distinguish between melanoma and non-melanoma without any invasive procedures. These CAD systems are based on the image processing techniques which include: image acquisition, image pre-processing, image segmentation, feature extraction and classification. Out of these steps, segmentation plays vital role as this is the step which credits to accurate differentiation of region of interest that is skin lesion from skin cancer images. Also, effective segmentation directly affects the accuracy of CAD systems. Despite of its importance, segmentation continued to be weakest step in

© Springer Nature Switzerland AG 2019
A. Abraham et al. (Eds.): IBICA 2018, AISC 939, pp. 452–462, 2019.
https://doi.org/10.1007/978-3-030-16681-6_45

melanoma classification task because shape of the melanoma lesions is irregular, ragged and blurry. This is the reason that different authors use different techniques to resolve this purpose of segmentation.

In this work, classification of melanoma is performed using four different types of segmentation techniques namely: otsu method, maximum entropy method, k-means clustering method and active contour method. Results got using these three approaches are compared and efficient segmentation technique for melanoma is decided for the classification task. Section 2 includes literature survey. Section 3 is dedicated to methodology which has been followed. Section 4 involves experimental results and Sect. 5 includes conclusion.

2 Literature Survey

In [6] Amaliah, Fatichah and Widyanto perform segmentation of images with region growing method. To filter noise and thus make images ready for segmentation median filter is used. Then ABCD parameters calculated with the help of these segmented images and on the basis of TDS score classification of melanoma has been performed. Jyothilakshmi and Jeeva [7] segment skin lesion from rest of image using active contour method. Before segmentation task images are refined by employing contrast enhancement and median filtering. By calculating ABCD features from segmented images melanoma has been detected. Bumrungkun, Chamnongthai and Patchoo [8] obtained segmentation of images using snake model in conjunction with SVM. SVM is provided with templates and thus initial information of these templates is used by snake algorithm to segment skin lesions. In [9] Eltayef, Li and Liu removed hairs and reflection artefacts from images. Then segmentation executed using particle swarm optimization and markov random field. Patel, Dhayal, Roy and Shah [10] segment skin cancer images using combination of k means clustering and thresholding based on entropy value. Agarwal, Issac and Dutta [11] performed pre-processing of images which includes XYZ space conversion, median filtering and histogram equalization. Afterwards region growing method used to apply segmentation. As a result of this segmentation two binary masks obtained and on the basis of solidity and extent best out of two binary masks is selected as a final segmented image. In [12] Diniz and Coderio use fuzzy numbers to segment skin lesions from cancerous images. To improve final segmentation results post-processing of images is carried out. Reshma and Shan [13] carried out distinction of skin lesions from images using sobel operator. Then stolz's rule used to classify melanoma and non-melanoma. In [14] Udrea and Mitra implemented segmentation task by utilizing Generative Adversarial neural network. Soumya, Neethu, Niju, Renjini and Aneesh [15] performed removal of hairs with 84 directional filters. Then segmentation of images done by employing active contour method. Afterwards, colour correlogram and texture features extracted to give as input to Bayesian classifier.

Alquran et al. [16] applied pre-processing techniques to refine images followed by lesion segmentation using otsu method. After segmentation step some morphological operations also applied to enhance the images. Then colour, texture and shape features extracted to classify cancerous and non-cancerous images using SVM classifier. Jafari, Samavi, Soroushmehr, Mohaghegh, Karimi and Najarian [17] segment skin lesions with help of k-means clustering. Then colour variation, spatial colour coordination, intensity and colour value extracted to feed as features to ANN classifier. Munia, Alam, Neubert and Fael-Rezai [18] executed segmentation using K-means clustering and Otsu method followed by post processing using morphological operations and guided filter. After this, colour, texture, asymmetry, border and non-linear features extracted and given to four different types of classifier SVM, KNN, Decision tree and Random forest for melanoma classification. Mustafa and Kimura [19] carried out segmentation using GrabCut algorithm. Then find out shape, geometry, border features and given to RBF-SVM for classification of cancerous and non-cancerous mole. In [20] Pathan, Lakshmi and Siddalingaswamy proposed algorithm for hair removal and perform segmentation using Fuzzy C-means clustering then extracted shape features, texture features, and colour variation. These features given to ANN with back propagation training for classification task. In [21] Takruri and Abubakar enhance images by removing hairs and other noise. Then k-means clustering employed for image segmentation followed by extraction of Wavelet transform, Curvelet transform and texture features. These features passed to three parallel SVM for classification of melanoma. Chatterjee, Dey and Munshi [22] carried out median filtering and hair removal using bottom hat for pre-processing. Segmentation is performed by employing morphological operations. Further shape, colour and texture features extracted. Then these features fed to SVM for classification purpose. In [23], Do et al. executed hierarchical segmentation by using combination of otsu and Minimum Spanning Tree method followed by extraction of lesion's colour, border, asymmetry and texture features to classify melanoma and non-melanoma from skin cancer images. Ansari and Sarode [24] carry out pre-processing steps for image refinement. Segmentation of these pre-processed images performed using maximum entropy method. Then GLCM features calculated and fed to SVM classifier for classification. In [25] Wiselin Jiji and Johnson Durai Raj segment skin lesions using CIE L * a * b colour space. After segmentation colour, texture, shape features extracted and given to SVM for classification purpose.

3   Methodology

Aim of this paper is to investigate that which segmentation technique among otsu, K-means clustering, maximum entropy and active contour method proves efficient in accurate classification of melanoma and non-melanoma images. Methodology being followed is mentioned in Fig. 1.

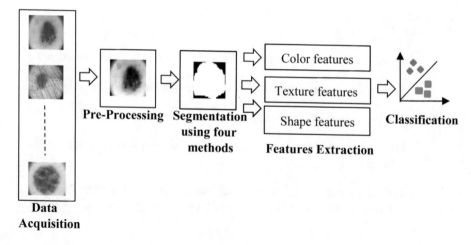

Fig. 1. Methodology

3.1 Segmentation of Affected Portion

Segmentation is used to locate skin lesion in the images that is the region of interest is attained. From literature, it has been seen that otsu, k-means clustering, maximum entropy and active contour methods are most prevalent.

Otsu Method. Firstly, otsu method was applied on the images. It is a threshold-based method in which threshold is chosen in such a way that the intra-class variance is minimum between foreground and background.

K-means Clustering Method. Secondly, we performed segmentation using k-means clustering. It is a kind of unsupervised learning in which groups in data are identified. Algorithm works in iterative fashion and assign each data point among one of the K groups based on distances.

Maximum Entropy-Based Method. Thirdly, we computed segmentation of images using this method. This is also a threshold-based method much similar to otsu technique, but here inter-class entropy is maximized rather than inter-class variance. Clearly it is based on the maximization of information between the foreground and background.

Active Contour Method. In this segmentation method, initial contour is taken and then this contour is adjusted to match to the actual lesion boundaries, resulting in separation of skin lesion from rest of the image.

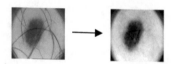

Fig. 2. Result of pre-processing steps. (a) Original skin lesion image, (b) Pre-processed image

---

Algorithm 1 Segmentation

---

1. To make image ready for segmentation pre-processing steps are carried out:

    1.1) First, to remove hairs from skin cancer images Dullrazor [26] is applied.

    1.2) Then convert RGB image to gray scale image.

    1.3) Carry out Contrast enhancement of gray scale image by adjusting the image.

    1.4) Then remove noise from resulting image using median filter. Resultant image obtained after application of pre-processing steps is given in Fig 2.

2. Apply segmentation using above four mentioned techniques. Results of all segmentation methods are given in Fig 3.

3. Results of segmentation are usually not refined, for this purpose utilize image filling Fig 4(c) followed by image opening operation Fig 4(d) to remove unwanted pixels which doesn't lie within the region of interest and to smooth lesion edges respectively.

4. Lastly, select the largest object within the image and crop it. Fig 4(e).

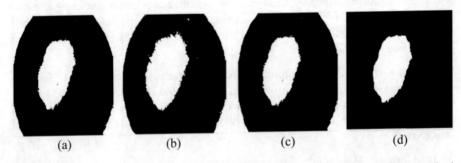

(a)   (b)   (c)   (d)

Fig. 3. Results of segmentation (without morphological operations). (a) Otsu method, (b) Maximum entropy method, (c) k-means clustering method, (d) Active contour method

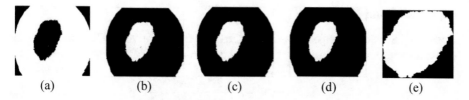

(a)   (b)   (c)   (d)   (e)

Fig. 4. Post segmentation steps. (a) Segmented image using otsu method, (b) Inverted image, (c) Image filling operation applied, (d) Image opening operation applied, (e) Cropped image

3.2 Feature Extraction

Despite of complete image, only those features are selected which result in fruitful classification of melanoma and non-melanoma. In this work, colour, texture and shape features are being extracted. For colour features RGB, HSV and CIE L * a * b colour spaces are utilized. From individual channels of all these three colour spaces four colour moments [27] that is mean, standard deviation, skewness and variance are generated. For texture features GLCM descriptor is used to obtain four prominent texture features [28] namely contrast, energy, homogeneity and correlation. Lastly for shape features area, perimeter, Major Axis Length, Minor Axis Length, circularity index, irregularity index A, irregularity index B, irregularity index C, irregularity index D [29], asymmetry index [30] and border irregularity [30] of lesion are computed. Using all these features feature vector is generated and passed to the classifying stage.

3.3 Classification

Classification is a process which make groups of data having similar properties. For classification purpose supervised classifier support vector machine is employed which finds the hyperplane that best classifies melanoma and non-melanoma images.

4 Experimental Results

4.1 Dataset Used

PH^2 database [31] is being used in this work. This dermoscopic dataset is publicly available and free of cost. Total of 200 images are included out of which 160 are benign and 40 are melanoma. Distribution of images present in database is given in Fig. 5.

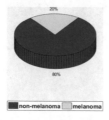

Fig. 5. Distribution of melanoma and non-melanoma images in PH^2 database

4.2 Evaluation Metrics

Accuracy, Sensitivity, Specificity and Area under the curve (AUC) are used for evaluation purpose given in Eqs. (1), (2) and (3) respectively.

$$Accuracy = \frac{TP + TN}{TP + TN + FP + FN} \tag{1}$$

$$Sensitivity = \frac{TP}{TP + FN} \tag{2}$$

$$Specificity = \frac{TN}{TN + FP} \tag{3}$$

- TP (true positive) = positive images recognised as positive.
- TN (true negative) = negative images recognised as negative.
- FP (false positive) = negative images recognised as positive.
- FN (false negative) = positive images recognised as negative.

4.3 Experimental Results

SVM with three-fold cross validation is used for classification purpose. Classification of melanoma and non-melanoma performed using four different types of segmentation techniques individually namely otsu, k-means clustering, maximum entropy and active contour method. Every time shape features are changed as they are extracted from segmented images, whereas texture and color features remain same for all the four-classification tasks. Classification results of these are compared in Table 1. It is evident that among all techniques Active contour-based method yields best results in terms of Accuracy (as seen in Fig. 7), Sensitivity and AUC whereas Maximum entropy method provides an edge when specificity comes into play. Confusion matrices obtained for all four classification tasks performed using different segmentation techniques are given in Fig. 6.

Table 1. Comparison of various segmentation techniques

Segmentation technique	Accuracy (%)	Sensitivity (%)	Specificity (%)	AUC
Otsu method	91.5	67.5	97.5	0.96
K-means clustering	91.5	67.5	97.5	0.96
Maximum entropy	93.5	75	**98.1**	**0.97**
Active contour	**94.5**	**82.5**	97.5	**0.97**

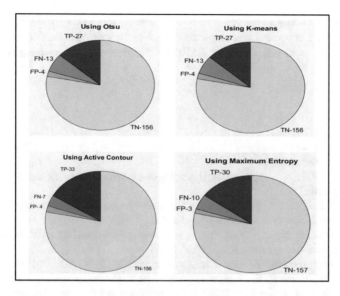

Fig. 6. Confusion matrices

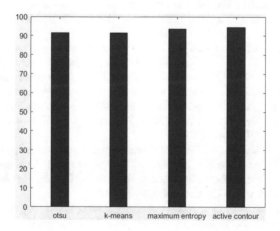

Fig. 7. Accuracy v/s segmentation techniques plot

5 Conclusion

Proper segmentation of skin lesion from dermoscopic images greatly affects the melanoma diagnosis rate of CAD systems. So in this work, classification of malignant melanoma and benign lesions has been carried out using four different segmentation techniques. These four techniques are: otsu method, k-means clustering method, maximum entropy method and active contour method. Classification results obtained using all these techniques are compared and it has been seen that active contour method yields higher accuracy of 94.5% which is more than any other mentioned techniques. Undoubtedly it can be said that shape features obtained using active contour technique in combination with other texture and color features elevate performance of CAD systems

References

1. American Cancer Society. https://www.cancer.org/cancer/melanoma-skin-cancer/about/what-is-melanoma.html
2. LIVESCIENCE. https://www.livescience.com/34783-uv-rays-increase-melanoma-skin-cancer-risk.html
3. MAYOCLINIC. https://www.pharmacytimes.com/perspectives/management-of-melanoma/burden-and-disease-characteristics-of-melanoma
4. Pharmacy Times. https://www.pharmacytimes.com/perspectives/management-of-melanoma/burden-and-disease-characteristics-of-melanoma
5. ScienceDaily. https://www.sciencedaily.com/releases/2017/03/170303091710.htm
6. Amaliah, B., Fatichah, C., Widyanto, M.R.: ABCD feature extraction of image dermatoscopic based on morphology analysis for melanoma skin cancer diagnosis. Jurnal Ilmu Komputer dan Informasi **3**, 82–90 (2012). https://doi.org/10.21609/jiki.v3i2.145
7. Jyothilakshmi, K.K., Jeeva, J.B.: Detection of malignant skin diseases based on the lesion segmentation. In: 2014 International Conference on Communication and Signal Processing, Melmaruvathur, pp. 382–386. IEEE Press (2014) https://doi.org/10.1109/iccsp.2014.6949867
8. Bumrungkun, P., Chamnongthai, K., Patchoo, W.: Detection skin cancer using SVM and snake model. In: 2018 International Workshop on Advanced Image Technology (IWAIT), Chiang Mai, pp. 1–4. IEEE Press (2018). https://doi.org/10.1109/iwait.2018.8369708
9. Eltayef, K., Li, Y., Liu, X.: Lesion segmentation in dermoscopy images using particle swarm optimization and markov random field. In: IEEE 30th International Symposium on Computer-Based Medical Systems (CBMS), Thessaloniki, pp. 739–744. IEEE Press (2017). https://doi.org/10.1109/cbms.2017.26
10. Patel, B., Dhayal, K., Roy, S., Shah, R.: Computerized skin cancer lesion identification using the combination of clustering and entropy. In: International Conference on Big Data Analytics and Computational Intelligence (ICBDAC), Chirala, pp. 46–51. IEEE Press (2017). https://doi.org/10.1109/icbdaci.2017.8070807
11. Agarwal, A., Issac, A., Dutta, M.K.: A region growing based imaging method for lesion segmentation from dermoscopic images. In: 4th IEEE Uttar Pradesh Section International Conference on Electrical, Computer and Electronics (UPCON), Mathura, pp. 632–637. IEEE Press (2017). https://doi.org/10.1109/upcon.2017.8251123
12. Diniz, J.B., Cordeiro, F.R.: Automatic segmentation of melanoma in dermoscopy images using fuzzy numbers. In: IEEE 30th International Symposium on Computer-Based Medical Systems (CBMS), Thessaloniki, pp. 150–155. IEEE Press (2017). https://doi.org/10.1109/cbms.2017.39
13. Reshma, M., Shan, B.P.: Two methodologies for identification of stages and different types of melanoma detection. In: Conference on Emerging Devices and Smart Systems (ICEDSS), Tiruchengode, pp. 257–259. IEEE Press (2017). https://doi.org/10.1109/icedss.2017.8073689

14. Udrea, A., Mitra, G.D.: Generative adversarial neural networks for pigmented and non-pigmented skin lesions detection in clinical images. In: 21st International Conference on Control Systems and Computer Science (CSCS), Bucharest, pp. 364–368 (2017). IEEE https://doi.org/10.1109/cscs.2017.56
15. Soumya, R.S., Neethu, S., Niju, T.S., Renjini, A., Aneesh, R.P.: Advanced earlier melanoma detection algorithm using colour correlogram. In: 2016 International Conference on Communication Systems and Networks (ComNet), Thiruvananthapuram, pp. 190–194 (2016). https://doi.org/10.1109/csn.2016.7824012
16. Alquran, H., Qasmieh, I.A., Alqudah, A.M., Alhammouri, S., Alawneh, E., Abughazaleh, A., Hasayen, F.: The melanoma skin cancer detection and classification using support vector machine. In: IEEE Jordan Conference on Applied Electrical Engineering and Computing Technologies (AEECT), Aqaba, pp. 1–5. IEEE Press (2017). https://doi.org/10.1109/aeect.2017.8257738
17. Jafari, M.H., Samavi, S., Soroushmehr, S.M.R., Mohaghegh, H., Karimi, N., Najarian, K.: Set of descriptors for skin cancer diagnosis using non-dermoscopic color images. In: 2016 IEEE International Conference on Image Processing (ICIP), Phoenix, pp. 2638–2642. IEEE Press (2016). https://doi.org/10.1109/icip.2016.7532837
18. Munia, T.T.K., Alam, M.N., Neubert, J., Fazel-Rezai, R.: Automatic diagnosis of melanoma using linear and nonlinear features from digital image. In: 39th Annual International Conference of the IEEE Engineering in Medicine and Biology Society (EMBC), Seogwipo, pp. 4281–4284. IEEE Press (2017). https://doi.org/10.1109/embc.2017.8037802
19. Mustafa, S., Kimura, A.: A SVM-based diagnosis of melanoma using only useful image features. In: 2018 International Workshop on Advanced Image Technology (IWAIT), Chiang Mai, pp. 1–4. IEEE Press (2018). https://doi.org/10.1109/iwait.2018.8369646
20. Pathan, S., Siddalingaswamy, P.C., Lakshmi, L., Prabhu, K.G.: Classification of benign and malignant melanocytic lesions: a CAD tool. In: International Conference on Advances in Computing, Communications and Informatics (ICACCI), Udupi, pp. 1308–1312. IEEE Press (2017). https://doi.org/10.1109/icacci.2017.8126022
21. Takruri, M., Abubakar, A.: Bayesian decision fusion for enhancing melanoma recognition accuracy. In: International Conference on Electrical and Computing Technologies and Applications (ICECTA), Ras Al Khaimah, pp. 1–4. IEEE Press (2017). https://doi.org/10.1109/icecta.2017.8252063
22. Chatterjee, S., Dey, D., Munshi, S.: Mathematical morphology aided shape, texture and colour feature extraction from skin lesion for identification of malignant melanoma. In: 2015 International Conference on Condition Assessment Techniques in Electrical Systems (CATCON), Bangalore, pp. 200–203. IEEE Press (2015). https://doi.org/10.1109/catcon.2015.7449534
23. Do, T.T., Hoang, T., Pomponiu, V., Zhou, Y., Chen, Z., Cheung, N.M., Koh, D., Tan, A., Hoon, T.: Accessible melanoma detection using smartphones and mobile image analysis. IEEE Trans. Multimedia. https://doi.org/10.1109/tmm.2018.2814346
24. Ansari, U.B., Sarode, T.: Skin cancer detection using image processing. Int. Res. J. Eng. Technol. **4**, 2875–2881 (2017). https://www.irjet.net/archives/V4/i4/IRJET-V4I4702
25. Wiselin Jiji, G., Johnson Durai Raj, P.: An extensive technique to detect and analyze melanoma: a challenge at the international symposium on biomedical imaging (ISBI) (2017). J CoRR. abs/1702.08717. http://arxiv.org/abs/1702.08717
26. Lee, T., Ng, V., Gallagher, R., Coldman, A., McLean, D.: DullRazor: a software approach to hair removal from images. Comput. Biol. Med. **27**, 533–543 (1997)
27. Almansour, E., Arfan Jaffar, M.: Classification of dermoscopic skin cancer images using color and hybrid texture features. Int. J. Comput. Sci. Netw. Secur. (2016). www.ijcrcst.com/iclsit17/IJCRCST-ICLSIT17-29

28. Kolkur, S., Kalbande, D.R.: Survey of texture-based feature extraction for skin disease detection. In: 2016 International Conference on ICT in Business Industry and Government (ICTBIG), Indore, pp. 1–6. IEEE Press (2016). https://doi.org/10.1109/ictbig.2016.7892649

29. Jain, S., Jagtap, V., Pise, N.: Computer aided melanoma skin cancer detection using image processing. In: International Conference on Computer, Communication and Convergence (ICCC 2015), Procedia Computer Science, pp. 735–740 (2015). https://doi.org/10.1016/j.procs.2015.04.209

30. Firmansyah, H.R., Kusumaningtyas, E.M., Hardiansyah, F.F.: Detection melanoma cancer using ABCD rule based on mobile device. In: International Electronics Symposium on Knowledge Creation and Intelligent Computing (IES-KCIC), Surabaya, pp. 127–131. IEEE Press (2017). https://doi.org/10.1109/kcic.2017.8228575

31. Mendonça, T., Ferreira, P.M., Marques, J.S., Marcal, A.R., Rozeira, J.: PH^2 - a dermoscopic image database for research and benchmarking. In: 35th International Conference of the IEEE Engineering in Medicine and Biology Society, Osaka, pp. 3–7. IEEE Press (2013). https://doi.org/10.1109/embc.2013.6610779

A Fuzzy Based Hybrid Firefly Optimization Technique for Load Balancing in Cloud Datacenters

M. Lawanya Shri[1], E. Ganga Devi[2], Balamurugan Balusamy[3],
Seifedine Kadry[4], Sanjay Misra[5(✉)], and Modupe Odusami[5]

[1] VIT University, Vellore, India
lawanyaraj@gmail.com
[2] Loyla College, Chennai, India
smgangadevi@yahoo.co.in
[3] Galgotia University, Greater Noida, India
kadavulai@gmail.com
[4] Department of Mathematics and Computer Science, Faculty of Science,
Beirut Arab University, Beirut, Lebanon
s.kadry@bau.edu.lb
[5] Department of Electrical and Information Engineering,
Covenant University, Ota, Nigeria
{misra.sanjay,
modupe.odusami}@covenantuniversity.edu.ng

Abstract. Cloud computing technology has massive inferences with the use of virtualization technologies. Most of the organizations have incorporated to practice the virtualization strategies to create and operate an effective dynamic data center. The growing maturity of the technologies and utilities of the cloud make the users hasten the adoption of the cloud. The dynamic demanding nature of cloud resources leads to an imbalance in virtual machine utilization and radically increases the energy consumption and operating cost of the data center. In this paper, we propose a fuzzy based hybrid load balancing algorithm for the optimal utilization of virtual machines. The proposed algorithm aim is to reduce the makespan, response time and cost with minimal energy usage and resource wastage. The fuzzy based hybrid optimization approach unveils better performance than existing metaheuristic load balancing algorithms.

Keywords: Cloud computing · Load balancing · Firefly algorithm ·
Fuzzy rules · Simulated annealing algorithm

1 Introduction

Cloud computing is an approach of new automation meant for composite systems with huge scale services which share with various users. In the area of information technology, cloud computing has become the buzzword which grants on-demand access to the combined pool of resources through the internet in a self-service, energetically scalable, and metered manner which results to achieve Green computing. Cloud Computing is known as the next generation's computing infrastructure, and it gives

A. Abraham et al. (Eds.): IBICA 2018, AISC 939, pp. 463–473, 2019.
https://doi.org/10.1007/978-3-030-16681-6_46

multiple advantages to its users in the area of electronic health record system [1]. Enormous adaptable storage and computing devices are also made available via the Cloud [2]. The significant merits of cloud computing lie in its flexibility, availability of service to users and. Those advantages operate the demand for cloud services and the request increments scientific controversies on the Internet of Services (IoS) and in Service Oriented Architectures like applications, which results in high availability and scalability. Several users are drawn towards cloud services as it has a massive pool of combined resources, software packages, and other applications.

The effective use of load balancing will ensure the energy distribution in an improved manner for high traffic situations. Load balancing grants cloud computing to scale up to increasing demands by accurately designating energetic workload among various nodes in the system. In our proposed technique, fuzzy rules are applied in firefly optimization algorithm to deal with the uncertainties that occur during the selection process of Firefly for virtual machines (VM). The workload of each VM is determined, and the VMs are ranked based on the loads. The VM having more loads or overloaded are selected, and the assigning of jobs to distinct VMs are ordered based on priority. The jobs with lower priority are removed from that separate VM to reduce the load. The removed job is called Firefly, and the VM which is having fewer loads are chosen. The disconnected jobs are allocated to that particular VM to align the system.

The paper is structured as follows. The related work is presented in Sect. 2. The methodology and experimentation are demonstrated in Sects. 3 and 4. Finally conclusion drawn is in Sect. 5.

2 Related Work

Load Balancing plays a vital role in developing the rhythm of cloud computing. It distributes the workload available on multiple computers through network links. Many researchers have presented the need of Load Balancing with the current issues while using the cloud. Different Load Balancing existing techniques are discussed in this section. [3] proposed a soft computing based Load Balancing approach using Stochastic Hill Climbing method to designate jobs that are entering into the servers or VMs and the performance is analyzed. This paper compares Round Robin and FCFS algorithm. [4] focused on two Load Balancing algorithms in the cloud which includes Min-Min and Max-Min algorithm. The result of their analysis proved that Max-Min functions better than Min-Min regarding makespan. [5] compared the controversies of existing Load Balancing algorithms on the basis of diverse qualitative metrics. [6] proposed an algorithm which grows to align the load of the cloud infrastructure while minimizing the makespan of a given set of task.

[7] addressed the Global Load Optimization approach which energetically allocates the resources using a load factor for Energy-efficient Load Balancing. [8] discussed the need for Load Balancing and presented several Load Balancing techniques. [9] proposed to decrease the makespan time and enhance the usage of resources and compared

the Min-Min algorithm, FCFS, and SJF in all conditions. [10] presented the Intercloud environment supports scaling of applications across numerous vendor clouds. [11] utilized a weighted signature for a VM level load balancing approach. The fundamental goal of their advanced method is to analyze three existing algorithms to minimize the response time of the user. [12] applied a load balancing for extra tasks which are migrated from overloaded VM to under-loaded VM at different host using particle swarm optimization algorithm.

[13] addressed energy efficient allocation method based on metadata heterogeneity for cloud storage. This paper associate the inactive nodes are switched to low energy mode to reduce the power consumption. [14] proposed an index table to maintain the availability of virtual servers and the sequence of the request. [15] designed a model to solve the load balancing problem by using an Adaptive Firefly Algorithm (ADF) in cloud computing by operating VM scheduling over datacenters. This paper proposed to apply a fuzzy based hybrid firefly optimization algorithm for load balancing in a cloud environment.

3 Methodology

The fireflies for load balancing algorithm acts like a collaborative agent, that balances the workload among cloud resources. The simulated annealing approach is used to hybrid with firefly algorithm to overcome the premature convergence Fuzzy rules entirely control the hybrid algorithm behavior.

3.1 Firefly Algorithm

The major objective of the proposed approach is to design a load balancing model using firefly algorithm, which in turn maximizes the resource usage and gives a good-aligned load among all the resources in cloud servers. Firefly algorithm is a meta-heuristic optimization algorithm that is energized by the flashing behavior of fireflies. The purpose of the flash in Firefly is to operate as a signal system to draw other fireflies. The three major idealized rules of Firefly algorithm are stated below:

Rule one state that all Fireflies are unisexual, and this enables any individual firefly to get attracted to all other fireflies.

Rule two states that attractiveness is proportional to their brightness. Considering any two fireflies, the less bright one will be attracted to the brighter one; however, the intensity decreases as their mutual distance increases.

Rule three states that in case there are no fireflies brighter than a given Firefly, it will move randomly and the brightness of a firefly is determined by the objective function.

Attractiveness and Light intensity are the striking aspects of firefly optimization algorithm. As the distance d increases, then the light intensity I decrease accordingly with respect to base on the inverse-square law. The variation of light intensity I(p) depends on the distance d exponentially.

$$I(d) = I_o e^{-\gamma d^2} \tag{1}$$

Where:

γ = the light absorption rate
I_o = the initial light intensity

Thus, the attractiveness of Firefly is defined by Eq. 2

$$\beta(d) = \beta_o e^{-\gamma d^2} \tag{2}$$

Where:

d = the distance among the fireflies
β_o = the attractiveness at the distance d = 0.
γ = the attractiveness variation.

The Cartesian distance is calculated as shown in Eq. 3

$$d_{ij} = \sqrt{\sum_{k=1}^{n} (x_{ik} - x_{jk})^2} \tag{3}$$

Where: i and j of x_i and x_j = the distance between two fireflies.
X_{ik} = spatial coordinator xi of the kth component.
N = number of dimensions
The attracted movement of one firefly with another is determined by Eq. 4.

$$x_j(v+1) = x_i + \beta_o e^{-\gamma d^2} (x_j - x_i) + \alpha \left(rand - \frac{1}{2}\right) \tag{4}$$

Where: α = the randomization parameter
d = the random number which is distributed uniformly.

3.1.1   Firefly Optimization Algorithm

```
INPUT: population size, maximum iteration
OUTPUT: Optimal solution
1.     Initialization phase: populate fireflies
       x(i=1,2,......n)
2.     Determine light absorption value
3.     Do
4.     For i in 1 to n
5.     For all j in 1 to i
6.     Find the light intensity Ii by f(xi)
7.     If light intensity Ii is greater than Ij then
8.     Move fly i towards the direction j based on the p
       dimensions
9.     End if
10.    Variation of attractiveness with distance d
11.    Calculate new solutions and renew the light
       intensity I
12.    End for i
13.    End for j
14.    Classify each fly and determine the current best
       solution
15.    While stopping condition is not outrun.
```

3.2   Simulated Annealing

A global optimization technique stirred by annealing process is referred to as simulated annealing. It is a metaheuristic technique that uses probabilistic rules. The heating and cooling strategy act like a controller to strengthen the crystal size. The defects in metals are reduced by energy reduction based on room temperature. An initial solution X, and the current solution X* is the starting point for simulated annealing. The solution is generated based on the procedure if and only if the f(X) is greater than f(X*).

$$Prob_i = \exp\left(\frac{-f(X^*) - f(X)}{T_M}\right) \qquad (5)$$

Where

f(X) = the fitness function of the primary solution X
f(X*) = the fitness function of the updated solution

The peak fitness rate of the current solution X* is considered with the well-defined probability $Prob_i$. This specific strategy avoids the solution to dive in local optima. The control parameter TM is used to identify the equilibrium state. The performance of the global search fully depends on the control parameter TM.

$$T_M = \chi^p + T_O + T_{final} \qquad (6)$$

Where χ^p is the descending rate of the control parameter, p is a number of stints. T_0 and T_{final} are the initial and final temperature value.

3.2.1 Simulated Annealing Optimization Algorithm

```
INPUT:  final  temperature,  initial  temperature,  and  Control
parameter or cooling parameter
OUTPUT: Optimal solution
1. Determine the initial solution X
2. Do
3. Develop a current solution X* based on the neighbor solution
   X.
4. Determine the probability
5.Decision for accepting or rejecting the solution based
   on step4
6. Identify the best solution by comparing the neighbor
   solutions.
7. Amend the best one
8. Minimize the Temperature
9.While stopping condition is not outrun.
```

3.3 Fuzzy Hybrid Firefly Optimization for Load Balancing

Load balancing in a cloud environment is addressed by a new approach called fuzzy hybrid firefly optimization algorithm. This approach utilizes a hybrid firefly optimization method based on simulated annealing optimization algorithm to enhance the optimization accuracy and the convergence rate. The algorithm not only balances the workload, it also reduces the energy consumption. This fuzzy approach is used during the selection procedure of virtual machines to allocate the unwanted task. The fuzzy rules are used to establish the control policies of Firefly during the selection process. The behavior of firefly is adopted for a load balancing technique. The dominant flies in the Firefly group are having numerous submissive flies and more brightness Here in the proposed approach, the dominant flies are considered as virtual machines in the data center and submissive flies are termed as jobs or task that is in the data center and submissive flies are termed as jobs or task that is allotted to the virtual machines. If the more number of submissive flies are employed to a particular dominant fly for searching the partner, then the other submissive flies are forwarded to another dominant firefly. So balancing of loads among virtual machines are taken care of by the behavior of the firefly. The simulated annealing algorithm is used in this context to update the solution in every iteration in order to obtain the global optimal solution.

3.4 Hybrid Firefly Optimization Algorithm Based on Simulated Annealing

```
INPUT:    maximum    generation,    population    size,    initial
temperature, final temperature and Control parameter /
cooling rate.
1.   Initialization phase: populate fireflies x(i=1,2,.......n)
2.   Determine light absorption value
3.   Do
4.   For i in 1 to n
5.   For all j in 1 to i
6.   Find the light intensity Ii by f(xi)
7.   If light intensity Ii is greater than Ij then
8.   Move fly i towards the direction j based on the p
     dimensions
9.   End if
10.  Variation of attractiveness with distance d
```

11. $\Delta f = f(X_i^*) - f(X_i)$

12. $U = \exp\left(-\dfrac{\Delta f}{T_M}\right)$

13. $if \; \Delta f \leq 0 \; or \; U > R(0,1)$

14. $X_i \leftarrow X_i^*$

```
15   End if
16.  R(0,1) is random number for Control parameter.
17.  Calculate new solutions and renew the light intensity I
18.  End for
19.  End for
20.  Classify each fly and determine the current best solution
21.  While stopping condition is not outrun.
```

3.5 Problem Formation

G with set P = {p1, p2, ...pg) signifies the real machine in the cloud data center, VMs is denoted as Q and R denotes the jobs. The jobs are appeased to the cloud data center by the cloud users. The datacenter broker makes appropriate decisions to allocate the appeased jobs among virtual machines based on a scheduling algorithm. The main aim of our proposed technique is to align the jobs among virtual machines. The initial task of our work has to deal with the identification of an overloaded virtual machine. The identified virtual machines are stored in a list in descending order. After identifying the virtual machines which are highly overloaded, the jobs are removed from the selected virtual machine. The jobs that are removed have to be allocated to the best suitable VMs that is low loaded. The resources that are not utilized will be considered to migrate the removed jobs. Load of the virtual machine is calculated in Eq. 7

$$Load_{VM} = \frac{Num\,(Job_j)}{VM_i} \tag{7}$$

$$Makespan = MAX\{Completion\_Time(Job_{ij})\} \tag{8}$$

In a VM Makespan is often referred to as the overall completion time of the jobs. The proposed algorithm reduces the completion time of jobs significantly. The selection process of VM to allot the removed task is based on the attractiveness of the firefly. The desirability for identifying the best virtual machine depends on the fuzzy values like low, high. The efficacies of Firefly on selecting the appropriate virtual machine can also term by the fuzzy sets (very very high, very high, high, medium, low, very low, and very very low). The triangular membership functions are formed based on the fuzzy set state variables. The following rules are considered for the firefly to choose an appropriate virtual machine to allocate the removed jobs.

> if β_{ij} is low and $Load_{VM}$ is very very low then
> efficacy of selecting VM is very very low.
> if β_{ij} is medium and $Load_{VM}$ is very low then
> efficacy of selecting VM is very low.
> if β_{ij} is high and $Load_{VM}$ is low then
> efficacy of selecting VM is low.
> if β_{ij} is low and $Load_{VM}$ is medium then
> efficacy of selecting VM is low.
> if β_{ij} is medium and $Load_{VM}$ is medium then
> efficacy of selecting VM is medium.
> if β_{ij} is high and $Load_{VM}$ is medium then
> efficacy of selecting VM is high.
> if β_{ij} is low and $Load_{VM}$ is high then
> efficacy of selecting VM is high.
> if β_{ij} is medium and $Load_{VM}$ is high then
> efficacy of selecting VM is very high.
> if β_{ij} is high and $Load_{VM}$ is high then
> efficacy of selecting VM is very very high.

The fuzzy rule defined in our proposed work not only contains the attractiveness, but also the antecedent part of the heuristic scheme.

4   Experimental Results

The efficacy of our proposed fuzzy-based hybrid firefly optimization algorithm for load balancing is evaluated using cloud sim. It is an effective simulation environment for investigating and simulating. The evaluation of our proposed algorithm based on fuzzy rules are simulated using fuzzy cloud in cloud sim. We have compared and evaluated our algorithm with existing algorithms like particle Swarm optimization (PSO) and Energy-aware Fruit fly algorithm (EFOA-LB). The simulation results reveal that our proposed approach outperforms well compared to another existing approach. Figure 1 depicts the makespan comparison of PSO, HBB-LB, EFOA-LB, and Fuzzy based Firefly Hybrid Algorithm based on Simulated Annealing (FFA-SA). Figure 2 depicts the Response time comparison of HBB-LB, PSO, EFOA-LB, and FFA-SA based on Simulated Annealing. The level of imbalance before and after load balancing is indicated in Fig. 3.

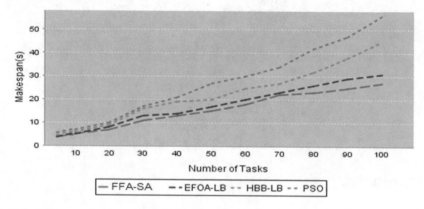

Fig. 1. Comparison of makespan

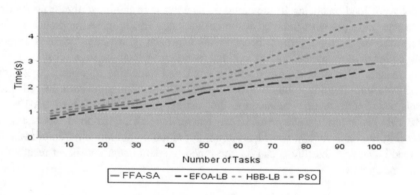

Fig. 2. Comparison of Response time

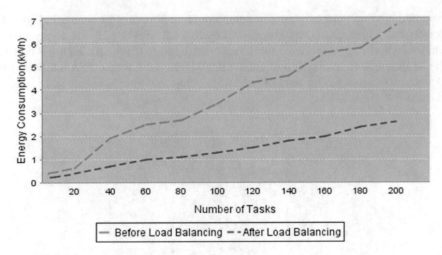

Fig. 3. Degree of imbalance

5 Conclusion

In this study, we have proposed a fuzzy based hybrid firefly optimization technique for load balancing in cloud data centers. The load on the cloud is increasing tremendously and load balancing a major controversy in the cloud computing paradigm. The main objective of load balancing in the cloud data center is to minimize the response time, makespan and cost with minimal resource wastage. The proposed approach is energized from the firefly algorithm because this approach is fast and decreases randomization during the search for an optimal solution, which in turn results in good performance. Also, this approach uses a simulated annealing algorithm to go for a better solution. The fuzzy based hybrid firefly optimization algorithm proved that it enacts better performance compared to the past approaches like PSO, EFOA-LB, and HBB-LB. As a future enhancement, we can include nature-inspired algorithms for balancing the loads amidst virtual machines for cloud computing environments. Also, the load balancing algorithms can be fine-tuned to attain better consistent results from all the perspectives.

Acknowledgments. We acknowledge the support and sponsorship provided by Covenant University through the Centre for Research, Innovation and Discovery (CUCRID).

References

1. Abayomi-Alli, A.A., Ikuomola, A.J., Robert, I.S., Abayomi-Alli, O.O.: An enterprise cloud-based electronic health records system. J. Comput. Sci. **2**(2), 21–36 (2014)
2. Odun-Ayo, I., Omoregbe, N., Odusami, M., Ajayi, O.: Cloud ownership and reliability–issues and developments. In: International Conference on Security, Privacy and Anonymity in Computation, Communication and Storage, pp. 231–240. Springer, Cham, December 2017

3. Mandal, B., Das Gupta, K., Dutta, P.: Load balancing in cloud computing using stochastic hill climbing. Procedia Technol. **4**, 783–789 (2012)
4. Geethu Gopinath, P.P., Vasudevan, S.K.: An in-depth analysis and study of Load balancing techniques in the cloud computing environment. Procedia Comput. Sci. **50**, 427–432 (2015)
5. Mishra, N.K., Mishra, N.: Load balancing techniques: need objectives and major challenges in cloud computing-a systematic review. Int. J. Comput. Appl. (0975-8887) **131**(18), 11–19 (2015)
6. Das Gupta, K., Mandal, B., Dutta, P.: A Genetic Algorithm (GA) based load balancing strategy for cloud computing. Procedia Technol. **10**, 340–347 (2013)
7. Shaji, D.S., Baburaj, E.: Green cloud: an energy efficient load balancing approach using global load optimization. Int. Rev. Comput. Softw. (IRECOS), 9, i8-2955 (2014)
8. Chamoli, N., Suyal, H., Panwar, A.: Load balancing technique in cloud computing: a review. Int. J. Comput. Appl. (0975-8887) **145**(15), 6–10 (2016)
9. Kumar, M., Sharma, S.C.: Dynamic load balancing algorithm for balancing the work load among virtual machine in cloud computing. Procedia Comput. Sci. **115**, 322–329 (2017)
10. Buyya, R., Ranjan, R.: Inter cloud: utility–oriented federation of cloud computing environments for scaling of application services. In: International conference on Algorithms and Architectures for parallel processing, pp. 13–31 (2010)
11. Ajit, M., Vidya, G.V.M.: Level load balancing in cloud environment. In: Proceedings of Fourth International Conference on Computing, Communications and Networking Technologies (ICCCNT), 4 July 2013, pp. 1–5 (2013)
12. Lu, Y., Xie, Q., Kliot, G., Geller, A., Larus, J.R.: Join-Idle-Queue: a novel load balancing algorithm for dynamically scalable web services. Perform. Eval. **68**(11), 1056–1071 (2011)
13. Karakoyunlu, C., Chandy, J.A.: Exploiting user metadata for energy-aware node allocation in a cloud storage system. J. Comput. Syst. Sci. **82**(2), 282–309 (2016)
14. Florence, A.P., Shanthi, V.: A load balancing model using firefly algorithm in cloud computing. J. Comput. Sci. **10**(7), 1156–1165 (2014)
15. Kaur, G., Kaur, K.: An adaptive firefly algorithm for load balancing in cloud computing. In: Proceedings of Sixth International conference on Soft Computing for Problem Solving, pp. 63–72 (2017)

A Fuzzy Expert System for Diagnosing and Analyzing Human Diseases

Nureni Ayofe Azeez[1], Timothy Towolawi[1], Charles Van der Vyver[1],
Sanjay Misra[2(✉)], Adewole Adewumi[2], Robertas Damaševičius[3],
and Ravin Ahuja[4]

[1] North-West University, Vaal Triangle Campus, Vanderbijlpark, South Africa
nurayhnl@gmail.com, timtowy@gmail.com,
Charles.VanDerVyver@nwu.ac.za
[2] Covenant University, Ota, Nigeria
{sanjay.misra,wole.adewumi}@covenantuniversity.edu.ng
[3] Kaunas University of Technology, Kaunas, Lithuania
robertas.damasevicius@ktu.lt
[4] University of Delhi, Delhi, India
ravinahujadce@gmail.com

Abstract. According to the World Health Organization (WHO), human disease results in at least 70% of deaths every year. Approximately, 56 million people died in 2012 and 68% of all deaths in 2012 were as a result of non-communicable diseases. The aim of this paper is to design and develop a web-based fuzzy expert system that would diagnose some of these diseases and provide users with expert advice and prescriptions based on the diagnosis generated by the system. The system would not only indicate if the disease is present but will also indicate the level at which the disease is present. The system is designed to diagnose five diseases which include asthma, diabetes, hypertension, malaria and tuberculosis. The system uses Mamdani inference method which has four phases: fuzzification, rule evaluation, rule aggregation and defuzzification. The fuzzy expert system was designed based on clinical observations and the expert knowledge. Having performed the experimentation and obtained relevant results, it is worthy of note that this approach of diagnosing human diseases has put the accuracy and reliability to 97%. It is the strong opinion of the authors that its full-scale implementation will assist in no small measure in carrying out same function in some of the hospitals and health institutions.

Keywords: Disease · Expert system · Fuzzy logic · Mamdani

1 Introduction

Consulting a medical doctor in a private or public health institution is very essential for anyone suffering from one ailment or the other. Doing this is important to get adequate medical attention within a reasonable period of time. This is however very challenging in developing nation across the globe. Aside from dearth of medical personnel, patients find it extremely difficult to consult with a medical doctor because of paucity of fund

© Springer Nature Switzerland AG 2019
A. Abraham et al. (Eds.): IBICA 2018, AISC 939, pp. 474–484, 2019.
https://doi.org/10.1007/978-3-030-16681-6_47

and distance to cover before reaching a hospital that is very close by. Consequently, this scenario has over the years, resulted in the death of people and therefore increase mortality rate across the globe [10].

Allowing medical support to be freely and easily available to everyone would help reduce the mortality rate and go a long way in ensuring a healthy living. One way of accomplishing this is to develop expert systems that can help patients diagnose various diseases and assist qualified medical personnel in making faster and reliable diagnosis [13]. In order to circumvent these challenges, computer experts have decided to come up with a solution being referred to as expert system. An expert system is a system that encodes the knowledge of an expert in a specific domain. Invariably, an expert system replicates the knowledge, decision making capability as well as proficiency of any human expert such as a medical doctor, engineer, pharmacist, accountant etc. [14].

Expert systems have played an important role in medicine. Rather than solely relying on medical doctors, expert diagnostics systems are now being developed to diagnose these diseases and offer expert advice to the patients [17]. These diagnostic systems are being used in medical centers and clinics while some are open source for use by anyone. Expert diagnostics systems aren't developed to eliminate the need for doctors but can even assist the doctors or medical personnel in making faster and reliable diagnosis [5]. It could also be of great use to patients who live in areas where access to doctor is limited and very expensive.

The paper is structured as follows. The second section provides the literature review. System architecture is given in Sect. 3. Methodology, result and discussion and conclusion are provided in Sects. 3, 4 and 5 respectively.

2 Literature Review

It is very clear that diagnosis of human diseases is a very complex procedure and service that demands a medical expert who is proficient in his chosen field. Any efforts made to come up with a fuzzy expert system should be able to proffer solutions to all the identified challenges earlier mentioned. We attempted to review few of the existing researches related to medical diagnostic solutions developed with fuzzy logic [2].

Hasan et al. [12], developed a web-enabled fuzzy [7] expert system for diagnosing diseases. The system was described and finally implemented by making use of various linguistic rules. The application was specifically designed to address clinical solution to some medical challenges by improving the quality of information exchange between patients and various health care practitioners. The performance of this application has been verified to be excellent [12].

A fuzzy expert system for the diagnosis of heart-related diseases was developed by Adeli and Neshat in [1]. The solution was based on Long Beach, and Cleveland Clinic Foundation database. The system has one output variable and thirteen input variables [11]. The solution uses Mamdani inference method. The results obtained after implementing and testing the solution give 94% accuracy in terms of effectiveness and efficiency [1].

A fuzzy expert system for diagnosing and analyzing prostate cancer was proposed and developed by Saritas et al. in [15]. The system takes prostate volume (PV), prostate specific antigen (PSA) and age as inputs parameters while prostate cancer risk (PCR) was used as output parameter [6]. This system gives room to determine if there is any need for the biopsy while it provides the user various risk of cancer diseases. The system is adjudged very useful, economical, risk-free and can be used as a learning tool by medical students [15].

Having acknowledged the prevalence and global effects of hypertension, Chandra and Singh in [8] developed a fuzzy expert system for the diagnosis and management of High Blood Pressure [4]. They used body mass index, blood pressure and age as input variables while the output is considered to be hypertension risk. This application has been confirmed to valuable and useful [8].

Consequently, having studied the existing researches in this area, we decided to develop a fuzzy expert system for the diagnosis of five different diseases [9]. The diseases are asthma, diabetes, hypertension, malaria and tuberculosis. The results obtained in this regard have proven to be reliable and efficient. It accommodates a wider domain due to its flexibility and dynamic nature.

3   System Architecture

The system was built using fuzzy logic with a domain for five diseases and has an inference system for each disease and a knowledge base for each inference system (Fig. 1). Each inference system uses the mamdani-style inference method.

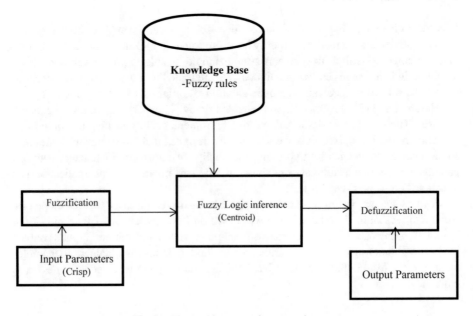

Fig. 1. The architecture of proposed system

Algorithm 1: Algorithm of the proposed system

1.   Retrieve data from the user in form of crisp input
2.   Change crisp input values to fuzzy values using the functions of membership (Fuzzification).
3.   Appraise the rules in the rule base (Inference).
4.   Merge the results of each rule (Inference).
5.   Convert the output data to non-fuzzy values (Defuzzification).
6.   Determine the diagnosis of the user based on the crisp input gotten from the deffuzifi cation phase.
7.   Output the fuzzy diagnosis back to the user.

The reason behind using multiple inference system (one for each disease) is for efficiency. If we took disease as a single domain and had just one inference system for disease, then disease could have output values of asthma, malaria, HIV, cholera etc. While this looks simple, the system would be very inefficient for a large knowledge base. This is because the system would require much unnecessary information from the end-user. Also, the system might not be able to indicate the level of presence of each disease or any other sort of information pertaining to that disease such as cause, type etc. Having multiple inference system allows us to incorporate some of these features. In this project, the inference systems developed indicate whether the disease is present and also the severity of the disease (if present).

4 Methodology

For the implementation phase, the mamdani inference system was used (as inference method) and it includes four phases; fuzzification, rule evaluation, rule aggregation and defuzification. The fuzzy operator (AND, OR) is used to evaluate multiple antecedents and obtain a single number. The *max* method was used for the OR operation while the *min* method was used for the AND operation. The max method was also used in the rule aggregation phase while the centroid method was used in the defuzzification phase. Diseases examined in this work are Malaria [10], Diabetes [3], Tuberculosis [18] Asthma [3] and Hypertension.

Activity Diagram of the System. The process and activity flow of information as proposed in this work is depicted in Fig. 2. The diagram has been used to explain the features and characteristics of actions and transitions within the system.

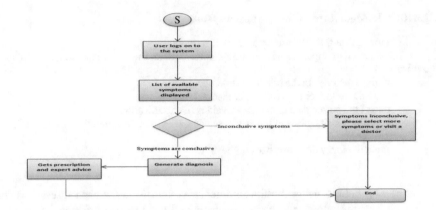

Fig. 2. Activity diagram for the Medical diagnostic system

Design of the Inference System. The system contains a knowledge base of 133 rules overall for all five diseases and uses triangular and trapezoidal membership function for each fuzzy set. Each phase of inference used in the system is described below:
Fuzzification- Fuzzification involves taking crisp inputs and determining the degree to which these inputs belong to each of the appropriate fuzzy sets. The crisp input is always a numerical value that is usually used to represent a linguistic object [16]. For example, if we say "temperature is 100 k", 100 is the crisp input while temperature is the linguistic object. The degree to which each crisp input belongs to a fuzzy set is called fuzzy value. In this work, triangular and trapezoidal membership functions are used to fuzzify crisp inputs.

Rule Evaluation. This involves making use of fuzzified inputs and making use of them to the fuzzy rules antecedents. In case there is more than one antecedent, either AND or OR fuzzy operator will be used to estimate a single value which represents the result of the antecedent evaluation. The truth value will thereafter be applied to the resulting membership function [16].

To evaluate the conjunction of rule antecedents, we use the AND fuzzy operation. Two methods can be used in AND operation: min and the product method. However, the min method is the most common method used in AND operation.

The min method is calculated as:

$$\mu_{A \cap B}(x) = \min[\mu_A(x), \mu_B(x)] \tag{1}$$

Rule Aggregation. This is the process combining and integrating the outcome of the rules. Invariably, the membership functions of all rule consequents are taken and combined into a single fuzzy set [14]. For example, let's consider the rule 1 and rule 2 from the asthma inference rule base:

RULE 1: IF PEFR is Normal and DSF is No and NSF is No and PEFRV is NO and OSL is No then asthma is No

RULE 2: IF PEFR is Moderate and DSF is No and NSF is No and PEFRV is Low then asthma is No

Now let's imagine both rules fire, what value will represent set 'No' of the consequent asthma in the defuzification phase? This is the essence of rule aggregation. We need to unify the outputs of all rules so that the set 'No' is of the consequent asthma is represented by a single value in the defuzification phase. The fuzzy OR operation is used in rule aggregation. Max method is the most common fuzzy OR operation used in rule aggregation phase.

The min method is calculated as:

$$\mu_{A \cup B}(x) = \max[\mu_A(x), \mu_B(x)] \tag{2}$$

Defuzzification. The final step in the fuzzification process is defuzzification. This phase assists to assess the guiding rules. The last output of any fuzzy system must be a crisp number. An aggregate output fuzzy set is considered the input while a single number is known to be output in any defuzification process. There are several methods for carrying out defuzification. The prominent among them is the centroid technique [14]. It looks for a point where a vertical line would divide the set into two equal parts. Centre of gravity (COG) can be mathematically represented as:

$$COG = \frac{\sum_{x=a}^{b} \mu_A(x)x}{\sum_{x=a}^{b} \mu_A(x)} \tag{3}$$

Where $\mu_A(x)$ = Membership value in the membership function and x = Center of membership function.

5   Results and Discussion

Several patients used the web diagnostic tool to test for probable diseases. The results of all tests are provided in Tables 1, 2, 3 and 4 and corresponding graphs are given in Figs. 3, 4, 5 and 6.

Table 1. Results of Asthma test for several patients

Patient No	PEFR	DSF	NSF	PEFRV	OSL	Asthma
1	93	4	3	17	95	20.2
2	57	13	9	27	30	72.1
3	77	14	9	28	93	53.2
4	92	1	1	7	96	17.8
5	60	25	23	30	99	65.8
6	85	2	2	10	95	26.5
7	83	10	5	24	93	46.1

PEFR => Peak Expiratory, Flow Rate, DSF => Daytime Symptom frequency, NSF => Nighttime Symptom frequency, PEFRV => Peak Expiratory Flow Rate Variability, OSL => Oxygen saturation level, Asthma => Asthma presence.

The system uses membership function to transform the crisp input into fuzzy values. For example, Rule which states that "IF PEFR is Moderate and DSF is Medium and NSF is Medium and PEFRV is Medium and OSL is Moderate then Asthma is Mild". Patient 1 crisp input can be transformed (using membership function) to:

$$\mu_{moderate}(PEFR) = \frac{95-93}{95-85} = \frac{2}{10} = 0.2$$

$$\mu_{medium}(DSF) = \frac{4-3}{10-3} = \frac{1}{7} = 0.143$$

$$\mu_{medium}(NSF) = \frac{3-2}{5-2} = \frac{1}{3} = 0.33$$

$$\mu_{medium}(PEFRV) = \frac{17-15}{24-15} = \frac{2}{9} = 0.22$$

$$\mu_{medium}(SaO_2) = 1$$

Since all antecedents have non-zero membership function, the consequent will fire with a certainty factor of:

$$\mu_{mild}(Asthma) = \min[\mu_A(x), \mu_B(x)] = \min[0.2, 0.143, 0.33, 0.22, 1] = 0.143$$

The fuzzy values for patient 1 that are represented in the rule evaluation phase are shown in Fig. 5 as Fuzzy values for patient 1.

Rule Aggregation. It can be observed that there are several values for μ_{mild} and $\mu_{intermittent}$. We therefore need to aggregate all consequents so only one value represents each consequent in the defuzzification phase.

$$\mu_{mild} = \max[\mu_A(x), \mu_B(x)] = \max[0.143, 0.2] = 0.2$$

$$\mu_{intermitent} = \max[\mu_A(x), \mu_B(x)] = \max[0.2, 0.3, 0.143] = 0.3$$

$$\mu_{no} = 0.6$$

$$\mu_{moderate} = 0, \ \mu_{severe} = 0$$

Defuzzification. The last thing to be done is to defuzzify the consequents into a single crisp input. We use the Centroid method with the formula:

$$COG = \frac{\sum_{x=a}^{b} \mu_A(x)x}{\sum_{x=a}^{b} \mu_A(x)}$$

So we can derive that

$$COG = \frac{0.6(0+5+10+15)+0.3(20+25+30+35+)+0.2(40+45+50+55)+0(60+65+70+75)+0(80+85+90+95+100)}{0.6+0.6+0.6+0.6+0.3+0.3+0.3+0.3+0.2+0.2+0.2+0+0+0+0+0+0+0+0} = 20.2$$

So from the value gotten from the defuzification phase, the doctor can determine the severity of the disease. The graphical representation of diagnosis for several patients who used the web diagnostic tool is shown in Fig. 3. Further results are provided in the tables showing patients numbers, the variables considered and their corresponding graphical representation. Figure 4 presents the graphical representation of the results of Diabetes test for several patients while Fig. 5 shows the graphical representation of the results of Hypertension test for several patients. Finally Fig. 6. Presents the graphical representation of the results of Malaria test for several patients.

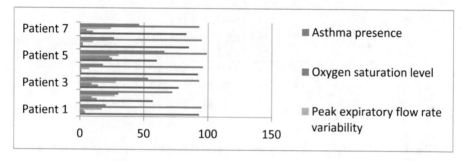

Fig. 3. Graphical representation of the results of Asthma test for several patients

Table 2. Results of Diabetes test for several patients

Patient No	Glucose	INS	BMI	BP	Age	Urine	Diabetes
1	150	80	40	160	57	115	51.1
2	56	18	14	93	15	70	22.9
3	40	20	30	100	46	1	37.5
4	110	50	18	90	10	45	31
5	146	98	60	95	70	5	55.4
6	126	118	40	135	30	200	46,4
7	109	38	33	140	50	80	43.6

Glucose => Fasting plasma glucose level, INS => Insulin Level, BMI => Body mass index, Bp => Blood pressure, Urine => Urine level, Diabetes =>Diabetes presence.

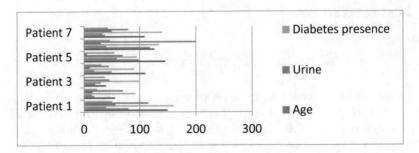

Fig. 4. Graphical representation of the results of Diabetes test for several patients

Table 3. Results of Hypertension test for several patients

Patient No	Glucose	DWH	BMI	BP	Age	Hr	Gender	Hypertension
1	57	7	37	200	45	116	1	47.9
2	115	12	44	180	65	160	1	69.7
3	64	3	35	120	25	100	0	17.8
4	90	1	20	100	15	90	0	30.4
5	198	18	67	200	70	160	1	88.7
6	113	13	40	170	40	150	1	71.3
7	150	10	42	90	40	150	1	56.1

Glucose => Fasting plasma glucose level, DWH => No of daily work hours for patient, BMI => Body mass index, Bp => Blood pressure, Urine => Urine level, HR => heart rate, Gender => female(0), male(1), Hypertension => .Hypertension presence

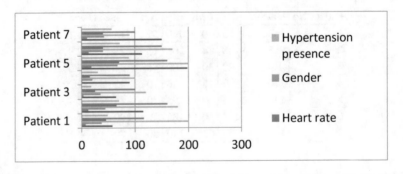

Fig. 5. Graphical representation of the results of Hypertension test for several patients

Table 4. Results of Malaria test for several patients

Patient No	Headache	Temperature	Nausea	Vomiting	Jaundice	Weakness	Mp	Appetite	Dizziness	**Malaria**
1	Mild	38	Mild	Mild	Mild	Moderate	Moderate	Severe	Severe	46.4
2	Mild	36	Mild	Mild	Mild	Mild	Mild	Mild	Mild	15.2
3	Moderate	39	Mild	Mild	Mild	Severe	Moderate	Severe	Severe	70.8
4	Mild	42	Mild	Mild	Mild	Severe	Severe	Mild	Mild	90.6
5	Moderate	38.8	Moderate	Moderate	Mild	Moderate	Moderate	Moderate	Moderate	45.3
6	Severe	40	Severe	Severe	Severe	Very severe	Severe	Severe	Severe	91.2

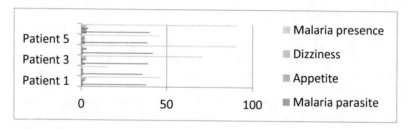

Fig. 6. Graphical representation of the results of Malaria test for several patients

6 Conclusion

Disease diagnosis is obviously not a luxury but a necessity and it's a practice that has been done for years but up until early 1970s, people have had to rely only on humans to test whether they have a disease or not. However, since the emergence of expert systems, people's reliance on humans has reduced, thereby leading to effortless diagnosis for patients as tasked performed only by humans can now be emulated by machines. The system proposed can diagnose several human diseases and even accommodate a wider domain due to its flexibility but would require sufficient data. It is worthy of note that this approach of diagnosing human diseases has put the accuracy and reliability to 97%. It is the strong opinion of the authors that its full-scale implementation will assist in no small measure for carrying same function in some of the hospitals and health institutions.

Acknowledgments. We acknowledge the support and sponsorship provided by Covenant University through the Centre for Research, Innovation and Discovery (CUCRID).

References

1. Adeli, A., Neshat, M.: A fuzzy expert system for heart disease. In: International Multiconference of Engineers and Computer Scientists, Hong Kong, pp. 134–139 (2010)
2. Ali, M.S., Sartakar, S.L.: A fuzzy expert system for pathological investigation and diagnosis of jaundice. Int. J. Adv. Res. Comput. Commun. Eng. **2**(6), 2431–2436 (2013)

3. Anand, S.K., Nandhini, K.M., Ajmalahamed, A.: Designing a rule based fuzzy expert controller for early detection and diagnosis of diabetes. ARPN J. Eng. Appl. Sci. **9**(5), 819–827 (2014)
4. Anand, S.K., Vijayalakshmi, S., Kalpana, R.: Design and implementation of a fuzzy expert system for detecting and estimating the level of asthma and chronic obstructive pulmonary disease. Middle-East J. Sci. Res. **14**(11), 1435–1444 (2013)
5. Ansari, G.A.: An adoptive medical diagnosis system using expert system with applications. J. Emerg. Trends Comput. Inf. Sci. **4**(3), 2009–2013 (2013)
6. Awad, E.M.: Building Expert systems, 1st edn. West Publishing Co., Eagan (1996)
7. Awotunde, J.B., Matiluko, O.E., Fatai, O.W.: Medical diagnosis system using fuzzy logic. Afr. J. Comput. ICT **7**(2), 99–106 (2014)
8. Chandra, V., Singh, P.: Fuzzy based high blood pressure diagnosis. Int. J. Adv. Res. Comput. Sci. Technol. **2**(2), 137–139 (2014)
9. Djam, X.Y., Wajiga, G.M., Kimbi, Y.H., Blamah, N.V.: A fuzzy expert system for the management of malaria. Int. J. Pure Appl. Sci. Technol. **5**(2), 84–108 (2011)
10. Fatumo, S.A., Adetiba, E., Onaolapo, J.O.: Implementation of XpertMalTyph: an expert system for medical diagnosis of the complications of malaria and typhoid. IOSR J. Comput. Eng. **8**(5), 34–40 (2013)
11. Giarratano, J., Riley, G.: Expert Systems, Principles and Programming, 4th edn. Course Technology, Boston (2004)
12. Hasan, M.A., Sher-E-Alam, K., Chowdhury, A.R.: Human disease diagnosis using a fuzzy expert system. J. Comput. **2**(6), 66–70 (2010)
13. Jackson, P.: Introduction to Expert Systems, 3rd edn. Addision-Wesley, Wokingham (1998)
14. Negnevitsky, M.: A Guide to Intelligent Systems, 2nd edn. Pearson Education Limited, Essex (2005)
15. Saritas, I., Torun, S., Allahverdi, N.: A fuzzy expert system design for diagnosis of prostate cancer. In: International Conference on Computer Systems and Technologies (2003)
16. Siler, W., Buckley, J.J.: Fuzzy Expert Systems and Fuzzzy Reasoning. Wiley, Hobeken (2005)
17. Simfukwe, M., Kunda, D., Zulu, M.: Addressing the shortage of medical doctors in Zambia: medical diagnosis expert system as a solution. IJISET - Int. J. Innovative Sci. Eng. Technol. **1**(5), 1–6 (2014)
18. Walia, N., Singh, H., Tiwari, S.K., Sharma, A.: A decision support system for tuberculosis diagnosability. Int. J. Soft Comput. (IJSC) **6**(3), 1–3 (2015)

Diagnosing Oral Ulcers with Bayes Model

Nureni Ayofe Azeez[1], Samuel O. Oyeniran[1], Charles Van der Vyver[1],
Sanjay Misra[2(✉)], Ravin Ahuja[3], Robertas Damasevicius[4],
and Rytis Maskeliunas[4]

[1] North-West University, Vaal Triangle Campus, Vanderbijlpark, South Africa
nurayhnl@gmail.com, so.oyeniran@gmail.com,
Charles.VanDerVyver@nwu.ac.za
[2] Covenant University, Ota, Nigeria
sanjay.misra@covenantuniversity.edu.ng
[3] Vishwakarma Skill University Gurugram, Gurugram, India
[4] Kaunas University of Technology, Kaunas, Lithuania
robertas.damasevicius@ktu.lt

Abstract. An oral ulcer is considered an ulcer that occurs on the mucous membrane of the oral cavity. Its treatment however poses a serious challenge as experts in this area of family health care are rare. In order to affirm this claim, questionnaire was designed and distributed to get up-to-date information concerning the ailment, particularly its treatment, causes, and solutions among others. An expert system seeks to mimic the functions and capabilities of a real life expert. This means that just as the expert is domain specific, the expert system as well is domain specific. It consists of a user interface, a knowledgebase, a fact-base, and an inference engine; all working together to mimic an expert. Bayesian inference was used in this work to factor in probability as it is well known that the fact that one is experiencing the symptoms of a disease does not absolutely mean that one has that disease. The domain of this expert system is oral ulcer and it focuses on four common oral ulcers which include: Cold sores, Gingivostomatitis, Herpangina as well as Neutropenia. The application which was designed with Object-Oriented Analysis and Design methodology was implemented with java programming language, provided on the Netbeans 8.0.2 IDE.

Keywords: Expert system · Oral ulcers · Diagnosis · Bayes' theorem

1 Introduction

Artificial intelligence [19] is the aspect of Computer Sciences that is concerned with solving problems using non-algorithmic, and symbolic methods. One of some of its branches is the Expert System. An expert system in itself is a system that is able to mimic an expert [23]. This implies that the system should be able to do what that expert can do. In other words, an expert system is an intelligent system that follows a theoretical knowledge of an expert, and gives inferences just like the expert does [13].

© Springer Nature Switzerland AG 2019
A. Abraham et al. (Eds.): IBICA 2018, AISC 939, pp. 485–494, 2019.
https://doi.org/10.1007/978-3-030-16681-6_48

Oral ulcers are common place disorders of the mouth. It is commonly referred to as a breach in the oral epithelium [24]. It exposes nerve endings in the underlying lamina propria which results in sores in the mouth preventing one from the want to eat something that contains pepper [20].

With advancement in technology, researchers, as well as industries, have been able to bring to us an advancement in artificial intelligence [29]. This work well done introduced expert systems into the world of dental informatics [22]. An era of advanced dentistry has now come upon us with the arrival of this new paradigm in the delivery of dental health care. The application of expert systems to dentistry cannot be overemphasized [30]. One of such is the subject of this work. This work aims at developing of an expert system that is capable of imitating the role of a dental expert in the diagnosis of oral ulcers.

With the advent of dental informatics, it is essential to begin improving dental research and treatment using information technology tools [27]. One of such tools is the expert system [25]. In the case of this project, the problem that is being solved is that of the development of an expert system that diagnoses common oral ulcers as well as provide useful information in treating them [26]. Also, the basic theories involved will be clearly expressed. Expert systems will be looked at closely vis a vis: history, basic concepts, techniques, as well as its application in the development of diagnostic systems.

The work is presented as follows. The background of oral ulcers and expert system are provided in next section. The methodology and data collection are given in Sect. 3. Implementation and testing is in Sect. 4 and finally conclusion drawn in Sect. 5.

2 Background of the Work

Oral ulcers – also known as mouth ulcers – are usually small, painful lesions that grow in the mouth or at the bottom of the gums [9, 15]. They are painful round or oval sores that appear in the mouth, commonly on the inside of the cheeks or lips [16]. They can make oral activities such as eating, talking, drinking, etc. uncomfortable [14]. Women (more likely than men), young adults, and people with family history of oral ulcers have a higher risk of mouth ulcers [9]. In this work, about four (4) different oral ulcers are considered. These ulcers include:

Cold Sores: Cold sores, are red blisters that are usually full of fluid that appear close to the mouth or on other parts of the face [21]. They are small blisters that develop on the lips or around the mouth [17]. In infrequent cases, they may appear on the finger, nose or inside the mouth. These sores are incurable, contagious and may reoccur without warning. The major cause of cold sores is the Herpes Simplex Virus type 1. Type 2 generally causes the genital herpes [21].

Gingivostomatitis: Gingivostomatitis (GM), is a common infection of the mouth and gums. This infection may be caused by a virus, bacteria, improper hygiene for the teeth and mouth [18]. It is a common condition among children. These children may drool and not eat or want any drink due to the discomfort that the sore causes [18]. Causes of

Gingivostomatitis include: Herpes Simplex Virus type 1, Coxsackie virus; often transmitted by touching a feces-contaminated surface and Poor oral hygiene [18].

Herpangina: Herpangina is an illness caused by virus that involves ulcers and sores inside the mouth, a sore throat, and fever [8].

Neutropenia
Neutropenia is a condition of the blood. This is caused by low neutrophil (type of white blood cell that fights infections) level [6]. This condition may be caused by any of the followings: Shwachman-Diamond syndrome, glycogen-storage disease type 1b, leukemia, viral illnesses [6].

Expert Systems: An expert system in itself is a system that is able to mimic the knowledge of an expert [5]. This implies that the system should be able to do what that expert can do. In other words, an expert system is an intelligent system that follows a theoretical knowledge of an expert, and gives inferences just like the expert does. Since these systems make use of knowledge, it is important to take a lot at what knowledge is [28].

Brief History of Expert Systems: In 1957, Herbert, Simon, Shaw, and Newell developed the first ever computer program that separated its knowledge of problems from its problem solving strategy. It was the general problem solver (G.P.S).

In the 1970s, expert systems began to surface [11]. Ted Shortliffe developed MYCIN, one of the first expert systems to demonstrate the power of rule-based architectures in 1974 [7]. Then in the 1980s, they began to go commercial encouraging more and more insight into the advances in the development of expert systems [4].

Concept and Design of Rule-Based Expert Systems: There are basically five main components for a rule-based expert system [3]. They are, the knowledge base, the inference engine, the database, the user interface and the explanation facilities [13]. This is depicted in Fig. 1 [13].

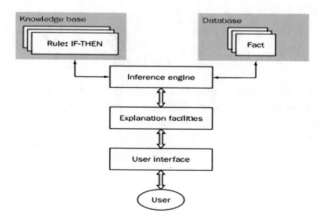

Fig. 1. Rule-based expert system architecture.

Bayes' Theorem-Bayes' theorem deals with conditional probabilities that are the reverse of themselves. It introduces the concept of prior probability and posterior probability; where the latter is the former, updated by information given from data [1]. Conditional probability refers to the probability of observing an event due to the observation of another event [12]. This is given in Eq. 1 [2].

$$P(B|A) = \frac{P(A \text{ and } B)}{P(A)} \tag{1}$$

Having looked at conditional probability, Bayes' theorem can now be described. Bayes' theorem can be given as shown in Eq. 2.

Mathematically,

$$P(B|A) = \frac{P(A|B)P(A)}{P(A|B)P(A) + P(A|\neg B)P(\neg A)} \tag{2}$$

Using a simple Bayes' model, proposed by Ledley and Lusted [10], we follow two assumptions:

1. That all results are provisionally and conditionally independent, given an ulcer instance. This implies that, if the true ulcer state of a patient is known, the possibility and likelihood of having any observation doesn't rely on observations made on any other characteristics and features. In essence, from the rules of probability:

$$P(s_i, s_{i+1}, s_{i+2}, \ldots, s_n | D_i) = P(s_i | D_i)P(s_{i+1} | D_i) \ldots P(s_n | D_i) \tag{3}$$

2. That the traditional entities of the ulcer are mutually exclusive and exhaustive.

Given these two (2) assumptions, the expert system can now calculate the posterior probabilities of the ulcer using:

$$P(D_i | s_i, s_{i+1}, s_{i+2}, \ldots, s_n) = \frac{P(s_i | D_i)P(s_{i+1} | D_i) \ldots P(s_n | D_i)}{\sum_i P(s_i | D_i)P(s_{i+1} | D_i) \ldots P(s_n | D_i)} \tag{4}$$

After the expert system does this calculation, it can thenceforth choose the optimum diagnosis; referring to the most suitable diagnosis given the symptoms that the patient is experiencing.

3   Methodology and Data Collection

The methodology used for this project work is the object oriented analysis and design (OOAD). OOAD can be broken down into parts.

System Implementation

System implementation may be seen as the building or developing of the system and its delivery into full business or organization operations. The implementation phase has one major activity: deploying the system. At this point in the software development life cycle, every bit of theoretical and conceptual design is converted into a working system; implying that the given system specifications are used to come up with a new system.

In this work, a questionnaire was used to check for the relevance of this work. It was shared to twenty people. The detail about the interpretation of questionnaire is included in the appendix.

Data Used in Bayesian Inference

Sample data was collected from an anonymous source. The data of people who came in for consultation on any of the four types of oral ulcers considered in this work were collected. We considered 400 of them; that is 100 for each case of ulcer to see what happens. Each patient narrated his symptoms and then the medical [10] personnel checks if they match the case for the ulcer, he then goes ahead to make conclusive tests to determine specifically which oral ulcer the patient had. The Tables 1, 2, 3, and 4 were the recorded observation of this setup.

Activity diagram for the system implementation is given in Fig. 2.

Activity Diagram

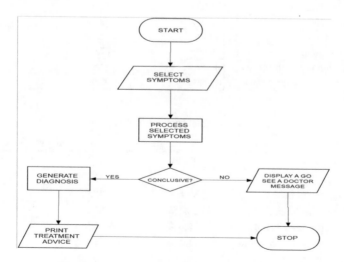

Fig. 2. Activity diagram for the implementation

Table 1. Statistics on Cold sores

Status	Cold sores (74%)	No cold sores (26%)
Tingling Sensation Positive	89.2%	15.2%
Tingling Sensation Negative	10.8%	84.8%
Red, fluid-filled blisters Positive	79.8%	5.4%
Red, fluid-filled blisters Negative	20.2%	94.6%
Fever Positive	99.24%	10.56%
Fever Negative	0.76%	89.44%
Muscle Aches Positive	69.25%	17.0%
Muscle Aches Negative	30.75%	83.0%
Swollen Lymph Nodes Positive	80.2%	15.4%
Swollen Lymph Nodes Negative	19.8%	84.6%

Table 2. Statistics on Gingivostomatitis

Status	Gingivostomatitis (79%)	No Gingivostomatitis (21%)
Tender Sores Positive	92.5%	19%
Tender Sores Negative	7.5%	81%
Bad Breath Positive	72%	21%
Bad Breath Negative	28%	79%
Fever Positive	91%	51%
Fever Negative	9%	49%
Bleeding Gum Positive	76.55%	21.23%
Bleeding Gum Negative	23.45%	78.77%
Swollen Lymph Nodes Positive	82%	15%
Swollen Lymph Nodes Negative	18%	85%
Drooling Positive	76.09%	11.04%
Drooling Negative	23.91%	88.96%
Malaise Positive	79.1%	12.1%
Malaise Negative	20.9%	87.9%
Discomfort in Mouth Positive	82.17%	0.91%
Discomfort in Mouth Negative	17.83%	99.09%

Table 3. Statistics on Herpangina

Status	Herpangina (70%)	No Herpangina (30%)
Fever Positive	99.7%	45.8%
Fever Negative	0.3%	54.2%
Headache Positive	89.6%	48.95%
Headache Negative	10.4%	51.05%
Appetite Loss Positive	88.6%	25.8%
Appetite Loss Negative	11.4%	74.2%
Sore Throat Positive	80.3%	44.2%
Sore Throat Negative	19.7%	55.8%
Similar Sores (HFB) Positive	78.6%	0.8%
Similar Sores (HFB) Negative	21.4%	99.2%

Table 4. Statistics on Neutropenia

Status	Neutropenia (71%)	No Neutropenia (29%)
Fever Positive	99.9%	67.5%
Fever Negative	0.1%	32.5%
Pneumonia Positive	78.65%	17.32%
Pneumonia Negative	21.35%	82.68%
Sinus Infection Positive	92.6%	1.5%
Sinus Infection Negative	7.4%	98.5%
Ear Infection Positive	88.6%	9.5%
Ear Infection Negative	11.4%	90.5%
Gum Inflammation Positive	92.53%	42.75%
Gum Inflammation Negative	7.47%	57.25%
Navel Infection Positive	98.6%	27.5%
Navel Infection Negative	1.4%	72.5%
Skin Abscesses Positive	90.6%	7.55%
Skin Abscesses Negative	9.4%	92.45%

4   Validation- Implementation and Testing

The implementation was done using java. The program starts by collecting data from the user. It collects the user's first name, last name, gender, as well as body temperature. With the body temperature, the system determines whether the patient has a fever (that is, temperature >37.5) or not. The other part of data collection involves the patient selecting the set of symptoms that he is experiencing. A screenshot of the input screen is demonstrated in Fig. 3.

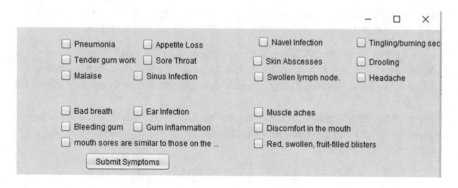

Fig. 3. Input screen

With the symptoms collected, the conditional probabilities of all the symptoms the user entered are aggregated and then converged using the assumptions proposed as the simple Bayes' model.

The expert system uses JEOPS, a java library that serves as an inference engine used in expert system. The rules were developed around the optimality of the probabilities. That implies that the ulcer with the highest probability is said to be the ulcer that the patient may be experiencing. The outcome of the test is then displayed on the outcome text area of the system. A screenshot of an example of an anonymous patient who used this system is given in Fig. 4.

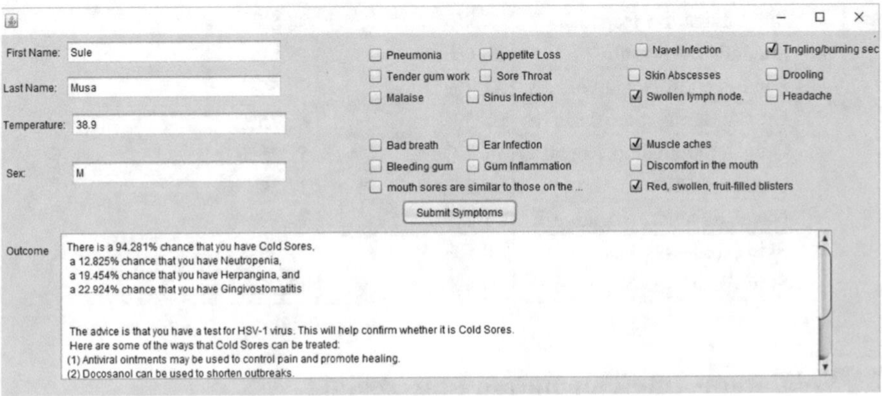

Fig. 4. Test case

The system also generates a printable Microsoft word document of the diagnosis which can be taken to the nearest oral health care specialist for confirmation and further examinations.

Recommendation

This work is a step in the right direction towards achieving good health care delivery system. With this system been made accessible to everyone to use, it could go on to reduce long incessant queues in the hospitals. Consequently, authors are of the following recommendations:

1. This system be made available to everyone so as to facilitate its use.
2. Workshops and training sessions should be set-up on the use of this system.
3. More systems be developed that can help bring medical care to the fingertips of the less privileged.
4. More research should be carried out on how artificial intelligence (in this case, expert systems) can be much more useful for the common good of man (especially in the medical fields).

5   Conclusion

This work has been able to demonstrate, to a large extent, that artificial intelligence can be a useful tool in the medical field. With the introduction of more expert systems, more people will be able to get access to health diagnosis without having to queue and wait long hours to see a physician. An efficient use of this system will result in reduction in cost and increase in the speed of oral health care delivery. Consequently, the application which was designed with the Object-Oriented Analysis and Design methodology was implemented with java programming language, provided on the Netbeans 8.0.2 IDE provides 98% degree of accuracy for diagnosing oral ulcers among the infected patients. It is the recommendation of the authors that its full scale implementation will assist medical personnel and experts in this domain of medical ailment.

Acknowledgments. We acknowledge the support and sponsorship provided by Covenant University through the Centre for Research, Innovation and Discovery (CUCRID).

References

1. Azad, K.: An Intuitive (and Short) Explanation of Bayes' Theorem (n.d.). http://betterexplained.com/articles/an-intuitive-and-short-explanation-of-bayes-theorem/
2. Bayes, T., Price, R.: An Essay Towards Solving a Problem in the Doctrine of Chances. By the late Rev. Mr. Bayes, communicated by Mr. Price, in a letter to John Canton, MA. and F. R.S. Philosophical Transactions of the Royal Society of London, pp. 370–418 (1763)
3. Blereau, R.P.: Viral infections in children and adolescents. Consultant for pediatrcians 360 (2013). http://www.pediatricsconsultant360.com/sites/default/files/Screen%252520Shot%2525202013-04-30%252520at%2525202.17.46%252520PM.png
4. Elstein, A.S., Schwarz, A.: Clinical problem solving and diagnostic decision making. Chicago (2002)
5. Franco, J.: Primary Herpetic Gingivostomatitis (n.d.). http://janellef.weebly.com/primary-herpetic-gingivostomatitis.html. Accessed 10 Oct 2017
6. Giorgi, A.Z.: Neutropenia, 27 August 2017. http://www.healthline.com/health/neutropenia
7. Jones, M.T.: Artificial intelligence: a systems approach. Infinity Science Press LLC, Hingham (2008)
8. Kaneshiro, N.K.: Retrieved from MedlinePlus, 22 August 2017. https://www.nlm.nih.gov/medlineplus/ency/article/000969.htm
9. Krucik, G., Johnson, S.: What causes mouth ulcers? 25 possible conditions, 25 August 2015. http://www.healthline.com/symptom/mouth-ulcers
10. Ledley, R.S., Lusted, L.B.: Reasoning foundations of medical diagnosis. Science **130**, 9–21 (1959)
11. Matarasso, S., Daniele, V., Siciliano, V.I., Mignogna, M.D., Andreuccetti, G., Cafiero, C.: The effect of recombinant granulocyte colony-stimulating factor on oral and periodontal manifestations in a patient with cyclic neutropenia: a case report. Int. J. Dent. **2009**, 6 (2009)
12. Meagher, P.: Implement Bayesian inference using PHP, Part 1, 5 January 2005. http://www.devshed.com/c/a/php/implement-bayesian-inference-using-php-part-1/
13. Negnevitsky, M.: Artificial Intelligence: A Guide to Intelligent Systems. Addison-Wesley (Pearson Education) (2005)
14. Newell, A., Simon, H.A.: Human Problem Solving. Prentice Hall, Upper Saddle River (1972)
15. Newell, A., Simon, H.A., Shaw, J.C.: Report on a general problem-solving program (1959)
16. NHS: Cold sore (herpes simplex virus), 10 April 2014. http://www.nhs.uk/conditions/cold-sore/pages/introduction.aspx

17. NHS: Mouth ulcers. Retrieved from NHS, 6 March 2014
18. Phillips, N.: Gingivostomatitis, 16 July 2012. http://www.healthline.com/health/gingivostomatitis
19. Ramsay, G.G.: Noam Chomsky on Where Artificial Intelligence Went Wrong (n.d.). http://www.theatlantic.com/technology/archive/2012/11/noam-chomsky-on-where-artificial-intelligence-went-wrong/261637/. Accessed 7 October 2017
20. Saudi, H.I., Ali, S.A.: An expert system for the diagnosis and management of oral ulcers. Tanta Dent. J. **11**, 42–46 (2014)
21. Selner, M.: Retrieved from Healthline (2012). http://www.healthline.com/health/herpes-labialis
22. Whitten, J.L., Bentley, L.D.: System Analysis and Design Methods, 7th edn. McGraw-Hill Companies Inc., New York (2007)
23. Azeez, N.A., Ademolu, O.: CyberProtector: identifying compromised URLs in electronic mails with Bayesian classification. In: International Conference Computational Science and Computational Intelligence, pp. 959–965 (2016)
24. Azeez, N.A., Olayinka, A.F., Fasina, E.P., Venter, I.M.: Evaluation of a flexible column-based access control security model for medical-based information. J. Comput. Sci. Appl. **22**(1), 14–25 (2015)
25. Denning, E.D.: An intrusion-detection model. IEEE Trans. Softw. Eng. **SE-13**, 222–232 (1987)
26. Azeez, N.A., Babatope, A.B.: AANtID: an alternative approach to network intrusion detection. J. Comput. Sci. Appl. **23**(1), 129–143 (2016)
27. Azeez, N.A., Venter, I.M.: Towards ensuring scalability, interoperability and efficient access control in a multi-domain grid-based environment. SAIEE Afr. Res. **104**(2), 54–68 (2013)
28. Nureni, A.A., Irwin, B..: Cyber security: challenges and the way forward. GESJ: Comput. Sci. Telecommun. **29**(6), 56–69 (2010). ISSN 1512-1232. 14p. 2
29. Azeez, N.A., Iliyas, H.D.: Implementation of a 4-tier cloud-based architecture for collaborative health care delivery. Niger. J. Technol. Dev. **13**(1), 17–25 (2016)
30. Ayofe, A.N., Adebayo, S.B., Ajetola, A.R., Abdulwahab, A.F.: A framework for computer aided investigation of ATM fraud in Nigeria. Int. J. Soft Comput. **5**(3), 78–82 (2010)

Design and Implementation of a Fault Management System

Abiola Salau[1], Chika Yinka-Banjo[1], Sanjay Misra[2(✉)],
Adewole Adewumi[2], Ravin Ahuja[3], and Rytis Maskeliunas[3,4]

[1] University of Lagos, Lagos, Nigeria
salauabiola@gmail.com, cyinkabanjo@unilag.edu.ng
[2] Covenant University, Ota, Nigeria
{sanjay.misra, wole.adewumi}@covenantuniversity.edu.ng
[3] Kaunas University of Technology, Kaunas, Lithuania
ravinahujadce@gmail.com, robertas.damasevicius@ktu.lt
[4] University of Delhi, Delhi, India

Abstract. Faults management is a daily process in every organization, what makes the difference is the approach each organization applies in managing such problems. The use of information technology to improve on reporting and management of faults is essential to the growth of an organization. For a telecommunication network that prides with over twenty five million subscribers and over seven thousand cell sites, the cases of faults are not any different. Faults are reported on the alarm page every second and the NOC (Network operations center) engineer is expected to log such faults and escalate to the field engineers to ensure quick and prompt resolution of the faults. This paper shows the designing and implementation of a centralized fault management system and compares the new system to previously existing systems. The designed system help transit from the traditional system of fault logging and escalating manually to a web portal system that Logs fault, tracks the fault, escalate the fault to the appropriate region, and generate a report.

Keywords: Alarm list · Fault detection · Fault management ·
Network management · NOC

1 Introduction

The Fault management system (FMS) involves management of faults within an organization. It will assist engineers in the Network operations center to log faults and be able to escalate faults to expected Field engineers [1]. The Field engineers are also expected to report observations and activities carried out in respect to a particular fault to ensure the fault is closed appropriately from the system [2]. The Aim of the project is to improve on the process of escalation, tracking and reporting of faults from a traditional method of typing in excel files to a more driven information technology process. This would be achieved by developing a system for managing faults in an operations environment, ensuring faults are logged promptly, ensuring quick and prompt escalation to all stakeholders, ensuring reports are delivered as well as proper

A. Abraham et al. (Eds.): IBICA 2018, AISC 939, pp. 495–505, 2019.
https://doi.org/10.1007/978-3-030-16681-6_49

management of the fault. The Network operations center of Globacom Nigeria log faults on Excel sheets on computer systems and these faults are escalated by SMS (short message service) to field engineers, reports are prepared as needed which could be cumbersome especially when there is a need to get updates from field engineers over the phone or via SMS (short message service) as regard failure on network elements [3]. Also when faults are detected via alarm systems to the network operations center engineer, the process of escalating this fault to the field engineer is a slow process as the engineers need to manually type the fault type into an SMS application platform and also manually input the contacts of the appropriate field Engineers before sending out the SMS [4].

Scope of Study: The Fault Management System (FMS) will be designed to be applicable to the Network Operations center of Globacom Nigeria. It is expected to improve the way data is stored and retrieved and greatly improves the process of reporting faults thereby, reducing the Mean time to restore (MTTR) faults.

The Fault management system will ensure faults are logged, escalated and reported as against different processes that are used to log, escalate and report faults on the network. It will also provide a platform for escalation through SMS to all stakeholders involved. It will also provide a wider query for reporting and provide better data analysis of faults resolution details.

Required Tools: The major software tools used in developing the application includes Spring, a Java framework, Hypertext Mark-up Language (HTML), cascading style sheet (CSS) and MySQL. The front end interface was designed using HTML and CSS while Java was used as back end scripting language to connect with the front end, MySQL was used as a relational database management system (RDBMS) which houses the entire materials of the warehouse for easy storage, update and retrieval.

2   Background Study on a Fault Management System

Operators in the telecommunications industry are saddled with increasing competition worldwide and their ability to provide uninterrupted services keeps them at the edge of the market flow [5]. To ensure the system are properly monitored most Telecommunication outfit have a Network operations center where all faults are monitored adequately, logged for management purpose and escalated to the field engineers for immediate actions [6]. The methods used by various telecommunications operations center to report and manage faults also vary depending on the scale of operations and the best way to provide management with informed decisions on the best way to improve a Network [7].

Network Operation Centre (NOC): A Network Operations center (NOC) can be referred to as a place where various teams or administrators supervise, monitor and maintain a telecommunications network in order to ensure availability and optimal performance per time [8, 9].

Fault Management: [10] said it is important to distinguish between faults and failures. A fault is any kind of defect that leads to an error. A failure is a state of a system when it deviates from specification and is unable to deliver its intended functions. Fault tolerance can be classified into four levels from a system point of view such as: hardware layer, software layer, network communication layer, and applications layer [11]. [12] defines Fault management as a component of the network management which is concerned with the detection, isolation and resolution of problems.

Fault Management Phases: To provide a resilient network in faulty situations [13–15], proffers three main actions (Fault detection, fault diagnosis and fault recovery) shown in the Fig. 1.

Fig. 1. Fault management and it three phases

Fault Detection: This is the first stage of a Fault management system, Fault detection in sensor networks depends on the type of failures that occur on a network.

Fault Diagnosis: Fault Diagnosis is the process of properly identifying the cause of problem or cause of the alarm. It is the process of properly troubleshooting the failure or fault in the system to know the root cause failure.

Fault Recovery: This is the way faults are treated, this is the stage the network is restructured or reconfigured such that failures do not affect or impact the network performance any further.

Network Map: The network map is a view on the network server application that displays the network being managed as well as the status of the network elements (NE) within [16, 17]. A network submap for an Example GSM Network is given in Fig. 2.

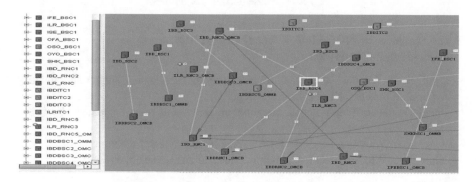

Fig. 2. Network submap for an example GSM network

3   System Analysis and Design

The System analysis and design is concerned with planning the development of information systems through understanding and specifying what a system should do and the way and manner each of the components of the system should be implemented to work together [18]. The System analysis primary goal is to improve and enhance the efficiency of the current system [19, 20]. This will feature a Use case diagram which will define the functional requirements.

3.1   Functional Requirements Using Use Case Analysis

Use Case is an important role analysis tool, it is essential to define the functional requirements of the system during software development [21]. It is said that to be used to evaluate the link between functional and non-functional requirements of a system. A use case diagram contains actors and use case which describes a specific usage of the system contains one or more actors [22–24]. Figure 3 present the use case diagram for all uses of fault management systems.

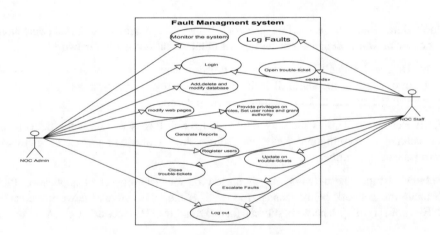

Fig. 3. Use case diagram for all users

Basic Scenario for NOC Staff- Primary Actor: NOC Staff
Stakeholders and Interests:

- NOC Staff: Wants accurate fault logging, prompt escalation, opening and updating of trouble-tickets, and generation of reports.
- Field Engineer: Wants fast notification on faults and proper updates of trouble-tickets.
- Customer: Wants reduced MTTR for reliable and efficient network
- Company: Wants to accurately report fault and satisfy customer interests.

  Wants to ensure that all faults are properly logged.

Wants some fault tolerance to allow network availability even if server components (e.g., Intelligent Network (IN) app for billing) are unavailable.

Preconditions: NOC Staff is identified and authenticated.

Success Guarantee (post-conditions): Fault is escalated.

Field Engineer successfully gets notification.

Trouble-tickets are opened and are updated.

Report is generated.

Main Success Scenario (or Basic Flow):

1. Fault occurs on a network element(s) and alarms pops up on server.
2. NOC staff acknowledges the alarms and saves them into a spreadsheet document.
3. NOC staff opens a trouble-ticket for every fault and generates a unique ID for each.
4. NOC staff escalates the faults using the trouble-ticket ID to the appropriate field engineer.
5. NOC staff gets updates from field engineers and update(or close) trouble-tickets accordingly.

   NOC staff repeats steps 2–5 for every new fault(s) developing.
6. System logs all faults escalated and generates report.

The Fig. 4 provide the context diagram of file management system.

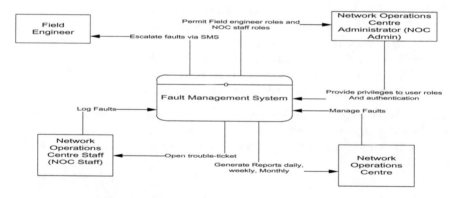

Fig. 4. Context diagram for the fault management system

3.2   Non Functional Requirements

Accessibility: The System should be accessible only to the Operations team of the organization to monitor the network performance activities; every other section will be accessible to the Network operations staff.

Usability: The system should be designed simple and be easy to navigate by all intended users within the organization and be available by all authorized users within the stated roles.

Availability: The System should operate 24 h every day 365 days in a year.

Accuracy: Errors should not be found in the existing system. The system will improve and ease the processes of logging faults, escalation and reporting thereby improving network availability and provide management with the real state of the network.

Maintainability: The System should have maintenance procedure so it can stand the test of time.

4   Fault Supervision and Escalation

The fault supervision and escalation is carried out by the NOC team who are at the back end supervising remote servers connected to the physical equipment at site locations. The remote servers are:

- ZTE NetNumen U31 Unified Management System
- HUAWEI iManager U2000 MBB Network Management System

These servers share the same working principle but differ in the interface designs and methodologies. They are connected through high speed transmissions links to the Network Elements (NE) on location to allow remote monitoring and troubleshooting of the NE. The server comprises many capabilities for effective troubleshooting and resolution of faults on the connected ZTE network elements.

The Alarm List Page: The alarm page displays a log of alarms of current and historical faults that has happened on the nodes. It displays the Node, the Raised Time as well as the corresponding Alarm Code (Fig. 5).

The Cleared Log Page: The cleared log page displays the log of all alarms that have been cleared from the alarm list page showing the Node, the Raised Time and the Cleared Time. This page is used when the NOC staff is working on the report (Fig. 6).

The Escalation Page: The escalation page allows a user to select the file(s) containing the alarm logs from the respective servers, processes the data by extracting the nodes, regions and corresponding phone contacts for escalation via SMS to the resolution team on field (Fig. 7).

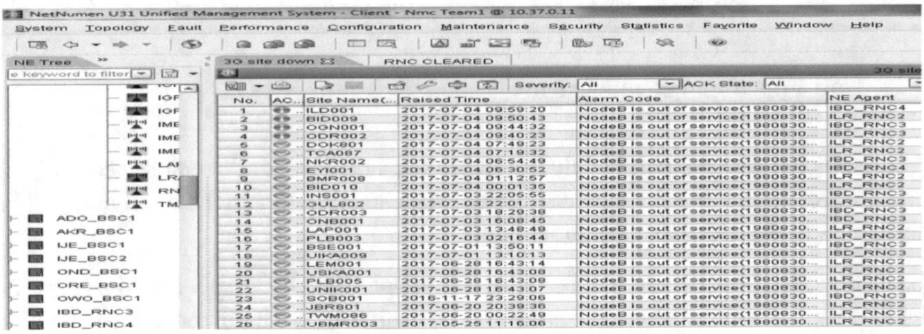

Fig. 5. ZTE server alarm list page

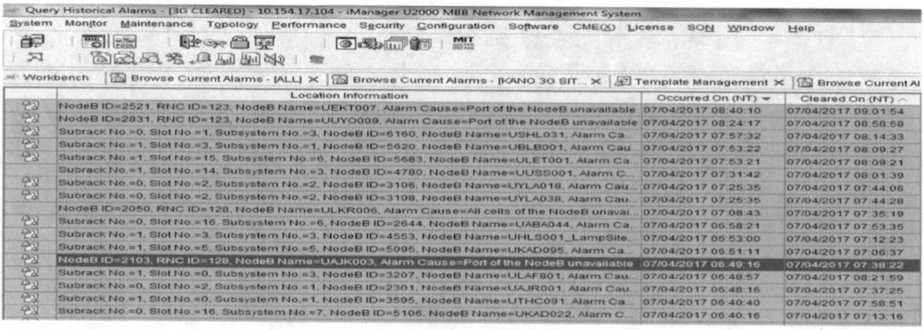

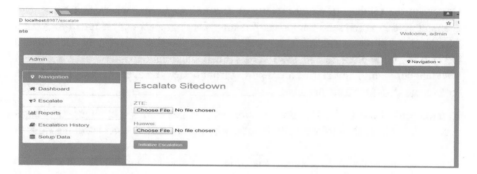

Fig. 6. Huawei server cleared log page

Fig. 7. Screenshot of escalation file selection page

Escalation Algorithm

> Steps
> 1. Highlight all alarms to be escalated from the server.
> 2. Right click and click on "save selected items"
> 3. On the "Escalation window" of the fault management system, select the saved file(s) using the appropriate file chooser button.
> 4. Click on "Initialize Escalation".
> Repeat steps 3-4 until all saved files has been chosen and processed
> 5. Click on "Escalate Now".

Reports Page: This page accepts three files as input and produces one updated report file in excel format. It accepts a copy of the current report and the saved excel files of the cleared logs of both the ZTE and Huawei servers as its input. The application loops through every opened ticket in the inputted report combining its fault node (Site ID) and the raised time to make a search in the cleared log files for a corresponding match. If a match is found for a particular ticket, the corresponding cleared time for that ticket is automatically inputted and the ticket is closed. If a match is not found, the trouble ticket is left opened and the next trouble ticket is processed (Fig. 8).

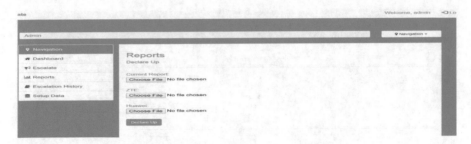

Fig. 8. Screenshot of reports file selection page

5 Results and Comparison

Currently, the Globacom Nigeria Network Operations Center where this fault management system is required comprises 3 teams of 4 engineers per team carrying out a 24 h technical surveillance and fault management on a total of 15041 network elements (7321 BTS, 6200 NodeBs and 1520 4G LTE cell sites) nationwide on a 12 h rotation schedule per team. Fault alarms are reported from these NEs every second resulting in an average of 500 alarms per hour.

Calculations Based on Existing System: Average alarms per hour per engineer = 500; Time taken to manually input alarm into formatted excel sheet per hour ≈ 30 s;

Latency in escalating to 1 cluster via SMS ≈ 15 s; Total number of clusters = 70;

Therefore, time taken to escalate to 70 clusters of field engineers per NOC staff = 70×15 = 1050 s;

Total time taken to complete escalation process per hour per engineer = Time taken to manually input alarm into formatted excel sheet per hour + Time taken to escalate to 70 clusters of field engineers per NOC staff = 30 + 1050 = 1080 s = 18 min

Therefore, total time saved per hour per engineer = 60 min − 18 min = 42 min

Calculations Based on Proposed System: Average alarms per hour per engineer = 500; Time taken to save alarm data into excel file per hour ≈ 35 s; latency in escalating to 1 cluster via SMS ≈ 15 s; Total number of clusters = 70; Time taken to escalate to 70 clusters of field engineers per NOC staff ≈ 15 s (Here, alarms are escalated simultaneously at the click of "Escalate Now" button);

Total time taken to complete escalation process per hour per engineer = time taken to save alarm data into excel file per hour + Time taken to escalate to 70 clusters of field engineers per NOC staff = 35 + 15 = 50 s ≈ 0.83 min.

Therefore, total time saved per hour per engineer = 60 min − 0.83 min ≈ 59 min

Overall percentage improvement from the old system to the proposed system =

$$\frac{59 - 42}{42} \times 100\% = \frac{17}{42} \times 100\% \cong 40.48\%$$

Comparison-Challenges of the Existing System: The challenges of the existing system are stated below:

- It is quite monotonous because it requires different activities to achieve a result
- Sites can be sent via SMS but not inserted in the report. That is, it is possible to delete any records.
- It takes a long time to prepare the reports.
- The reports are not accurate.
- Time is largely wasted preparing reports there by slowing down escalation processes.

Benefits of the Proposed System: The Fault management system is expected to provide the following benefits to the process in the Network operations center (NOC):

- All processes are integrated into one application.
- Escalations are done immediately the Faults are logged on the application system.
- Reports are prepared in little time.
- Faults are escalated in a prompt manner.

A comparison between existing systems and the proposed system is given Table 1.

Table 1. Comparison between existing system and proposed system

	Old system	Proposed system
Approach	Distributed: The systems involves a combination of four applications working together to achieve the goal of fault management i.e. fault identification, diagnosis and recovery. Notification to field engineers via SMS is done using a bulk SMS platform	Centralized: This system uses only one application to achieve the fault management goals. Notification to field engineers is also achieved within the system
Latency	High: The process is much delayed as a lot of manual input is required to achieve each step	Low: The latency here is very much reduced as all activities are done with only one application and the escalation steps are semi-automatic
Overhead required	Due to the cumbersome steps involved, the overhead cost is high as it requires more human resources and applications	The overhead is low as only one NOC staff is sufficient to handle the entire processes and only one application is used
Fault logging	Every escalation done via SMS is logged onto an excel sheet which could be easily manipulated or even deleted by any user of the system	Every escalation done is logged onto the application with a ticket ID and only the NOC admin has the rights to delete a ticket
Results	Reports are generated manually by copying the logs onto an excel sheet and the alarm occurrence time as well as the cleared time are inputted on corresponding rows	Reports are generated on requested based on the need of the NOC staff/admin from the application and can be exported directly from the system onto excel spreadsheet or .pdf formats

6 Conclusion and Recommendation

This research work is the design of a fault management system developed for a Network Operating Centre which facilitates the reporting, logging, tracking and monitoring of faults from the start to finish within in less than one minute. Before now for most Network Operating Centers, when faults are reported, the NOC staff either notifies the appropriate individuals by a call and SMS, records the fault in a book or an excel file and waiting till the fault is resolved. This system has been developed to aggregate all the functions of the Network operations center into a system the logs the fault, tracks the fault, notifies via an SMS and generates a report accordingly thereby improving the efficiency of the NOC by 40.48% in comparison to the old system. It is however recommended that a mobile phone version of the application is designed to enable field engineers login and input updates directly to faults which they resolve rather than making phone calls to the NOC Engineer.

References

1. Akinbinu, A.: New telecommunication technologies and the financial health of the Nigeria telecommunications Ltd. Nigerian J. Eng. Manag. 14–22 (2001)
2. Kim, H.-S., Baek, C.H.: A quality assurance process model on fault management. Int. J. Inf. Process. Syst. **2**(3), 163–169 (2006)
3. Abubakar, A.N., Garba, B.B.: Strengths, weaknesses, opportunities and threats (SWOT) analysis on globacom ltd. Int. J. Inf. Technol. Bus. Manag. **16**(1), 83–91 (2013)
4. Folarin, D.A., Abdul-Hameed, T.A.: Developing a software package for forecasting mobile communications network subscribers in Nigeria. Int. J. Sci. Technol. **1**(6), 289–298 (2011)
5. Chukwudebe, G., Chika, I.E.: Sustainable national telecommunication development: an overview critical requirements. In: Proceedings of the National Engineering Conference of the Nigerian Society of Engineers, pp. 269–278 (2005)
6. Wojuade, J.I.: Impact of global system for mobile telecommunication on Nigerian economy: a case of some selected local government areas in Oyo State, Nigeria. A Med thesis, University of Ibadan, Nigeria, Ibadan (2005)
7. Maduka, V.I.: Telecommunications in NEEDS. In: Proceedings of COREN 13th Engineering Assembly, Abuja, pp. 68–75 (2004)
8. Raman, G.L.: Fundamentals of Telecommunications Network Management. Wiley, Hoboken (1999)
9. Laghari, K.R., Ben Yahia, I.G., Crespi, N.: Analysis of telecommunication management. Int. J. Comput. Sci. Inf. Technol. (IJCSIT) **1**(2), 152–166 (2009)
10. Rahul, T.: Techno Trice (2016). http://www.technotrice.com/what-is-spiral-model-software-engineering/
11. Tutorialspoint. Tutorialspoint (2015). https://www.tutorialspoint.com/sdlc/sdlc_spiral_model.htm
12. Kirk, J.S.: An introduction to TMN. J. Netw. Syst. **3**(1) (1995). http://www.cs.stevens.edu/ ~ sghosh/courses/cs669/old-classnotes/
13. Hajela, S.: HP OEMF: alarm management in telecommunications networks. Hewlett-Packard J. **3**, 1–11 (1996)

14. Saravanan, P., Emmanuel, R., Sekhar, V.: Enhancing enterprise network management using SMART. In: Annual IEEE India Conference, 11–13 December 2008, vol. 2, pp. 343–348 (2008). (secured mobile agents for heterogeneous environment)
15. Kamoun, F.: Toward best maintenance practices in communication network. Int. J. Netw. Manag. **15**, 321–334 (2005)
16. Strahonja, V.: Proactive approach to the incident and problem. J. Inf. Organisational Sci. **31** (1), 5–6 (2007)
17. Erick, S.: MSP Business runbook www.n-able.com. In: MSP Business runbook www.n-able. com.: MSP University, pp. 45–50 (2013)
18. Richard, F., Darren, C., Dave, R., Robert, S., Jesse, V.: RT Essentials. O'Reilly Media, Inc., Sebastopol (2005)
19. Khan, M.Z., et al.: Centralized schemes of fault management in wireless sensor networks. GESJ: Comput. Sci. Telecommun. **4**(36), 66–68 (2012)
20. Liu, H., et al.: Fault-tolerant algorithms/protocols in wireless sensor networks. In: Guide to Wireless Ad Hoc Networks. Springer, London, pp. 265–295 (2009)
21. Margret, R.: searchnetworking.techtarget.com, September 2006. http://searchnetworking. techtarget.com/definition/fault-management
22. Adda, M., Al-Kasassbeh, M.: Network fault detection with wiener filter-based agent. J. Netw. Comput. Appl. **32**, 824–833 (2009)
23. Bhute, A.N., Meshram, B.B.: System analysis and design for multimedia retrieval systems. Int. J. Multimed. Appl. (IJMA) **5**(6), 25–44 (2013)
24. Chon, A., Iris, J., Mike, W.: System analysis and design for services oriented architecture projects: a case study at the federal financial institutions examination council (FFIEC). J. Emerg. Trends Comput. Inf. Sci. **2**(1), 1–15 (2011)

Optimized Path Selection in Oceanographic Environment

N. V. Sobhana[1]([⊠]), M. Rahul Raj[2], B. Gayatri Menon[3],
and Elizabeth Sherly[3]

[1] Rajiv Gandhi Institute of Technology, Kottayam, Kottayam, India
sobhananv.rit@gmail.com
[2] Illahia College of Engineering and Technology,
Muvattupuzha, Mulavoor, India
rahulraj2kl6.m@gmail.com
[3] Indian Institute of Information Technology and Management-Kerala,
Thiruvananthapuram, India
{gayatri.menon, sherly}@iitmk.ac.in

Abstract. Wireless sensor networks (WSN) react to events in specified circumstances by sensing, computing and communicating with thousands of sensors arranged at different locations operating in different modes. Typical applications include, but are not limited to, data collection, military operations, surveillance, and medical telemetry. The sensors are battery powered devices and hence their lifetime is very limited. It may not be possible to recharge or replace the battery depending upon the application environment. Communication overhead has to be reduced because energy is a very valuable resource for these sensor nodes. Long distance communication among sensors will cause large amount of energy drain which may reduce the lifetime of the network. In this work we propose Genetic Algorithm (GA) and Gravitational Search based methods to address sensor network optimization problem. The GA, GSA and PSO based clustering of WSN can greatly minimize the total communication distance, thus lengthening the network lifespan. Kerala, on the west coast of India, holds a vital role in India's fishing industry that gives a sustainable steady income. A new technique based on WSN technology using GA, GSA and PSO methods has been presented in this paper which helps to provide protection to the fishermen while they are in the deep sea. The results obtained from the proposed work shows that the network optimization based on cluster head using GSA has better performance and less energy consumption than the network without clustering.

Keywords: Wireless sensor networks · Energy of nodes ·
Optimization problem · Clustering · Genetic algorithm ·
Gravitational search algorithm

1 Introduction

Wireless sensor networks contain sensor-nodes having restricted resources that are composed randomly on a medium-large geographic area. These networks assemble data from the environment and forward it to the base station. Sometimes it is not

© Springer Nature Switzerland AG 2019
A. Abraham et al. (Eds.): IBICA 2018, AISC 939, pp. 506–517, 2019.
https://doi.org/10.1007/978-3-030-16681-6_50

possible to vary or recharge these sensors; because of they are situated in hardly accessible places. Therefore, energy preservation might be a fundamental concern, in order to boost the system life span. This could be accomplished with worthy design of the routing algorithms, thinking about of the distances between the nodes and their energy levels.

During the last decade, WSNs are widely utilized in a very kind of application areas associated with water observance, forest observance, industrial observance, agriculture observance, battlefield police   work,   intelligent   transportation, sensible homes, animal behavior observance, and disaster interference. This technology can definitely be applied to the observance of marine environments. In the case of the oceanic environment position of the nodes depends on many factors. The full network keeps on changing owing to tide effects, temperature and pressure variations, wind effects and horizon problems. The changes may well be a cause of difficulties in communication. Hence it is very necessary to address these issues. In literature Genetic Algorithms are used to optimize the network [1].

Wireless sensor network (WSN) of spatially distributed autonomous sensors is utilized to observe physical or environmental conditions, like temperature, sound, pressure, and so on and to hand and glove pass their knowledge through the network to a main location. Here, we proposed an approach to solve the sensor node network optimization problem supported genetic algorithms (GAs), Gravitational Search Algorithms and PSO to predict the movements of nodes supported some parameters and cluster head re-election. Long distance communication between sensors and a sink (or destination) during a sensor network will highly drain the energy of sensors and scale back the period of a network. By clustering a sensor network into variety of independent clusters utilizing a GA [2], GSA and PSO, we will greatly minimize the full communication distance, so prolonging the network period. This approach is recommended for shortest distance optimization issues.

Instead of a direct connection from the nodes to the sink, the clusters may be used for sending information to   a   base   station that successively leverages the   benefits of tiny transmit distances for many nodes, requiring solely a number of nodes to transmit way distances to the bottom station [3, 4]. By using the clustering formation, we tend to might partition the network into variety of independent clusters, every of that feature a cluster-head that collects information from all nodes in its cluster. These cluster-heads then compress the information and send it on to the sink. Clustering will greatly cut back communication prices of most nodes as a result of they solely got to send information to the closest cluster-head, instead of on to a sink that will be more away is static. Sensors are deployed in a very remote, inhospitable atmosphere and are far away from the sink that is sometimes positioned in a very safe place. All nodes are assumed to own the capabilities of a cluster-head and therefore the ability to   regulate their.   transmission   power supported transmission   distance. Every sensor's position is exactly measured by GPS (Global Position System) devices.

2 Related Work

The related works are divided into two categories. They are works that based on (i) GA based optimization (ii) cluster routing (iii) GSA based optimization.

Mansouri et al. [5] proposed a technique to prolong battery life by upgrading the quantization and picking the best set of sensors that take part in information accumulation in the WSN for target tracking application. In wireless sensor network the problem of Node localization is critical for applications, for example, military reconnaissance, ecological observing, mechanical autonomy, and numerous others.

Routing in wireless sensor network (WSN) is more difficult than that of wired network. WSN consists of a large number of tiny connected sensor nodes. There is no fixed infrastructure and it is self-organizing structure. The wireless links are not reliable and has to meet many energy saving constraints [10]. Major routing protocols for WSN are categorized in to seven [11]. Singh et al. [11] focused on clustering routing or hierarchical routing methods. Each cluster has a cluster head (CH) and various sensors called member nodes (MN) associated with it through a remote media. The member nodes sense nature and convey to group head. The cluster head process the information and transmit to sink or base station (BS). Clustering can be utilized to actualize vitality productive directing strategies. High energy hubs are doled out as CH and low energy nodes are allocated as MN. The CHs likewise can be organized in progressive request to transmit information to BS. Singh et al. [12] audited different energy efficient conventions like Low Energy Adaptive Clustering Hierarchy (LEACH), Power-Efficient Gathering. Delavar et al. [13] proposed a cluster routing algorithm called SLGC. This algorithm divides the WSN into grids. In each grid computes the center-gravity and threshold of energy for selecting the node that has the best condition base on these parameters in grid for selecting Cluster-Head in current round, also SLGC selecting Cluster-Heads for next rounds thereby this CHs reduce the volume of controlling messages for next rounds and inform nodes for sending data into CH of respective round. This algorithm prolong network lifetime and decrease energy consumption by selecting CH in grid and sending data of grid to sink by this CH.

3 Clustering and Optimized Path Selection Using GSA, GA and PSO

In this proposed model all nods are assumed to have the capacities of a cluster head and the capacity to adjust their transmission power in view of transmission distance. Every sensor's position can be figured by GPS gadgets. The proposed system can be divided into four tasks: 1. Node placement 2. Cluster formation 3. Sensing and data transmitting 4. Network optimization.

3.1 Node Placement

Node placement is a vital phase before establishment of a WSN. Different types of node distributions in WSNs including regular, random, and grid distributions [13]. In this work sensor nodes are placed in fishing boats. Node placement comes under random distribution, where the nodes can cover the entire area. Sensor nodes are deployed in the fishing region and base station or sink is far away from this area.

3.2 Cluster Formation

Custer formation could be divided into two phases; setup phase and clustering phase. In the setup phase after the node deployment, the base station broadcasts a packet which contains the following details: Coordinates of BS, No of nodes, Beginning time and Time slot. When a sensor node gets this packet it will compute its direct distance from base station and energy level. The sensor nodes are location aware by using GPS system and the distance from the BS could be calculated using normal distance formula

$$dist(x, y) = \sqrt{\left((x2 - x1)^2 + (y2 - y1)^2\right)}$$

And the energy level is calculated by

$$E(i) = lciel\left(residual\frac{en(i)}{\propto}\right)rceil$$

Where E is the energy parameter [9] and it is an estimated value 'i' is the ID of each node α is the minimum energy unit, a constant. After this every node sends its packet around with a specific radius during its own time slot For instance, the node whose ID is i will send out its packet. This packet contains preamble and the information, such as coordinates and EL of node i. The packet made by every node contains the following. In the ith time slot, for node i, the packet contains a preamble and the information such as coordinates, Energy level of node I and DD of BS is the direct distance from base station (Table 1).

Table 1. Packet created by each node

Preamble	Coordinates of i	E (i)	DD from BS

All the other nodes during this time slot will screen the channel by sending and accepting packets and every node records a table in their memory which contains the data of every one of its neighbors. In the clustering phase after the initial phase, each node knows information about its neighbors. So, all the sensor nodes have a connection with their nearest neighbors. For the first clustering, the node which is at the shortest distance will be elected as CH. After the selection of cluster head, all cluster heads will keep a bit (either 0 or 1) to represent whether it is a cluster head or not. Clustering is formed with its neighbors. Cluster heads are represented by 1 and cluster members are represented by 0.

3.3   Sensing and Data Transmitting

Once clustering is done each node senses its environment and continuously senses the details such as Latitude, Longitude, Wind direction, Wind speed, Sea level pressure, Pressure change, Air temperature, Sea surface temperature, Signal strength. They send this data to the cluster head. Cluster head will remove the duplicates and send data to the base station. So this information transmitting phase is divided into a number of time slots. After each time slot each node recalculates its energy level and checks whether it falls below a particular threshold. This threshold is dependent on the distance and signal strength. If it shows that energy level $E\ (i)$ of cluster head is below this particular threshold, then cluster head re-election is needed. For this cluster head re-election, we use Genetic Algorithm thereby optimizing the network.

3.4   Network Optimization

3.4.1   Network Optimization Using Genetic Algorithm

(a) *Chromosome design:* Genetic Algorithm performs fitness test on each chromosome on the basis of a fitness function. The chromosome structure on a rule based system is defined here. Each chromosome consists of four features, namely position, signal strength, sea temperature and wind speed (Fig. 1).

Position	Signal Strength	Sea Temperature	Wind speed

Fig. 1. Chromosome for proposed scheme

Position is the current position of the wireless node. Signal strength is measured from a reference node and which is an important parameter. Sea temperature is the temperature of the sea surface and wind speed is the speed of the wind in km/h. It means not all four features are having same importance. For example, the sea temperature is almost common for all nodes because we are considering only small distance communication, but energy of the node is very much depends on the working criteria of node, so we cannot give equal importance to these two features. The fitness function will try to match each chromosome towards a perfect chromosome, i.e. all fields are 1 in this case which the best case is. But the ordering will be different for different parameters. The order in which a chromosome is said to have the qualities to be chosen as best is also given. The parameters of the fitness [3] and their gene can be developed as follows. Position is computed based on GPS location, Direct distance from base station (DD), Cluster distance (C).

Choosing the best chromosome in this case is according to the priorities. In this case the first preference will go to the signal strength. That particular field should be 1 to consider the other factors. Then its order will be position, wind, temperature, and then pressure. Cluster distance (C) is the sum of distances from nodes to the cluster head and distance from cluster head to the base station [4].

$$C = \sum Dih + Dhs$$

Dih is the distance of a cluster member to a cluster head. *Dhs* is the distance from cluster head to base station If C = DD of CH, then C = 1 or C = 0. This cluster distance should be minimum for the best case. The cluster distance standard deviation will follow the order, first position, then signal strength, wind, pressure. This chromosome will try to find out a maximum match towards the best case. The other parameters are Signal Strength, Sea Temperature and Wind Speed. From these factors we will select the best chromosome and the node which has maximum number of best chromosomes is chosen as next CH. Since the current CH has the genes of all of its cluster members, the fitness function is applied on those genes and it chooses new CH. We have assigned different weights to each feature. The overall fitness of chromosome is determined by the equation: $\sum feature \cdot priority$. For each feature, a range of values are already calculated and divided them in to four sector based on quality. They are bad, average and good correspondingly binary values 00, 01, 11 are given. Here onwards each chromosome is defined by these five groups of two bit sequences i.e. each chromosome is represented by a 10 bit pattern. We use two bit representation for each of these parameters. The codes assigned are as follows:

Bad – 00 Average – 01 Good – 11

(b) Cluster head election: Clustering and cluster head election are the processes which make smooth communication between every node and the base station. The coverage of base station is in the scope of 20 nautical miles. In such case, long communication between nodes and base station can significantly deplete the energy of sensors and decrease the life expectancy of the system. In the proposed framework, we group the sensor network into a number of independent clusters by choosing the head node. The cluster head node is chosen by analyzing the strength of various parameters using fitness function in GA. Cluster heads act as an intermediate nodes from the base station. It can greatly minimize the total distance of communication and better lifespan of the network. In the cluster head selection algorithm using genetic algorithm, each node is represented as a ten bit pattern based on the corresponding feature values. The chromosome selection for doing cross over was Roulette-Wheel selection which is a common selection method. The cross over is used was one point cross over and mutation was 10 percentage of the population. The best chromosome is stored in a variable called max_fitness for each iteration. The best chromosome is the chromosome with the highest fitness value. After a number of iterations a best chromosome i.e., a node with best fitness value, is selected as the cluster head.

3.4.2 Network Optimization Using GSA

The gravitational search algorithm (GSA) [28] is an algorithm used for optimization. It is based on the principle of Newton's laws of motion. Based on the principle of Newton's laws of motion the objects with large masses attract more. According to this movement or next position of the object is determined. GSA is an algorithm to find the best node to be elect as the cluster head. GSA is an adapted concept from Newton's law of attraction. The mass calculation of each node is based on its fitness value. For computing fitness value we have to consider certain features. The score given by these features contributes to the mass of each node. The five different features are considered. They are position, signal strength, sea temperature and wind speed. For each mass i.e.; node the algorithm computes the force of attraction with other all masses. According to the law, larger masses will attract each other. So if the effective force which acts on a masses are found, it will help us to find the node which is more suitable or desirable as the cluster head. Next section deals with the cluster head election algorithm. The last portion contain the algorithm for cluster merging.

(a) *Cluster head election*

Clustering and cluster head decision are the procedures which make smooth communication between every node and the base station. The coverage of base station is in the scope of 20 nautical miles. In such case, long communication between nodes and base station can greatly deplete the energy of sensors and diminish the life expectancy of the system by dividing sensor network into a number of independent clusters and by choosing the head node by analyzing the strength of different parameters utilizing a GSA. Group heads go about as an intermediate between nodes and base station. It can greatly limit the aggregate communication distance and better system lifetime.

Algorithm.1. Cluster-head selection
1. For all nodes under consideration do the following:
2. Calculate the score of each node by summing the two bit binary patterns of features. Considering it as the initial mass of each node.
3. For all masses do the following until termination condition reaches:
4. Calculate position, velocity and acceleration of each mass.
5. Find force of attraction between each pair of masses.
6. Calculate effective force acting on each mass.
7. Calculate next generation mass.
8. The node with greater total force will be the new cluster head.

In the cluster head selection algorithm using GSA each node is represented as a ten bit pattern based on the corresponding feature values. After the execution of cluster head election algorithm, the best node is stored in a variable called max_score. The best node is the node with the highest feature value. After a number of iteration a best node i.e., a node with the best overall score is selected as the cluster head. One advantage of GSA is that it requires less number of iterations to obtain a solution, but in the case of the genetic algorithm it requires more number of iterations than GSA. One thing we cannot deny the accuracy of genetic algorithms especially for a large number of inputs. So we can make use of a new strategy GSA when the nodes to be clustered are considerably less i.e., if we have to cluster more number of nodes to be clustered, genetic algorithm strategy can be used for cluster head election

(b) *Cluster merging*

Need for cluster merging is that, the head of a cluster is not having a better score or not efficient enough to make effective communication between nodes and base station. If it is unable to elect a cluster head based on optimal overall score, then the messages from those nodes would travel more time and more nodes. If it is not able to elect new cluster head, then a region containing several nodes cannot be able to communicate and there is a chance of losing an entire region. Obviously, this will lead to attenuation and erroneous transmission. In this case, we need to merge the cluster with the nearest one having better cluster head. This will prevent the above mentioned scenario. From the evaluation, it is observed that the system after cluster merging nodes has less energy consumption.

Algorithm.2. Cluster_merging

1. For all neighboring clusters of current cluster repeat
   1.1. Find fitness calculation for each node
2. Find five number summary for each neighboring cluster based on the fitness value
3. Apply pareto-optimal ranking in the five number summary vector
4. Select first three top ranked clusters from the group
5. Select a cluster with more number of nodes and cluster head node nearer to base station
6. Merge current cluster the selected cluster.

For cluster merging, the cluster under consideration is i^{th} cluster and totally there are n numbers of clusters. In this situation we can merge i^{th} cluster with any one of the $n - 1$ cluster. In the Algorithm 1, a method for finding a cluster with better cluster head is described. First of all for each cluster the fitness value for each node is calculated also for each cluster arrange the nodes in the ascending order of their fitness value. Next a suitable cluster to be merged with cluster-i is selected. For that we are adopting a statistical concept called 'Five number summary'. Five number summary gives an overall behavior of a group of data. It will give you a vector with five values which are the sample minimum (here smallest fitness value of a cluster), first quartile (i.e. 25% of the fitness values of the cluster), median (middle fitness value of the cluster), third quartile (75% of the fitness value of the cluster) and sample maximum (the maximum fitness value in the cluster). So for each cluster, we have created a five element vector. Eg: [12.4,26.5, 57.2, 81.3, 95.1]. Now apply pareto-optimal ranking [29] in the five number summary vectors. After the application of this algorithm, we will get the best cluster in the rank-1 position and select that as the cluster to be merged with cluster-i. After cluster merging, cluster head is elected using Gravitational Search Algorithm GSA. Here we adopted Gravitational search algorithm (GSA) [33] again for cluster head selection because in this case cluster head election is based on less number of iterations. If we apply genetic algorithm it takes more time. Using this concept the cluster head re-election is also done.

3.4.3 Network Optimization Using PSO

Particle swarm optimization (PSO) is inspired by the collective behavior of a group of animals or birds. Example: Social behavior of a school of fish, flock of birds. Consider a flock of bird as example. Each birds in the flock would try to achieve the speed of its neighboring bird (If neighbor has comparatively more speed) also tries to achieve the speed of the bird which is having highest speed among the flock. That is each agent tries to achieve local maximum and global maximum. Each particle (birds in this case) called candidate solution. In each of the iteration, particles would get new position, according to its speed. After a number of iteration the global maximum would be selected as the optimum solution.

The PSO optimization is the concept adapted from a swarm of words flying in a direction. Here each particle is allowed to fly through the plane with some velocity. According to the velocity of a particle the position of each particle is updated using the below equations.

v[] = v[] + c1 * rand() * (pbestpl[] - present[]) + c2 * rand() * (gbestpl[] - present[])

Position [] = Position[] + v[]

In each cycles, every particle is updated by following two "best" values. The first is the best solution (fitness) it has accomplished up until this point. The fitness value is also stored. This value is called pbestpl. Furthermore, in light of the neighbor value it would enhance its own value. Another "best" value is tracked by the particle swarm optimizer. It is the best value acquired by any particle in the population. This best value is a global best and called gbestpl. PSO is used to find the best node to be elect as the cluster head. The particle is considered based on the fitness value. For computing fitness value we have to consider certain features. The score given by these features contributes to the particle. The five different features are considered. They are position, signal strength, sea temperature and wind speed. In each iteration. velocity of the particle is calculated based on the equation given. After computing velocity, position is also calculated i.e. each neighbor tries to achieve the position compared to other nodes, if the position is a better one. This will help us to find the node which is more suitable or desirable as the cluster head. Next section deals with the cluster head election algorithm. The last portion contains the algorithm for cluster merging.

(a) *Cluster head election*

Using a PSO clustering the a sensor network into a number of independent clusters is done by electing the head node and analyzing the strength of various parameters. In the cluster head selection algorithm using PSO each node is represented as a ten bit pattern based on the corresponding feature values. After the execution of cluster head election algorithm, the best node is stored in a variable called max_score. The best node is the node with the highest feature value. After a number of iteration a best node i.e., a node with the best overall score, is selected as the cluster head. One advantage of PSO is that it requires less number of iterations to obtain a solution, but in the case of the genetic algorithm it requires more number of iterations than PSO.

4   Experimental Results and Analysis

The entire data collection is divided into several groups on the basis of GA, GSA and PSO Algorithm application. In the literature GA and GSA Algorithm, is used for improving network lifetime and some of them use to optimize the entire network. The

NS2 simulation of proposed system is done for both case 1 and case 2. Several sensor nodes are placed in a 750 * 750 field and one node is selected as base station for the scenario. Case 1 is the situation when the clustering technique is not used. Each node is directly connected to this base station and sends data packets to it. More amount of energy is consumed and wasted in this scenario.

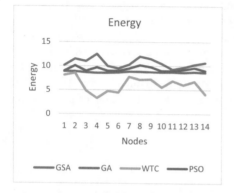
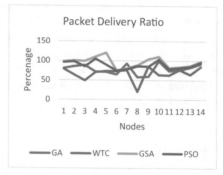

Fig. 2. Energy of nodes with and without cluster head

Fig. 3. Packet delivery ratio of nodes with and without cluster head

Case 1 is the situation when the clustering technique is not used. Each node is directly connected to this base station and sends data packets to it. This is possible nodes lies within 20 nautical miles. More amount of energy is consumed and wasted in this scenario. Case 2 has predefined number of clusters and cluster heads. There will be one cluster head for a cluster (number of sensor nodes), cluster head communicates to nearest cluster head of the clusters and finally the last cluster head communicates to the base station. So each cluster head collects information from all of its cluster members and sends it to the base station. The total energy consumed by each node is lesser in this case and the network lasts for a long time compared to case 1. The comparative graph is given in Fig. 2 shows Energy level of nodes with cluster head based on GA and GSA and without cluster-head.

The graph consists of energy consumption of each node in the network with cluster head and without cluster head. So this clearly indicates that network, which has cluster head based on GSA, outperforms the others. So clustering technique is applicable in the oceanic environment. The graph plotted with node number in X coordinate and energy at time T in each node. Energy is in the range of Micro Joules. It is assumed that initially, all nodes have the energy of 10 Micro Joules. The energy is the remaining energy in each node it difference of initial energy and the consumed energy.

Figure 3 shows the packet delivery ratio. It is the ratio between numbers of packet sent to the number of a packet received. From the evaluation, we found that the packet delivery ratio is higher in cluster head based on GSA mechanism rather than the normal case. It is because of the reason that in normal scenario the packets want to pass through several nodes because of that reason packet drop may also be increased.

5 Conclusion

When fishermen want to communicate from the deep sea to the shore or neighboring boats it becomes a difficult task. A solution to this is yet to be identified by means of new development of communication technology to ensure the wellbeing and protection while they are in deep sea. A new method for making the deep sea fishing less troublesome is proposed. This system using Genetic Algorithm, Gravitational Search Algorithm, and Particle Swarm Optimization can determine the most influential parameters by considering Signal Strength, Energy and wind speed etc. which helps to track, classify and disseminate information to the fishing fleets while they are in the deep sea. This technique employs wireless communication to impart safety and security to the fishermen and to equip them with the current status of the fisheries sector while they are in the deep sea fishing.

References

1. Heinzelman, W.R., Chandrakasan, A., Balakrishnan, H.: Energy-efficient communication protocol for wireless micro-sensor networks. In: Proceedings of the Hawaii International Conference on System Science, Maui, Hawaii (2000)
2. Cheng, X., Xu, J., Pei, J., Liu, J.: Hierarchical distributed data classification in wireless sensor networks. In: Proceedings of the IEEE International Conference on Mobile Ad-hoc and Sensor Systems (MASS) (2009)
3. Lai, C.-C., Ting, C.-K., Ko, R.-S.: An effective genetic algorithm to improve wireless sensor network lifetime for large-scale surveillance applications. In: IEEE Congress on Evolutionary Computation (CEC 2007) (2007)
4. Hussain, S., Matin, A.W., Islam, O.: Genetic algorithm for energy efficient clusters in wireless sensor networks. J. Netw. 2(5) (2007)
5. Mansouri, M., Nounou, H., Nounou, M.: Genetic algorithm-based adaptive optimization for target tracking inwireless sensor networks. J. Sig. Process. Syst. 74, 189–202 (2014)
6. Chagas, S.H., Martins, J.B., de Oliveira, L.L.: Genetic algorithms and simulated annealing optimization methods in wireless sensor networks localization using artificial neural networks, pp. 928–931. IEEE (2012)
7. Jourdan, D.B., de Weck, O.L.: Layout optimization for a wireless sensor network using a multi-objective genetic algorithm. Published in: 2004 IEEE 59th Vehicular Technology Conference, VTC 2004-Spring (2004)
8. Fonseca, C.M., Fleming, P.J.: Genetic algorithms for multi objective optimization: formulation, discussion and generalization. In: Genetic Algorithms: Proceedings of Fifth International Conference, pp. 416–423. Morgan Kaufmann (1993)
9. Sara, G.S., Devi, S.P., Sridharan, D.: A genetic-algorithm-based optimized clustering for energy-efficient routing in MWSN. ETRI J. 34(6), 922–931 (2012)
10. Misra, S., et al. (eds.): Guide to Wireless Sensor Networks, Computer Communications and Networks. Springer, London (2009). https://doi.org/10.1007/978-1-84882-218-4
11. Singh, S.K., Singh, M.P., Singh, D.K.: Routing protocols in wireless sensor networks – a survey. Int. J. Comput. Sci. Eng. Surv. (IJCSES) 1(2) (2010). https://doi.org/10.5121/ijcses.2010.1206
12. Singh, P., Bhatia, M., Kaur, R.: Energy-efficient cluster based routing techniques: a review. Int. J. Eng. Trends Technol. 4(3) (2013). ISSN 2231-5381

13. Delavar, A.G., Shamsi, S., Mirkazemi, N., Artin, J.: SLGC: a new cluster routing algorithm in wireless sensor network for decrease energy consumption. Int. J. Comput. Sci. Eng. Appl. (IJCSEA) **2**(3) (2012). https://doi.org/10.5121/ijcsea.2012.2304

14. Goyal, R.: A review on energy efficient clustering routing protocol in wireless sensor network. IJRET: Int. J. Res. Eng. Technol. **03**(06) (2014). eISSN: 2319-1163, pISSN: 2321-7308

15. Kumar, S., Prateek, M., Ahuja, N.J., Bhushan, B.: MEECDA: multihop energy efficient clustering and data aggregation protocol for HWSN. Int. J. Comput. Appl. (0975 – 8887) **88** (9) (2014)

16. Kaur, A., Buttar, A.S.: Energy efficient clustering techniques using genetic algorithm in wireless sensor network: a survey. IJIRST –Int. J. Innov. Res. Sci. Technol. **2**(09) (2016). ISSN (online): 2349-6010

17. Hussain, S., Matin, A.W., Islam, O.: Genetic algorithm for hierarchical wireless sensor networks. J. Netw. **2**(5) 87–97 (2007)

18. Singh, V.K., Sharma, V.: Elitist genetic algorithm based energy efficient routing scheme for wireless sensor networks. Int. J. Adv. Smart Sensor Netw. Syst. (IJASSN) **2**(2) (2012). https://doi.org/10.5121/ijassn.2012.2202

19. Zahmatkesh, A., Yaghmaee, M.H.: A genetic algorithm-based approach for energy- efficient clustering of wireless sensor networks. Int. J. Inf. Electron. Eng. **2**(2), 165 (2012)

20. Peiravi, A., Mashhadi, H.R., Hamed Javadi, S.: An optimal energy-efficient clustering method in wireless sensor networks using multi-objective genetic algorithm. Int. J. Commun. Syst. 26, 114–126 (2013). Published online 24 August 2011 in Wiley Online Library (wileyonlinelibrary.com). https://doi.org/10.1002/dac.1336

21. Zitzler, E.: Evolutionary algorithms for multiobjective optimization: methods and applications. Ph.D. dissertation, Swiss Federal Institute of Technology Zurich, November 1999

22. Khalil, E.A., Attea, B.A.: Energy-aware evolutionary routing protocol for dynamic clustering of wireless sensor networks. Swarm Evol. Comput. **1**, 195–203 (2011). https://doi.org/10.1016/j.swevo.2011.06.004

23. Chagas, S.H., Martins, J.B., de Oliveira, L.L.: An approach to localization scheme of wireless sensor networks based on artificial neural networks and genetic algorithms. IEEE (2012). 978-1-4673-0859-5

24. He, S., Dai, Y., Zhou, R., Zhao, S.: A clustering routing protocol for energy balance of WSN based on genetic clustering algorithm. IERI Procedia **2** 788–793 (2012). Elsevier

25. Peng, B., Li, L.: An improved localization algorithm based on genetic algorithm in wireless sensor networks. Cogn. Neurodyn. **9**, 249–256 (2015). https://doi.org/10.1007/s11571-014-9324-y

26. Nicolescu, D., Nath, B.: DV based positioning in ad hoc networks. J. Telecommun. (2003)

27. Rostami, A.S., Bernetty, H.M., Hosseinabadi, A.R.: A novel and optimized algorithm to select monitoring sensors by GSA. In: 2nd International Conference on Control, Automation and Instrumentation. IEEE (2011). 978-1-4673-1690-3

28. Rafsanjani, M.K., Dowlatshahi, M.B.: Using gravitational search algorithm for finding near-optimal base station location in two-tiered WSNs. Int. J. Mach. Learn. Comput. 2(4), 377 (2012)

29. Huynh, T.T., Dinh-Duc, A.-V., Tran, C.-H., Le, T.-A.: Balance particle swarm optimization and gravitational search algorithm for energy efficient in heterogeneous wireless sensor networks. In: 2015 IEEE International Conference on Computing and Communication Technologies Research, Innovation, and Vision for Future (RIVF (2015). 978-1-4799-8044-4/15

Author Index

Printed in the United States
By Bookmasters